METHODS IN CELL BIOLOGY

VOLUME 32

Vesicular Transport

Part B

Series Editor

LESLIE WILSON

Department of Biological Sciences
University of California, Santa Barbara
Santa Barbara, California

METHODS IN CELL BIOLOGY

Prepared under the Auspices of the American Society for Cell Biology

VOLUME 32

Vesicular Transport
Part B

Edited by

ALAN M. TARTAKOFF
INSTITUTE OF PATHOLOGY
CASE WESTERN RESERVE UNIVERSITY SCHOOL OF MEDICINE
CLEVELAND, OHIO

ACADEMIC PRESS, INC.
Harcourt Brace Jovanovich, Publishers
San Diego New York Berkeley Boston
London Sydney Tokyo Toronto

Sci
QH 585 P92 v.32
Methods in cell biology
v. 32
University of the Pacific
Library
Received on: 11-09-89

COPYRIGHT © 1989 BY ACADEMIC PRESS, INC.
All Rights Reserved.
No part of this publication may be reproduced or transmitted in any form or by any means, electronic or mechanical, including photocopy, recording, or any information storage and retrieval system, without permission in writing from the publisher.

ACADEMIC PRESS, INC.
San Diego, California 92101

United Kingdom Edition published by
ACADEMIC PRESS LIMITED
24-28 Oval Road, London NW1 7DX

LIBRARY OF CONGRESS CATALOG CARD NUMBER: 64-14220

ISBN 0-12-564132-X (alk. paper)

PRINTED IN THE UNITED STATES OF AMERICA
89 90 91 92 9 8 7 6 5 4 3 2 1

For Paola Ymayo,
Daniela Helen Elizabeth,
and Joseph Michael

CONTENTS

CONTRIBUTORS

Numbers in parentheses indicate the pages on which the authors' contributions begin.

MELVIN BERGER, Departments of Pediatrics and Pathology, Case Western Reserve University School of Medicine, Cleveland, Ohio 44106 (351)

JOHN E. BERGMANN, Department of Anatomy and Cell Biology, College of Physicians and Surgeons, Columbia University, New York, New York 10032 (85)

PHILIP P. BREITFELD, Department of Pediatrics, University of Massachusetts, Medical School, Worcester, Massachusetts 01655 (329)

JAMES E. CASANOVA, Whitehead Institute for Biomedical Research, Cambridge, Massachusetts 02139 (329)

RICHARD D. CUMMINGS, Department of Biochemistry, The University of Georgia, Athens, Georgia 30602 (141)

JEFFREY D. ESKO, Department of Biochemistry, Schools of Medicine and Dentistry, University of Alabama at Birmingham, Birmingham, Alabama 35294 (387)

MARY-JANE GETHING, Howard Hughes Medical Institute and Department of Biochemistry, University of Texas Southwestern Medical Center, Dallas, Texas 75235 (185)

KAREN S. GIACOLETTO, Howard Hughes Medical Institute and Division of Rheumatology, Department of Medicine, Washington University School of Medicine, St. Louis, Missouri 63110 (207)

LUTZ GRAEVE, Cornell University Medical College, Department of Cell Biology and Anatomy, New York, New York 10021 (37)

ROBERT HAAS, Department of Pharmacology, Case Western Reserve University School of Medicine, Cleveland, Ohio 44106 (231)

JEANNE M. HARRIS, Whitehead Institute for Biomedical Research, Cambridge, Massachusetts 02139 (329)

LAWRENCE HOBBIE, Department of Biology, Massachusetts Institute of Technology, Cambridge, Massachusetts 02139 (57)

INHAE JI, Department of Molecular Biology, University of Wyoming, Laramie, Wyoming 82071 (277)

TAE H. JI, Department of Molecular Biology, University of Wyoming, Laramie, Wyoming 82071 (277)

LIAN-WEI JIANG, Department of Biochemistry, Michigan State University, East Lansing, Michigan 48824 (423)

DAVID KINGSLEY, Department of Biology, Massachusetts Institute of Technology, Cambridge, Massachusetts 02139 (57)

KAREN KOZARSKY,[1] Department of Biology, Massachusetts Institute of Technology, Cambridge, Massachusetts 02139 (57)

MONTY KRIEGER, Department of Biology, Massachusetts Institute of Technology, Cambridge, Massachusetts 02139 (57)

ESA KUISMANEN, Department of Biochemistry, University of Helsinki, Helsinki, Finland (257)

CLAIRE LANGLET, Centre d'Immunologie, INSERM-CNRS de Marseille-Luminy, 13288 Marseille CEDEX 9, France (447)

[1] *Present address:* University of Michigan, Ann Arbor, Michigan 48109.

ANDRE LEBIVIC, Cornell University Medical Center, Department of Cell Biology and Anatomy, New York, New York 10021 (37)

LEE LESERMAN, Centre d'Immunologie, INSERM-CNRS de Marseille-Luminy, 13288 Marseille CEDEX 9, France (447)

MICHAEL LISANTI, Cornell University Medical Center, Department of Cell Biology and Anatomy, New York, New York 10021 (37)

PATRICK MACHY, Centre d'Immunologie, INSERM-CNRS de Marseille-Luminy, 13288 Marseille CEDEX 9, France (447)

KAREN MCCAMMON, Howard Hughes Medical Institute, University of Texas Southwestern Medical Center, Dallas, Texas 75235 (185)

ROBERTA K. MERKLE, Department of Biochemistry, The University of Georgia, Athens, Georgia 30602 (141)

PHILIP A. MORTON, Hematology–Oncology Division, Departments of Pediatrics and Pharmacology, Washington University School of Medicine, St. Louis, Missouri 63110 (305)

KEITH E. MOSTOV, Whitehead Institute for Biomedical Research, Cambridge, Massachusetts 02139 (329)

RYUICHIRO NISHIMURA, Department of Molecular Biology, University of Wyoming, Laramie, Wyoming 82071 (277)

SJUR OLSNES, Institute for Cancer Research, The Norwegian Radium Hospital, Montebello, 0310 Oslo 3, Norway (365)

DWAIN A. OWENSBY, The Cardiovascular Division, Department of Medicine, Washington University School of Medicine, St. Louis, Missouri 63110 (305)

MARSHA PENMAN, Department of Biology, Massachusetts Institute of Technology, Cambridge, Massachusetts 02139 (57)

OLE W. PETERSEN, Structural Cell Biology Unit, Department of Anatomy, The Panum Institute, University of Copenhagen, DK-2200 Copenhagen N, Denmark (365)

THOMAS H. PLUMMER, JR., Wadsworth Center for Laboratories and Research, New York State Department of Health, Albany, New York 12201 (111)

PRANHITHA REDDY,[2] Department of Biology, Massachusetts Institute of Technology, Cambridge, Massachusetts 02139 (57)

WILLIAM L. ROBERTS, Department of Pharmacology, Case Western Reserve University School of Medicine, Cleveland, Ohio 44106 (231)

ENRIQUE RODRIGUEZ-BOULAN, Cornell University Medical College, Department of Cell Biology and Anatomy, New York, New York 10021 (37)

TERRONE L. ROSENBERRY, Department of Pharmacology, Case Western Reserve University School of Medicine, Cleveland, Ohio 44106 (231)

JONATHAN ROTHBLATT, Department of Biochemistry, University of California at Berkeley, Berkeley, California 94720 (3)

TARA RUMBARGER, Howard Hughes Medical Institute and Division of Rheumatology, Department of Medicine, Washington University School of Medicine, St. Louis, Missouri 63110 (207)

PEDRO J. SALAS,[3] Cornell University Medical Center, Department of Cell Biology and Anatomy, New York, New York 10021 (37)

JOE SAMBROOK, Department of Biochemistry, University of Texas Southwestern Medical Center, Dallas, Texas 75235 (185)

[2] *Present address:* 1130 18th Avenue East, Seattle, Washington 98112.
[3] *Present address:* Instituto de Investigaciones Bioquimicas, Fundacion Campomar, 1405 Buenos Aires, Argentina.

YULA SAMBUY,[4] Cornell University Medical Center, Department of Cell Biology and Anatomy, New York, New York 10021 (37)

KIRSTEN SANDVIG, Institute for Cancer Research, The Norwegian Radium Hospital, Montebello, 0310 Oslo 3, Norway (365)

JAAKKO SARASTE, Ludwig Institute for Cancer Research, Karolinska Institute, Stockholm, Sweden (257)

MASSIMO SARGIACOMO, Cornell University Medical Center, Department of Cell Biology and Anatomy, New York, New York 10021 (37)

RANDY SCHEKMAN, Department of Biochemistry, University of California at Berkeley, Berkeley, California 94720 (3)

MELVIN SCHINDLER, Department of Biochemistry, Michigan State University, East Lansing, Michigan 48824 (423)

ANNE-MARIE SCHMITT-VERHULST, Centre d'Immunologie, INSERM-CNRS de Marseille-Luminy, 13288 Marseille CEDEX 9, France (447)

ALAN L. SCHWARTZ, Hematology–Oncology Division, Departments of Pediatrics and Pharmacology, Washington University School of Medicine, St. Louis, Missouri 63110 (305)

BENJAMIN D. SCHWARTZ, Howard Hughes Medical Institute and Division of Rheumatology, Department of Medicine, Washington University School of Medicine, St. Louis, Missouri 63110 (207)

NEIL E. SIMISTER, Whitehead Institute for Biomedical Research, Cambridge, Massachusetts 02139 (329)

MARTIN D. SNIDER, Department of Biochemistry, School of Medicine, Case Western Reserve University, Cleveland, Ohio 44106 (339)

NANCY L. STULTS, Department of Biochemistry, The University of Georgia, Athens, Georgia 30602 (141)

MARK SWAISGOOD, Department of Biochemistry, Michigan State University, East Lansing, Michigan 48824 (423)

ANTHONY L. TARENTINO, Wadsworth Center for Laboratories and Research, New York State Department of Health, Albany, New York 12201 (111)

ALAN M. TARTAKOFF, Department of Pathology, Case Western Reserve University School of Medicine, Cleveland, Ohio 44106 (351)

JEAN-PIERRE TOUTANT,[5] Department of Pharmacology, Case Western Reserve University School of Medicine, Cleveland, Ohio 44106 (231)

ROBERT B. TRIMBLE, Wadsworth Center for Laboratories and Research, New York State Department of Health, Albany, New York 12201 (111)

JERROLD R. TURNER, Department of Pathology, Case Western Reserve University School of Medicine, Cleveland, Ohio 44106 (351)

BO VAN DEURS, Structural Cell Biology Unit, Department of Anatomy, The Panum Institute, University of Copenhagen, DK-2200 Copenhagen N, Denmark (365)

DORA VEGA-SALAS,[6] Cornell University Medical Center, Department of Cell Biology and Anatomy, New York, New York 10021 (37)

MARGARET H. WADE, Meridian Instruments, Inc., Okemos, Michigan 48864 (423)

[4] *Present address:* Instituto Nazionale della Nutrizione, 00179 Rome, Italy.

[5] *Permanent address:* Laboratoire de Physiologie Animale, INRA, Place Viala, 34060 Montpellier Cedex, France.

[6] *Present address:* Instituto de Investigaciones Bioquimicas, Fundacion Campomar, 1405 Buenos Aires, Argentina.

PREFACE

The elucidation of the events of vesicular transport along the secretory and endocytic paths has grown out of the combined efforts of electron microscopy, cyto- and immunocytochemistry, autoradiography, genetics, and biochemistry. The traditional cell types investigated exhibited macroscopic (i.e., visible in the light microscope) evidence of their transport activities—for the secretory path, so-called "regulated" secretory cells were frequent objects of study and for the endocytic path an accessible and conspicuous model was provided by phagocytic cells. We now realize that these were both special cases—much as the study of skeletal muscle was a special case which provided essential background for investigating the contractile and cytoskeletal elements of non-muscle cells. Today, the attention of cell biologists has broadened to include cells engaged in constitutive protein transport and cells engaged in pinocytosis and receptor-mediated endocytosis of soluble ligands.

The diversification of choice of cell type has been matched by a diversification of the macromolecules under study since the availability of highly sensitive and specific immune reagents has made it possible to study the transport of essentially any macromolecule. The result has been a major increase in interest in this area of membrane cell biology and the development of an "applied" or "protein-specific" cell biology. The dividends have not, however, been exclusively applied—the insistence of scientists upon investigation of their favorite objects of study has inevitably brought to light basic phenomena of far-reaching importance, such as the numerous observations of general importance for understanding vesicular transport which have resulted from the study of transport of viral envelope glycoproteins or from the study of transport in yeast.

These volumes of *Methods in Cell Biology* highlight procedures of general interest, some of which are of use for investigation of the basic mechanisms operating in intracellular transport and some of which will be most valuable for descriptive studies by investigators monitoring the transport of a particular macromolecule. A limited number of other publications include procedural detail of the sort which is included in these chapters. When possible, these cross-references are given either in the prefatory remarks or in the chapters themselves. Nevertheless, cell biological methods have often not been systematized—they must be witnessed first-hand in order to be exactly reproduced. As motivation for clarity and completeness of exposition, I have urged the contributors to these volumes to consider that their chapters provide an opportunity to reduce their expenditure of time in explaining their methods first-hand to others.

Despite the established importance of nucleic acid-based procedures (transfection, quantitation of mRNA, manipulation of the structure both of the cell's transport apparatus and of the structure of molecules undergoing transport, etc.) for study of vesicular transport, these procedures are not emphasized in these volumes. This is because they are still in a state of development, because they are covered in several up-to-date texts and symposia (Spandidos and Wilkie, 1984; Chirikjian, 1985; Hooper, 1985; Glover, 1985, 1987; Miller and Calos, 1987), and because most of these methods are not designed primarily for the study of vesicular transport.

Methods for the study of vectorial transport to the cisternal space of the RER have also been covered before (Fleischer and Fleischer, 1983), although the newest yeast protocols, which make use of permeabilized cells and post-translational vectorial transport, are presented in Section I,A.

Because of the remarkable genetic accessibility of yeast (Schekman, 1985; Botstein and Fink, 1988), it has come to serve as a minimal model for studies of vesicular transport. I have therefore included several chapters on methods specific to yeast (an overview of the secretory path in the context of SEC mutants, permeabilization procedures, subcellular fractionation, and EM immunocytochemical and fluorescent methods). Two earlier volumes in this series have been entirely devoted to yeast. Their tables of contents are included in the Contents of Recent Volumes.

Apart from the effort to provide a precise description of the covalent and conformational maturation of macromolecules undergoing transport, major issues which I hope these volumes will help address are: (1) the identification of structural determinants which govern the destination and itinerary of individual macromolecules and (2) the description of vesicular carriers and the mechanisms which underlie their functions. It is especially this latter area which is obscure, despite definite progress in development of cell-free models of transport and first analysis of cellular mutants defective in transport.

Outstanding open questions concern: (1) the nature of specific membrane–membrane recognition; (2) the control of membrane fusion; (3) the possible roles of "cytoskeletal" components; (4) the issue of how individual vesicular carriers can participate in cyclic transport; and (5) the issue of the extent to which membrane–membrane interactions are stochastic.

Many well-studied models for specific membrane–membrane interaction concern the approximation and fusion of the ectodomain of membranes, for example, sperm–egg interaction, recognition between *Chlamydomonas* mating types, enveloped virus-cell surface or -endosome fusion. By contrast, events of vesicle fusion along the secretory and endocytic paths must be initiated at the endodomain (e.g., endosome–endosome fusion; secretion granule–plasma membrane fusion). Vesicle departure from a closed surface may, however, involve fusion between the apposed ectodomains of a single continuous surface (e.g., budding from the transitional elements of the RER, endocytosis).

Although fusion of isolated vesicles has been accomplished *in vitro*, and although there are specific proteins on the endodomain of vesicular carriers, it is not known to what extent these components are responsible for the specificity of membrane–membrane interaction. Considering the very large thermodynamic barrier which opposes bilayer fusion, a powerful catalyst must intervene. The extent of resolution of the events of fusion has not yet reached a level which makes it possible to generalize with regard to parameters which regulate fusion, e.g., ATP and/or specific ions. Recent studies of exocytosis argue that although ATP may be important for a priming function, it is not needed for exocytosis itself.

An important consideration in comparing the validity of reconstitution experiments using isolated organelles with experiments on permeabilized cell models is the extent to which the cytoskeleton (conceived in its broadest sense) is involved. Many pharmacologic and anatomic observations have pointed to intimate relations between microtubules and vesicular traffic; nevertheless, much of this information is consistent with tubules playing a passive rather than an active role and the most recent studies of yeast argue against an essential role for tubulin in protein secretion. Since the organization of the cytoplasm is clearly nonrandom, it would nevertheless be surprising if no cytoskeletal proteins were obligatorily involved in membrane traffic.

Ongoing studies of vesicular traffic have identified several examples of membrane recycling (in receptor-mediated endocytosis, in recapture of "exocytic membrane" contributed to the cell surface at the moment of exocytosis, in exit from the RER). These striking examples of membrane economy raise a basic question of vesicle targeting specificity—apparently a single carrier can either fuse with a given partner or separate from it. This situation suggests that the parameters which initially lead to fusion or fission decay, so that the transport options of a given vesicle change through time.

Related to this issue is the question of the extent to which membrane–membrane interactions are stochastic. Studies of the *N*-glycans of glycoproteins which exit from the RER argue strongly that secretory glycoproteins do not undergo exocytosis until they have traversed the Golgi stack; nevertheless, it is by no means clear that they traverse the Golgi in altogether sequential fashion—indeed the considerable kinetic dispersal of a cohort of pulse-labeled newly synthesized secretory proteins suggests that the itinerary may be far from direct. Moreover, along the endocytic path (which appears to have considerable overlap with terminal steps of the constitutive secretory path) growing evidence indicates that a significant portion of endocytic tracers returns to the plasma membrane rather than being uniformly delivered to lysosomes.

A broad overview of the biological issues which are addressed by the procedures described in these two volumes is given in several recent books, reviews, and symposia (Silverstein, 1978; Evered and Collins, 1982; Gething, 1985; Kelly, 1985; Pfeffer and Rothman, 1987; Pastan and Willingham, 1987; Tartakoff, 1987).

I thank Sonya Olsen for help in preparation of these volumes, Michael Lamm and the National Institutes of Health for their support, and Leslie Wilson and the inspiring Mexican countryside for the impetus to edit these volumes.

ALAN M. TARTAKOFF

REFERENCES

Botstein, D., and Fink, G., (1988). Yeast, an experimental organism for modern biology. *Science* **240**, 1439–1444.

Chirikjian, J., ed. (1985). "Gene Amplification and Analysis," Vol. 3. Elsevier, New York.

Evered, D., and Collins, G., eds (1982). "Membrane Recycling" (Ciba Foundation Symposium, New Series #92). Pittman, London.

Fleischer, S., and Fleischer, B., eds (1983). "Methods in Enzymology," Vol. 96. Academic Press, Orlando, Florida.

Gething, M. J., ed. (1985). "Protein Transport and Secretion." Cold Spring Harbor Laboratory, Cold Spring Harbor, New York.

Glover, D., ed. (1985). "DNA Cloning, A Practical Approach," Vol. II. IRL Press, Oxford.

Glover, D., ed. (1987). "DNA Cloning, A Practical Approach," Vol. III. IRL Press, Oxford.

Hooper, M. (1985). "Mammalian Cell Genetics." Wiley and Sons, New York.

Kelly, R. (1985). Pathways of protein secretion in eukaryotes. *Science* **189**, 347–358.

Miller, J., and Calos, M., eds. (1987). "Gene Transfer Vectors for Mammalian Cells." Cold Spring Harbor Laboratory, Cold Spring Harbor, New York.

Pastan, I., and Willingham, M., eds. (1987). "Endocytosis." Plenum, New York.

Pfeffer, S., and Rothman, J. (1987). Biosynthetic protein transport and sorting by the endoplasmic reticulum and Golgi. *Ann. Rev. Biochem.* **56**, 829–852.

Schekman, R. (1985). Protein localization and membrane traffic in yeast. *Ann. Rev. Cell Biol.* **1**, 115–143.

Silverstein, S., ed. (1978). "Transport of Macromolecules in Cellular Systems." Dahlem Conferenzen, Berlin.

Spandidos, D., and Wilkie, N. (1984). *In* "Transcription and Translation" (B. Hames and S. Higgins, eds.). IRL Press, Oxford.

Tartakoff, A. (1987). "The Secretory and Endocytic Paths." Wiley and Sons, New York.

Part I. Monitoring and Regulating the Progress of Transport—Secretory Path

A wide range of tissues and cells have been used for study of the secretory path. Certain of these preparations have been described in previous volumes (Hauri, 1988; Pitot and Sirica, 1980; Quaroni and May, 1980; Scheele 1983). A very large portion of recent research interest has centered on cells in culture, especially cells infected with enveloped viruses or transfected with genes of choice. In this volume, Chapters 2, 3, and 4 by Rodriguez-Boulan *et al.*, Krieger *et al.*, and Bergmann, respectively, Chapter 13 by Breitfeld *et al.* (Part II), and in Volume 31, Chapter 9 by Ellens *et al.* (Part I), exemplify the impressive potential of such systems. Chapter 4 by Bergmann describes the use of temperature-sensitive mutants of vesicular stomatitis virus for studies of the secretory path. Previous articles have described temperature-sensitive transport mutants of other enveloped viruses (Kääriäinen *et al.*, 1983).

An introductory advertisement (hitchhiker's guide) for the use of yeast is included as Chapter 1. The most technically complete introduction to exploitation of *Xenopus* oocytes for studies of transport is Colman (1984).

In addition to the numerous noncovalent modifications that accompany intracellular transport (assembly of oligomeric proteins, binding of lipids or other prosthetic groups, etc.), a large number of covalent modifications have been identified (Wold and Moldave, 1984). In only a limited number of cases are corresponding inhibitors available.

Because of the extremely widespread occurrence of N and O glycosylation, and because the progress of transport can often be inferred from study of glycan structure (see also Snider, Chapter 14, Part II, this volume), two of the following chapters (Tarentino *et al.*, Chapter 5; Cummings *et al.*, Chapter 6) provide procedural detail for enzymologic and chemical analysis of oligosaccharide maturation. Two other recently discovered covalent modifications that may accompany transport of membrane proteins are detailed in Chapter 8 by Giacoletto *et al.* and Chapter 9 by Rosenberry *et al.* The reader's attention is also called to protocols employing heavy isotopes, which are suitable for monitoring transport (Devreotes *et al.*, 1977; Krupp *et al.*, 1983).

Temperature reduction or addition of pharmacologically active agents provide versatile means for studying mechanisms of transport between compartments, for interrupting transport along both the secretory and endocytic paths, and for identifying the subcellular site of structural (covalent and noncovalent) modifications. Such procedures are of widespread and easy application, and therefore have come to be broadly used. Vesicular traffic can also be manipulated by modification of the cell cycle (Birky, 1983; Warren, 1985) or cytoplasmic pH (see Sandvig *et al.*, Chapter 16, Part II, this volume).

References

Birky, C. (1983). The partitioning of cytoplasmic organelles at cell division. *Int. Rev. Cytol.* **15,** Suppl., 49–89.

Colman, A. (1984). Translation of eukaryotic mRNA in *Xenopus* oocytes. *In* "Transcription and Translation" (D. Hames and S. Higgins, eds.). IRL Press, Oxford.

Devreotes, P., Gardner, J., and Fambrough, D. (1977). Kinetics of biosynthesis of acetylcholine receptor and subsequent incorporation into the plasma membrane of cultured chick skeletal muscle. *Cell (Cambridge, Mass.)* **10,** 365–373.

Hauri, H.-P. (1988). Biogenesis and intracellular transport of intestinal brush border membrane hydrolases: Use of antibody probes and tissue culture. *Subcell. Biochem.* **12,** 155–220.

Kääriäinen, L., Virtanen, I., Saraste, J., and Keranen, S. (1983). Transport of virus membrane glycoproteins, use of temperature-sensitive mutants and organelle-specific lectins. *In* "Methods in Enzymology" (S. Fleischer and B. Fleischer, eds.), Vol. 96, pp. 453–465. Academic Press, New York.

Krupp, M., Knutson, V., Ronnett, G., and Lane, M. (1983). Use of heavy isotope density-shift method to investigate insulin receptor synthesis, turnover and processing. *In* "Methods in Enzymology" (S. Fleischer and B. Fleischer, eds.), Vol. 96, pp. 423–433. Academic Press, New York.

Pitot, H., and Sirica, A. (1980). Methodology and utility of primary cultures of hepatocytes from experimental animals. *In* "Methods in Cell Biology" (C. Harris, B. F. Trump, and G. Stoner, eds.), Vol. 21B, pp. 441–456. Academic Press, New York.

Quaroni, A., and May, R. (1980). Establishment and characterization of intestinal epithelial cell cultures. *In* "Methods in Cell Biology" (C. Harris, B. F. Trump, and G. Stoner, eds.), Vol. 21B, pp. 403–428. Academic Press, New York.

Scheele, G. (1983). Pancreatic lobules in the vitro study of pancreatic acinar cell function. *In* "Methods in Enzymology" (S. Fleischer and B. Fleischer, eds.), Vol. 96, pp. 17–27. Academic Press, New York.

Warren, G. (1985). Membrane traffic and organelle division. *Trends Biochem. Sci.* **10,** 439–443.

F. Wold and K. Moldave, eds. (1984). "Methods in Enzymology," Vols. 106 and 107. Academic Press, Orlando, Florida.

Chapter 1

A Hitchhiker's Guide to Analysis of the Secretory Pathway in Yeast

JONATHAN ROTHBLATT AND RANDY SCHEKMAN

Department of Biochemistry
University of California at Berkeley
Berkeley, California 94720

I. Introduction

The highly compartmentalized organization of cellular activities in the budding yeast, *Saccharomyces cerevisiae,* as in higher eukaryotes, requires a mechanism that ensures the correct localization of protein and lipid. Aside from a small number of mitochondrial DNA-encoded proteins, all cellular protein biosynthesis operates in the cytoplasm. Proteins to be assembled into the nucleus, mitochondrion, or peroxisome must be distinguished from those requiring entry into the secretory pathway for

Copyright © 1989 by Academic Press, Inc.
All rights of reproduction in any form reserved.

their distribution. Secretory proteins are synthesized on polyribosomes that preferentially associate with the endoplasmic reticulum (ER) shortly after initiation and discharge the nascent polypeptides into the ER. Within the secretory pathway there is a selective modification and transport (or retention) of proteins en route to the cell surface, the vacuole, or one of a number of compartments (including the ER, Golgi apparatus, and secretory vesicles) composing the pathway. Both *in vivo* and *in vitro* analyses, using a variety of cell systems, have shown that secretory precursors, as well as mitochondrial, peroxisomal, and nuclear precursors, have signals or conformational features that specifically target polypeptides to their correct destination (Garoff, 1985; Schekman, 1985). Recent studies suggest that secretory protein transport to the cell surface is a default process unless there is specific retention of ER- or Golgi-specific constituents (Wieland *et al.*, 1987; Poruchynsky *et al.*, 1985; Munro and Pelham, 1987; Pääbo *et al.*, 1987).

N-Glycosidically linked precursor oligosaccharides are essentially identical in all eukaryotic glycoproteins. In yeast secretory proteins, the N-linked core oligosaccharides are trimmed to $Man_8GlcNAc_2$ and then are extensively modified with additional mannose residues during their transit through the Golgi appartus, producing a carbohydrate structure referred to as the outer chain (Ballou, 1982; Kukuruzinska *et al.*, 1987). Although the outer-chain mannose structure provides yeast cells with distinctive antigenic determinants, carbohydrate does not appear to serve a role in yeast protein sorting. Invertase secretion is not affected by mutations interfering with outer-chain glycosylation (Ballou *et al.*, 1980). Targeting of hydrolytic enzymes to the lysosome of animal cells is determined by mannose-6-phosphate (Man6P) added to the core oligosaccharide of lysosomal precursors. Interaction of Man6P with its cognate receptor in the Golgi apparatus redirects the hydrolases to lysosomes (von Figura and Hasilik, 1986). In contrast, at least two vacuolar proteins, carboxypeptidase Y (CPY) and alkaline phosphatase, are synthesized and activated normally in the absence of oligosaccharide synthesis (Schwaiger *et al.*, 1982; Onishi *et al.*, 1979). By examining mutations in the structural gene for CPY that result in missorting of an otherwise normal proenzyme, the signal responsible for vacuolar targeting has been localized to an amino-terminal domain of the propeptide (Valls *et al.*, 1987; Johnson *et al.*, 1987).

Protein secretion and plasma membrane assembly in *S. cerevisiae* are directed topologically to the bud portion of a growing cell, it being the region of active cell surface growth. Both invertase and acid phosphatase are transported into the bud (Tkacz and Lampen, 1973; Field and Schekman, 1980). Enzyme cytochemistry has localized acid phosphatase reaction product to vesicles in the bud, as well as to the ER and a Golgi-like structure (Linnemans *et al.*, 1977). The *sec1* strain specifically accumulates mature secretory proteins within vesicles which fail to release their contents at the plasma membrane. These cells also fail to continue bud growth as a consequence of a block in plasma membrane and cell wall assembly (Novick and Schekman, 1979). Holcomb *et al.* (1988) and Brada and Schekman (1988) demonstrated biochemically and immunocytochemically in *sec1* cells that a constitutive integral protein of the plasma membrane, the vanadate-sensitive ATPase, and two catabolite-repressible periplasmic glycoproteins, acid phosphatase and invertase, are transported to the cell surface in the same secretory vesicles.

Molecular dissection of the mechanism(s) of protein sorting and intercompartmental transport within the yeast secretory pathway has been facilitated by the isolation and characterization of a large number of conditionally lethal, temperature-sensitive secretion, *sec* (Novick *et al.*, 1980; Ferro-Novick *et al.*, 1984a; Deshaies and Schekman, 1987) and *bet* (Newman and Ferro-Novick, 1987), mutants. Biochemical analyses of the *sec* mutations showed that secretion and growth are blocked at the restrictive temperature (37°C), leading to the accumulation of soluble secretory and vacuolar precursors, as well as integral membrane proteins residing in the vacuole, plasma membrane, and organelles of the secretory pathway (reviewed in Schekman and Novick, 1982).

The mutants, which define a minimum of 25 genes acting at different stages in the secretory pathway, are of three types. The earliest step in secretion, namely, protein translocation into the ER, is blocked when cells from one group, *sec61, sec62,* and *sec63,* are shifted to the nonpermissive temperature (Deshaies and Schekman, 1987; J. Rothblatt, R. Deshaies, S. Sanders, G. Daum, and R. Schekman, unpublished observations). Mutants in the second group, comprising *sec53* and *sec59,* fail to glycosylate and assemble polypeptides properly in the ER lumen (Ferro-Novick *et al.*, 1984a). Members of the last group are defective in interorganellar transport, leading to the accumulation of secretory intermediates in atypical or exaggerated forms of the ER *(sec18),* Golgi apparatus *(sec7* and *sec14),* or secretory vesicles *(sec1* and *sec4)* Schekman and Novick, 1982).

Analysis of wild-type and secretion-defective *S. cerevisiae* strains by pulse-radiolabeling of spheroplasts and intact cells will continue to provide important information about the temporal and spatial organization, as well as the molecular components, of the secretory pathway. The protocols and tables presented in this chapter are intended as a guide to follow (or hitchhike along with) secretory proteins during their modification and transport en route to their respective destinations.

II. Yeast Strains and Growth Conditions

A. Yeast Strains

Table I provides a list of commonly used *S. cerevisiae* strains bearing *sec* mutations affecting distinct points in the yeast secretory pathway. All *sec* mutants are derived from X2180-1A *(gal2 MAT a;* Yeast Genetics Stock Center, University of California, Berkeley) (Novick *et al.*, 1980), except RDM 15-5B, which was obtained by mutagenesis of FC2-12B (Deshaies and Schekman, 1987). Yeast strains are maintained on YPD–agar plates (1% Bacto-Yeast extract, 2% Bacto peptone, 2% D-glucose, 2% Bacto-agar). For long-term storage, strains are preserved as frozen stocks. Cells are grown to stationary phase in YPD (1% Bacto-Yeast extract, 2% Bacto-peptone, 5% D-glucose), and 0.8 ml of the culture is transferred to a sterile 1.2-ml freezing vial. The cell suspension is supplemented with 0.2 ml of sterile 75% glycerol, and the vial is transferred to a −80°C freezer. Cells from frozen stocks are streaked onto YPD–agar plates prior to use.

For studies of CPY transport to the vacuole, it is useful to work with a strain that has a mutation at (or deletion of) the *PEP4* locus. *PEP4* encodes a processing protease (protease A) required for propeptide cleavage and activation of CPY and other vacuolar hydrolases. Relative to the primary translation product, prepro-CPY, the increase in mass due to oligosaccharide addition is offset by the release of the 8-kDa propeptide. Thus, the mature, vacuolar form of CPY cannot be resolved from prepro-CPY on sodium dodecyl sulfate (SDS)–polyacrylamide gels. The *pep4-3* mutation blocks the final maturation step, and as a result, in pulse–chase experiments, it is possible to distinguish between prepro-CPY and the fully glycosylated species.

B. Growth Conditions for Radiolabeling

For ^{35}S or ^{32}P radiolabeling, the appropriate growth conditions will depend largely on the protein(s) to be examined. Expression of a variety of regulated secretory and membrane proteins requires growth conditions that starve yeast of a particular catabolite. Whereas derepression of extracellular invertase is very rapid in low-glucose medium, high-level expression of extracellular acid phosphatase or plasma membrane sulfate permease requires longer periods of derepression. Two commonly studied yeast proteins—the α-cell-specific mating pheromone, α-factor, and the soluble vacuolar protease, CPY—are produced constitutively, and thus no special growth conditions are required to examine the biogenesis

TABLE I

SITE OF SECRETORY PROTEIN ACCUMULATION IN AND GENOTYPE OF COMMONLY USED *sec* MUTANTS

Strain name (lab collection no.)	Site of accumulation	Genotype	Reference
RDM 15-5B (RSY 283)	Cytoplasm[a]	*sec61-2 ura3-52 leu2-3 leu2-112 ade2 pep4-3 MATα*	Deshaies and Schekman (1987)
SEY 5536 (RSY 12)	ER	*sec53-6 ura3-52 leu2-3 leu2-112 gal2 MATα*	Ferro-Novick *et al.* (1984a)
SEY 5186 (RSY 11)	ER	*sec18-1 ura3-52 leu2-3 leu2-112 gal2 MATα*	Emr *et al.* (1984)
SEY 5076 (RSY 9)	Golgi apparatus	*sec7-1 ura3-52 leu2-3 leu2-112 gal2 MATα*	Emr *et al.* (1984)
SF590-3B (RSY19)	Golgi apparatus	*sec14-3 pep4-3 gal2 MATα*	Stevens *et al.* (1982)
SEY 5016 (RSY 10)	Secretory vesicles	*sec1-1 ura3-52 leu2-3 leu2-112 gal2 MATα*	Emr *et al.* (1984)

[a] Although accumulated precursor molecules are protease-sensitive, they sediment with the membrane fraction. It is not known whether the precursor molecules specifically associate with ER membranes.

of these proteins. However, growth conditions necessary for derepression of one gene can indirectly lead to reduced expression of another gene. We have observed that the low concentrations of glucose that derepress synthesis of invertase result in reduced levels of α-factor synthesis. In general, though, for ^{35}S radiolabeling of proteins, yeast cultures are grown for several generations under conditions that deplete the intracellular sulfate and methionine pools and induce the plasma membrane sulfate permease activity. For ^{32}P radiolabeling and studies of acid phosphatase processing and transport, strains are grown in low-phosphate medium.

C. Preparation of Synthetic Growth Medium

Synthetic minimal media (MV) (Wickerham, 1946) suitable for growth and radiolabeling of yeast are prepared from the following components according to experimental needs: 10× minimal salts (sulfate- and/or phosphate-free), 1000× trace elements (sulfate-free), 200× vitamins, 50% (w/v) D-glucose, 100× supplements (uracil, adenine HCl, and/or required amino acids), 10 m*M* ammonium sulfate, 10 m*M* potassium phosphate, monobasic. All of these components with the exception of 200× vitamins can be prepared and autoclaved individually and stored at room temperature. The 200× vitamins should be filter-sterilized (0.22-μm pore diameter) and stored refrigerated.

10× Minimal Salts (Sulfate-Free)

	per 500 ml H_2O
Ammonium chloride	10.1 g
Potassium phosphate, monobasic	5.0 g
Magnesium chloride · $6H_2O$	4.15 g
Sodium chloride	0.5 g
Calcium chloride, anhydrous	0.5 g

Note: For SO_4^{2-}-rich medium, the ammonium and magnesium chloride salts are replaced by 25 g of $(NH_4)_2SO_4$ and 2.5 g of anhydrous $MgSO_4$. For PO_4^{3-}-free medium, the potassium phosphate salt is replaced by 2.75 g potassium chloride and 5.6 g sodium acetate and the pH adjusted to pH 5.5.

1000× Trace Elements (Sulfate-Free)

	per 100 ml H_2O
Boric acid	50 mg
Copper chloride · $2H_2O$	3 mg
Potassium iodide	10 mg
Ferric chloride	20 mg
Manganese chloride · $4H_2O$	46 mg

Sodium molybdate	20 mg
Zinc chloride	34 mg

Note: If a precipitate forms at room temperature, it is simply resuspended by mixing prior to use.

200× Vitamins

	per 100 ml H_2O
D-Biotin and folic acid	1 ml
D-Pantothenic acid, hemicalcium salt	8 mg
Myoinositol	40 mg
Nicotinic acid	8 mg
p-Aminobenzoic acid	4 mg
Pyridoxine HCl	8 mg
Riboflavin	4 mg
Thiamine HCl	8 mg

Note: D-Biotin and folic acid are prepared together as a 20,000× stock (4 mg of each per 100 ml H_2O), of which 1 ml is added per 100 ml of 200× vitamins.

100× Amino Acid and Nitrogen Base Supplements (Sulfate-Free)

	per 100 ml H_2O
Adenine HCl	200 mg
Uracil	200 mg
L-Tryptophan	200 mg
L-Histidine	200 mg
L-Leucine	300 mg

Note: These are auxotrophies routinely used in our lab strains as selectable markers. A complete listing of amino acid supplements for other auxotrophies can be found in Sherman *et al.* (1983).

III. *In Vivo* Radiolabeling and Lysis of Yeast Cells

A. Protocol for $^{35}SO_4^{2-}$ Radiolabeling of Intact Yeast Cells

A protocol for radiolabeling of secretory and vacuolar proteins, such as invertase and CPY, as well as low-abundance membrane proteins in wild-type and *sec* mutant strains is described here.

A stationary-phase YPD culture is prepared in advance of a series of radiolabeling experiments. A single colony is inoculated into 5 ml of YPD broth (1% Bacto-Yeast extract, 2% Bacto-peptone, 5% D-glucose) in a sterile 80 × 150-mm culture tube and grown at permissive temperature with good aeration to stationary phase (OD_{600} ~40–50). The stationary culture is usually stable for at least 2 weeks at 4°C.

To obtain logarithmic-phase cells (OD_{600} = 0.2–0.8; 1 OD_{600} unit = 1 × 10^7 cells) for ^{35}S radiolabeling, it is convenient to start an MV culture the evening before the radiolabeling experiment. The exact volume of inoculum for overnight growth (roughly five generations) is dependent on the growth rate of the particular strain in use, which, in turn, will depend on the phenotype of the strain, the growth temperature, and the titer of viable cells in the stationary culture. In practice, the doubling time of a wild-type or *sec* strain at 24°C (permissive for *sec* growth) is roughly 3–3.5 hours in MV-low sulfate. Hence, a 1 : 2000 dilution of the stationary culture into minimal medium is a safe estimate for the inoculation. A 25-ml liquid culture of MV-low sulfate (1× each of sulfate-free minimal salts, trace elements, vitamins, and supplements, 5% glucose, and 200 μ*M* ammonium sulfate) in a sterile 125-ml Erlenmeyer flask is inoculated with 12.5–25 μl of the stationary YPD culture. A gyrotory water bath shaker (New Brunswick Scientific, model G76), fitted with flask clamps and a test tube rack is suitable for both growing and radiolabeling up to nine different cultures.

Once cells have reached midlogarithmic phase, growth conditions can be adjusted for the purpose of imposing a *sec* mutation block in the secretory pathway, derepressing the expression of extracellular invertase, or introducing drugs that interfere with protein modification, such as the inhibitor of lipid-linked core oligosaccharide precursor synthesis, tunicamycin.

To establish the *sec* block, the required number of cells, expressed as OD_{600} units, is transferred to each of two sterile flasks (or culture tubes if the volumes are small). One culture is shifted to a gyrotory water bath adjusted to the restrictive temperature while the other culture is kept at the permisive temperature for the appropriate length of time.

Derepression of the secretory form of invertase is initiated by harvesting the overnight culture, resuspending the cells in low-sulfate medium containing 0.1% glucose, and continuing the temperature shift for 15–30 minutes before labeling. Since CPY synthesis is not affected by invertase derepression, both proteins can be examined in derepressed cells.[1] The efficiency of invertase derepression is modestly sensitive to growth at 37°C, so that it may be advisable to transfer cells briefly to derepressing

[1]Analyzing α-factor and invertase in the same experiment is somewhat more laborious in that it requires that an aliquot of cells from each culture be shifted to low-glucose medium for invertase derepression and another be kept under conditions of invertase repression.

conditions before the shift to elevated temperature required to impose a Sec phenotype. The suitability of this largely depends on the length of time required to establish a completely restrictive *sec* block. The duration of the temperature shift necessary for a tight block can vary from 15 minutes at 30°C for *sec18* to 2 hours at 37°C for *sec61* (Deshaies and Schekman, 1987). For long temperature shifts (i.e., ~1–2 hours), we usually shift to the restrictive temperature first and then begin the invertase derepression 15–30 minutes before radiolabeling. For periods of up to 3 hours of invertase derepression, minimal medium supplemented with 2% sucrose and 0.05% glucose has been used (Novick and Schekman, 1979).

Assuming a 2-hour shift to 37°C before radiolabeling, the OD_{600} of the culture is measured after 90 minutes at the restrictive temperature. The required number of OD_{600} units of cells[2] is transferred from each culture to small sterile culture tubes (13 × 100-mm disposable) and harvested by centrifuging 2–3 minutes at top speed in a tabletop clinical centrifuge at room temperature. The cell pellet is washed once with MV-low sulfate, 0.1% glucose (prewarmed to the permissive or restrictive temperature) and resuspended in the same medium at a density of ≤4 OD_{600} units/ml (maximum volume of 3 ml in 13 × 100-mm tubes). The cultures are returned to the 24° or 37°C shaking water bath for 15–30 minutes' additional incubation.

Addition of asparagine-linked oligosaccharides to polypeptides entering the ER can be blocked by the addition of tunicamycin, an inhibitor of dolichol-oligosaccharide precursor biosynthesis (Kuo and Lampen, 1974), to the medium for the last 15 minutes of the prelabeling temperature shift and for the duration of the radiolabeling. A 10-mg/ml stock suspension of tunicamycin (prepared in 100% ethanol and stored at −20°C) is added to the culture to a final concentration of 10 μg/ml.

At the end of the 2-hour prelabeling temperature shift, the cells are harvested as before, washed with 1 ml of MV-no sulfate (prewarmed to the appropriate temperature), and resuspended to 2 OD_{600} units/ml in MV-no sulfate[3] in the same 13 × 100-mm culture tubes. The cultures are preincubated for an additional 5 minutes at the appropriate temperature. Ra-

[2]To detect radiolabeled proteins by autoradiography of SDS–polyacrylamide gels within 1–2 days, immunoprecipitations are performed on the following amounts of cell extract: α-factor and acid phosphatase, 0.5 OD_{600} unit of cells; invertase, 1.0 OD_{600} unit of cells; CPY, 0.25–0.5 OD_{600} unit of cells; low-abundance membrane proteins (e.g., dipeptidylaminopeptidase B, Kex2 endopeptidase, and α-factor receptor), 2–3 OD_{600} units of cells.

[3]In order to protect small peptides secreted into the medium, such as the mature form of α-factor, the labeling is performed in MV-no sulfate containing 200 μ/ml bovine serum albumin (BSA) and 2 m*M* TAME (*N*α-*p*-tosyl-L-arginine methyl ester, Sigma Chemical Co.) (Ciejek and Thorner, 1979).

diolabeling is initiated with the addition of 200–400 μCi of $Na_2{}^{35}SO_4$ (ICN Radiochemicals, cat. no. 64041) per OD_{600} unit of cells.[4] For ease of handling, the radioisotope, which is usually supplied at ~400 μCi/μl, is diluted with sterile distilled water to a concentration of 100 μCi/μl. Typically, the radiolabeling is allowed to proceed for 30 minutes at the permissive or restrictive temperature. For pulse–chase experiments, shorter labeling times may be appropriate. In the latter case, the chase is initiated with the addition of 1/100 volume of a solution containing 100 m*M* $(NH_4)_2SO_4$, 0.3% L-cysteine, and 0.4% L-methionine. After the desired period of time, the radioisotope pulse and chase are terminated with the addition of an equal volume of ice-cold 20 m*M* sodium azide and rapid chilling in an ice-water bath. To determine the efficiency of radiolabel uptake, a 2-μl aliquot is removed at this stage, spotted onto a Whatman GF/C filter disk, and counted in a nonaqueous scintillant to measure the total input of counts. An equivalent aliquot is taken after cell lysis to determine cell-associated counts. In general, >80% of the sulfate label is found associated with cells. The amount of label incorporated into protein can be obtained by measuring trichloroacetic acid (TCA)-precipitable counts, essentially as described by Erickson and Blobel (1983), for cell-free translation products. An aliquot is spotted onto a Whatman GF/C filter and immersed in ice-cold 10% TCA. After 10 minutes, the 10% TCA is replaced with fresh ice-cold 10% TCA. After an additional 10 minutes, the 10% TCA is replaced with 5% TCA and and heated to boiling for 5–10 minutes to destroy [^{35}S] methionyl-tRNA. The filters are then washed sequentially in ethanol–ether (1 : 1) and 100% ether, air dried, and counted in nonaqueous scintillant.

The cells are collected by centrifugation and washed once with ice-cold 10 m*M* sodium azide. The culture supernatant may be saved, if necessary, for later immunoprecipitation of polypeptides that are released into the medium. Rapid lysis of radiolabeled intact cells for the analysis of total cell protein (intracellular and periplasmic) is achieved by glass bead disruption in the same small culture tubes in which the radiolabeling was performed. Cells are resuspended at a density of 5–20 OD_{600} units/ml in 1% SDS–1 m*M* phenylmethylsulfonyl fluoride (PMSF)–50 m*M* Tris-HCl, pH 7.4. For reasons of convenience in setting up immunoprecipitations, the lysis is done in a volume of 200 μl, although multiples of this volume would be appropriate if more than one immunoprecipitation from each sample is anticipated. Approximately 200 mg (filling the conical portion of a 0.5-ml microcentrifuge tube) of chromic–sulfuric acid-washed 0.5-

[4] It is also possible to label yeast efficiently with Tran^{35}S-label (ICN Radiochemicals, cat. no. 51006), an *Escherichia coli* hydrolysate containing L-[^{35}S]methionine. Cells are labeled with 25 μCiperOD$_{600}$ unit at a cell density of 10 OD_{600} units/ml in MV-no sulfate.

mm glass beads (Biospec Products, Bartlesville, OK; cat. no.11078-105) are added and the samples are vortexed twice for 30 seconds at top speed. In general, samples are denatured in a boiling-water bath for 5 minutes prior to immunoprecipitation. However, certain hydrophobic (membrane) proteins aggregate at 100°C, and it may be necessary to solubilize the samples at lower temperatures: 75°C for vacuolar dipeptidylaminopeptidase B (C. Roberts and T. Stevens, personal communication), or as low as 37°C in the presence of 8 *M* urea for the α-factor receptor (Blumer *et al.*, 1988).

B. Preparation and Homogenization of Yeast Spheroplasts

In pulse–chase experiments or experiments examining the thermoreversibility of a *sec* block, it may be desirable to distinguish between the intracellular and periplasmic forms of invertase or acid phosphatase. This can be achieved by removing the cell wall after radiolabeling to release periplasmic proteins into the spheroplasting medium or by radiolabeling yeast spheroplasts so that the polypeptides of interest are secreted into the labeling medium. In order to perform subcellular fractionation or proteolysis assays of polypeptide topography, prior removal of the yeast cell wall allows gentle and efficient homogenization of spheroplasts.

The yeast cell wall is removed by enzymatic digestion of the 1,3-β-glucans that compose much of the cell wall structure. A number of commercial preparations are available, including Zymolyase (Miles Pharmaceuticals), which is a lyophilized powder of partially purified *Arthrobacter luteus* culture supernatant, and Glusulase (DuPont Pharmaceuticals) or β-glucuronidase (Sigma), both of which are prepared from the gut juices of the snail *Helix pomatia*. These enzyme preparations work very well, but they possess high levels of contaminating nuclease and protease activity in the glucanase-enriched fraction. Lyticase (Scott and Schekman, 1980), which is prepared from the culture supernatant of *Oerskovia xanthineolytica* (reclassification of *A. luteus*) has a very high 1,3-β-glucanase activity relative to protease and nuclease and has been used effectively for rapid and reliable spheroplast formation. A commercial preparation of lyticase, termed Oxalyticase, is available as a lyophilized powder (Enzogenetics, Corvallis, OR 97333).

1. Preparation of Lyticase

Preparation of lyticase is a two-stage procedure that entails (1) preparing yeast glucan to be used as a substrate for induction of lytic activity by *O. xanthineolytica*, and (2) purification of the glucanase activity from the culture supernatant. Lyticase fraction II is prepared, with minor modifications, essentially as described by Scott and Schekman (1980).

a. Growth Conditions. Oerskovia xanthineolytica is grown at 30°C in M63 medium (Sistrom, 1958) containing (per liter): 100 ml of 10× M63 salts [136 g/liter KH_2PO_4, 20 g/liter $(NH_4)_2SO_4$, 42 g/liter KOH, 0.01 g/liter $Fe_2(SO_4)_3$], 1 ml of $MgSO_4 \cdot 7H_2O$ (0.2 g/ml), 1 ml of thiamine (1 mg/ml), 5 ml of biotin (0.2 mg/ml), and made to 1 liter with sterile distilled H_2O. The stock solutions of 10× M63 salts and magnesium sulfate can be autoclaved and stored at room temperature, whereas the thiamine and biotin solutions should be sterilized by filtration and stored refrigerated. For *O. xanthineolytica* growth and induction, M63 minimal medium is supplemented with 0.4% glucose, 0.4% yeast glucan, or 3% washed autoclaved yeast as carbon source.

Autoclaved yeast cells are prepared from yeast strain X2180 grown in 500 ml of YPD to stationary phase. Cells are harvested (14,000 *g* for 10 minutes in GSA or GS-3 rotor) and washed once with H_2O. The cell pellet is resuspended in 100 ml H_2O and autoclaved for 30 minutes. The cells are collected by centrifugation. The supernatant is discarded and the wet weight of the cell pellet noted.

To prepare yeast β-glucan, 4.8 liters of 6% NaOH are added to 12 pounds of Red Star yeast (until used, the yeast can be stored as cakes at −20°C and thawed overnight at room temperature just before use). The yeast is stirred into suspension and heated at 75°–80°C while stirring. After 1 hour, 8 liters of cold H_2O are added. The suspension is centrifuged in a Sorvall GS-3 rotor at 14,000 *g* for 20 minutes. The pellets are resuspended in 4 liters of H_2O and the pH adjusted to pH 7.0 with HCl. The glucan is sedimented as before, and the pellets are resuspended in 5 liters H_2O. The H_2O wash is repeated once. The pellets are resuspended in 5 liters of 95% ethanol and centrifuged as before. The ethanol wash is repeated once. The final pellet is resuspended as a paste in H_2O. A small aliquot is removed, weighed, then dried and weighed again to determine the wet weight–dry weight ratio of the yeast glucan. The recovery should be ~24 g of glucan (dry weight) per pound of yeast. The glucan is stored as a paste at −80°C. Before use, the glucan is thawed and washed once with H_2O.

b. Trial Lyticase Induction. Oerskovia xanthineolytica is stored at 4°C on a MAY slant [minimal medium, 3% (wet weight) autoclaved yeast, 2% Bacto-agar]. The strain is streaked out onto a MAY plate and incubated at 30°C. Four single colonies that show a distinct halo around the colony (the halo is due to secreted lyticase, which solubilizes the yeast cell wall material in the MAY plate) are picked and inoculated into four cultures, each containing 5 ml of MGS (minimal medium, 0.4% glucose).

The cultures are grown at 30°C to stationary phase (OD_{600} ~2), which requires ~2 days. Each culture is diluted 1 : 100 into 50 ml of MGN [minimal medium, 0.4% (dry weight) glucan] and grown at 30°C with good aeration. The MGS stationary cultures can be kept temporarily at 4°C for inoculation of the large-scale cultures after completing the trial induction. Lyticase activity in the culture supernatant is monitored by removing 1 ml and centrifuging at 14,000 *g* in a Sorvall SS-34 rotor for 10 minutes. The supernatant fraction is assayed for lyticase activity as described later. Activity should appear ~15–20 hours after inoculation and should peak between 30 and 50 hours at an activity of 700–1500 units/ml. A fresh MAY slant of *O. xanthineolytica* is prepared from the culture that produced the most lyticase activity and stored at 4°C.

c. Large-Scale Induction and Purification of Lyticase. A 1-liter flask containing 200 ml of MGN is inoculated with 2 ml of the MGS stationary that produced the most lyticase activity and incubated overnight at 30°C. The 200-ml MGN culture is diluted into 10–12 liters of MGN and grown with maximum aeration in a fermenter. After 18–19 hours, activity is assayed every 1-5 hours and the supernatant fraction harvested when the activity exceeds 1000 units/ml or when the activity begins to drop off. The level of activity usually peaks between 24 and 30 hours.

The culture supernatant is collected by sedimenting the cells at 14,000 *g* for 20 minutes in a GS-3 rotor. A Sharples centrifuge is not satisfactory for harvesting the culture fluid from a fermenter. The supernatant fraction is transferred into 1 $\frac{7}{8}$-in. dialysis tubing, prepared by boiling in 2% sodium bicarbonate, 1 m*M* EDTA. A 12-liter culture will require about 12 pieces of tubing, each 80 cm long. The filled dialysis tubes are placed into large plastic tubs, covered with 2.5 kg of polyethylene glycol (PEG) 8000 powder (Sigma), and kept at 4°C. After ~16 hours, the H_2O-saturated PEG is poured off, the dialysis tubing is retied so that the bag remains taut, and 1.5 kg of PEG are added. This results in a 10-fold reduction of the volume, which usually takes about 1–2 days. The concentrate is pooled into one dialysis tube and dialyzed against 5–7 liters of 10 m*M* sodium succinate, pH 5.0 (10× sodium succinate stock: 35.4 g succinic acid and 15.2 g NaOH per 3 liters). The dialysis buffer is changed every 10–15 hours until the pH of the concentrated supernatant is ~5.0 and the conductivity is 0.5–1.0 mΩ. Dialysis should be complete in 2–3 days. When the dialysis is complete, the concentrate is transferred to a beaker.

In the purification stage of the preparation, it is usually sufficient to carry the preparation through CM-cellulose chromatography. CM-52 cellulose (Whatman) is preequilibrated in 10 m*M* sodium succinate, pH 5.0. Approximately 300 g (wet weight) of equilibrated CM-52 are added to the beaker containing the concentrated supernatant, and the mixture is stirred gently for 2 hours at 4°C. A small aliquot is removed, the CM-52 is sedimented, and the lyticase activity in the supernatant fraction is measured to ensure that most of the activity is removed. After the activity is adsorbed, the CM-52 is poured into a 5.5-cm-diameter column. The flowthrough is collected and the lyticase activity measured once again. Typically, ~20% of the activity does not bind to CM-52. The column is washed with 300–400 ml of sodium succinate buffer and the lyticase eluted with 0.25 *M* NaCl–10 m*M* sodium succinate, pH 5.0, at a rate of 3–7 ml/minute. Column fractions of 7 ml are collected. Every fifth fraction is assayed for activity, and then every other fraction across the peak of lyticase activity is assayed. The peak fractions are combined and the activity of the pooled material is measured. The final yield should be $\sim 10^7$ units from a 12-liter culture. The lyticase is sterilized by filtration, or sodium azide is added to a final concentration of 10 m*M*. Stored at 4°C, lyticase is stable for as long as 1 year with no more than a 50% decrease in glucanase activity.

d. Lyticase Assay. A suitable yeast strain (e.g., X2180) is grown overnight in 100 ml YPD to an OD_{600} of 1–4. The cells are harvested and washed once with H_2O. The cell pellet is resuspended in 50 m*M* Tris-HCl (pH 7.5), 10 m*M* sodium azide at a density of 20 OD_{600} units/ml. This cell suspension can be stored at 4°C for 2–3 days. For each assay, 1 ml of 2× assay buffer (80 m*M* β-mercaptoethanol, 100 m*M* Tris-HCl, pH 7.5) is mixed with 0.92 ml of H_2O and 0.04 ml of yeast cells, giving an OD_{600} of ~0.4. An aliquot (0–100 μl containing ~5 units of activity to be in the linear range of the assay) of culture supernatant is added. Immediately 1 ml is removed and the OD_{600} is measured. The remaining portion is incubated at 30°C. After 30 minutes, the OD_{600} is measured. One unit of lyticase activity is defined as the amount required to cause a 10% decrease in the OD_{600} after 30 minutes at 30°C.

2. Preparation of Yeast Spheroplasts

To convert radiolabeled whole cells to spheroplasts, the sodium azide-washed cells are resuspended at 20 OD_{600} units/ml in 10 m*M* dithiothreitol (DTT), 100 m*M* Tris-SO_4, pH 9.4, and incubated for 5–10 minutes at 30°C with gentle shaking. This step has been reported to facilitate enzymatic digestion by weakening the yeast cell wall (Daum *et al.*, 1982). The cells are collected in a tabletop clinical centrifuge and resuspended at ~20 OD_{600} units/ml in spheroplast buffer (1.4 *M* sorbitol, 60 m*M* β-mercaptoethanol, 10 m*M* sodium azide, 2 m*M* $MgCl_2$, 25 m*M* Tris-HCl, pH 7.5). Lyticase (25 units/OD_{600} units of cells) is added and the cell suspension is incubated for 30–60 minutes at 30°C with gentle shaking. Spheroplast conversion is monitored by following the decrease in absorbance at OD_{600} when 10 μl of spheroplast suspension is diluted into 1 ml of water. The conversion is complete when the OD_{600} is 5–10% of the initial OD_{600}. Under a phase-contrast light microscope, spheroplasts display a rounded morphology in contrast to the ovoid shape of intact yeast.

The efficiency of spheroplast conversion by lyticase is affected by cell growth conditions. Lyticase works poorly on stationary-phase cells, and growth in minimal medium results in less efficient spheroplast conversion. However, spheroplast formation does not appear to be sensitive to the density of yeast cells in spheroplast buffer. Depending on the needs of the experiment, spheroplasting has been performed at concentrations ranging from 5 to 100 OD_{600} units/ml of buffer. The exact conditions for rapid and efficient spheroplast preparation, including the sorbitol concentration necessary for osmotic stabilization of spheroplasts and the amount of lyticase required, is strain-dependent and should be optimized under conditions of the proposed experiment. Reducing agent (β-mercaptoethanol or DTT) is essential for 1,3-β-glucanase activity. Spheroplast conversion has been observed to level off prematurely, after only 10–15 minutes of treatment with lyticase. Simply adding more lyticase at this stage does not appear to restore activity. Since the pH optimum of lyticase fraction II is 7.5, and activity drops off rapidly below that, the decrease in spheroplast formation may be a consequence of cellular metabolites that acidify the surrounding medium. Thus, it may be necessary to check and readjust the pH of the buffer during spheroplast conversion if the digestion has not proceeded to completion. Alternatively, larger quantities of lyticase are added initially so that conversion is complete within 15 minutes.

Simply to examine the distribution of intracellular forms of radiolabeled proteins (e.g., invertase) that are retained within the spheroplast and extracellular forms released into the medium during spheroplast formation, spheroplast conversion is performed in a microcentrifuge tube. Following lyticase digestion, the spheroplasts are sedimented and the supernatant fraction is transferred to another microcentrifuge tube. Sample buffer for SDS–polyacrylamide gel electrophoresis (SDS–PAGE) is added to the pellet and supernatant fractions, and the samples are denatured in a boiling-water bath for 5 minutes and subjected to immunoprecipitation.

3. Radiolabeling of Yeast Spheroplasts

Extended incubations under the conditions described earlier for the preparation of spheroplasts after radiolabeling could adversely affect the analysis of an unstable secretory protein or the isolation of intracellular organelles by subcellular fractionation. In such circumstances, it may be desirable to convert intact yeast cells to spheroplasts before radiolabeling with $^{35}SO_4^{2-}$. For efficient incorporation of radioactivity into cell protein, it is essential that the cells remain metabolically active during removal of the cell wall. Hence, spheroplasts are prepared in buffered minimal growth medium supplemented with a carbon source and with sorbitol for osmotic support.

The role of secretory vesicles in the transport of plasma membrane and cell wall proteins has been examined by subcellular fractionation of radiolabeled *sec1* spheroplasts (Holcomb *et al.*, 1988). The *sec1* cells are grown in MV-low sulfate medium–2% glucose to an OD_{600} of 0.5. Cells are harvested, washed once in 100 m*M* Tris-SO_4, pH 9.4, and incubated in 10 m*M* DTT, 100m*M* Tris-SO_4, pH 9.4, for 5 minutes at 24°C with shaking. After washing twice with spheroplasting medium (MV-low sulfate–2% glucose–0.7 *M* sorbitol, adjusted to pH 7.4 with potassium phosphate), the cells are resuspended in spheroplast medium at 50 OD_{600} units/ml and treated with 20 units of lyticase per OD_{600} unit for 30 minutes at 24°C with gentle shaking. The spheroplasts are sedimented at 1000 *g* for 5 minutes, washed twice in labeling medium [MV–20 μM $(NH_4)_2SO_4$–2% glucose–0.7 *M* sorbitol] and resuspended at 50 OD_{600} units/ml in labeling medium. After a 30-minute shift to the restrictive temperature to impose the *sec1* block, the spheroplasts are labeled for 60 minutes with $^{35}SO_4^{2-}$. Since spheroplasts are very sensitive to breakage, the spheroplast pellets are resuspended gently with a rubber Pasteur pipet bulb attached to a glass rod. Buffer is added dropwise while mixing to avoid formation of cell aggregates.

4. Homogenization of Spheroplasts

A variety of lysis and centrifugal fractionation procedures have been described for the analysis of protein targeting to the yeast mitochondrion (Gasser, 1983) and ER (Rothblatt and Meyer, 1986; Hansen *et al.*, 1986; Waters and Blobel, 1986), and within the secretory pathway of yeast (Holcomb *et al.*, 1987; Baker *et al.*, 1988). The sedimentation characteristics of organelles in yeast homogenates appear to differ somewhat from those of membrane fractions prepared from homogenates of mammalian tissues. For instance, in yeast the ER marker enzyme NADPH–cytochrome-*c* reductase largely sediments at low relative centrifugal force, whereas much higher centrifugal force is required to sediment ER-derived membranes (microsomes) from homogenates of rat liver. Whether the tendency of yeast membranes to sediment at low *g*- forces is due to differences in organelle structure, organelle–cytoskeleton interactions, or simply a predisposition toward nonspecific aggregation of membranes during homogenization has not been systematically examined. Nevertheless, the key step in efficient lysis and good organelle resolution during fractionation is the optimization of homogenization and homogenization buffer conditions.

It is beyond the scope of this chapter to present a critical review of yeast subcellular fractionation. However, a rapid and efficient means of preparing homogenates from spheroplasts is crucial for studying the subcellular localization of polypeptides in the secretory pathway of wild-type or *sec* strains. One protocol that has been particularly effective for examining the susceptibilty of secretory precursors to exogenously added proteases (Feldman *et al.*, 1987; Deshaies and Schekman, 1987) gives efficient spheroplast lysis and minimal aggregation of membranes. Briefly, spheroplasts are harvested by layering the cell suspension onto a cushion composed of 1.9 *M* sorbitol, 10 m*M* sodium azide, 2 m*M* $MgCl_2$, 25 m*M* Tris-HCl (pH 7.5) and sedimenting in a Sorvall HB-4 rotor at 4000 *g* for 5 minutes. The spheroplast pellet is resuspended at a density of ~20–25 OD_{600} units/ml in 0.3 *M* mannitol, 100 m*M* KCl, 1 m*M* EDTA, 50 m*M* Tris-HCl, pH 7.5. A motor-driven Potter–Elvehjem tissue grinder fitted with a serrated Teflon pestle is used to homogenize the spheroplasts. Three 1-minute bursts on ice are usually sufficient to complete the cell lysis. The efficiency of cell lysis is generally ~75–80% with this method, but strain differences can affect efficiency over the range of 50–100%. We have also tested a Dura-grind stainless-steel Dounce-type homogenizer (Wheaton Scientific) with a clearance of 0.0005 in. (12.7 μm, which approximates the 10-μm average diameter of spheroplasts) under modified conditions for the preparation of microsomes for yeast *in vitro* protein translocation assays. The microsomes prepared in this manner are as active in protein translocation as microsomes prepared with a standard glass Dounce homogenizer. This type of tissue grinder may provide greater reproducibility of lysis than the somewhat looser (0.1–0.15 mm clearance) Potter–Elvehjem homogenizer.

C. Immunoprecipitation of Radiolabeled Proteins

Immunoprecipitation is the most direct means of retrieving particular polypeptides from whole extracts of labeled cells for characterization by SDS–PAGE or quantitation by liquid scintillation counting. A protocol for immunoprecipitation under denaturing conditions is summarized here. A useful discussion of various parameters pertaining to immunoprecipitation can be found in an article by Anderson and Blobel (1983).

1. Buffers for Immunoprecipitation (IP)

IP Dilution Buffer
1.25% Triton X-100
190 m*M* NaCl
6 m*M* EDTA
60 m*M* Tris-HCl, pH 7.4
IP Buffer
1% Triton X-100
0.2% SDS
150 m*M* NaCl
5 m*M* EDTA
50 m*M* Tris-HCl, pH 7.4
Urea Wash Buffer
1% Triton X-100
0.2% SDS
2 *M* Urea,
150 m*M* NaCl
5 m*M* EDTA
50 m*M* Tris-HCl, pH 7.4
High-Salt Wash Buffer
1% Triton X-100
0.2% SDS
500 m*M* NaCl
5 m*M* EDTA
50 m*M* Tris-HCl, pH 7.4
Detergent-Free Wash Buffer
150 m*M* NaCl
5 m*M* EDTA
50 m*M* Tris-HCl, pH 7.4

10% (v/v) IgGsorb (fixed *Staphylococcus aureus* cells; The Enzyme Center, Inc., Malden, MA) in IP buffer.

20% (v/v) protein A–Sepharose CL-4B (Pharmacia Biochemicals) in IP buffer.

2. Immunoprecipitation Protocol

1. Four volumes (i.e., 800 μl) of IP dilution buffer are added to the cooled SDS-denatured sample and the sample (without the glass beads) is transferred to a 1.5-ml microcentrifuge tube.
2. Fifty microliters of 10% IgGsorb are added to the sample and the

mixture is incubated at room temperature with continuous rotation. Optionally, ~25 μl of unlabeled "delete" extract (300–400 OD_{600} equivalents of cells per milliliter), prepared from a yeast strain with a chromosomal deletion for the protein of interest (or from a *MATa* strain for immunoprecipitation of α-factor), are added together with the IgGsorb to the sample. The IgGsorb suspension is included to adsorb any material that might bind nonspecifically to protein A and to form a tight pellet together with the yeast cell debris during the clearing centrifugation. "Delete" extract helps to reduce nonspecific background arising from antiserum cross-reactivity or "stickiness." "Delete" extract is prepared from an unlabeled YPD stationary culture by glass bead lysis in 1% SDS–1 m*M* PMSF–50 m*M* Tris-HCl, pH 7.4 (cell pellet–lysis buffer ratio of 1 : 1) and heat denaturation. Efficient lysis requires sufficient glass beads to fill the glass culture tube to the top of the cell–lysis buffer suspension. After heat denaturation, the cell debris is sedimented by centrifugation and the supernatant fraction is stored at −80°C until needed.

3. After 10 minutes, the samples are centrifuged in a microcentrifuge for 5 minutes at top speed, and the supernatant fractions are transferred to fresh tubes.

4. A preliminary experiment is performed to titrate the amount of antibody per OD_{600} unit of cell equivalent required to immunoprecipitate 100% of the protein of interest. The optimum amount of antibody is added and immunoprecipitation continued for 12 hours at 4°C with continuous rotation.

5. Five microliters of 20% protein A–Sepharose suspension per microliter of antiserum are added and incubated for 1–2 hours at room temperature with continuous rotation.

6. The Sepharose beads are allowed to settle by gravity or sedimented very briefly (i.e., 5 seconds) in a microcentrifuge, and the supernatant solution is either aspirated with a 22-gauge hypodermic needle into radioactive waste or transferred to a fresh tube for immunoprecipitation of a different protein (provided that any added "delete" extract does not compete with the labeled species).

7. The beads are washed twice with 1 ml of IP buffer, twice with urea wash buffer, and once with high-salt wash buffer, with vortexing each time.

8. After a final wash with 1 ml of detergent-free wash buffer, the supernatant fraction is aspirated with a 25-gauge needle, which will not draw up packed Sepharose beads.

9. At this stage, the sample is either prepared for SDS–PAGE or reimmunoprecipitated to reduce the level of contaminating protein species. For SDS–PAGE, sample buffer containing DTT or β-mercaptoethanol is

added and the sample heated in boiling water (or lower temperature, if necessary) for 5 minutes. The beads are sedimented and the supernatant loaded onto a SDS–polyacrylamide gel.

10. If serial immunoprecipitation is necessary, 200 μl of 1% SDS–50 m*M* Tris-HCl (pH 7.4) are added to the Sepharose beads and the immunoprecipitate dissociated at 100°C (or lower temperature) for 5 minutes. After cooling, step 1 (with the addition of 150 μ of BSA as a carrier protein) and steps 2–9 are repeated.

IV. Analysis of Intercompartmental Protein Transport

A variety of posttranslational protein modifications accompany the transit of polypeptides along the secretory pathway. These reactions include proteolytic processing, glycosylation, fatty acid acylation, phosphorylation, and oligomerization. The processing of nascent polypeptides is not only site-specific with regard to polypeptide domains that are subject to modification; many of the processing reactions are themselves compartmentalized within known steps of the secretory pathway. As such, these modifications are diagnostic of, or landmarks for, the organelle(s) performing the protein modifications. From a genetic point of view as well, the extent of modification to a polypeptide, and by inference its location in the pathway, could reveal important information about the site(s) of action of *SEC* gene products. Table II summarizes the specific modifications experienced by a representative group of yeast secretory and vacuolar proteins during transport through the secretory pathway. A summary of genetic mutations affecting secretory protein modification is provided in Table III.

A. Tests of Precursor Entry into the ER

Sequestration of nascent precursor molecules within the ER lumen, or assembly into the ER membrane, represents the first step in protein transport to the yeast vacuole or cell suface. Entry into the ER can be examined biochemically in spheroplast homogenates as the passage of the nascent polypeptide from a soluble, protease-sensitive compartment (the cytoplasm) to a membrane-bounded, protease-resistant compartment (the ER). Alternatively, one of several ER-specific, posttranslational protein modifications can be analyzed.

1. Assay of Exogenous Protease Sensitivity

As an example, susceptibility to exogenously added protease has been used to characterize two different types of *sec* mutants, *sec53* and *sec61*, both of which are defective at very early steps in protein secretion (Feldman *et al.*, 1987; Deshaies and Schekman, 1987). In order to test whether the *sec61* defect defined a step in protein translocation across the ER membrane, the accessibility of the precursors of α-factor and CPY to proteolytic digestion was examined. If the *sec61* mutation represents an ER translocation defect, then the nascent secretory polypeptides accumulated in *sec61* cells are expected to remain fully (or partially) exposed in the cytoplasm as protease-sensitive, unmodified species at the nonpermissive temperature. The proteolytic analysis was facilitated by the construction of a *sec61 sec18* haploid strain. Shifting the strain to 30°C, which is semirestrictive for the *sec61* defect and fully restrictive for *sec18*, permits the accumulation of any translocated precursor molecules within the ER. Thus, the *sec18* precursors of α-factor and CPY serve as internal controls for properly sequestered, protease-resistant molecules.

Pulse-labeled cells are converted to spheroplasts and homogenized in a motor-driven Potter–Elvehjem tissue grinder as described earlier. The homogenate is subjected to brief centrifugation at 650 *g* for 4 minutes in a Sorvall HB-4 rotor at 4°C. To avoid unnecessary handling of the radioactive homogenate, the lysate is centrifuged directly in the 2-ml homogenizer fitted into a modified rubber adaptor for the 15-ml Corex tube. After sedimentation of the unbroken cells, the supernatant fraction is divided into two aliquots. One sample is adjusted to 0.4% Triton X-100 in order to solubilize the microsomal (ER) membrane. To examine the α-factor precursor species prior to the addition of exogenous protease, a 40-μl aliquot of each sample is transferred to a 1.5-ml microcentrifuge tube containing 0.56 ml of 20% TCA. Proteinase K is added to the remainder of both samples at a final concentration of 0.3 mg/ml and incubated on ice. Immediately after adding protease and at time increments up to 20 minutes, aliquots are removed and proteolysis terminated by transferring to 20% TCA. The TCA precipitates are sedimented by centrifugation in a microcentrifuge for 10 minutes and washed once with −20°C acetone. The TCA pellets are solubilized in 200 μl of 1% SDS–50 m*M* Tris-HCl (pH 7.4) at 95°C and immunoprecipitated as described before.

In the presence of Triton X-100, both the 26-kDa, ER form of α-factor and the 19-kDa, unglycosylated species are digested by proteinase K. In the absence of detergent, the 19-kDa species is degraded progressively

TABLE II

POSTTRANSLATIONAL MODIFICATIONS TO REPRESENTATIVE SOLUBLE AND MEMBRANE-ASSOCIATED SECRETORY PROTEINS

			ER			
Protein (gene)	Destination	Percursor MIW	Signal sequence cleavage	N-Linked core addition/ trimming	O-Linked addition	Phosphorylation
α-Factor (*MFα1*)	Medium	18.6 kDa	+	+3 Cores (26 kDa)	–	–
Killer Toxin (ScV-M$_1$)[c]	Medium	32–35 kDa	+	+3 Cores (42 kDa)	–	–
Invertase (*SUC2*)	Periplasm/ cell wall	61 kDa	+	+8–13 Cores (79–83 kDa)	–	–
Acid phosphatase (*PHO5*)	Periplasm/ cell wall	58–60 kDa	+	+8 (?) Cores (~80 kDa)	?	?
Processing endopeptidase (*KEX2*)	Golgi apparatus	90 kDa	?	+ (95 kDa)	+	?
CPY (*PRC1*)	Vacuole	59 kDa	+	+4 (67 kDa)	?	+
α-Factor receptor (*STE2*)	Plasma membrane	43 kDa	–	+2–3 (49 kDa)	?	+, ≥ 5

[a]Kex2 enzyme, prohormone processing endopeptidase; Ste13 enzyme, dipeptidylaminopeptidase A; Kex1 enzyme, carboxypeptidase B-like peptidase.

[b]References: 1. Julius *et al.* (1983, 1984a,b); Waters *et al.* (1988); Wagner and Wolf (1987); Dmochowska *et al.* (1987). 2. Bussey *et al.* (1983); Lolle and Bussey (1986); Dmochowska *et al.* (1987). 3. Perlman and Halvorson (1981); Byrd *et al.* (1982); B. Esmon *et al.* (1984); Trimble and Atkinson (1986); P.C. Esmon *et al.* (1987); Reddy *et al.* (1988). 4. Bostian *et al.* (1980); Barbarić *et al.* (1984); Schönholzer *et al.* (1985). 5. Fuller *et al.* (1988). 6. Hasilik and Tanner (1978a,b); Müller and Müller (1981); Stevens *et al.* (1982); Hemmings *et al.*

with time (cf. Fig. 8 in Deshaies and Schekman, 1987), whereas the 26-kDa glycosylated species is resistant to proteolysis throughout the incubation period.[5] Likewise, the low molecular weight precursor of CPY accumulated at the *sec61* block is rapidly degraded by proteinase K, whereas the ER species accumulated at the *sec18* block are resistant in the absence of detergent. These results suggest that the *sec61* mutation causes a defect in translocation of secretory and vacuolar proteins into the ER (Deshaies and Schekman, 1987). In contrast, α-factor precursor accumulated in *sec53* showed no such protease susceptibility (Feldman *et al.*, 1987)

[5]A small percentage of the ER form is proteolysed as a result of microsome leakage during homogenization (Deshaies and Schekman, 1987).

TABLE II *(continued)*

POSTTRANSLATIONAL MODIFICATIONS TO REPRESENTATIVE SOLUBLE AND MEMBRANE-ASSOCIATED SECRETORY PROTEINS

	Golgi apparatus						Vacuole	
	Outer-chain elongation			Proteolytic processing[a]				
Oligomerization	Man_{9-14}	Branched α-1,6	O-Linked mannose elongation	Kex2	Ste13	Kex1	Propeptide cleavage	References[b]
–	+[d]	+[d]	–	+	+	+	NA[e]	1
–	?	?	–	+	–	+	NA	2
Octamer (~600 kDa)	+	+ (100–150 kDa)	–	–	–	–	NA	3
Dimer (~250 kDa)	+	+ (100–150 kDa)	–	–	–	–	NA	4
?	?	–	+ (~135 kDa)	–	–	–	NA	5
?	+ (69 kDa)	–	–	–	–	–	+[f] (61 kDa)	6
?	?	–	?	–	–	–	NA	7

(1981); Zubenko *et al.* (1983); Ammerer *et al.* (1986); Woolford *et al.* (1986). 7. Blumer *et al.* (1988); Reneke *et al.* (1988).

[c]Killer toxin precursor is encoded by a linear double-stranded RNA encapsidated in a viruslike particle (Tipper and Bostian, 1984).

[d]α-Factor precursor containing outer chain oligosaccharides is observed only in *sec* strains defective in transport from the Golgi apparatus (e.g., *sec7*; Julius *et al.*, 1984a).

[e]Not applicable.

[f]Proteinase A (*PEP4, PRA1*)-dependent processing.

2. ENDOPLASMIC RETICULUM-SPECIFIC PROTEIN MODIFICATIONS

a. Signal Sequence Cleavage. The earliest modification of a nascent polypeptide as it crosses the ER membrane is the endoproteolytic cleavage of the amino-terminal signal sequence, which encodes information for targeting the mRNA–ribosome–polypeptide translation complex to the ER membrane (Hortsch and Meyer, 1986). Efficient processing of the signal peptide on invertase and acid phosphatase is essential for rapid transport from the ER to the Golgi apparatus (Schauer *et al.*, 1985; Haguenauer-Tsapis and Hinnen, 1984). Signal sequence cleavage can usually be assessed on SDS–polyacrylamide gels by direct comparison of the primary translation product, synthesized *in vitro*, with unglycosylated precursor molecules that accumulate at the restrictive temperature in *sec18* cells in the presence of tunicamycin. However, depending on the com-

TABLE III

MUTATIONS AFFECTING SECRETORY PROTEIN MODIFICATION

Mutation	Defect	Substrate	Site of defect	Reference
sec11	Signal sequence cleavage	Signal peptide	ER	Bohni *et al.* (1988)
sec53/alg4	Phosphomannomutase activity	Man6P	Cytoplasm	Kepes and Schekman (1988)
sec59	Mannose addition	Dolichol-linked oligosaccharide	ER membrane	Bernstein *et al.* (1989)
gls1	Trimming of core glucoses	Asn-linked oligosaccharide	ER	Esmon *et al.* (1984)
alg1–3, 5–8 dpg1	Lipid-linked core olizgosaccharide assembly	Dolichol-linked monosaccharides and oligosaccharides	ER	Huffaker and Robbins (1983); Runge *et al.* (1984); Runge and Robbins (1986a); Ballou *et al.* (1986)
kex1	Carboxy-terminal processing of prohormones	α-Factor*[a] and killer toxin	Late Golgi or secretory vesicles	Wagner and Wolf (1987); Dmochowska *et al.* (1987)
kex2	Endoproteolytic cleavage after Lys-Arg residues	Pro-α-factor and killer toxin	Late Golgi or secretory vesicles	Julius *et al.* (1984a); Fuller *et al.* (1988)
ste13	Amino-terminal processing of prohormones	α-Factor*[a]	Late Golgi	Julius *et al.* (1983)
pep4, pra1	Proteolytic processing of vacuolar hydrolases	Vacuolar hydrolases	Vacuole	Hemmings *et al.* (1981); Zubenko *et al.* (1983); Ammerer *et al.* (1986)
mnn1-9	Core and outer-chain mannosylation	N-Linked core oligosaccharide	Golgi apparatus	Ballou (1982)

[a] α-Factor* refers to the Kex2 (prohormone) processing endopeptidase cleavage product.

position of the SDS–polyacrylamide gel, unglycosylated pro-α-factor migrates ahead, behind, or in parallel with prepro-α-factor. The original observation that tunicamycin-treated *sec18* cells accumulate a precursor that comigrates with the *in vitro*-synthesized species led to the wrong conclusion that prepro-α-factor does not undergo signal sequence cleavage (Julius *et al.*, 1984a). *In vitro* analysis of prepro-α-factor processing (Waters *et al.*, 1988), has shown that an 18% polyacrylamide–4 *M* urea gel provides good resolution of the two species, with unglycosylated pro-α-factor migrating, as expected, faster than prepro-α-factor. Using the 18% polyacrylamide–4 *M* urea gel system, the α-factor precursor species accumulated in *sec61* cells at the restrictive temperature has been shown to comigrate with *in vitro*-synthesized prepro-α-factor (R. Deshaies and P. Böhni, unpublished observations).

b. Carbohydrate Addition and Trimming. Among the various protein modifications that occur in the yeast secretory pathway, glycosylation is the best characterized (reviewed in Kukuruzinska *et al.*, 1987). Enzymatic activity of at least two externalized glycoproteins, invertase and acid phosphatase, depends on asparagine-linked glycosylation (Arnold and Tanner, 1982). Conformational changes arising from carbohydrate addition presumably direct correct folding and oligomerization during protein synthesis (Esmon *et al.*, 1987). Core oligosaccharides of the form $Glc_3Man_9GlcNAc_2$ are assembled on a dolichol lipid intermediate in the ER membrane and attached in N-glycosidic linkage to nascent polypeptides emerging at the luminal surface of the membrane. The three glucoses and one α-1,2-mannose of the core oligosaccharide are removed by specific glycosidases within the ER (Byrd *et al.*, 1982; Esmon *et al.*, 1984). In addition to N-linked carbohydrates, both intracellularly localized proteins (for example, the Golgi-associated *Kex2* endopeptidase) and cell wall mannoproteins contain O-linked oligosaccharides (Fuller *et al.*, 1988; Ballou, 1982). O-Linked glycosylation of serine and threonine residues is initiated in the ER by transfer of mannose from a dolichol carrier (Haselbeck and Tanner, 1983).

Invertase accumulated in a *sec18* strain contains core oligosaccharides which are a single structural isomer of $Man_8GlcNAc_2$ Asn (Byrd *et al.*, 1982). Of the 14 available N-linked glycosylation sites in invertase, 8 are always or almost always glycosylated and another 5 sites are utilized ~50% or less of the time (Reddy *et al.*, 1988). In mammalian cells there is a stringent requirement for core glycosylation with complete Glc_3Man_9-$GlcNAc_2$ oligosaccharides. In contrast, analysis of yeast asparagine-linked glycosylation (*alg* and *dpg*) mutants, which are defective in core oligosaccharide synthesis, showed that incomplete oligosaccharides (as small as $GlcNAc_2$ or $Man_{1-2}GlcNAc_2$) are transferred to asparagine residues (Huffaker and Robbins, 1983; Runge *et al.*, 1984; Ballou *et al.*,

1986). However, $Glc_3Man_9GlcNAc_2$ is the preferred donor oligosaccharide in yeast (Parodi, 1981).

A direct role for *SEC* gene products in protein glycosylation has been established with the observation that an intermediate in GDP-mannose biosynthesis can complement the *sec53* defect in an *in vitro* system (Kepes and Schekman, 1988). It appears that the *sec59* mutation falls into this category of glycosylation defects as well. The *sec59* mutation affects the addition of mannose to dolichol-linked oligosaccharides (Bernstein *et al.*, 1989). Whereas the *sec53* and *sec59* mutations results in accumulation of inactive invertase in the ER lumen at 37°C, the *alg* and *dpg* mutations have no effect on the secretion of incompletely glycosylated invertase. Even completely unglycosylated invertase, synthesized in the presence of tunicamycin, is secreted at 25°C, albeit at a reduced rate (Ferro-Novick *et al.*, 1984b).

A variety of biochemical approaches are available to distinguish between N-linked and O-linked carbohydrate addition and to characterize the structure of oligosaccharides present on proteins in defined compartments. Proteins containing N-linked and O-linked sugars can be isolated from radiolabeled cell extracts by concanavalin A (Con A)–Sepharose affinity purification, essentially as described earlier for immunoprecipitation (see also Baker and Schekman, Chapter 7, Volume 31), following one cycle of immunoprecipitation. The presence of N- and O-linked oligosaccharides can be detected by SDS–PAGE. Tunicamycin, which blocks the synthesis of lipid-linked oligosaccharide precursors for asparagine-linked glycosylation (Kuo and Lampen, 1974), has no effect on the assembly of O-linked mannose donors. Likewise, O-linked structures are resistant to digestion with mannosyl-glycoprotein endo-β-*N*-acetylglucosaminidase H, which cleaves the di-(*N*-acetyl)-chitobiose residue linked to asparagine (for a more extensive discussion of glycosidases and glycosylation inhibitors, see Tarentino *et al.*, Chapter 5, this volume). [^{3}H]Mannose radiolabeling, in conjunction with Bio-gel P4 chromatography, has been used to characterize the oligosaccharide structure of glycoproteins accumulating at the *sec18* block (Esmon *et al.*, 1984; Runge and Robbins, 1986b).

c. Glycoprotein Acylation and Phosphorylation. Fatty acid acylation of yeast secretory proteins has been observed when transport from the ER is blocked (Wen and Schlesinger, 1984). Four minor glycoproteins label with [^{3}H]palmitate in *sec18* cells shifted to the restrictive temperature and are recovered in membrane fractions sedimented through a sucrose density gradient. Inhibition of core glycosylation by tunicamycin does not interfere with acylation. Whether the membrane association of these proteins is a consequence of acylation or is due to the presence of a hydrophobic membrane anchor domain within the polypeptides is unknown. These proteins are further glycosylated during transit through the Golgi apparatus, but the palmitate label is lost at some stage during or after transport to their final destination. Since the transit time from the ER to the cell surface is very rapid, palmitate-labeled proteins are not observed in wild-type cells. The function of glycoprotein acylation in yeast remains unclear. Palmitoylation of the soluble (nonsecretory) yeast Ras1 and Ras2 proteins and *ras*-related yeast Ypt1 protein results in membrane association (Fujiyama and Tamanoi, 1986; Molenaar *et al.*, 1988). Although a mutation (known as *ste16, ram,* or *dpr1*) affecting the acylation of yeast *RAS* gene products and secretion of **a**-factor has been identified (Powers *et al.*, 1986; Fujiyama *et al.*, 1987; Wilson and Herskowitz, 1987), no mutants defective in glycoprotein acylation have been isolated.

To examine fatty acid acylation of proteins in yeast, cells are grown to early-logarithmic growth phase in YPD (or in MV-low sulfate medium if both [^{3}H]palmitate and $^{35}SO_4^{2-}$ radiolabeling is intended). Cells are harvested and washed, and then resuspended at a density of 2 OD_{600} units/ml in YPD, pH 6.8. Fatty acid radiolabeling is usually carried out in YPD medium and at higher pH (the pH of normal YPD is ~5.8) in order to keep the fatty acid in solution. $^{35}SO_4^{2-}$ labeling in minimal medium is carried out in parallel with cells from the same overnight culture. Cerulenin (Calbiochem), an inhibitor of fatty acid synthetase, is added to a final concentration of 2 μg/ml. Preincubation with cerulenin improves incorporation of radiolabeled fatty acids ~5-fold (Towler and Glaser, 1986). After 15 minutes, [^{3}H]palmitate is added at a concentration of 0.1–1.0 mCi/ml. Sufficient incorporation for a pulse–chase experiment can be obtained with a 20-minute period of labeling. Since [^{3}H]palmitate is shipped in toluene, it should be dried down and resuspended at 20–50 mCi/ml in 100% ethanol prior to use.

Both the vacuolar glycoprotein CPY and extracellular invertase are phosphorylated after entering the secretory pathway. Yeast mutants defective in phosphomannose addition to N-linked oligosaccharides show no alteration in proper localization of either CPY or invertase. Stevens *et al.* (1982) examined the subcellular site of phosphate addition in the various *sec* mutants. As expected for phosphate incorporated into outer-chain carbohydrate, endoglycosidase H-releasable phosphate is present on invertase accumulated in *sec7* cells, which are defective in protein transport from the Golgi apparatus. Core-glycosylated invertase accumulated in *sec18* cells does not contain endoglycosidase H-releasable phosphate. In contrast, core-glycosylated CPY does contain endoglycosidase H-releasable phosphate, indicating carbohydrate phosphorylation in the ER. However, no further addition of phosphate is observed during transit through the Golgi apparatus. Approximately 20% of CPY-associated phosphate is covalently attached to the polypeptide as phosphoserine (R. Feldman and R. Schekman, unpublished observations). Polypeptide-associated phosphate on invertase is seen only on unglycosylated invertase accumulated in tunicamycin-treated *sec18* cells. What significance, if any, is attributable to the differential localization of phosphate addition to vacuolar and secretory proteins remains to be elucidated.

To study phosphate modification of secretory proteins or proteins regulating events in the secretory pathway, ^{32}P radiolabeling of intact yeast can be employed. The conditions for cell growth, radiolabeling, and lysis are essentially identical to those for ^{35}S radiolabeling, except that cultures are grown in MV-low phosphate (i.e., 100 μM KH_2PO_4) and labeled in phosphate-free MV medium. Whereas yeast reach stationary-growth phase at a cell density of ~40–50 OD_{600} in YPD medium and of ~4–8 OD_{600} in complete MV medium, stationary phase in low-sulfate or low-phosphate minimal medium is attained at a cell density of 2–3 OD_{600}. For this reason it is particularly important to harvest and label cells at very low density (OD_{600} ≤ 0.3) in order to get efficient radioisotope uptake and incorporation. After the temperature shift and a 30-minute prelabeling incubation in MV-no phosphate medium at a density of 2 OD_{600} units/ml, $H_3{}^{32}PO_4$ (HCl-free, ICN Radiochemicals) is added at a concentration of 250–500 $\mu Ci/OD_{600}$ unit. Under these conditions, a labeling period of 60 minutes is adequate for efficient labeling of phosphate residues in both invertase and CPY (Stevens *et al.*, 1982).

d. Oligomerization. The process of protein folding and assembly in the ER is emerging as an important factor in transport through the secretory pathway. Protein disulfide-isomerase, which catalyzes the formation and breakage of disulfide bonds common in cell surface and extracellular proteins, is associated with the luminal surface of the ER (Freedman, 1984). The quaternary structure of a variety of cell surface molecules is completed before exiting the ER (Kvist *et al.*, 1982; Gething *et al.*, 1986; Doms *et al.*, 1987). What little information is known about the role and timing of these reactions in yeast has been facilitated by the analysis of *sec* mutants. Whereas the constitutively expressed, unglycosylated cytoplasmic invertase exists as a dimer, the extracellular form of invertase exists as a homo-octamer (Chu *et al.*, 1983). Esmon *et al.* (1987) showed that secretory invertase exists in its octameric form within the ER and at all definable steps thereafter in the secretory pathway. The data further suggest that asparagine-linked carbohydrate addition promotes octamer formation or stability. In the presence of tunicamycin, cells secrete unglycosylated dimeric invertase rather than an unglycosylated octamer. As noted before, the rate of invertase transport at 25°C is significantly reduced in the absence of core glycosylation. Thus, oligomerization may be one of several factors contributing to efficient sorting and transport of proteins in yeast.

Native gel electrophoresis is used to examine the oligomeric state of invertase. Enzymatically active dimer, tetramer, hexamer, and octamer are resolved in a nondenaturing polyacrylamide gel system. A protocol for native gel analysis of invertase is described in the legend to Fig. 2 in the chapter by Baker and Schekman (Chapter 7, Volume 31).

B. Interorganellar Protein Transport

Transit of polypeptides through the Golgi apparatus en route to the vacuole and the cell surface, and arrival of some vacuolar hyrdolases in the vacuole are accompanied by further modifications, including carbohydrate elongation (reviewed in Ballou, 1982; Kukuruzinska *et al.*, 1987) and proteolytic processing (reviewed in Schekman, 1985; Fuller *et al.*, 1988).

1. Carbohydrate Elongation

In a compartment(s) distal to the *sec18* block, but before the *sec7* and *sec14* defects that block the exit of proteins from the Golgi apparatus, both asparagine-linked oligosaccharides and O-linked mannoses are extended by the addition of mannose residues. Both invertase and acid phosphatase accumulate in their mature forms in *sec7* and later pathway mutants. Very large electrophoretic mobility shifts on SDS–polyacrylamide gels are observed as a consequence of branched outer-chain glycosylation of the core-glycosylated ER intermediates. As many as 150 mannose residues may be present in each N-linked oligosaccharide. Some of the side chains are substituted with diester-linked mannose-phosphate or mannobiose-phosphate. O-linked mannose, on the other hand, receives 2–3 additional mannose residues in the Golgi apparatus. The order of events in outer-chain synthesis has been characterized by both genetic and biochemical means (Ballou *et al.*, 1980; Trimble and Atkinson, 1986; reviewed in Kukuruzinska *et al.*, 1987).

Unlike mammalian glycoproteins, which are modified by the addition of *N*-acetylglucosamine, galactose, fucose, and sialic acid to produce so-called complex sugars, yeast N- and O-linked outer-chain oligosaccharides consist entirely of mannose. For studies of glycoprotein transport in mammalian cells, endoglycosidase H is commonly used to distinguish between high-mannose core oligosaccharides, which are released, from the Golgi-modified complex oligosaccharides, which resist hydrolysis. Yeast outer-chain glycosylation does not alter the susceptibility of the di-(*N*-acetyl)-chitobiosyl linkage to endoglycosidase H. Hence, acquisition of endoglycosidase H resistance cannot be used as a landmark for transport from a compartment(s) containing core-glycosylated proteins to a later compartment(s) performing outer-chain mannose addition. However, immunological probes, directed against different yeast cell wall mannan structures, have been generated by immunizing rabbits with either wild-type strains or mutant *(mnn)* yeast strains defective at various steps in outer-chain carbohydrate addition (Ballou, 1982). Baker *et al.* (1988; see also Baker and Schekman, Chapter 7, Volume 31) have used a specific antibody, detecting the presence of mannose in α-1,6 linkage in the outer chain, to assay transport of α-factor precursor from the ER to the Golgi apparatus *in vitro*.

Unlike the periplasmic glycoproteins invertase and acid phosphatase, CPY shows only limited outer-chain elongation. The oligosaccharides observed on mature CPY are usually of the form $Man_{11-14}GlcNAc_2$. Whereas invertase accumulated at the *sec7* block is fully glycosylated, the principal species of CPY is the core-glycosylated, or so-called p1, form. *sec14* accumulates CPY at equal amounts in the p1 form and in its outer chain-glycosylated, or p2, form. Although both *sec* mutations block transport to the vacuole and the cell surface, this results suggests the possibility that CPY, and perhaps other vacuolar proteins, are sorted from invertase somewhere before the *sec7* block. Studies on vacuolar acid trehalase, which has been shown to undergo extensive outer-chain glycosylation (Mittenbühler and Holzer, 1988), in the *sec7* and *sec14* mutants may provide additional information on the modification and sorting of vacuolar and secretory proteins within the Golgi apparatus.

2. Proteolytic Processing

Another series of proteolytic cleavages occur in the Golgi apparatus. Both the precursor of α-factor and the precursor of killer toxin are processed to mature forms at some point during transport through the Golgi stack (Julius *et al.*, 1984a; Bussey *et al.*, 1983). In the case of α-factor, the sequence of the gene predicts four tandem repeats of the pheromone, separated by nearly identical spacer peptide segments. Analysis of α-factor precursor forms accumulated at various stages of secretion showed that some processing at the spacer peptides occurred before the *sec7* block, and was essentially complete by the time α-factor appeared in secretory vesicles accumulated in *sec1*. Three enzymes involved in the maturation of α-factor in a late Golgi compartment have been identified biochemically and by genetic analysis. The initial endoproteolytic cleavage is made after a pair of basic residues (Lys-Arg). An enzyme with this specificity is detectable in wild-type strains and is absent in the mutant *kex2* strain that secretes intact α-factor precursor (Julius *et al.,* 1984b). The *kex2* mutation also interferes with the processing of a Lys-Arg site in killer toxin (Bostian *et al.,* 1984; Leibowitz and Wickner, 1976). After the initial cleavage, a group of four or six amino acids on the amino terminus of each peptide (termed α-factor*) is removed by the action of membrane-bound dipeptidylaminopeptidase A (*STE13,* Julius *et al.,* 1983). The Lys-Arg residues present on three copies of processed α-factor are removed by a carboxypeptidase B-like protease, which is deficient in the *kex1* mutant strain (Dmochowska *et al.,* 1987).

Proteolytic processing also plays an important role in the activation of precursor forms of vacuolar hydrolases. Various hydrolases require the proteolytic action of active *PEP4* gene product (Hemmings *et al.,* 1981). In the cases of CPY and proteinases A and B, *pep4* mutant cells produce higher molecular weight precursors which, at least for CPY, are detected in the vacuole (Stevens *et al.,* 1982; Zubenko *et al.,* 1983). The *PEP4* gene product has been shown to be proteinase A (Ammerer *et al.,* 1986; Woolford *et al.,* 1986). The precursor of CPY is also detected in *sec* mutants blocked in transport from the ER or from the Golgi apparatus, whereas normal localization to and processing within the vacuole occurs when secretion is blocked after the Golgi apparatus (i.e., *sec1*). Hence, secretory and vacuolar precursors must be sorted from each other in the Golgi apparatus, and final transport to the vacuole is independent of functions required for discharge of secretory vesicles at the cell surface.

Acknowledgments

We would like to acknowledge the contributions of the members of our laboratory, past and present, to the methods and research reviewed in this chapter, and their helpful discussions during its preparation. Special thanks to Rachel Sterne for providing details about fatty acid radiolabeling, to Greg Payne for the lyticase protocol, and to Rick Feldman for information on phosphate radiolabeling. Thanks also to Colin Stirling for details on the use of Tran^{35}S label. Last, we would like to thank Paul Atkinson, Ray Deshaies, Alex Franzusoff, and Francois Kepes for comments on the manuscript, and Peggy McCutcheon for help with its preparation. J. R. has been supported by a National Institute of General Medical Sciences Postdoctoral Fellowship and a Senior Postdoctoral Fellowship from the American Cancer Society, California Division. Research in our laboratory has been supported by grants from the National Institutes of Health and the National Science Foundation to R. S.

References

Ammerer, G., Hunter, C. P., Rothman, J. H., Saari, G. C., Valls, L. A., and Stevens, T. H. (1986). *Mol. Cell. Biol.* **6,** 2490–2499.

Anderson, D. J., and Blobel, G. (1983). *In* "Methods in Enzymology" (S. Fleischer and B. Fleischer, eds.), Vol. 96, pp. 111–120. Academic Press, New York.

Arnold, E., and Tanner, W. (1982). *FEBS Lett.* **148,** 49–53.

Baker, D., Hicke, L., Rexach, M., Schleyer, M., and Schekman, R. (1988). *Cell (Cambridge, Mass.)* **54,** 335–344.

Ballou, C. E. (1982), *In* "The Molecular Biology of the Yeast *Saccharomyces:* Metabolism and Gene Expression" (J. Strathern, E. Jones, and J. Broach, eds.), pp. 335–360. Cold Spring Harbor Lab., Cold Spring Harbor, New York.

Ballou, L., Cohen, R. E., and Ballou, C. E. (1980). *J. Biol. Chem.* **255,** 5986–5991.

Ballou, L., Supal, P., Krummel, B., Markku, T., and Ballou, C. E. (1986). *Proc. Natl. Acad. Sci. U.S.A.* **83,** 3081–3085.

Barbarić, S., Kozulić, B., Ries, B., and Mildner, P. (1984). *J. Biol. Chem.* **259,** 878–883.

Bernstein, M. Kepes, F., and Schekman, R. (1989). *Mol. Cell. Biol.* **9,** 1191–1199.

Blumer, K. J., Reneke, J. E., and Thorner, J. (1988). *J. Biol. Chem.* **263,** 10,836–10,842.

Böhni, P. C., Deshaies, R. J., and Schekman, R. W. (1988). *J. Cell Biol.* **106,** 1035–1042.

Bostian, K. A., Lemire, J. M., Cannon, L. E., and Halvorson, H. O. (1980). *Proc. Natl. Acad. Sci. U.S.A.* **77,** 4504–4508.

Bostian, K. A., Elliot, Q., Bussey, H., Burn, V., Smith, A., and Tipper, D. J. (1984). *Cell (Cambridge, Mass.)* **36,** 741–751.

Brada, D., and Schekman, R. (1988). *J. Bacteriol.* **170,** 2775–2783.

Bussey, H., Saville, D., Greene, D., Tipper, D. J., and Bostian, K. A. (1983). *Mol. Cell. Biol.* **3,** 1362–1370.

Byrd, J. C., Tarentino, A. L., Maley, F., Atkinson, P. H., and Trimble, R. B. (1982). *J. Biol. Chem.* **257,** 14657–14666.

Chu, F., Watorek, W., and Maley, F. (1983). *Arch. Biochem. Biophys.* **223,** 543–555.

Ciejek, E., and Thorner, J. (1979). *Cell (Cambridge, Mass.)* **18,** 623–635.

Daum, G., Böhni, P. C., and Schatz, G. (1982). *J. Biol. Chem.* **257,** 13028–13033.

Deshaies, R. J., and Schekman, R. (1987). *J. Cell Biol.* **105,** 633–645.

Dmochowska, A., Dignard, D., Henning, D., Thomas, D. Y., and Bussey, H. (1987). *Cell (Cambridge, Mass.)* **50,** 573–584.

Doms, R. W., Keller, D. S., Helenius, A., and Balch, W. E. (1987). *J. Cell Biol.* **105,** 1957–1969.

Emr, S. D., Schauer, I., Hansen, W., Esmon, P., and Schekman, R. (1984) *Mol. Cell. Biol.* **4,** 2347–2355.

Erickson, A., and Blobel, G. (1983). *In* "Methods in Enzymology" (S. Fleischer and B. Fleischer, eds.), Vol. 96, pp. 38–50. Academic Press, New York.

Esmon, B., Esmon, P. C., and Schekman, R. (1984). *J. Biol. Chem.* **259,** 10322–10327.

Esmon, P. C., Esmon, B. E., Schauer, I. E., Taylor, A., and Schekman, R. (1987). *J. Biol. Chem.* **262,** 4387–4394.

Feldman, R. I., Bernstein, M., and Schekman, R. (1987). *J. Biol. Chem.* **262,** 9332–9339.

Ferro-Novick, S., Novick, P., Field, C., and Schekman, R. (1984a). *J. Cell Biol.* **98,** 35–43.

Ferro-Novick, S., Hansen, W., Schauer, I., and Schekman, R. (1984b). *J. Cell Biol.* **98,** 44–53.

Field, C., and Schekman, R. (1980). *J. Biol. Chem.* **254,** 796–803.

Freedman, R. (1984). *Trends Biochem. Sci.* **9,** 438–441.

Fujiyama, A., and Tamanoi, F. (1986). *Proc. Natl. Acad. Sci. U.S.A.* **83,** 1266–1270.

Fujiyama, A., Matsumoto, K., and Tamanoi, F. (1987). *EMBO J.* **6,** 223–228.
Fuller, R. S., Sterne, R. E., and Thorner, J. (1988). *Annu. Rev. Physiol.* **50,** 345–362.
Garoff, H. (1985). *Annu. Rev. Cell Biol.* **1,** 403–445.
Gasser, S. M. (1983). *In* "Methods in Enzymology" (S. Fleischer and B. Fleischer, eds.), Vol. 97, pp. 329–336. Academic Press, New York.
Gething, M.-J., McCammon, K., and Sambrook, J. (1986). *Cell (Cambridge, Mass.)* **46,** 939–950.
Haguenauer-Tsapis, R., and Hinnen, A. (1984). *J. Mol. Cell Biol.* **4,** 2668–2675.
Hansen, W., Garcia, P. D., and Walter, P. (1986) *Cell (Cambridge, Mass.)* **45,** 397–406.
Haselbeck, A., and Tanner, W. (1983). *FEBS Lett.* **158,** 335–338.
Hasilik, A., and Tanner, W. (1978a). *Eur. J. Biochem.* **85,** 599–608.
Hasilik, A., and Tanner, W. (1978b). *Eur. J. Biochem.* **91,** 567–575.
Hemmings, B. A., Zubenko, G. S., Hasilik, A., and Jones, E. W. (1981). *Proc. Natl. Acad. Sci. U.S.A.* **78,** 435–439.
Holcomb, C. L., Etcheverry, T., and Schekman, R. (1978). *Anal. Biochem.* **166,** 328–334.
Holcomb, C. L., Hansen, W. J., Etcheverry, T., and Schekman, R. (1988). *J. Cell Biol.* **106,** 641–648.
Hortsch, M., and Meyer, D. I. (1986). *Int. Rev. Cytol.* **102,** 215–242.
Huffaker, T. C., and Robbins, P. W. (1983). *Proc. Natl. Acad. Sci. U.S.A.* **80,** 7466–7470.
Johnson, L. M., Bankaitis, V. A., and Emr, S. D. (1987). *Cell (Cambridge, Mass.)* **48,** 875–885.
Julius, D., Blair, L., Brake, A., Sprague, G., and Thorner, J. (1983). *Cell (Cambridge, Mass.)* **32,** 839–852.
Julius, D., Schekman, R., and Thorner, J. (1984a). *Cell (Cambridge, Mass.)* **36,** 309–318.
Julius, D., Brake, A., Blair, L., Kunisawa, R., and Thorner, J. (1984b). *Cell (Cambridge, Mass.)* **37,** 1075–1089.
Kepes, F., and Schekman, R. (1988). *J. Biol. Chem.* **263,** 9155–9161.
Kukuruzinska, M. A., Bergh, M. L. E., and Jackson, B. J. (1987). *Annu. Rev. Biochem.* **56,** 915–944.
Kuo, S. C., and Lampen, J. O. (1974). *Biochem. Biophys. Res. Commun.* **58,** 287–295.
Kvist, S., Wiman, K., Claesson, L., Peterson, P. A., and Dobberstein, B. (1982). *Cell (Cambridge, Mass.)* **29,** 61–69.
Leibowitz, M. J., and Wickner, R. W. (1976). *Proc. Natl. Acad. Sci. U.S.A.* **73,** 2061–2065.
Linnemans, W. A. M., Boer, P., and Elbers, P. F. (1977). *J. Bacteriol.* **131,** 638–644.
Mittenbühler, K., and Holzer, H. (1988). *J. Biol. Chem.* **263,** 8537–8543.
Molenaar, C. M. T., Prange, R., and Gallwitz, D. (1988). *EMBO J.* **7,** 971–976.
Müller, M., and Müller, H. (1981). *J. Biol. Chem.* **256,** 11962–11965.
Munro, S., and Pelham, H. R. B. (1987). *Cell (Cambridge, Mass.)* **48,** 899–907.
Newman, A. P., and Ferro-Novick, S. (1987). *J. Cell Biol.* **105,** 1587–1594.
Novick, P., and Schekman, R. (1979). *Proc. Natl. Acad. Sci. U.S.A.* **76,** 1858–1862.
Novick, P., Field, C., and Schekman, R. (1980). *Cell (Cambridge, Mass.)* **21,** 205–215.
Onishi, H. R., Tkacz, J. S., and Lampen, J. O. (1979). *J. Biol. Chem.* **254,** 11943–11952.
Pääbo, S., Bhat, B. M., Wold, W. S. M., and Peterson, P. A. (1987). *Cell (Cambridge, Mass.)* **50,** 311–317.
Parodi, A. J. (1981). *Arch. Biochem. Biophys.* **210,** 372–382.
Perlman, D., and Halvorson, H. (1981). *Cell (Cambridge, Mass.)* **25,** 525–536.
Poruchynsky, M. S., Tyndall, C., Both, G. W., Sato, F., Bellamy, A. R., and Atkinson, P. H. (1985). *J. Cell Biol.* **101,** 2199–2209.
Powers, S., Michaelis, S., Broek, D., Santa Anna-A., S., Field, J., Herskowitz, I., and Wigler, M. (1986). *Cell (Cambridge, Mass.)* **47,** 413–422.

Reddy, V. A., Johnson, R. S., Biemann, K., Williams, R. S., Ziegler, F. D., Trimble, R. B., and Maley, F. (1988). *J. Biol. Chem.* **263,** 6978–6985.
Reid, G. A. (1983). *In* "Methods in Enzymology" (S. Fleischer and B. Fleischer, eds.), Vol. 97, pp. 324–329. Academic Press, New York.
Reneke, J. E., Blumer, K. J., Courchesne, W. E., and Thorner, J. (1988). *Cell* **55,** 221–234.
Rothblatt, J. A., and Meyer, D. I. (1986). *Cell (Cambridge, Mass.)* **44,** 619–628.
Runge, K. W., and Robbins, P. W. (1986a). *J. Biol. Chem.* **261,** 15582–15590.
Runge, K. W., and Robbins, P. W. (1986b). *Microbiology (Washington, D.C.)* pp. 312–316.
Runge, K. W., Huffaker, T. C., and Robbins, P. W. (1984). *J. Biol. Chem.* **259,** 412–417.
Schauer, I., Emr, S., Gross, C., and Schekman, R. (1985). *J. Cell Biol.* **100,** 1664–1675.
Schekman, R. (1985). *Annu. Rev. Cell Biol.* **1,** 115–143.
Schekman, R., and Novick, P. (1982). *In* "The Molecular Biology of the Yeast *Saccharomyuces:* Metabolism and Gene Expression" (J. Strathern, E. Jones, and J. Broach, eds.), pp. 361–393. Cold Spring Harbor Lab., Cold Spring Harbor, New York.
Schönholzer, F., Schweingruber, A.-M., Trachsel, H., and Schweingruber, M. E. (1985). *Eur. J. Biochem.* **147,** 273–279.
Schwaiger, H., Hasilik, A., von Figura, A., Wiemken, A., and Tanner, W. (1982). *Biochem. Biophys. Res. Commun.* **104,** 950–956.
Scott, J. H., and Schekman, R. (1980). *J. Bacteriol.* **142,** 414–423.
Sherman, F., Fink, G. R., and Hicks, J. B. (1983). "Methods in Yeast Genetics: A Laboratory Manual" (rev. ed.). Cold Spring Harbor Lab., Cold Spring Harbor, New York.
Sistrom, W. R. (1958). *Biochim. Biophys. Acta* **29,** 579–587.
Stevens, T. H., Esmon, B., and Schekman, R. (1982). *Cell (Cambridge, Mass.)* **30,** 439–448.
Tipper, D. J., and Bostian, K. A. (1984). *Microbiol. Rev.* **48,** 125–156.
Tkacz, J. S., and Lampen, J. O. (1973). *J. Bacteriol.* **113,** 1073–1075.
Towler, D., and Glaser, L. (1986). *Proc. Natl. Acad. Sci. U.S.A.* **83,** 2812–2816.
Trimble, R. B., and Atkinson, P. H. (1986). *J. Biol. Chem.* **261,** 9815–9824.
Valls, L. A., Hunter, C. P., Rothman, J. H., and Stevens, T. H. (1987). *Cell (Cambridge, Mass.)* **48,** 887–897.
von Figura, K., and Hasilik, A. (1986). *Annu. Rev. Biochem.* **55,** 167–193.
Wagner, J.-C., and Wolf, D. H. (1987). *FEBS Lett.* **221,** 423–426.
Waters, M. G., and Blobel, G. (1986). *J. Cell Biol.* **102,** 1543–1550.
Waters, M. G., Evans, E. A., and Blobel, G. (1988). *J. Biol. Chem.* **263,** 6209–6214.
Wen, D., and Schlesinger, M. J. (1984). *Mol. Cell. Biol.* **4,** 688–694.
Wickerham, L. S. (1946). *J. Bacteriol.* **52,** 293–301.
Wieland, F. T., Gleason, M. L., Serafini, T. A., and Rothman, J. E. (1987). *Cell (Cambridge, Mass.)* **50,** 289–300.
Wilson, K. L., and Herskowitz, I. (1987). *Genetics* **155,** 441–449.
Woolford, C. A., Daniels, L. B., Park, F. J., Jones, E. W., Van Arsdell, J. N., and Innis, M. A. (1986). *Mol. Cell. Biol.* **6,** 2500–2510.
Zubenko, G. S., Park, F. J., and Jones, E. W. (1983). *Proc. Natl. Acad. Sci. U.S.A.* **80,** 510–514.

Chapter 2

Methods to Estimate the Polarized Distribution of Surface Antigens in Cultured Epithelial Cells

ENRIQUE RODRIGUEZ-BOULAN, PEDRO J. SALAS,[1] MASSIMO SARGIACOMO, MICHAEL LISANTI, ANDRE LEBIVIC, YULA SAMBUY,[2] DORA VEGA-SALAS,[1] AND LUTZ GRAEVE

Cornell University Medical College
Department of Cell Biology and Anatomy
New York, New York 10021

I. Polarity of Epithelial Cells

The polarized distribution of surface proteins between the apical and the basolateral surfaces of epithelial cells is the basis of their vectorial

[1]Present address: Instituto de Investigaciones Bioquimicas, Fundacion Campomar, 1405 Buenos Aires, Argentina.
[2]Present address: Instituto Nazionale della Nutrizione, 00179 Rome, Italy.

Copyright © 1989 by Academic Press, Inc.
All rights of reproduction in any form reserved.

function. Work in the last decade has begun to elucidate the nature of the cellular mechanisms responsible for this polarized distribution (Rodriguez-Boulan, 1983a; Simons and Fuller, 1985). Epithelial cell lines have proved excellent tools for this class of studies. A variety of them is available; the best studied is the dog kidney line MDCK (Misfeldt *et al.*, 1976; Cerejido *et al.*, 1978). Table I shows a partial list of the polarized epithelial lines available.

In this paper we review the methods employed in our laboratory to transfect and express foreign genes into epithelial cells and to determine the polarized surface expression of both endogenous and exogenous plasma membrane proteins.

II. Expression of Foreign Genes in Polarized Epithelial Lines

Recent work has characterized the polarized surface distribution of viral glycoproteins or receptor molecules introduced into epithelial cells via infection or transfection (Table II) (for reviews, see Rodriguez-Boulan, 1983a; Simons and Fuller, 1985; Garoff, 1985; Matlin, 1986). A review of the methods used in our laboratory to study the polarized assembly of viruses by epithelial cells can be found elsewhere (Rodriguez-Boulan, 1983b).

TABLE I

FREQUENTLY USED POLARIZED EPITHELIAL CELL LINES

Line	Origin	Transmonolayer electrical resistance (Ω/cm^2)	References
MDCK	Dog kidney	Type I: 1000–3000	Cerejdo *et al.* (1978); Richardson *et al.* (1981)
		Type II: 100–300	Simons and Fuller (1988)
LLC-PK1	Pig kidney	200–400	Saier *et al.* (1986)
MA-104	Monkey kidney	600–700	Roth *et al.* (1987)
Caco-II	Human colonic tumor	110–200	Grasset *et al.* (1984); Pinto *et al.* (1983)
HT-29	Human colonic tumor	—	Pinto *et al.* (1982)
T-84	Human colonic tumor	100	Dharmsathaphorn *et al.* (1984)
A6	Toad bladder epithelium	7000	Perkins and Handler (1981)

TABLE II

EXPRESSION OF FOREIGN PLASMA MEMBRANE PROTEINS IN EPITHELIAL CELLS

Method of introduction	Cell line	Transient (T) or permanent (P)	Protein expressed	Apical (A) or basolateral (B)	References
Viral infection					
Vesicular stomatitis	MDCK	T	G protein	B	Rodriguez-Boulan and Pendergast (1980)
Influenza	MDCK	T	HA	A	Rodriguez-Boulan and Pendergast (1980)
		T	NA	A	Fuller *et al.* (1985a); Jones *et al.* (1985)
Semliki Forest	MDCK	T	E_2	B	Fuller *et al.* (1985b)
Sendai	MDCK	T	F, HN	A	Rodriguez-Boulan and Pendergast (1980)
Herpes simplex	MDCK, Vero	T	gB, gC, gD, gE, gG	B	Srinivas *et al.* (1986)
Murine retrovirus	MDCK	T	$p15^{env}$	B	Roth *et al.* (1983a)
Recombinant Virus Infection					
SV40	Monkey kidney	T	HA	A	Roth *et al.* (1983b)
	MA-104	T	HA	A	Roth *et al.* (1987)
	MA-104, Vero	T	NA	A	Jones *et al.* (1985)
Vaccinia	MDCK	T	HA	A	Stephens *et al.* (1986)
	MDCK	T	G protein	B	Stephens *et al.* (1986)
	MDCK	T	$gp70/p15^{env}$	B	Stephens and Compans (1986)
Retrovirus	MDCK	P	IgA receptor	B→A	Mostov and Deitcher (1986)
Transfection					
SV40 promoter	MDCK	P	HA	A	Gottlieb *et al.* (1986)
SV40 promoter	MDCK	P	G protein	B	Gottlieb *et al.* (1986); Puddington *et al.* (1987)
SV40 promoter	MDCK	P	p62	B	Roman and Garoff (1986)
MuLV promoter	MDCK	P	ASGP-R[a]	B	Graeve *et al.* (1989)

[a] Asialoglycoprotein receptor.

DNA Transfection of MDCK Cells and Establishment of Stably Transformed Cell Lines

Retroviral promoters, specifically Moloney murine leukemia virus (MuLV) and Rous sarcoma virus long terminal repeats (LTR), have resulted in high levels of expression of exogenous plasma membrane proteins in our hands (see also Mostov and Deitcher, 1986). Other laboratories have used Simian virus 40 (SV40) promoters successfully (Gottlieb *et al.*, 1986; Roman and Garoff, 1986; Puddington *et al.*, 1987). However, SV40 promoters require pretreatment of the monolayers with sodium butyrate (Gottlieb *et al.*, 1986; Roman and Garoff, 1986; Puddington *et al.*, 1987) to include high levels of expression. For transfection, the cDNA is introduced into MDCK cells as a calcium phosphate precipitate (Graham and van der Eb, 1973). Since the efficiency of transformation is usually very low (1 out of 10^4–10^6 cells), cells are cotransfected with a selectable marker. In our hands the neomycin-resistance marker coupled with selection with G418 has been very useful for the selection of stable transfectants.

1. Transfection of MDCK Cells

Transfection of MDCK cells is more efficient with cells in suspension. Normally, 1–2 $\times$ 10^6 cells are transfected with 10 μg of the nonselectable gene and 1 μg of the marker gene. If both genes are located on the same plasmid, 1–2 μg of the specific DNA plus 10 μg of carrier DNA are used. Nevertheless, in our experience it is preferable that the selectable and the nonselectable genes are on two separate plasmids. The DNA is purified in $CsCl_2$ gradients by standard procedures. The plasmids are not linearized.

2. Solutions (Sterile-Filtered, 0.22 μm)

70 m*M* sodium phosphate, pH 6.8
2 *M* $CaCl_2$
2× HBS (HEPES-buffered saline): 50 m*M* HEPES, 280 m*M* NaCl. Adjust pH to 7.10 ± 0.05 with NaOH. This pH has to be exact and should be readjusted before each use.
15% (w/v) Glycerol in HBS
10 m*M* Tris-HCl, pH 7.5
10 mg/ml Chloroquine in H_2O (Sigma) (stored frozen at −20°C)
G418 (Gibco, Grand Island, NY), stock solution 125 mg/ml in H_2O (aliquots stored at −20°C)
Dulbecco's minimal essential medium (DME) (Gibco)

3. Procedure

1. Plate 1×10^6 cells per transfection in a 100-mm Petri dish in DME–10% fetal calf serum (FCS). Use cells of low passage number that are not confluent. Incubate overnight.

2. Mix 20 μg of nonselectable and 2 μg of selectable DNA, ethanol-precipitate, centrifuge, and remove ethanol and water quantitatively under sterile hood. Let pellet dry and dissolve in 440 μl of 10 m*M* Tris-HCl, pH 7.5.

3. Take sterile 3-ml tubes and add 500 μl of $2 \times$ HBS to each tube. Add 10 μl 70 m*M* sodium phosphate (pH 6.8) to each tube.

4. Add 60 μl 2*M* $CaCl_2$ to the dissolved DNA, mix, and add dropwise to the tube containing $2\times$ HBS, while gently blowing sterile air through the solution. A white precipitate should form (check against black background). Let the precipitate stand at room temperature for 20 minutes.

5. Trypsinize cells, collect by centrifugation, and resuspend them in 1 ml fresh DME–10% FCS. Pipet cells into a new 10-cm Petri dish and add dropwise 500 μl of the DNA precipitate (mix it before). Gently agitate the dish to distribute cells and precipitate uniformly. Let stand for 20 minutes at room temperature.

6. Add 3.5 ml of DME–10% FCS containing 100 μg/ml chloroquine. Let cells attach for 6–9 hours at 37°C.

7. Remove medium, add 15% glycerol solution for 1 minute (37°C), wash cells twice with DME and add 10 ml DME–10% FCS. Let cells grow for 2–3 days until they reach confluency.

4. Selection and Screening of Resistant Clones

1. Trypsinize cells and replate them 1 : 8 in a 150-mm dish (or 1 : 20 in a 100-mm dish) in selection medium (DME–10% horse serum containing 500 μg/ml G418). Change medium every 3–4 days. After 10–14 days all nontransformed cells should have detached from the plate and 10–100 resistant colonies per dish should be visible.

2. Resistant colonies are isolated by trypsinization with cloning rings and replated in 24-well tissue culture dishes; an aliquot is plated on glass coverslips for screening of expression by indirect immunofluorescence. As an alternative, the whole dish with the resistant colonies is trypsinized and cells are cloned by limiting dilution in microtiter plates. By this strategy one can screen on a single coverslip the resistant cells for expression of the transfected gene. Furthermore, this is the method of choice when the colonies are grown too close to each other or too many colonies are present on the dish.

3. Colonies that show expression are expanded, frozen down, and further characterized.

III. Estimation of Polarity of Epithelial Surface Molecules

The determination of the extent of surface polarization of a given molecule is becoming increasingly important as investigators in the field address the mechanisms responsible for its biogenesis. Ultimately, polarity can be expressed as a ratio between the amounts of a given molecule in apical and basolateral plasma membrane domains. An important consideration in the analysis of this ratio is the fact that the area of the apical domain is usually a fraction of the area of the basolateral domain. In the MDCK cell line, this ratio is 1 : 4 for the type II (low-resistance) cells and 1 : 7 for type I (high-resistance) cells (von Bonsdorff *et al.*, 1985). Therefore, it is important to attempt to express polarity as a *density ratio,* that is, a ratio between the amounts of the molecule per unit surface area in each surface domain. This is, in fact, a relatively difficult task; the main difficulty resides in the fact that apical and basolateral membranes have differential degrees of accessibility to probes, and this difference becomes more apparent as the size of the probe increases. Table III summarizes the various types of procedures utilized to estimate the surface polarity of a molecule in expressed epithelial cells.

A. Video-Enhanced Analysis of Fluorescence in Semithin Frozen Sections

1. Solutions

Rat tail collagen (3 mg/ml in 20 m*M* acetic acid)
Sterile water
10× Dulbecco's MEM (10× DME)
0.34 *M* NaOH (sterile)
Phosphate-buffered saline, 1 m*M* $MgCl_2$, 0.1 m*M* $CaCl_2$ (PBS/CM)
2% Formaldehyde (prepared from solid paraformaldehyde) in PBS/CM
10% Gelatin in PBS/CM

2. Plating of Monolayers on CollagenCoated Coverslips

Substrates for cell attachment are prepared by mixing 1 ml of rat tail collagen with 400 μl sterile water, 200 μl 10× DME and 200 μl sterile NaOH. Clean and sterile coverslips are rapidly soaked in the collagen mixture and placed in a 24-well Petri dish. The collagen gels form in ~10 minutes upon neutralization with NaOH. However, we normally allow gelation to proceed for 1 hour at 37°C. For some experiments, "fixed" gels are used. The coverslips are dipped into collagen in acetic acid, exposed to ammonium hydroxide vapors in a humid chamber and fixed in 1% glutaraldehyde. These coverslips are then rinsed three times in sterile saline bicarbonate buffer (Moscona) or PBS and three times in DME before plating the cells.

MDCK or LLC-PK1 cells are plated at immediate confluency on collagen-coated coverslips (1–2 × 10^5 cells/cm^2). MDCK or MA-104 cells attach well to native or fixed collagen gels. LLC-PK1 cells attach more efficiently to fixed gels. Two hours after plating, unattached cells are removed with washes (four times) in DME.

3. Semithin Frozen Sections

The monolayers are fixed with 2% formaldehyde in PBS containing 1 m*M* Mg^{2+} and 0.1 m*M* Ca^{2+} (PBS/CM) (freshly prepared from paraformaldehyde) for 1 hour, scraped with a razor blade from the coverslip, embedded in 10% gelatin in PBS/CM, and infused with 1.8 *M* sucrose overnight at 4°C. Frozen sections (0.5 μm) are obtained at −80°C with a Dupont MT-5000 ultramicrotome equipped with a FS-1000 frozen-sectioning attachment and collected on glass coverslips (previously treated with 1 mg/ml polylysine MW 150,000–300,000 in PBS). After extensive washing in PBS/CM containing 1% bovine serum albumin (BSA), the sections are processed for indirect immunofluorescence. When fluoresceinated goat second antibodies are used, BSA is replaced with 50 mg/ml preimmune goat IgG.

After a final short rinse in double-distilled water, the coverslips are inverted and mounted on a drop of 20% polyvinylalcohol (Vinol 205, Air Products, Allentown, PA)–15% glycerol in PBS on a glass microscope slide, allowed to harden at room temperature for 2 hours, and stored in the cold. Samples are photographed with a Leitz Ortholux epifluorescence microscope using 400 ASA Kodak Tri-X film developed with Diafine for 10 minutes at 20°C. Exposure times for positive results are in the range of 20–50 seconds.

TABLE III

METHODS TO ESTIMATE SURFACE POLARITY IN CULTURED EPITHELIAL CELLS

Methods[a]	Advantages	Disadvantages	References
Physiological methods			
Isotopic uptake or binding Electrophysiologic procedures	Fast procedures; uptake and binding most reliable for small molecules and ions	Applicable only to functional molecules; access limited from the basal side for large molecules	Murer and Kinne (1980)
Immunological procedures			
En Face IF Monolayers on coverslips Monolayers on filters	Fast procedure	Qualitative; access of antibody to basolateral surface poor	Rodriguez-Boulan and Pendergast (1980) Fuller *et al.* (1984)
EM immunocytochemistry; preembedding staining Scraped monolayers Monolayers on filters	More quantitative than *en face* IF when colloidal gold or ferritin is used	Slow; access to lateral surface incomplete	Rodriguez-Boulan and Pendergast (1980) Roth *et al.* (1987) Herzlinger and Ojakian (1984)
IF on 1-μm (semithin) frozen sections	Semiquantitative; equal access of antibody to all surfaces; faster than EM immunocytochemistry	Does not discriminate accurately between membrane and submembrane fluorescence	Vega-Salas *et al.* (1987) Salas *et al.* (1988)
Postsection colloidal gold immunocytochemistry Ultrathin frozen sections Plastic sections (Iowicryl, etc.)	Quantitative; equal access of antibody to all surfaces	Ultrathin frozen section is difficult procedure; poor sensitivity (plastic sections)	Griffiths *et al.* (1983) Roth *et al.* (1978)

RIA on filter-grown monolayers	Semiquantitative	Access of antibody to basolateral surface poor (whole antibodies), somewhat better for Fab fragments	Pfeiffer *et al.* (1985) Salas *et al.* (1986)
Confocal microscopy	Fast, maybe semiquantitative; provides spatial distribution of the antigen	When carried out with antibodies, same as RIA	van Meer *et al.* (1987)
Biochemical Procedures			
Trypsin digestion	Semiquantitative, fast	Requires trypsin sensitivity of antigen; accessibility of basolateral membrane may be limited	Matlin and Simons (1984)
Biotin procedure Sulfo-NHS-biotin	Semiquantitative, fast	Accessibility of basolateral membrane may be limited	Sargiacomo *et al.* (1989)
Exogalactosylation	Semiquantitative, fast	Can only be performed on adequate glycosyltransfer-deficient lines; accessibility of enzyme may be limited	Brandli *et al.* (1988)
Subcellular Fractionation			
Open apical membrane sheets	Fast, allows enrichment of apical markers	Qualitative (since allows only purification of apical membrane)	Sambuy and Rodriguez-Boulan (1988)

[a] IF, Immunofluorescence; RIA, radioimmunoassay.

4. Digital Analysis of Fluorescence

For storage of digitized images of the fluorescent sections the same microscope is used, but the photographic camera is replaced with a silicon-intensified video camera (SIT 66, DAGE-MTI, Michigan City, IN). The camera output is digitized with a PCVISION frame grabber (Imaging Technology, Woburn, MA) and stored in an IBM AT personal computer. Nine images per field are averaged, but this number might be higher with a faster frame grabber that would allow real-time averaging (32 images per second). The averaging "filters" background noise—usually important with silicon-intensified cameras—and results in sharper images. The averaged image is stored in the hard disk of the computer. With non-real-time frame grabbers (as old versions of PCVISION), it may be helpful to define an "area of interest," centralize the desired field within this area, and average only the pixels in this area rather than all the pixels in the screen. This also reduces the size of images to be stored and results in better utilization of the hard disk. The use of mass-storage devices may be helpful, since each image may be 16–325 kbytes. It is key to the interpretation of the data that all measurements should be within the linear range of the video camera. Two criteria can be used to determine this: (1) the camera and digitizer manufacturers provide ranges of linearity of their products, usually for 8-bit systems (i.e., pixel values 0–255); pixel values >10 and <245 are safely linear. (2) An experimental calibration of the system can be obtained in the same way that immunogold is calibrated (Posthuma *et al.*, 1987). Increasing concentrations of mouse IgG in the range 0.1–20 mg/ml are mixed with freshly prepared gelatin at 44°C. Thin layers of each gelatin–IgG mixture are poured and allowed to gel one after the other. The gels are fixed in paraformaldehyde and processed in sucrose as described before. Semithin frozen sections of up to five layers can be obtained, and processed for immunofluorescence with fluorescent second antibody alone. Several pixels in the image from each layer are averaged, and the linearity of the system is calibrated. Many cameras provide "automatic" gain. It can be useful unless intensities from different fields are compared, in which case constant gain has to be used throughout the experiment.

The actual size of the image represented in each pixel varies with the total number of pixels and the magnification. For our system it is 0.31×0.23 μm at 630×. Therefore, the method cannot resolve single membranes from membrane foldings, and cannot resolve membrane fluorescence from cytoplasmic fluorescence. Both problems can be controlled through an estimation of the folding factor F.

5. Estimation of the Folding Factor F

In order to correct for "membrane density" in the pixel area, it is necessary to estimate the contribution of microvilli and basal folds. Because this is not possible at the optical microscope level (in the same frozen sections), we use morphometric procedures at the electron-microscopic (EM) level. Although these values are obtained in thinner sections, they yield significant estimates under the assumption that the fluorescence intensity in each spot is proportional to the density of membranes S_v. This assumption is justified because semithin frozen sections are ~0.5 μm, that is, 5–10 times thicker than Epon sections, and 5 times thicker than single microvilli. In parallel controls we have studied the penetration of molecules substantially larger than IgG, such as ferritin (M_r ~800,000) and found that they penetrate several microns in gelatin gels processed exactly as for the experimental sections (fixed in paraformaldehyde, and incubated with ferritin for 1 hour). In other words, differently from ultrathin frozen sections where immunogold particles only label the sections' surface, fluorescent antibodies can freely penetrate the entire volume of the section, and therefore, it is reasonable to assume they label the whole membrane area ("density") in the section.

The morphometric determination of $F = S_v$ apical domain/S_v basal domain has been described elsewhere (Vega-Salas *et al.*, 1987). Corrections for finite section thickness are particularly important for areas with microvilli or other membrane foldings with dimensions close to the section thickness. Corrections for anisotropy are also important in the estimation of S_v for individual epithelial cell membrane domains. This morphometric procedure may be tedious, but once an average folding factor is obtained it can be assumed to be constant for the same cell line under the same culture conditions (media, serum, substratum, and time after plating). Typically we have found F to be 1.4–1.6 for MDCK cells and 2.0–2.3 for LLC-PK1 cells confluent on collagen gels.

6. Calculation of Polarity Ratios

Once the folding factor is available, a semiquantitative estimation of apical to basal polarity is immediate: for each cell in the section, three random profiles are taken; usually two of them skip the nucleus and the third one contains the nucleus. For each profile 5 pixel values are recorded: apical background *(ab)*, apical membrane *(a)*, cytoplasmic fluorescence *(c)*, basal membrane *(b)*, and basal background *(bb)*. Because of nonspecific binding to the collagen substrate, often $bb > ab$ by up to 10 pixel units. Since the method cannot distinguish cytoplasmic from membrane fluorescence, it is important that the signal of the positive membrane (i.e., a for an apical antigen) is much greater than the extracellular and the cytoplasmic background; that is, $a >> aa, c$. The measurements are usually collected in computer spreadsheet for easier statistical manipulations. Polarity ratios are then calculated as

$$\text{Polarity ratio (apical antigen)} = (a - K)/F(b - K)$$
$$\text{Polarity ratio (basal antigen)} = (b - K)F/(a - K)$$

where $K < (ab + bb)/2$; usually $K < 70$ pixel units.

The cytoplasmic fluorescence is often the limit for the accuracy of the method. For poorly polarized membrane antigens, $a >> c$ and $b >> c$, the polarity ratios are clearly accurate. For very well-polarized antigens, on the other hand, polarity ratios calculated as just described are an underestimation of the real concentration gradient between both domains; that is, for a very well-polarized apical antigen (see, for example, the 184-kDa apical antigen described in Vega-Salas *et al.*, 1987) $a >> c$ but $b \cong c$. This problem is common to any ratio when the denominator approaches the background level. Conceptually, though, there is still one question that cannot be answered at the optical microscope level: Is there any concentration of antigen containing intracellular membranes very closely associated to the plasma membrane? Only ultrathin frozen sections with immunogold labeling may answer this question.

B. Biotin Assay for Polarity

A water-soluble biotin analog, sulfo-*N*-hydroxy-succinimido-biotin (sulfo-NHS-biotin) can be employed to label selectively the apical or basolateral surfaces of epithelial cell monolayers. When employed on MDCK cells or on other polarized epithelial cell lines, this procedure reveals strikingly different protein compositions of the apical and basolateral membranes (Sargiacomo *et al.*, 1989).

1. Materials

Polycarbonate filter chambers, 24.5 mm diameter, 0.4 μm pore size, tissue culture-treated (Transwell, Costar, Inc., Cambridge, MA)

Triton X-114 (TX-114) (Sigma), three times precondensed according to Bordier (1981)

[^{3}H]Inulin or [^{3}H]ouabain (NEN, Boston, MA)

Sulfo-NHS-biotin (Pierce, Rockford, IL). Stock solution: 200 mg/ml in dimethyl sulfoxide (DMSO). 10 μl aliquots are stored at −20°C.

^{125}I-Labeled streptavidin (^{125}I-streptavidin) radiolabeled with the chloramine-T procedure (Greenwood *et al.*, 1963) to a specific activity of 10 μCi/μg
PBS/CM
Lysis buffer I: 0.15 *M* NaCl, 1 m*M* MEDTA, 1% TX-114, 10 m*M* Tris, pH 7.4
BSA–TGG: 3% BSA in PBS containing 0.5% Tween 20 (Sigma), 10% glycerol, 1 *M* glucose

2. Cell Plating—Monolayer Integrity

After dissociation with trypsin–EDTA, cells are plated at high density on polycarbonate filter chambers. For MDCK cells, medium is changed every 2–3 days and experiments are performed 6–8 days after plating.

Monolayer integrity and the tightness of tight junctions are assayed by measuring the permeability of the monolayer to [^{3}H]inulin or [^{3}H]ouabain. Briefly, 1 ml complete medium containing ~0.2–0.5 μCi of the substances just listed is added to the apical compartment of the filter chamber. After 2 hours incubation at 37°C, samples are collected from the basolateral compartment and counted. Monolayer permeability is expressed as a percentage of the total counts per minute added to the apical compartment that leak to the basolateral side. Monolayers showing permeability >1% in 2 hours at 37°C are discarded.

3. Apical and Basolateral Biotinylation

To biotinylate the apical or the basolateral surfaces, 5–7 days confluent monolayers on filter chambers are washed four times with PBS/CM and agitated for 30 minutes at 4°C in PBS/CM. An aliquot of frozen (−20°C) stock sulfo-NHS-biotin solution (200 mg/ml in DMSO) is thawed just before use and diluted to a final concentration of 0.5 mg/ml in ice-cold PBS/CM. One milliliter is added to either the apical or the basolateral compartment of the filter chamber. Compartments not receiving sulfo-NHS-biotin are filled with an equivalent volume of PBS/CM. After 20–30 minutes of agitation at 4°C, filter chambers are washed once with serum-free medium and three times with PBS/CM.

To visualize the pattern of biotinylated proteins, filters are excised from the chamber with a scalpel and extracted with 1 ml of ice-cold lysis buffer for 1 hour with intermittent mixing. Samples are then transferred to Eppendorf tubes and clarified by centrifugation (14,000 *g* for 10 minutes at 4°C). Supernatants are collected and subjected to phase separation, as described by Bordier (1981). Briefly, they are incubated at 37°C for 3 minutes, and the aqueous and detergent phases are separated by centrifugation in a Beckman microcentrifuge (14,000 *g* for 10 minutes at room temperature).

Phase separation with TX-114 allows separation of hydrophilic "peripheral" proteins (which partition with the aqueous phase) from hydrophobic "integral" membrane proteins, which are collected in the detergent phase (Bordier, 1981). Some membrane proteins of intermediate hydrophobicity, however, are found to partition with the aqueous phase, which contains 0.7 m*M* TX-114 under the conditions described earlier (Pryde, 1986).

4. Detection of Biotinylated Proteins (Electrophoresis and Electroblotting)

Cell extracts (aqueous or detergent TX-114 phases) are precipitated with five volumes of acetone (−20°C for 30 minutes), spun (14,000 *g*, 10 minutes), resuspended in 100 μl Laemmli sample buffer, boiled for 2–3 minutes, and electrophoresed under reducing conditions in sodium dodecyl sulfate (SDS)–polyacrylamide slab gels (Laemmli, 1970). After electrophoresis, proteins are transferred to nitrocellulose as described by Towbin *et al.* (1979) at a constant voltage (60 V) for 14–16 hours. Nitrocellulose sheets are incubated with ^{125}I-streptavidin under conditions that reduce nonspecific binding to nitrocellulose (Birk and Koepsell, 1987). Briefly, blots are blocked with BSA–TGG for 1 hour at room temperature. ^{125}I-Streptavidin (1–2 $\times$ 10^6 cpm/ml in TGG–0.3% BSA) is allowed to bind for 2 hours at room temperature, followed by washing in PBS–0.5% Tween 20 (4×, 15 minutes each). Blots are dried and autoradiographed (2–12 hours at −70°C, with an intensifying screen) on Kodak XAR-5 film.

5. Use of the Biotinylation Procedure to Determine the Polarity of a Given Antigen

The biotin-labeling procedure can also be used to determine the polarized distribution of a given antigen. After labeling of the cell monolayers with sulfo-NHS-biotin as described earlier, cells are lysed in an appropriate immunoprecipitation buffer (e.g., lysis buffer II, Section III,C) and immunoprecipitated with the respective antiserum. Immunoprecipitates are analyzed by SDS–polyacrylamide gel electrophoresis (SDS–PAGE) and biotinylated antigen is detected with ^{125}I-streptavidin as described before. The use of antiserum for the immunoprecipitation may yield problems due to nonspecific binding of ^{125}I-streptavidin to IgG heavy chain. This problem can be overcome by either using affinity-purified antibodies, by coupling the antibodies to Sepharose beads and using a one-step immunoprecipitation procedure, or by including 1% (w/v) nonfat dry milk in the BSA–TGG buffer during the initial blocking step.

C. Trypsin Assay on Filter-Grown MDCK Cells

To determine the polarized surface expression of certain antigens, a trypsin assay developed by Matlin and Simons (1984) can be employed. This assay makes use of the observation that trypsinization of MDCK monolayers at 4°C in medium free of EDTA or EGTA leads to cleavage of surface proteins (e.g., influenza HA) without affecting transepithelial resistance or tightness of the monolayers. Polarity can be assayed by adding trypsin to either the apical or basolateral side of filter-grown cells and measuring either the appearance of cleavage products (in the case of HA/HA_0: $HA_1 + HA_2$) or the loss of the antigen. A prerequisite is, of course, a trypsin-sensitive antigen. Initial studies on subconfluent cells should be carried out to determine the amount of trypsin and the incubation time needed to cleave or remove the antigen quantitatively from the cell surface (determined, for example, by surface radioimmunoassay).

1. Materials

TPCK–Trypsin (Boehringer): stock solution 10 mg/ml in PBS (stored in aliquots at −20°C), is diluted to the appropriate concentration in PBS/CM.

Trypsin inhibitor (soybean, lima bean, etc.) in PBS/CM (usually 100 μg/ml)

Transwells, 24.5 mm diameter

DME containing one-tenth of normal concentration of methionine

Lysis buffer II: 1% Nonidet P-40, 0.4% Sodium deoxycholate, 66 m*M* EDTA, 10 m*M* Tris-HCl, pH 7.4

[^{35}S]Methionine (Translabel, ICN, Costa Mesa, CA)

2. Procedure

Cells are seeded at high density in polycarbonate filter chambers and grown for 5–7 days, with daily changes of medium. The tightness of the monolayers is determined by measuring the permeability of [^{3}H]ouabain or [^{3}H]inulin, as described in Section B, 2. Cells are labeled for 20–24 hours with [^{35}S]methionine (150 μCi/ml) in DME containing one-tenth the normal concentration of methionine. Labeling medium is added only to the basolateral side. Cells are washed extensively at 4°C with PBS/CM and TPCK–trypsin is added either to the apical or basolateral side. Trypsin inhibitor is added to the opposite side. Incubation is on ice for 15 minutes to 1 hour. Control cells receive only PBS/CM. After aspiration of the trypsin solution, monolayers are washed 2 × 5 minutes with 100 μg/ml trypsin inhibitor added to both sides of the filter and then several times with PBS/CM. Lysis is in lysis buffer II containing 100 μg/ml trypsin inhibitor. After immunoprecipitation, SDS–PAGE, and fluorography, the amount of antigen digested on the apical and on the basolateral side is determined by densitometry of the respective bands on the fluorogram.

D. Method to Obtain Apical Membrane Sheets from Polarized MDCK Cells

Cell fractionation procedures do not yield very pure fractions when applied to cultured cells. We have developed a fast and efficient (60% recovery) solid-phase procedure that yields open sheets of highly purified apical plasmalemma from MDCK monolayers in culture. The plasma membrane sheets have the cytoplasmic side exposed, which makes them potentially useful for reconstitution of transport between the Golgi apparatus and the cell surface.

1. Solutions

Millipore filters, HTAF, 0.45 μm pore size, 137 mm diameter
PBS/CM
MES–Saline: 130 m*M* NaCl, 20 m*M* 2 [*N*-morpholino]ethanesulfonic acid, 1 m*M* $MgCl_2$, pH 6.5
1% Cationized colloidal silica (obtained from Dr. Bruce S. Jacobson, Amherst, MA) in MES–saline
1 mg/ml Polyacrylic acid (M_r 50,000; Polyscience) in MES–saline, pH 6.5
1 mg/ml polylysine, MW >300,000 (Sigma)
2.5 m*M* Imidazole buffer, pH 7.3
100 m*M* Sodium carbonate, pH 11

2. Procedure

MDCK cells are seeded at high density ($1.5–2 \times 10^5$ cells/cm^2) on nitrocellulose filters (Millipore) in plastic culture dishes (Falcon, 150 mm diameter) in standard culture medium and grown for 4–6 days with daily medium changes.

The cell monolayer is washed twice with PBS and twice with MES–saline at 4°C; all solutions used throughout the procedure contain 1 mM $MgCl_2$ and 0.1 mM $CaCl_2$ to maintain the integrity of the tight junctions.

The monolayer is treated for 10 seconds with 1% cationized colloidal silica in MES–saline, treated for 10 seconds with 1 mg/ml polyacrylic acid (M_r 50,000; Polyscience) in MES–saline (pH adjusted to 6.5), and washed extensively with MES–saline. The coating procedure is repeated once and the cells on the filter are overlaid with a glass plate that was previously coated with 1 mg/ml polylysine (Sigma) in double-distilled water. Prior to polylysine treatment, the glass plates are acid-washed and extensively washed with double-distilled water and allowed to dry. Polylysine treatment is carried out for 1 hour to overnight and the plates are washed in water and allowed to dry.

The glass surfaces and the filters are pressed together under a weight or with the aid of a heavy rolling pin (marble) and the surfaces carefully peeled apart and transferred to the extraction medium of choice for further processing. What appears to be critical to obtain large sheets of apical membrane is a good adherence between the coat overlaying the cells and the polycation-coated glass plate; this is achieved by applying a uniform pressure and by removing excess water between the layers by blotting the filter well before peeling it away from the plate. A brief hypotonic shock with 2.5 mM imidazole buffer (pH 7.3) at 4°C improves the recovery of apical membrane at the cost of higher cytoplasmic and lateral membrane contamination of the fraction.

For further purification, the apical fraction is placed (still attached to the glass plate) in 100 mM sodium carbonate (pH 11) at 4°C and agitated in an orbital shaker for 60 minutes with two changes of the washing solution. The purified membranes are collected in the buffer of choice by scraping from the glass plate with a razor blade.

IV. Conclusions

Procedures to quantitate the polarized distribution of surface molecules in epithelial cells suffer from a basic limitation: the basolateral and the apical membranes have very different accessibility to external probes. In this regard, the most reliable procedures for intact monolayers are those that use very small probes (such as ions or small molecules, for example, [^{3}H]ouabain, to measure Na,K-ATPase, etc.). Molecules for which a functional assay is not available can be measured only by immunological or biochemical procedures. The most quantitative immunological procedure is, perhaps, immunoelectron microscopy on ultrathin frozen sections (Griffiths *et al.*, 1983). However, this method is relatively insensitive and therefore is limited to relatively abundant antigens that keep their antigenicity under the quite stringent fixation conditions required. Postembedding procedures are usually even less sensitive than ultrathin frozen sections.

Semithin frozen sections provide a relatively fast qualitative estimate of the polarity of surface antigens. When combined with video-enhanced microscopy, they give a semiquantitative determination of surface polarity. Accessibility of antibodies is probably similar for both surfaces, but the contribution of intracellular fluorescent structures cannot be easily measured.

Biochemical procedures are relatively fast and simple. When applied to individual molecules, they require a parallel method to identify them (such as immunoprecipitation). Because they are carried out on filter-grown monolayers, they may underestimate polarity because of reduced access to basal membrane. For the procedures described here, this is minimized by the fact that trypsin probably digests its way up to the tight junction and by the small size of the biotin analog.

To conclude, there is no perfect method to estimate polarity of a surface molecule. Judicious selection of the procedures to be utilized depends on the availability of functional assays, on the behavior of the antigen upon fixation, on the availability of good antibodies to the molecule, and on the technical expertise available to the researcher. Because of the uncertainty associated with most procedures, it is usually advisable to use more than one of them.

Acknowledgments

We thank Ms. Francine Sanchez for typing the manuscript. M. S. was supported by the Istituto Superiore di Sanita (Italy), M. L. by a Medical Scientist Training grant from the Cornell University M.D./Ph.D. program, L. G. by a fellowship from the Deutsche Forschungsgemeinschaft (Federal Republic of Germany), A. L. B. by a fellowship from Association pour la Recherche sur le Cancer (France), E. R. B. was a recipient of an Established Investigator Award from the American Heart Association. P. S. and D. V. S. are currently career investigators from the CONICET (Argentina). This research was supported by grants from NIH (R01 GM-34107) and from the American Heart Association (New York Branch).

References

Birk, H., and Koepsell, H. (1987). *Anal. Biochem.* **164,** 12–22.

Bordier, C. (1981). *J. Biol. Chem.* **256,** 1604–1607.

Brandli, A. W., Hansson, G. C., Rodriguez-Boulan, E., and Simons, K. (1988). *J. Biol. Chem.* **263,** 16283–16290.

Cerejido, M. E., Robbins, E. S., Dolan, W. J., Rotunno, C. A., and Sabatini, D. D. (1978). *J. Cell Biol.* **77,** 853–880.

Dharmsathaphorn, K., McRoberts, J. A., Mandel, K. G., Tisdale, L. D., and Masui, H. (1984). *Am. J. Physiol.* **246,** G204–6208.

Fuller, S. D., von Bonsdorff, C. H., and Simons, K. (1984). *Cell (Cambridge, Mass.)* **38,** 65–77.

Fuller, S. D., Bravo, R., and Simons, K. (1985a). *EMBO J.* **4,** 297–304.

Fuller, S. D., von Bonsdorff, C. H., and Simons, K. (1985b) *EMBO J.* **4,** 2475–2485.

Garoff, H. (1985). *Annu. Rev. Cell Biol.* **1,** 403–446.

Gottlieb, T. A., Gonzalez, S., Rizzolo, L., Rindler, M. J., Adesnik, M., and Sabatini, D. (1986). *J. Cell Biol.* **102,** 1242–1255.

Graeve, L., Patzak, A., Drickamer, K., and Rodriguez-Boulan, E. (1989). Submitted for publication.

Graham, F., and van der Eb, A. (1973). *Virology* **52,** 456.

Grasset, E., Pinto, M., Dussaulx, E., Zweibaum, A., and Desjeux, J. F. (1984). *Am. J. Physiol.* **247,** C260–C267.

Greenwood, F. C., Hunter, W. M., and Glover, J. S. (1963). *Biochem. J.* **89,** 289–300.

Griffiths, G., Simons, K., Warren, G., and Tokuyasu, K. T. (1983). *In* "Methods in Enzymology" (S. Fleischer and B. Fleischer, eds.), Vol. 96, pp. 435–450. Academic Press, New York.

Herzlinger, D. A., and Ojakian, G. K. (1984). *J. Cell Biol.* **98,** 1777–1787.

Jones, L. V., Compans, R. W., Davis, A. R., Bos, T. J., and Nayak, D. P. (1985). *Mol. Cell. Biol.* **5,** 2181–2189.

Laemmli, U. K. (1970). *Nature (London)* **227,** 680–685.

Matlin, K. S. (1986). *J. Cell Biol.* **103,** 2565–2568.

Matlin, K. S., and Simons, K. (1984). *J. Cell Biol.* **99,** 2131–2139.

Misfeldt, D. S., Hamamoto, S. T., and Pitelka, D. R. (1976). *Proc. Natl. Acad. Sci. U.S.A.* **73,** 1212–1216.

Mostov, K. E., and Deitcher, D. L. (1986). *Cell (Cambridge, Mass.)* **46,** 613–621.

Murer H., and Kinne, R. (1980). *J. Membr. Biol.* **55,** 81–95.

Perkins, F. M., and Handler, J. S. (1981). *Am. J. Physiol.* **241,** C154–C159.

Pfeiffer, S., Fuller, S. D., and Simons, K. (1985). *J. Cell Biol.* **101,** 470–476.

Pinto, M. S., Appay, M. D., Simmon-Assmann, P., Chevalier, G., Dracopoli, N., Fogh, J., and Zweibaum, A. (1982). *Biol. Cell.* **44,** 193–196.

Pinto, M. S., Robine, L., Appay, M. D., Kedinger, M., Triadou, N., Dussaulx, E., LaCroix, B., Simon-Assmann, P., Haffen, K., Fogh, J., and Zweibaum, A. (1983). *Biol. Cell.* **47,** 323–330.

Posthuma, G., Slot, J. W., and Geuze, H. J. (1987). *J. Histochem. Cytochem.* **35,** 405–410.

Pryde, J. G. (1986). *Trends Biochem. Sci.* **11,** 160–163.

Puddington, L., Woodgett, C., and Rose, J. K. (1987). *Proc. Natl. Acad. Sci. U.S.A.* **84,** 2756–2760.

Richardson, J. C. W., Scalera, V., and Simmons, N. L. (1981). *Biochim. Biophys. Acta* **673,** 26–36.

Rodriguez-Boulan, E. (1983a). *Mod. Cell Biol.* **1,** 119–170.

Rodriguez-Boulan, E. (1983b). *In* "Methods in Enzymology" (S. Fischer and B. Fleischer, eds.), Vol. 98, pp. 486–500. Academic Press, New York.

Rodriguez-Boulan, E., and Pendergast, M. (1980). *Cell (Cambridge, Mass.)* **20,** 45–54.

Roman, L. M., and Garoff, H. (1986). *J. Cell Biol.* **103,** 2607–2618.

Roth, J., Bendayan, M., and Orci, L. (1978). *J. Histochem. Cytochem.* **26,** 1074–1081.

Roth, M. G., Srinivas, R. V., and Compans, R. W. (1983a). *J. Virol.* **45,** 1065–1073.

Roth, M. G., Compans, R. W., Giusti, L., Davis, A. R., Nayak, D. P., Gething, M. J., and Sambrook, J. (1983b). *Cell (Cambridge, Mass.)* **33,** 435–443.

Roth, M. G., Gundersen, D., Patil, N., and Rodriguez-Boulan, E. (1987). *J. Cell Biol.* **104,** 769–782.

Saier, M. H., Jr., Boerner, P., Grenier, F. C., McRoberts, J. A., Rindler, M. J., Taub, M., and U Sang, Hoi (1986). *Miner. Electrolyte Metab.* **12,** 42–50.

Salas, P. J. I., Misek, D. E., Vega-Salas, D. E., Gundersen, D., Cereijido, M., and Rodriguez-Boulan, E. (1986). *J. Cell Biol.* **102,** 1853–1867.

Salas, P. J. I., Vega-Salas, D. E., Hochman, J., Rodriguez-Boulan, E., and Edidin, M. (1988). *J. Cell Biol.* **107,** 2363–2376.

Sambuy, Y., and Rodriguez-Boulan, E. (1988). *Proc. Natl. Acad. Sci. U.S.A.* **85,** 1529–1533.

Sargiacomo, M., Lisanti, M., Graeve, L., LeBivic, A., and Rodriguez-Boulan, E. (1989). *J. Membrane Biol.* In press.

Simons, K., and Fuller, S. D. (1985). *Annu. Rev. Cell Biol.* **1,** 243–288.

Srinivas, R. V., Balachandran, N., Alonso-Caplen, F. V., and Compans, R. W. (1986). *J. Virol.* **58,** 689–693.

Stephens, E. B., and Compans, R. W. (1986). *Cell (Cambridge, Mass.)* **47,** 1053–1059.

Stephens, E. B., Compans, R. W., Earl, P., and Moss, B. (1986). *EMBO J.* **5,** 237–245.

Towbin, H., Staehelin, T., and Gordon, J. (1979). *Proc. Natl. Acad. Sci. U.S.A.* **76,** 4350–4354.

van Meer, G., Stelzer, E. H. K., Wijnaendts-van-Resandt, R. W., and Simons, K. (1987). *J. Cell Biol.* **105,** 1623–1635.

Vega-Salas, D. E., Salas, P. J. I., and Rodriguez-Boulan, E. (1987). *J. Cell Biol.* **104,** 905–916.

von Bonsdorff, C. H., Fuller, S., and Simons, K. (1985). *EMBO J.* **4,** 2781–2792.

Chapter 3

Analysis of the Synthesis, Intracellular Sorting, and Function of Glycoproteins Using a Mammalian Cell Mutant with Reversible Glycosylation Defects

MONTY KRIEGER, PRANHITHA REDDY, KAREN KOZARSKY,[1]
DAVID KINGSLEY,[2] LAWRENCE HOBBIE, AND MARSHA PENMAN

Department of Biology
Massachusetts Institute of Technology
Cambridge, Massachusetts 02139

---

[1]Present address: University of Michigan, Ann Arbor, Michigan 48104.
[2]Present address: National Cancer Institute—Frederick Cancer Research Facility, Frederick, Maryland 21701.

Copyright © 1989 by Academic Press, Inc.
All rights of reproduction in any form reserved.

I. Introduction

The mechanisms of the intracellular sorting, secretion, and endocytosis of macromolecules by eukaryotic cells are complex and involve interconnected multicompartmental pathways. Classic biochemical genetics combined with recombinant-DNA technology provides a powerful approach for identifying and characterizing the genes, gene products, and biochemical and cellular functions that underlie these pathways. Several investigators have applied the power and elegance of yeast genetics to study these pathways (See Rothblatt and Schekman, Chapter 1 in this volume). In addition, mammalian genetic systems, especially those for the low-density lipoprotein (LDL) receptor pathway (reviewed in Krieger *et al.*, 1985; Goldstein *et al.*, 1985) and for posttranslational protein glycosylation (Stanley, 1985), have been particularly useful for the molecular genetic and biochemical analysis of these problems.

In the course of studying LDL receptor-mediated endocytosis, we and our colleagues isolated and characterized a Chinese hamster ovary (CHO) cell mutant that exhibits reversible defects in the synthesis of protein- and lipid-linked oligosaccharides (Krieger *et al.*, 1981; Krieger, 1983; Kingsley *et al.*, 1986a). These defects dramatically influence the cells' ability to add galactose (Gal) and *N*-acetylgalactosamine (GalNAc) to all cellular glycoconjugates. In this chapter we will summarize the properties of this mutant, called *ldlD,* review recent experiments that exploit its novel characteristics, and provide detailed suggestions for the experimental manipulation of these cells.

II. Synthesis and Function of Protein-Linked Oligosaccharides

The asparagine-linked (N-linked) and serine/threonine-linked (O-linked) oligosaccharides on proteins and the oligosaccharides of glycolipids (see Fig. 1 for representative structures) are synthesized by sequential

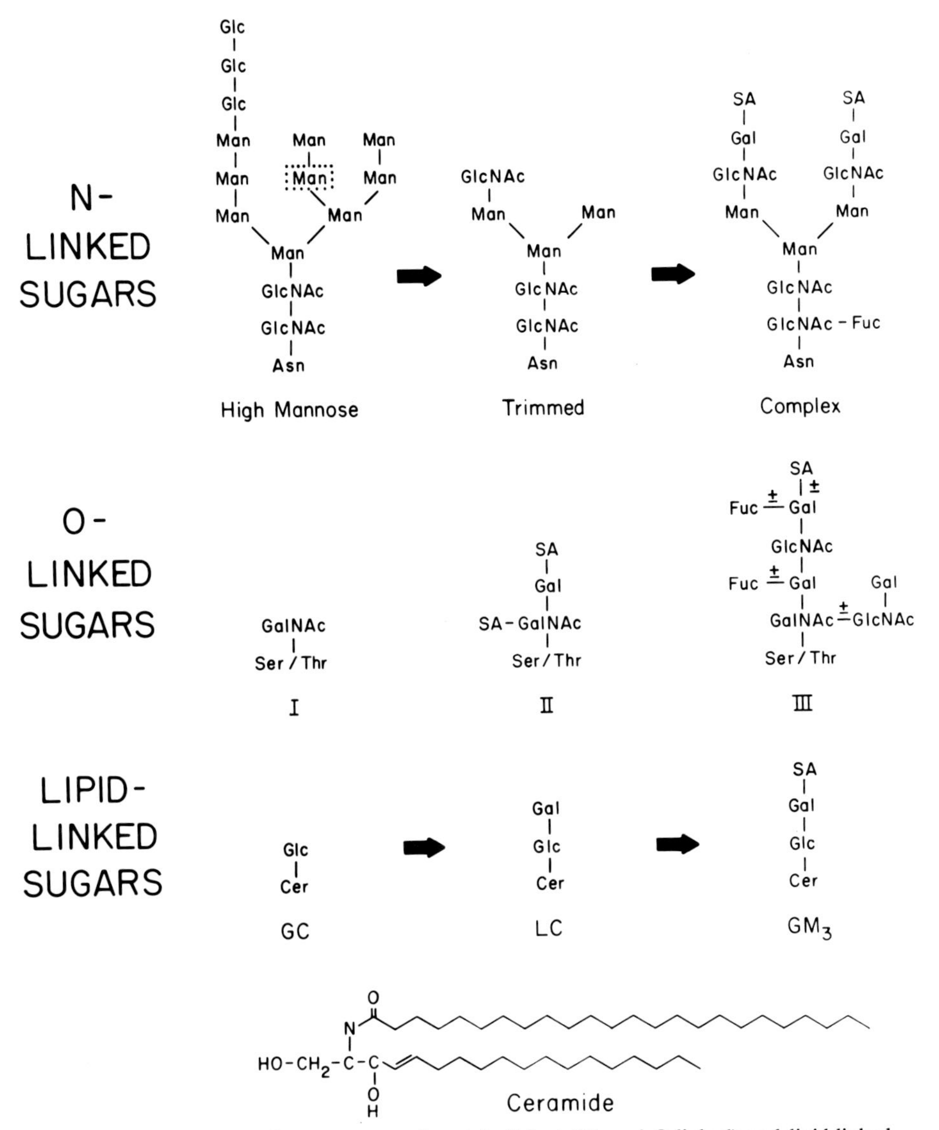

FIG. 1. Representative structures of protein-linked (N- and O-linked) and lipid-linked oligosaccharide chains. In the rough endoplasmic reticulum and the Golgi complex, the initial high-mannose type of N-linked oligosaccharide can be added to proteins and then trimmed and subsequently converted to the complex type (Hubbard and Ivatt, 1981; Kornfeld and Kornfeld, 1980; Stanley, 1985) (top). A wide variety of O-linked chains can be attached to Ser and Thr residues (Kornfeld and Kornfeld, 1980; Sadler, 1984). N-Acetylgalactosamine (GalNAc) is normally the first sugar added in O-linked chains (structure **I**; Cummings *et al.*, 1983), onto which a variety of sugars can be attached (e.g, structures **II** and **III**). The major glycolipid in wild-type CHO cells is GM_3 (Yogeeswaran *et al.*, 1974), which is synthesized by the sequential addition of glucose (Glc), galactose (Gal), and sialic acid (SA) to ceramide. The synthesis of all three of these types of oligosaccharides is disrupted in *ldlB*, *ldlC*, and *ldlD* mutants. Asn, Asparagine; Ser/Thr, serine or threonine; Man, mannose; GlcNAc, *N*-acetylglucosamine; Fuc, fucose). Reprinted from Kingsley *et al.* (1986b).

additions of sugars to growing oligosaccharide chains (Kornfeld and Kornfeld, 1980). The glycosyl transferases that catalyze the additions of these sugars use nucleotide sugar derivatives as substrates (UDP-glucose, CMP-sialic acid, UDP-galactose, etc.). In cultured cells, glucose in the culture medium is usually the primary precursor for the synthesis of these nucleotide sugars (e.g., see Fig. 2).

A. Asparagine-Linked (N-Linked) Sugars

The pathway for N-linked chain synthesis and processing, and the distribution of enzymes in this pathway have been worked out in great detail and are reviewed elsewhere (Kornfeld and Kornfeld, 1980). The distinctive intracellular compartmentalization of these enzymes and the characteristic properties of their oligosaccharide substrates and products (e.g.,

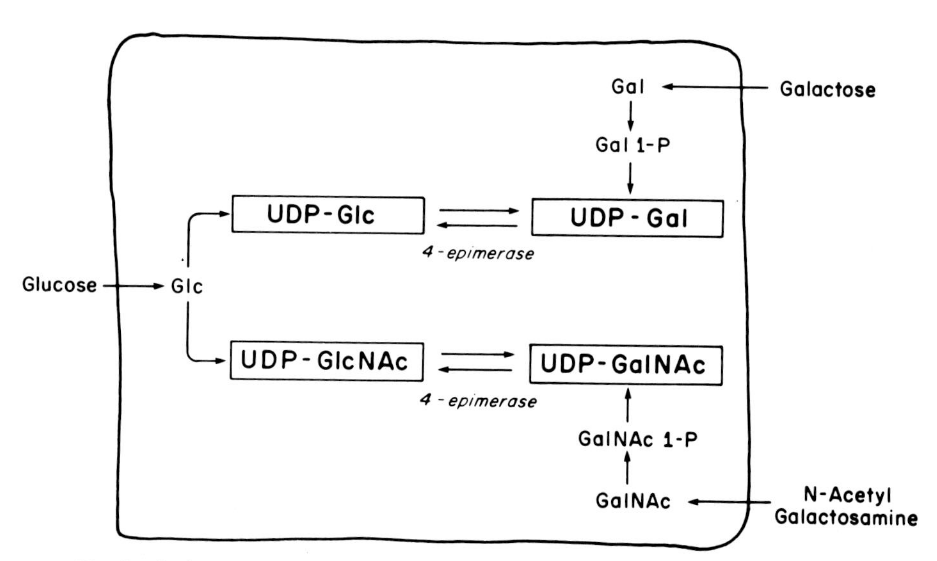

FIG. 2. Pathways of nucleotide sugar synthesis. Nucleotide sugars are the donors for the synthesis of the oligosaccharides of a variety of glycoconjugates, including glycoproteins and glycolipids (Schachter and Roseman, 1980). Under normal cell culture conditions glucose is the sole pure hexose in the culture medium, and its uptake and metabolic conversion can provide all of the nucleotide sugars (Schachter and Rodén, 1973). Under these conditions, the UDP-Gal/UDP-GalNAc 4-epimerase enzyme is required for the synthesis of UDP-Gal and UDP-GalNAc from their corresponding glucose isomers. These nucleotide sugars may be synthesized by alternative "salvage" pathways in which the unconjugated sugars are taken up from the medium and then converted to their phosphate derivatives (kinase reaction) and then the nucleotide sugars (uridyl transferase reaction). P, Phosphate; UDP, uridine diphosphate; Glc, glucose; Gal, galactose; GalNAc, N-acetylgalactosamine; GlcNAc, N-acetylglucosamine.

endoglycosidase H sensitivity) have provided powerful tools for the dissection of the pathways of intracellular protein sorting and secretion. Many advances in studying and exploiting the compartmentalization of N-linked sugar synthesis have relied on the availability of appropriate mutants, purified enzymes, and corresponding antibodies, as well as drugs such as tunicamycin (Hubbard and Ivatt, 1981), which specifically inhibit a variety of steps in N glycosylation. The availability of a variety of different types of mutants and drugs also has permitted the analysis of N-linked chain function, including effects on processing and cellular distribution of particular glycoproteins (Schlesinger *et al.*, 1984; Machamer and Rose, 1988a,b; Doyle *et al.*, 1985). From these and other studies one can conclude that the functions of N-linked chains vary enormously from protein to protein. With the exception of the lysosomal enzyme receptor recognition marker mannose 6-phosphate (Man6P), it has not been possible to deduce general rules that describe the function of N-linked chains. When N-linked chains are critical for processing or transport, their importance is usually attributed to their contribution to the overall folding of the protein.

B. Serine/Threonine-Linked (O-Linked) Sugars

Mucin-type O-linked oligosaccharide chains (attachment to serine or threonine via GalNAc) have been observed on many proteins (reviewed by Sadler, 1984), including cell surface receptors (Cummings *et al.*, 1983; Leonard *et al.*, 1985; Wasserman, 1987), polypeptide hormones (e.g., Lentz *et al.*, 1984; Wasley *et al.*, 1989), immunoglobulins (Mellis and Baenziger, 1983), clotting factors, apolipoproteins (Rall *et al.*, 1985), other serum proteins (Bock *et al.*, 1986; Edge and Spiro, 1987), and viral envelope proteins (Morgan *et al.*, 1984; Wertz *et al.*, 1985; Johnson and Spear, 1983). These oligosaccharides have been classified by Schachter and Williams (1982) into four core groups: Galβ1–3GalNAc, Galβ1–3(GlcNAcβ1–6)GalNAc, GlcNAcβ1–3GalNAc, and GlcNAcβ1–3(GlcNAcβ1–6)GalNAc. The sugar chains of proteoglycans also are often attached to their protein cores via Ser/Thr O-links, and recently Hart and colleagues have identified a new class of O-linked structures, GlcNAc–Ser/Thr (Holt and Hart, 1986). Unless noted otherwise (see Section V,G), we will restrict our consideration of O-linked sugars in this paper to the mucin-type chains and will not discuss proteoglycans.

We know less about the mechanisms of synthesis and the functions of mucin-type O-linked sugars than about those of N-linked chains, because far fewer experimental tools have been available for studying O-linked sugars and there has been a tendency for investigators to pursue studies of the more well-defined N-linked sugar pathway. Nevertheless, a number of methods have been used to make considerable progress in studying O-glycosylated proteins (Sadler, 1984). For example, O-linked chains can be removed from proteins by chemical or enzymatic treatments (see Section V,A) and sequenced (e.g., Cummings *et al.*, 1983; Mellis and Baenziger, 1983; Podolsky, 1985). Unfortunately, specific inhibitors of O glycosylation analogous to those that block N glycosylation [e.g., tunicamycin, deoxynorjirimycin (Burke *et al.*, 1984), etc.] have not been identified. In addition, among the few mutant mammalian cells with defects affecting O glycosylation (e.g., addition of galactose or sialic acid), most exhibit associated defects in N glycosylation and glycolipid synthesis (Table I, Stanley, 1985; Kingsley *et al.*, 1986b). Thus, until recently, it was not possible specifically to disrupt O glycosylation and examine the consequences of such disruptions on protein structure, processing, sorting, and function as well as cell metabolism.

There has been uncertainty about whether there are general rules that govern the function of O-linked chains for different classes of glycoproteins (e.g., membrane-associated versus secreted) or whether the functions of these sugars vary dramatically from protein to protein as appears to be the case for N-linked chains. The isolation of *ldlD* cells, a mutant line of CHO cells that express a reversible defect in O glycosylation, has provided a new tool for the study of the function of O-linked oligosaccharides.

III. CHO Mutants with Defects in Oligosaccharide Biosynthesis

A. Isolation and Phenotypes

ldlD cells were initially isolated in selections and screens designed to identify CHO mutants expressing defects in the endocytosis of LDL (Krieger *et al.*, 1981, 1983, 1985). To date, we have identified seven recessive complementation groups *(ldlA–ldlG),* which define genes required for the normal expression of LDL receptor activity (reviewed in Krieger *et al.*, 1985; Kingsley and Krieger, 1984; L. Hobbie and M. Krieger, unpublished data, 1988). One of these, *ldlA,* is the diploid structural gene for the LDL receptor (Sege *et al.*, 1986), and some *ldlA* alleles are analogous to mutant alleles described in cells from patients with familial hypercholesterolemia (Kozarsky *et al.*, 1986). The *ldlE, ldlF,* and *ldlG* cells are recently isolated, temperature-sensitive, conditional-lethal mutants that are currently under study (L. Hobbie and M. Krieger, unpublished data, 1988).

The *ldlB, ldlC,* and *ldlD* mutants exhibit global, pleiotropic glycosylation defects affecting protein- and lipid-linked oligosaccharides (Kingsley *et al.,* 1986b,d; Krieger *et al.,* 1985). These glycosylation abnormalities are directly responsible for the LDL receptor-deficient phenotypes of these cells. The *ldlB* and *ldlC* mutations affect medial and trans-Golgi-associated glycosylation reactions. The key biochemical defect in *ldlD* cells is described in the next section. In addition to the *ldlB–ldlD* cells, 17 other genetically or phenotypically different classes of general glycosylation mutations have been identified in CHO cells; some have been characterized in great detail (Stanley, 1984, 1985), and some have been used to study the posttranslational processing and sorting of glycoproteins (e.g., Balch *et al.,* 1984). Table I summarizes the properties of some

TABLE I

SURVEY OF CHO GLYCOSYLATION MUTANTS[a]

Mutant class	Established carbohydrate defects[b]		
	N-Linked	O-Linked	Lipid-linked
ldlB, ldlC, ldlD, Lec2, Lec8	X	X	X
Lec3	X	?	X
Lec1, Lec4, Lec5, Lec9, LEC10, LEC11, LEC12, Lec13, LEC14, Lec15, Lec16, Lecl.Lec6	X	—	—

Mutant class	Established biochemical defects
ldlD	UDP-Gal/UDP-GalNAc 4-epimerase deficient
Lec1 (clone 15b)	N-Acetylglucosaminyltransferase I deficient
Lec2 (clone 1021)	Golgi CMP-sialic acid transporter deficient
Lec8 (clone 13)	Golgi UDP-Gal transporter deficient
Lec4	β-1,6-N-Acetylglucosaminyltransferase deficient
Lec15	Dolichol-phosphate-mannose synthetase deficient
LEC10	N-acetylglucosaminyltransferase III acquired
LEC11	α-1,3-Fucosyltransferase I acquired
LEC12	α-1,3-Fucosyltransferase II acquired
Lec13	GDP-Mannose 4,6-dehydratase

[a]For detailed descriptions of CHO glycosylation mutants, see the comprehensive reviews by Stanley (1984, 1985). Additional references include Deutscher *et al.* (1986), Kingsley *et al.* (1986a,b), and Ripka *et al.* (1986). The names of frequently studied Lec1, Lec2, and Lec8 clones are given in parentheses.

[b]Only those defects that have been established directly are listed. Note that it is probable that the LEC11, LEC12, and Lec13 mutants also have defects in lipid-linked or O-linked chains containing fucose, and that Lec3 probably synthesizes defective O-linked glycans.

of these mutants. Studies with these mutants have highlighted not only the importance of carbohydrate chains for the expression and function of some cell surface glycoproteins (e.g., the lysosomal enzyme and LDL receptors; see Krieger *et al.*, 1985), but also the variety of effects that particular carbohydrate alterations can have on the functions of different glycoproteins.

B. Biochemical Defect in *ldlD* Cells: UDP-Gal/UDP-GalNAc 4-Epimerase Deficiency

The defective synthesis of N-linked, O-linked, and lipid-linked glycoconjugates and the LDL receptor deficiency of *ldlD* cells are due to a virtual absence of UDP-Gal/UDP-GalNAc 4-epimerase enzymatic activity (Kingsley *et al.*, 1986a). This enzyme normally catalyzes the reversible isomerizations of UDP-glucose to UDP-galactose (Gal) and UDP-*N*-acetylglucosamine to UDP-*N*-acetylgalactosamine (GalNAc) (Fig. 2, Maley and Maley, 1959; Piller *et al.*, 1983). UDP-Gal and UDP-GalNAc are the donor groups for transfer of galactose and GalNAc to glycoproteins and glycolipids (Schachter and Roseman, 1980). Galactose is added to all of the three major glycoconjugates in CHO cells (N-linked and O-linked sugars on glycoproteins, including proteoglycan sugars, and the lipid-linked ganglioside GM_3), while GalNAc usually is added only to O-linked glycans (mucin-type chains and chondroitin-type proteoglycans; Fig. 1) (Kornfeld and Kornfeld, 1980). Under normal culture conditions, glucose is the only sugar directly provided in growth medium. Wild-type cells convert glucose to UDP-glucose and UDP-*N*-acetylglucosamine, which are in turn isomerized to UDP-Gal and UDP-GalNAc by the 4-epimerase enzyme (Fig. 2). Because of the 4-epimerase deficiency, under normal culture conditions, the intracellular pools of UDP-Gal and UDP-GalNAc in *ldlD* cells are very low and cannot sustain normal glycoconjugate synthesis.

IV. Reversibility of the Glycosylation Defects in *ldlD* Cells

There are three different ways to restore the UDP-Gal and UDP-GalNAc pools in *ldlD* cells, and thereby permit normal protein and lipid glycosylation in spite of the 4-epimerase deficiency: addition of pure Gal and GalNAc to the culture medium (correction via sugar salvage pathways); cocultivation with 4-epimerase-positive cells (correction via sugar transfer through intercellular junctions), or addition of Gal- and GalNAc-bearing glycoproteins to the culture medium (correction via endocytosis, intracellular degradation, and salvage pathways).

A. Addition of Gal and GalNAc to the Culture Medium

As indicated in Fig. 2, *ldlD* cells express sugar salvage pathways by which they can take up pure Gal and GalNAc from the culture medium and convert these sugars to their UDP derivatives (Kingsley *et al.*, 1986a). As a consequence, addition of Gal and GalNAc to the culture medium of *ldlD* cells restores normal glycoconjugate synthesis. Correction of glycosylation abnormalities by sugar additions is fast, occurring within a few minutes after addition. In pulse–chase metabolic labeling experiments (Kozarsky *et al.*, 1988b), normal glycosylation of newly synthesized proteins can be achieved even when sugars are added to the chase medium after protein synthesis is initiated (e.g., 15 minutes after pulse-labeling). As a consequence, *ldlD* cells can be used for a variety of kinetic studies, including examination of the recycling of proteins through cellular compartments containing galactosyl and *N*-acetylgalactosaminyl transferases (see later).

The differential effects on N- and O-linked sugar structures of separately adding to the culture medium only Gal or only GalNAc are striking (Fig. 3) and suggest novel applications of *ldlD* cells for glycoprotein analysis (see later). The consequences of these effects on glycoprotein structure have been examined in detail by analysis of the LDL receptor (Kingsley *et al.*, 1986a).

The LDL receptor is an integral membrane protein composed of a short COOH-terminal cytoplasmic domain, a single membrane-spanning domain, and a large NH_2-terminal extracellular domain (reviewed in Goldstein *et al.*, 1985). The receptor undergoes distinctive and rapid processing from a precursor form ("p," ~120 kDa) to the functional mature form ("m," ~160 kDa) found on the cell surface (see Fig. 4). Approximately 20 oligosaccharide chains are added to the receptor protein. Two are N-linked and the rest are O-linked. The ~40-kDa shift in apparent molecular mass between the precursor and mature forms is due to posttranslational glycosylation, including the processing of high-mannose N-linked sugars to the complex type (see Fig. 1, top row) and completion of O-linked chain synthesis. Much of the shift is due to the addition of the terminal sialic acids to N- and O-linked chains.

1. Effects of Sugar Additions on Glycoconjugate Structures and the Electrophoretic Mobility of Receptors

a. No Additions. Under culture conditions in which glucose in the medium is the only source of sugar, neither galactose nor sialic acid can

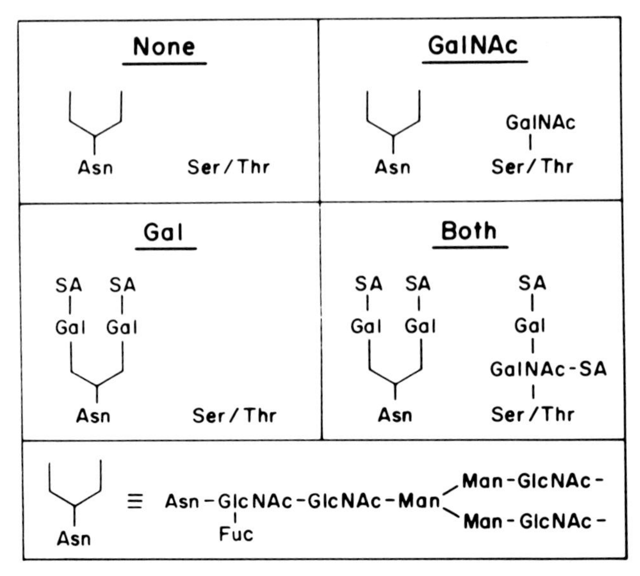

FIG. 3. Proposed effects of sugar additions on the structures of N-linked and O-linked oligosaccharides in *ldlD* cells. When exogenous GalNAc is *not* available (left panels), mucin-type O-linked chains cannot be made. When exogenous Gal is *not* available (top panels), full-length N- and O-linked chains (containing galactose and sialic acid) cannot be made. See Fig. 1 for abbreviations.

be added to typical complex-type N-linked oligosaccharide chains. As a consequence, the most mature forms of complex chains are truncated endoglycosidase H-resistant oligosaccharides (Fig. 3, upper left panel, None). Similarly, the major lipid-linked oligosaccharide in CHO cells, GM_3 (ceramide-glucose-galactose-sialic acid, Fig. 1), is also truncated in the absence of UDP-Gal. Usually, GalNAc is not found on N-linked chains, and therefore the presence or absence of GalNAc does not affect N-linked structures. Because GalNAc is the first sugar added to the side-chain hydroxyl groups of Ser and Thr residues on mucin-type O-linked chains, these chains are not synthesized in *ldlD* cells in the absence of an exogenous source of GalNAc (or UDP-GalNAc). In the absence of sugar additions, the most mature form of the human LDL receptor synthesized in *ldlD* cells has approximately the same apparent mass as the precursor form (Fig. 4, None).

b. Galactose Addition. When galactose alone is added to the culture medium, normal N- and lipid-linked chains are synthesized; however, the absence of UDP-GalNAc still prevents the addition of any sugars to muc-

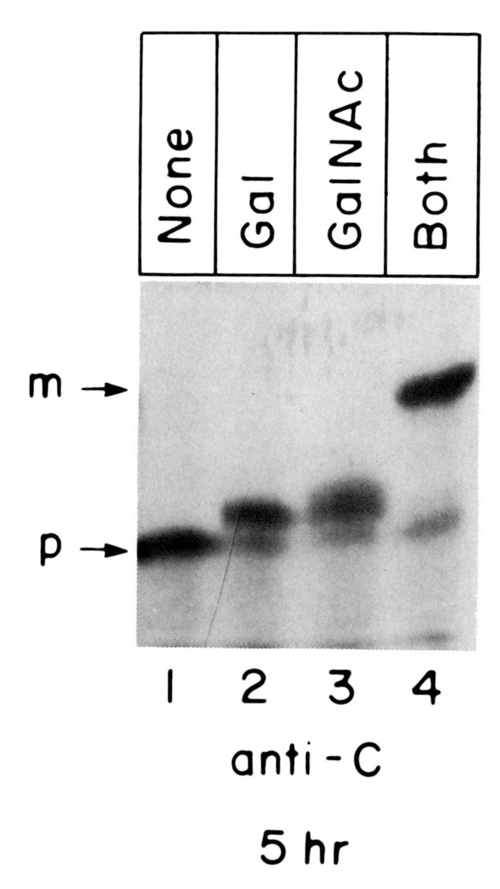

FIG. 4. Effects of Gal and GalNAc on the structure of the human LDL receptor synthesized in *ldlD* cells. An expression vector for the human LDL receptor (pLDLR4) was transfected into a derivative of *ldlD* cells that cannot express endogenous hamster LDL receptors (M. Penman, L. Hobbie, and M. Krieger, unpublished data). The transfected cells were plated and grown following the NCLPDS/ITS+/McCoy's protocol described in Section VI,B,2, and labeled for 5 hours with 100 μCi/ml of [^{35}S]methionine in methionine-free ITS+/Ham's medium in the presence of the indicated sugars as previously described (Kozarsky *et al.*, 1986). The cells were then harvested and cell extracts subjected to immunoprecipitation with an anti-LDL receptor antibody (anti-C), electrophoresis, and autoradiography as previously described (Kozarsky *et al.*, 1986). The positions of the precursor (p, 120 kDa) and mature (m, 160 kDa) forms of the human LDL receptor are indicated.

in-type O-linked chains (Fig. 3, lower left, Gal). The most mature form of the LDL receptor under these conditions appears to be somewhat larger than the precursor because of the addition of galactose and sialic acid to the N-linked chains (Fig. 4). Under the conditions of the experiment shown in Fig. 4, a small amount of unprocessed receptor precursor is also observed (lanes 2–4).

c. GalNAc Addition. Addition of GalNAc alone has no apparent effects on the structures of N- and lipid-linked chains: they remain truncated; however, it does permit the synthesis of truncated O-linked chains without their typical galactose and sialic acid residues (Fig. 3, upper right, GalNAc). As a consequence, GalNAc addition alone slightly decreases the electrophoretic mobility of LDL receptors in sodium dodecylsulfate (SDS)–polyacrylamide gels (Fig. 4).

d. Both Additions. When both Gal and GalNAc are added together, there is virtually full correction of the defects in the synthesis of all classes of glycoconjugate (Fig. 3, lower right, Both), and the mature LDL receptor exhibits normal electrophoretic mobility (Fig. 4).

2. Reversible Defect in O Glycosylation

Figure 3 clearly shows that culture medium can be adjusted so there is either normal synthesis of N-linked and mucin-type O-linked chains (both additions) or complete inhibition of O-linked chain synthesis while N-linked chains are made normally (addition of Gal alone). Thus, *ldlD* cells can be used to study O-linked glycosylation in a fashion analogous to that of using tunicamycin to study N-linked glycosylation, because in both cases there is a complete block in glycan addition. The *ldlD* cells have been used to study the consequences of O glycosylation on the structure (including dimerization), sorting (secretion, cell surface expression, rate of Golgi passage), recycling, stability, and function of several membrane and secreted proteins. For example, stable cell surface expression and endocytic activity of LDL receptors requires O glycosylation of the receptor protein (Kingsley *et al.*, 1986a; Kozarsky *et al.*, 1988a). Additional examples are summarized in the following paragraphs.

B. Formation of Intercellular Junctions by Cocultivation with 4-Epimerase-Positive Cells

When *ldlD* cells are cocultivated with any one of a variety of 4-epimerase-positive lines of cells (e.g., wild-type and mutant CHO, human fibroblasts) with which they can form functional communicating junctions, the glycosylation defects of the *ldlD* cells are corrected by the junctional transfer of sugars (UDP-Gal and UDP-GalNAc: Krieger, 1983; Hobbie *et al.*, 1987). Intercellular-junctional transfer of sugars to *ldlD* cells is easily detected because restoration of O-linked glycosylation is accompanied by restoration of normal LDL receptor glycosylation (Fig. 5, lane 4, compare to lanes 2 and 3) and receptor activity. When junctional communication is blocked by retinoic acid or when the cocultivated cells cannot efficiently form functional junctions with *ldlD* cells (e.g., mouse LTA or hu-

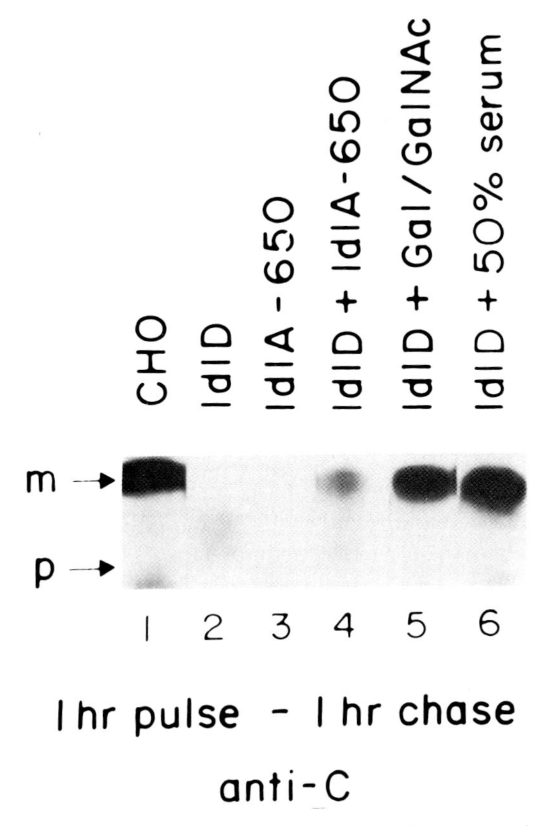

FIG. 5. Correction of the abnormal glycosylation of LDL receptors in *ldlD* cells by sugar additions, serum glycoproteins, or intercellular-junctional communication. Wild-type CHO cells (lane 1), *ldlD* cells (lanes 2, 5, and 6), the LDL receptor-negative *ldlA* mutant ldlA-650 (lane 3), or a 4:1 mixture of *ldlD* and ldlA-650 cells (lane 4) were grown in 3% NCLPDS media, labeled with [^{35}S]methionine for 1 hour, chased in unlabeled medium for 1 hour, and processed for immunoprecipitation, electrophoresis, and autoradiography as previously described (see Fig. 1 of Hobbie *et al.*, 1987). The pure cultures of *ldlD* cells contained no additions (lane 2), 10 μM Gal and 100 μM GalNAc (lane 5), or 50% (v/v) human lipoprotein-deficient serum (lane 6). The molecular weight standards m and p refer to the normal precursor and mature forms of the hamster LDL receptor.

man A431 cells), induction of LDL receptor activity by cocultivation does not take place. There are many simple, rapid, quantitative or qualitative assays for LDL receptor activity, and we have developed a number of genetic selections for LDL receptor-deficient cells. These methods can be used with *ldlD* cells for the genetic and biochemical analysis of intercellular-junctional communication.

C. Addition of Glycoproteins to the Culture Medium

When *ldlD* cells are cultured in medium containing high concentrations of serum, apparently normal glycosylation of the LDL receptor (Fig. 5, lane 6) and essentially normal LDL receptor activity are restored (Krieger, 1983). Receptor activity can also be induced when *ldlD* cells are incubated with high concentrations of a single glycoprotein, fetuin (K. Kozarsky, K. Malmstrom, L. Hobbie, D. Kingsley, and M. Krieger, unpublished data). These and other results suggest that *ldlD* cells can internalize and degrade glycoproteins from the culture medium and that the galactose and GalNAc released by degradation can be used to populate cellular UDP-Gal and UDP-GalNAc pools, thereby bypassing the 4-epimerase defect. As a consequence, the results of studies using *ldlD* cells can depend critically on the nature of the cell culture media (see Section VI,B). As the concentration of serum in the culture medium is increased, oligosaccharide chains are more completely processed. Figure 6 shows the effects of different concentrations of newborn calf lipoprotein-deficient serum (NCLPDS) and fetal bovine serum (FBS) on the electrophoretic mobility of the hamster LDL receptor in *ldlD* cells. When *ldlD* cells were cultured in Ham's F-12 medium supplemented with 10% FBS, normal N- and O-linked glycosylation of the LDL receptor were almost fully restored, whereas there was far less correction in media containing $\leq 3\%$ NCLPDS. Fetal bovine serum is apparently a better source of the sugars than newborn calf lipoprotein-deficient serum, perhaps because of its high content of fetuin.

V. Glycoprotein Synthesis, Sorting, and Function Studied Using *ldlD* Cells

A. Electrophoretic Detection of Mucin-Type O-Linked Sugars

Addition of galactose- and sialic acid-containing sugar chains to proteins usually affects their electrophoretic mobilities in SDS–polyacrylamide gels (e.g., Cummings *et al.*, 1983; Kingsley *et al.*, 1986a; Reddy *et al.*,

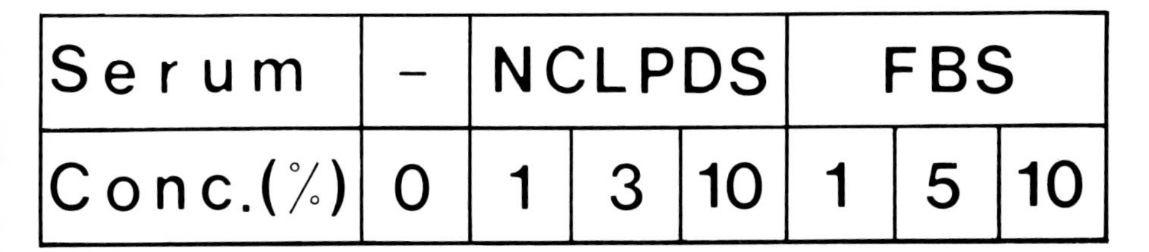

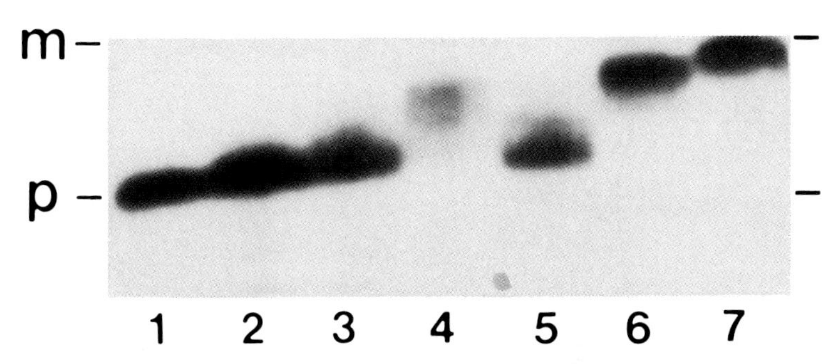

FIG. 6. Effects on LDL receptor structure of varying the concentration and type of serum used to supplement the growth medium of *ldlD* cells. On day 0, *ldlD* cells were seeded into the wells of six-well dishes at a concentration of 150,000 cells per well in 3% NCLPDS medium. On day 1, the cells were washed once with Ham's F-12 medium and refed with medium A containing the indicated serum supplements. On day 2, the cells were labeled for 5 hours with 80 μCi/ml of [^{35}S]methionine, harvested, and subjected to immunoprecipitation (anti-C antibody), electrophoresis, and autoradiography as previously described (Kozarsky *et al.*, 1986). The molecular weight standards m and p refer to the normal precursor and mature forms of the hamster LDL receptor.

1988). Mucin-type O-linked chains often contain one or both of these two sugars (Sadler, 1984). Thus, proteins expressed in *ldlD* cells, which normally contain O-linked chains, should exhibit altered electrophoretic mobilities when O glycosylation is inhibited. Therefore, comparison of the electrophoretic mobilities of proteins synthesized by *ldlD* cells in the presence of Gal and either with or without GalNAc provides a simple diagnostic test for their O glycosylation (see Fig. 3).

For example, Fig. 7 compares the effects of different sugar additions on the electrophoretic mobilities of the LDL receptor and the *neu* oncogene protein (Drebin *et al.*, 1984). In contrast to the mobility of the O-glycosylated LDL receptor, the mobility of the *neu* oncogene protein, which was sensitive to Gal addition, was unaffected by GalNAc addition. Similar results have been observed for the rabbit polymeric IgA receptor (P. Reddy, K. Mostov, and M. Krieger, unpublished data), hamster fibronectin (J. Paul, D. Kingsley, M. Krieger, and R. Hynes, unpublished data), and the gp120/41 envelope protein of human immunodeficiency virus (Kozar-

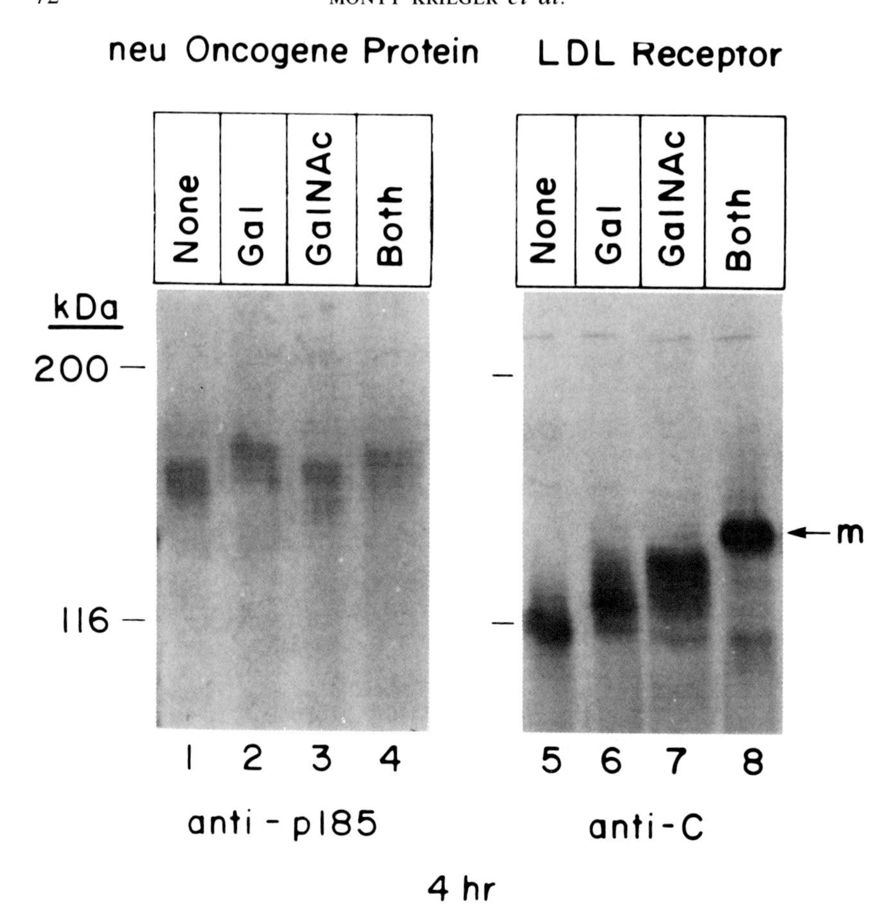

FIG. 7. Effects of Gal and GalNAc on the structures of the *neu* oncogene protein and the hamster LDL receptor in *ldlD* cells. An expression vector for the transforming form of the *neu* oncogene (pDOL/neu, Bargmann and Weinberg, 1988) was transfected into *ldlD* cells (P. Reddy, C. Bargmann, and M. Krieger, unpublished data). The transfected cells were grown in 3% NCLPDS and labeled for 4 hours with 80 μCi/ml of [^{35}S]cysteine in the presence of the indicated sugars (20 μ*M* Gal, 200 μ*M* GalNAc) as previously described (Kingsley *et al.*, 1986a). The cells were then harvested and cell extracts subjected to immunoprecipitation with either an anti-LDL receptor antibody (anti-C) or an anti-*neu* oncogene protein antibody, anti-p185 antibody 16.4 (Drebin *et al.*, 1984), electrophoresis, and autoradiography as previously described (Kozarsky *et al.*, 1986). The GalNAc independence of the electrophoretic mobility of the *neu* oncogene protein suggests that this molecule is not O-glycosylated in *ldlD* cells.

sky *et al.*, 1989). These observations suggest that, when these proteins are expressed in *ldlD* cells, they are N-glycosylated, but not significantly O-glycosylated.

There are a number of intrinsic limitations to this approach. First, it is possible (although it has not been observed) that some forms of mucin-type O-linked chains in *ldlD* cells may not influence the electrophoretic mobility of proteins. Second, in SDS–polyacrylamide gels, small glycosylation-dependent changes in mobility may not be easy to distinguish, particularly in the low-mobility (high molecular weight) regions where resolution can be poor. Thus, two-dimensional gel electrophoresis should enhance the sensitivity of this assay. Nevertheless, our experience using this approach has thus far been encouraging. We have had no difficulty detecting GalNAc-dependent shifts in mobility in seven of seven glycoproteins known by independent criteria to be O-glycosylated. In addition, we have not seen GalNAc-dependent changes in the electrophoretic mobilities of proteins known to contain only N-linked carbohydrate chains [e.g., the vesicular stomatitis virus G protein (Kingsley *et al.*, 1986a) and the α subunit of human chorionic gonadotropin (hCG; Matzuk *et al.*, 1987)]. Thus, the GalNAc dependence of the electrophoretic mobility of proteins expressed in *ldlD* cells provides strong evidence, but not definitive proof, for the presence or absence of mucin-type O-linked chains.

This application of *ldlD* cells complements previously described enzymatic and chemical deglycosylation techniques and immunochemical approaches. O-linked chains can be removed from proteins by β elimination or trifluoromethanesulfonic acid treatment (Herzberg *et al.*, 1985). The enzyme O-Glycanase also can be used to detach O-linked chains, but this enzyme is highly specific and only removes Gal-GalNAc (Lamblin *et al.*, 1984). Before this enzyme may be used, terminal sialic acids must be removed by sialidase, and even then, many chains cannot be recognized. In addition, O-linked chains can be identified by an antibody, AH8-28, which specifically recognizes Gal-GalNAc and lectins, which specifically bind to GalNAc or Gal-GalNAc (Nichols *et al.*, 1986).

B. Effects of O Glycosylation on the Stability and Cell Surface Expression of Membrane Glycoproteins

By appropriately adjusting the culture conditions, one can use the 4-epimerase defect in *ldlD* cells to, in essence, inhibit protein O glycosylation without disrupting normal N glycosylation (addition of Gal but not GalNAc). We have examined the effects of such an inhibition on the stability and cell surface expression of four types of membrane proteins that are normally O-glycosylated: the human and hamster LDL receptors (Kingsley *et al.*, 1986a; Kozarsky *et al.*, 1988b; D. Kingsley, M. Penman, and M. Krieger, unpublished data), decay-accelerating factor (Reddy *et al.*, 1989), the major envelope glycoprotein antigen of Epstein–Barr virus (EBV) (K. Kozarsky, M. Silberklang, and M. Krieger, unpublished data), and the gp55 subunit of the human interleukin-2 (IL-2) receptor (Kozarsky *et al.*, 1988a). With the exception of the endogenous hamster LDL receptor, these proteins were studied after transfection of corresponding cDNA expression vectors into the wild-type CHO and *ldlD* cells using standard calcium phosphate precipitation methods (see Section VI,D).

In all four cases, there was a dramatic decrease in cell surface expression when O glycosylation was inhibited by omitting GalNAc from the media of the transfected *ldlD* cells. For the human and hamster LDL receptors, decay-accelerating factor, and the EBV envelope protein, we found that newly synthesized O-linked deficient proteins (O^d) were transported to the cell surface. However, unlike their normally glycosylated counterparts, these O^d proteins were subject to rapid proteolysis. As a consequence, the bulk of their extracellular domains were released into the culture media and the cellular steady-state levels of these O^d proteins were very low (e.g., see Fig. 8). The properties of the critical proteolytic activity(s), including its cellular location, are currently under investigation. The rapid reversibility of the O-glycosylation defect has been used to show that the absence of O-linked sugars on the LDL receptor itself, and not other cellular glycoproteins, is responsible for the observed proteolysis–release (Kozarsky *et al.*, 1988b).

In contrast to these three surface glycoproteins, the O^d IL-2 receptor subunit appears to be missorted at a site in or beyond the trans-Golgi compartment, and, as a consequence, it does not reach the cell surface (Kozarsky *et al.*, 1988a). The stability of the missorted O^d IL-2 receptor subunit is similar to its normally glycosylated counterpart. As a control, we examined the effects of inhibiting O glycosylation on the gp120/41 envelope protein of human immunodeficiency virus expressed in *ldlD* cells (Kozarsky *et al.*, 1989). This viral envelope protein is not O-glycosylated in CHO cells. As anticipated, the stability, cell surface expression, and cell fusion activity of gp120/41 were unaffected by inhibition of O glycosylation.

These findings suggest the following generalization: O glycosylation of cell surface proteins may frequently play an important role in determining their cell surface expression, often, but not always, by controlling protein stability. Additional studies with other glycoproteins will help determine whether or not this finding is broadly applicable.

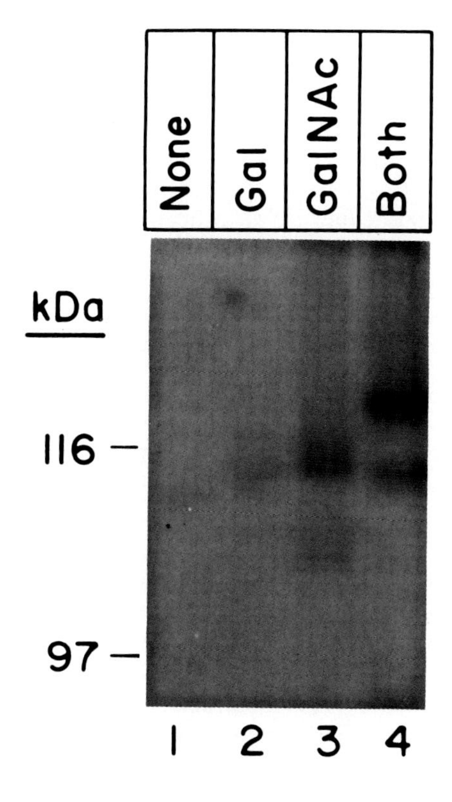

FIG. 8. Effects of Gal and GalNAc on the steady-state levels of LDL receptors in *ldlD* cells as measured by Western blotting. Membrane proteins from a derivative line of *ldlD* cells were isolated and subjected to gel electrophoresis (samples not reduced), transfer to cellulose nitrate paper, immunoblotting with an anti-LDL receptor antibody (anti-C), staining with ^{125}I-labeled Protein A, and visualization by autoradiography as previously described (Kozarsky *et al.*, 1986; Reddy *et al.*, 1989). The derivative line, ldlD[DAF], expresses human decay-accelerating factor (Reddy *et al.*, 1989) and was grown in 0.5% NCLPDS with the indicated sugar additions (20μM Gal, 200 μM GalNAc). The bands of lower apparent molecular weight in lanes 3 and 4 represent cell-associated degradation products of the LDL receptor that have been described previously (Kozarsky *et al.*, 1986); these were also seen in wild-type CHO cells (K. Kozarsky, P. Reddy and M. Krieger, unpublished data). In the absence of GalNAc (lanes 1 and 2), the steady-state levels of the O^d LDL receptors were very low.

C. Effects of O Glycosylation on Glycoprotein Secretion

Unlike the dramatic effects of O glycosylation seen for membrane-associated proteins, we and our colleagues have observed little or no effect of inhibiting O glycosylation on the rate or extent of secretion of four normally secreted, O-glycosylated proteins: hCG (Matzuk *et al.,* 1987), apolipoprotein E (Zanni *et al.,* 1987), a truncated and secreted form of the EBV envelope protein (K. Kozarsky, M. Silberklang, and M. Krieger, unpublished data), and erythropoietin (Wasley *et al.,* 1989). These findings suggest that *ldlD* cells may prove useful for the production of a wide variety of secreted glycoproteins bearing either normal or abnormal oligosaccharide chains.

D. Rapid Reversibility of the Glycosylation Defects in *ldlD* Cells and Intracellular Glycoprotein Sorting and Recycling

The distinctive compartmentalization and properties of the enzymes involved in oligosaccharide processing can be used to study the pathways of intracellular protein sorting, recycling, and secretion. For example, Snider and Rogers (1985, 1986) have shown that after initial synthesis, processing, and transport to the cell surface, mammalian cell surface glycoproteins can recycle to Golgi compartments. The reversible glycosylation defects of *ldlD* cells provide additional approaches for analyzing intracellular protein transport. For example, after abnormally glycosylated proteins are synthesized in the absence of Gal and/or GalNAc, these sugars can be added to the culture medium, rapidly repopulating the sugar–nucleotide pools. The rates and extents of subsequent endoplasmic reticulum (ER)/Golgi-associated addition of these sugars to the glycoproteins can then be determined and used to estimate protein recycling through late- or trans-Golgi (galactosyltransferase) or late-ER/early-Golgi (GalNAc transferase) compartments. This approach can be used only when the initial abnormal glycosylation of a protein neither interferes with its normal intracellular transport nor prevents its subsequent utilization as a substrate by the appropriate glycosyl transferases.

We have exploited the ability rapidly to fill the UDP-GalNAc pools of *ldlD* cells to study the kinetics and function of O glycosylation of the LDL receptor and to examine LDL receptor recycling (Kozarsky *et al.,* 1988b). We have also used the reversible defect in galactosylation of N-linked chains to study the pathway of maturation of the human immunodeficiency virus envelope protein (Kozarsky *et al.,* 1989). Duncan and

Kornfeld (1988) have used *ldlD* and other glycosylation mutants to study the intracellular cycling of Man6P receptors. Specifically, they used both surface [^{3}H]galactose labeling of agalactoreceptors and endogenous galactosylation with [^{3}H]galactose added to the culture medium to generate radiolabeled asialoreceptor. They then monitored the cycling of Man6P receptors between the cell surface and the trans-Golgi network by determining the rate and extent of Golgi-associated sialylation of the receptor.

E. *ldlD* Cells and Galactosemia

The *ldlD* cells provide a useful cell culture model of human galactosemia. The galactosemias (reviewed in Segal, 1983) are a set of autosomal recessive diseases that arise as consequences of defects in the enzymes of the Gal salvage pathway (galactokinase, UTP–hexose-1-phosphate uridylyltransferase, and UDP-Gal/UDP-GalNAc 4-epimerase, see Fig. 2). Transferase-deficiency galactosemia and galactokinase-deficiency galactosemia are the most common forms of galactosemia; there is only one confirmed case of epimerase-deficiency galactosemia (Holton *et al.,* 1981; Kingsley *et al.,* 1986c).

Galactose added to the medium of fibroblasts cultured from transferase-deficient patients is toxic. Similarly, moderate-to-high concentrations of Gal (>75 μ*M*) are toxic to *ldlD* cells while lower concentrations (≤ 20 μ*M*) are benign and, in fact, correct the Gal-associated glycosylation defects in *ldlD* cells (Kingsley *et al.,* 1986a). Concentrations of Gal as high as 10 m*M* have no apparent deleterious effects on wild-type CHO cells. Thus, *ldlD* cells may be useful for studying the biochemistry and genetics of Gal toxicity associated with galactosemia.

F. High-Specific-Activity Labeling of Gal-Containing Glycoconjugates

Because the pools of endogenously derived UDP-Gal in *ldlD* cells are very low, addition of radiolabeled Gal (e.g., [^{3}H]galactose) to the culture medium results in the very efficient labeling of Gal-containing glycoconjugates (Kingsley *et al.,* 1986a).

G. Proteoglycans

Proteoglycans usually contain Gal, and some classes of proteoglycans (e.g., chondroitin sulfate) contain GalNAc (Rodén, 1980). Thus, *ldlD* cells offer the potential of studying the synthesis, processing, and function of proteoglycans. For example, Esko *et al.* (1988) have used *ldlD* cells to study the effects of altered proteoglycan synthesis on the tumorigenicity of CHO cells in nude mice.

VI. Care and Feeding of *ldlD* Cells

A. Stock Cultures and Standard Techniques

ldlD cells are derived from the proline auxotrophic line CHO-Kl. *ldlD* cells are usually grown as stock monolayer cultures in medium A (Ham's F-12 containing 100 U/ml penicillin, 100 μg/ml streptomycin, and 2 m*M* glutamine) supplemented with 5% (v/v) fetal bovine serum (5% FBS medium) using standard mammalian cell culture conditions and techniques (e.g., 5% CO_2 atmosphere, 37°C, trypsin–EDTA harvesting). A detailed survey of the characteristics of CHO cells can be found in the comprehensive review edited by Gottesman (1985).

B. Plating and Growth for Experiments

After harvesting cells with standard cell culture trypsin–EDTA (1–2 minutes at 37°C), the trypsin should be quenched by dilution (at least 1 : 4) into serum containing medium before counting and plating cells into experimental dishes. If necessary, cells should be collected by centrifugation and resuspended into serum-containing medium to ensure removal of active trypsin. Selection of the type and amount of serum used to supplement the culture medium is critical for most experiments. As described earlier, *ldlD* cells can scavenge Gal and GalNAc from glycoproteins in the culture medium. Fetal bovine serum should be avoided if possible, because it is an efficient source of these sugars. We have previously used two types of conditions for plating and growing cells for studies of glycoprotein processing and function.

1. NCLPDS and Calf Serum-Supplemented Media

We often use 3% NCLPDS [medium A supplemented with 3% (v/v) newborn calf lipoprotein-deficient serum] for plating and growing cells. Lipoprotein-deficient serum has been used for studies of the LDL receptor because LDL receptor expression is induced when cells are grown in medium that does not contain LDL. The isolation of NCLPDS has been described previously (Goldstein *et al.*, 1983; Krieger, 1986). The glycoprotein content of 3% NCLPDS is sufficiently low to allow examination of many phenomena of interest. In general, the cells are plated on day 0 into this medium with or without Gal (20 μ*M*) and/or GalNAc (200 μ*M*), and experiments are performed on day 2 or 3. Despite the relatively low serum concentration in 3% NCLPDS, there are still high enough concentrations of Gal/GalNAc-containing proteins such that the cells synthesize glycoproteins containing somewhat heterogeneous oligosaccharide chains. This heterogeneity is manifested by fuzzy or diffuse glycoprotein bands of reduced electrophoretic mobility seen in gel electrophoretograms (Fig. 6). The heterogeneity is significantly reduced if the concentration of NCLPDS is lowered to 0–0.5% (v/v) (Reddy *et al.*, 1989). We have also obtained satisfactory results using medium A supplemented with 1% calf serum (Matzuk *et al.*, 1987). To completely eliminate this source of heterogeneity, protocols using ITS+ medium have been developed.

2. NCLPDS/ITS+ Media

ITS+ is a mixture of insulin (0.625 mg/ml), transferrin (0.625 mg/ml), selenium (0.625 μg/ml), linoleic acid (0.535 mg/ml), and BSA (0.125 g/ml), which is available from Collaborative Research Inc. (Two Oak Park, Bedford, MA 01730). CHO and *ldlD* cells can grow in either medium A supplemented with ITS+ (10 ml/liter) (ITS+/Ham's) or medium B (modified McCoy's medium containing 100 U/ml penicillin, 100 μg/ml streptomycin, and 2 m*M* glutamine) supplemented with ITS+ (10 ml/liter, ITS+/McCoy's) (see Chen *et al.*, 1988, and M. Penman, A. Fisher, and M. Krieger, unpublished data). In our experience, cells freshly harvested in trypsin do not attach well to the culture plastic when they are seeded directly into ITS+ media. Therefore, we plate cells on day 0 into 3% NCLPDS medium, change the medium on day 1 to ITS+/Ham's or ITS+/McCoy's with or without Gal (20 μ*M*) and/or GalNAc (200 μ*M*), and perform experiments on day 3. *ldlD* cells grow faster in ITS+/McCoy's than in ITS+/Ham's. The use of an ITS+ supplement, or its equivalent, is highly recommended over using serum-supplemented medium for glycosylation experiments with *ldlD* cells.

C. Sugar Supplements

When Gal or GalNAc is added to the culture medium at the time of plating or at least 2 days before processing (e.g., metabolic labeling with [^{35}S]methionine), they can be added at concentrations of 20 μ*M* and 200 μ*M*, respectively. For experiments requiring the rapid filling of the nucleotide–sugar pools (e.g., additions after pulse-labeling of cells), 20 μ*M* Gal and 400 μ*M* GalNAc should be used (Kozarsky *et al.*, 1988b).

Concentrations of Gal significantly higher than 20 μ*M* may prove to be toxic to the cells (see earlier). The optimal concentrations of these sugars may vary somewhat from subclone to subclone and may depend on the nature of the experimental protocol.

After growing *ldlD* cells for 2 days in 3% NCLPDS plus 10 μ*M* Gal and 100 μ*M* GalNAc followed by washing and additional growth in 3% NCLPDS without the sugars, we found that return to the sugar nucleotide-depleted state took at least 48 hours. Recovery to the basal state after incubation with high concentrations of serum (50% v/v human lipoprotein-deficient serum) appeared to be even slower.

D. Transfection Using the Calcium Phosphate Technique

It is relatively simple to use *ldlD* cells to study proteins not normally expressed in these cells by transfecting the cells with appropriate expression vectors (Matzuk *et al.,* 1987; Kozarsky *et al.,* 1988a,c; K. Kozarsky, M. Silberklang, and M. Krieger, unpublished data, Reddy *et al.,* 1989). The following protocol for calcium phosphate transfection is a slightly modified version of previously described procedures (Sege *et al.,* 1984; Parker and Stark, 1979; Graham and van der Eb, 1973; Kingsley *et al.,* 1986d).

Day 0

Plate *ldlD* cells: 750,000 cells per 100-mm plate in 10 ml of 5% FBS medium.

Day 2: Transfection

1. Thaw HBS (HEPES-buffered saline, 137 m*M* NaCl, 5 m*M* KCl, 0.7 m*M* Na_2HPO_4, 6 m*M* dextrose, 21 m*M* HEPES, pH 7.10, filter-sterilize, store aliquots at −70°C), check pH (must be 7.05–7.15).
2. Prepare expression vector DNA (plasmid)–HBS mixture (2 ml per tube, 1 ml per transfection plate, 0.5–5.0 μg expression vector DNA per plate); the DNA should encode a dominant selectable marker (e.g., neomycin resistance) on the vector itself or on a cotransfected vector mixed with the expression vector at 10–20% of the total DNA.
3. Prepare $CaPO_4$–DNA precipitate by adding 80–120 μl of 2 *M* $CaCl_2$ to each tube while gently vortexing (to avoid high local [$CaPO_4$]). A bluish-white haze should form almost immediately. Incubate at room temperature for 15–45 minutes while haze intensifies.

4. Treat cells (two plates at a time) with DNA–$CaPO_4$ mixtures as follows:

a. Aspirate off growth medium, and immediately (no washing) add 1 ml per plate DNA–HBS–$CaPO_4$ mixture.
b. Incubate 10 minutes at room temperature, rock once, and incubate for 10 more minutes.
c. Gently add 10 ml of 5% FBS medium.

5. Incubate 4–6 hours at 37°C in CO_2 incubator. (If a precipitate that looks like pepper is not visible on and around cells, repeat the experiment using different HBS.) Glycerol shock is not necessary when transfecting *ldlD* cells.

6. Aspirate DNA–5% FBS medium, wash cells three times with Ham's F-12 medium, and refeed with 5% FBS medium.

Day 3

Refeed cells 20–24 hours after transfection.

Day 4

Trypsinize cells and replate in selection medium (e.g., medium containing G418 if vector DNA or cotransfected DNA contains the neomycin resistance gene).

Acknowledgments

We are grateful to M. Matzuk, I. Boime, I. Caras, M. Silberklang, S. Dower, V. Zannis, K. Mostov, J. Paul, R. Hynes, J. Sodroski, W. Haseltine, J. Holton, L. Wasley, and R. Kaufman for stimulating collaborations; to D. Russell, M. Brown, and J. Goldstein for the human LDL receptor expression vector pLDLR4, and C. Bargmann and R. Weinberg for the *neu* oncogene vector pDOL/neu; to our colleagues K. Malmstrom, R. Jackman, H. Brush, R. Sege, and L. Couper for their contributions to this work; to J. Esko and colleagues for informing us of their results prior to publication, and to R. Rosenberg and P. Stanley for helpful discussions. The work in our laboratory was supported by grants from the National Institutes of Health. K. K. was a Whitaker Health Sciences Fund fellow, D. K. was a Johnson & Johnson Associated Industries Fund fellow, and M. K. was a National Institutes of Health career development awardee. This work was done during the tenure of a research fellowship to P. R. from the American Heart Association, Massachusetts Affiliate, Inc. M. K. was also supported by a Latham Family Professorship.

References

Balch, W. E., Dunphy, W. G., Braell, W. A., and Rothman, J. E. (1984). *Cell (Cambridge, Mass.)* **39,** 405–416.

Bargmann, C., and Weinberg, R. (1988). *EMBO J.* **7,** 2043–2052.

Bock, S. C., Skriver, K., Nielsen, E., Thogersen, H.-C., Wiman, B., Donaldson, V. H., Eddy, R. L., Marrinan, J., Radziejewska, E., Huber, R., Shows, T. B., and Magnusson, S. (1986). *Biochemistry* **25,** 4292–4301.

Burke, B., Matlin, K., Bause, E., Legler, G., Peyrieras, N., and Ploegh, H. (1984). *EMBO J.* **3,** 551–556.

Chen, H. W., Leonard, D. A., Fischer, R. T., and Trazaskos, J. M. (1988). *J. Biol. Chem.* **263,** 1248–1254.

Cummings, R. D., Kornfeld, S., Schneider, W. J., Hobgood, K. K., Tolleshaug, H., Brown, M. S., and Goldstein, J. L. (1983). *J. Biol. Chem.* **258,** 15261–15273.

Deutscher, S. L., and Hirschberg, C. B. (1986). *J. Biol. Chem.* **261,** 96–100.

Doyle, C., Roth, M. G., Sambrook, J., and Gething, M.-J. (1985). *J. Cell Biol.* **100,** 704–714.

Drebin, J. A., Stein, D. F., Link, V. L., Weinberg, R. A., and Green, M. I. (1984). *Nature (London)* **312,** 545–548.

Duncan, J. R., and Kornfeld, S. (1988). *J. Cell Biol.* **106,** 617–628.

Edge, A. S. B., and Spiro, R. G. (1987). *J. Biol. Chem.* **262,** 16135–16141.

Esko, J. D., Rostand, K. S., and Weinke, J. L. (1988). *Science* **241,** 1092–1096.

Goldstein, J. L., Basu, S. K., and Brown, M. S. (1983). *In* "Methods in Enzymology" (S. Fleischer and B. Fleischer, eds.), vol. 98, pp. 241–260. Academic Press, New York.

Goldstein, J. L., Brown, M. S., Anderson, R. G. W., Russell, D. W., and Schneider, W. J. (1985), *Annu. Rev. Cell Biol.* **1,** 1–39.

Gottesman, M. M., ed. (1985). "Molecular Cell Genetics: The Chinese Hamster Cell." Wiley, New York.

Graham, F. L., and van der Eb, A. J. (1973). *Virology* **52,** 456–467.

Herzberg, V. L., Grigorescu, F., Edge, A. S., Sprio, R. G., and Kahn, C. R. (1985). *Biochem. Biophys. Res. Commun.* **129,** 789–796.

Hobbie, L., Kingsley, D. M., Kozarsky, K. F., Jackman, R., and Krieger, M. (1987). *Science* **235,** 69–73.

Holt, G. D., and Hart, G. W. (1986). *J. Biol. Chem.* **261,** 8049–8057.

Holton, J. B., Gillett, M. G., MacFaul, R., and Young, R. (1981). *Arch. Dis. Child.* **56,** 885–887.

Hubbard, S. C., and Ivatt, R. J. (1981). *Annu. Rev. Biochem.* **50,** 555–583.

Johnson, D. C., and Spear, P. G. (1983). *Cell (Cambridge, Mass.)* **32,** 987–997.

Kingsley, D. M., and Krieger, M. (1984). *Proc. Natl. Acad. Sci. U.S.A.* **81,** 5454–5458.

Kingsley, D. M., Kozarsky, K. F., Hobbie, L., and Krieger, M. (1986a). *Cell (Cambridge, Mass.)* **44,** 749–759.

Kingsley, D. M., Kozarsky, K. F., Segal, M., and Krieger, M. (1986b). *J. Cell Biol.* **102,** 1576–1585.

Kingsley, D. M., Krieger, M., and Holton, J. B. (1986c). *N. Engl. J. Med.* **314,** 1257–1258.

Kingsley, D. M., Sege, R. D., Kozarsky, K., and Krieger, M. (1986d). *Mol. Cell. Biol.* **6,** 2734–2737.

Kornfeld, R., and Kornfeld, S. (1980). *In* "The Biochemistry of Glycoproteins and Proteoglycans" (W. J. Lennarz, ed.), pp. 1–34. Plenum, New York.

Kozarsky, K., Brush, H., and Krieger, M. (1986). *J. Cell Biol.* **102,** 1567–1575.

Kozarsky, K., Call, S. M., Dower, S. K., and Krieger, M. (1988a). *Mol. Cell. Biol.* **8,** 3357–3363.

Kozarsky, K., Kingsley, D., and Krieger, M. (1988b). *Proc. Natl. Acad. Sci. U.S.A.* **85,** 4335–4339.

Kozarsky, K., Penman M., Basiripour, L., Haseltine, W., Sodroski, J., and Krieger, M. (1989). *J. Acquired Immune Deficiency Syndromes,* **2,** 163–169.

Krieger, M. (1983). *Cell (Cambridge, Mass.)* **33,** 413–422.

Krieger, M. (1986). *In* "Methods in Enzymology" (J. J. Albers and J. P. Segrest, eds.), Vol. 129, pp. 227–237. Academic Press, Orlando, Florida.

Krieger, M., Brown, M. S., and Goldstein, J. L. (1981). *J. Mol. Biol.* **150,** 167–184.

Krieger, M., Martin, J., Segal, M., and Kingsley, D. (1983). *Proc. Natl. Acad. Sci. U.S.A.* **80,** 5607–5611.

Krieger, M., Kingsley, D. M., Sege, B. S., Hobbie, L., and Kozarsky, K. F. (1985). *Trends Biochem. Sci.* **10,** 447–452.

Lamblin, G., Lhermitte, M., Klein, A., Roussel, P., Van Halbeek, H., and Vliegenthart, J. F. G. (1984). *Biochem. Soc. Trans.* **12,** 599–600.

Lentz, S. R., Birken, S., Lustbader, J., and Boime, I. (1984). *Biochemistry* **23,** 5330–5337.

Leonard, W. J., Depper, J. M., Kronke, M., Robb, R. J., Waldmann, T. A., and Green, W. C. (1985). *J. Biol. Chem.* **260,** 1872–1880.

Machamer, C. E., and Rose, J. K. (1988a). *J. Biol. Chem.* **263,** 5948–5954.

Machamer, C. E., and Rose, J. K. (1988b). *J. Biol. Chem.* **263,** 5955–5960.

Maley, F., and Maley, G. F. (1959). *Biochim. Biophys. Acta* **31,** 577–578.

Matzuk, M. M., Krieger, M., Corless, C. L., and Boime, I. (1987). *Proc. Natl. Acad. Sci. U.S.A.* **84,** 6354–6358.

Mellis, S. J., and Baenziger, J. U. (1983). *J. Biol. Chem.* **258,** 11557–11563.

Morgan, A. J., Smith, A. R., Barker, R. N., and Epstein, M. A. (1984). *J. Gen. Virol.* **65,** 397–404.

Nichols, E. J., Fenderson, B. A., Carter, W. G., and Hakamori, S-I. (1986). *J. Biol. Chem.* **261,** 11295–11301.

Parker, B. A., and Stark, G. R. (1979). *J. Virol.* **31,** 360–369.

Piller, F., Hanlon, M. H., and Hill, R. L. (1983). *J. Biol. Chem.* **258,** 10774–10778.

Podolsky, D. K. (1985). *J. Biol. Chem.* **260,** 8262–8271.

Rall, S. C., Weisgraber, K. H., Innerarity, T. L., and Mahley, R. W. (1985). *Circulation* **72,** 111–143.

Reddy, P., Caras, I., and Krieger, M. (1989). Submitted.

Ripka, J., Adamany, A., and Stanley, P. (1986). *Arch. Biochem. Biophys.* **249,** 533–545.

Rodén, L. (1980). *In* "The Biochemistry of Glycoproteins and Proteoglycans" (W. J. Lennarz, ed.), pp. 267–371. Plenum, New York.

Sadler, J. E. (1984). *In* "The Biology of Carbohydrates" (V. Ginsburg and P. W. Robbins, eds.), Vol. 2. pp. 199–288. Wiley (Interscience), New York.

Schachter, H., and Rodén, L. (1973). *In* "Metabolic Conjugation and Metabolic Hydrolysis" (W. H. Fishman, ed.), Vol. 3, pp. 2–149. Academic Press, New York.

Schachter, H., and Roseman, S. (1980). *In* "The Biochemistry of Glycoproteins and Proteoglycans" (W. J. Lennarz, ed.), pp. 85–160. Plenum, New York.

Schachter, H., and Williams, D. (1982). *In* "Mucus in Health and Disease" (E. N. Chantler, J. B. Elder, and M. Elstein, eds.), Vol. 2, pp. 3–28. Plenum, New York.

Schesinger, S., Malfer, C., and Schlesinger, M. J. (1984). *J. Biol. Chem.* **259,** 7597–7601.

Segal, S. (1983). *In* "The Metabolic Basis of Inherited Disease" (J. B. Stanbury, J. B. Wyngaarden, D. S. Fredrickson, M. S. Brown, and J. L. Goldstein, eds.), 5th ed., pp. 167–192. McGraw-Hill, New York.

Sege, R. D., Kozarsky, K., Nelson, D. L., and Krieger, M. (1984). *Nature (London)* **307,** 742–745.

Sege, R. D., Kozarsky, K., and Krieger, M. (1986). *Mol. Cell. Biol.* **6,** 3268–3277.

Snider, M. D., and Rogers, O. C. (1985). *J. Cell Biol.* **100,** 826–834.

Snider, M. D., and Rogers, O. C. (1986). *J. Cell Biol. 103,* 265–275.

Stanley, P. (1984). *Annu. Rev. Genet.* **18,** 525–552.

Stanley, P. (1985). *In* "Molecular Cell Genetics: The Chinese Hamster Cell" (M. M. Gottesman, ed.), pp. 745–772. Wiley, New York.

Wasley, L., Horgan, P., Timony, G., Stoudemier, J., Krieger, M., and Kaufman, R. J. (1989). Submitted.

Wasserman, P. M. (1987). *Science* **235,** 553–560.

Wertz, G. W., Collins, P. L., Huang, Y., Gruber, C., Levine, S., and Ball, L. A. (1985). *Proc. Natl. Acad. Sci. U.S.A.* **82,** 4075–4079.

Yogeeswaran, G., Murray, R. K., and Wright, J. A. (1974). *Biochem. Biophys. Res. Commun.* **56,** 1010–1016.

Zanni, E. E., Krieger, M., Hadzopoulou-Cladaras, M., Forbes, G., and Zannis, V. I. (1987). *Circulation* **76,** 223.

Chapter 4

Using Temperature-Sensitive Mutants of VSV to Study Membrane Protein Biogenesis

JOHN E. BERGMANN

Department of Anatomy and Cell Biology
College of Physicians and Surgeons
Columbia University
New York, New York 10032

I. Introduction

Small lipid-enveloped viruses such as vesicular stomatitis virus (VSV) have proved to be important tools in the investigation of the biogenesis

Copyright © 1989 by Academic Press, Inc.
All rights of reproduction in any form reserved.

of integral membrane proteins. These viruses have extremely small genomes and rely totally on cellular machinery for the translation, post-translational processing, and intracellular transport of their membrane proteins. For example, the VSV genome codes for only five proteins: an integral membrane glycoprotein (G), a peripheral membrane protein (M), and nucleocapsid-associated proteins (N, NS, and L). As the virus shuts off the synthesis of the host cell's proteins, the total translational capacity of the cell is devoted to the synthesis of these proteins. It has therefore proved considerably easier to study the biogenesis of the G protein than that of an endogenous cellular glycoprotein.

During the initial investigations of the biogenesis of the G protein, investigators assumed that its biogenesis was not materially affected either directly by the other viral proteins or indirectly by the cytopathic effect of the virus on its host cell. More recent investigations have decisively proven these assumptions correct. The normal G protein when expressed from cloned cDNA is transported from the rough endoplasmic reticulum (RER) to the Golgi apparatus with the same kinetics as the G protein in virus-infected cells (Rose and Bergmann, 1982, 1983). The asparagine-linked oligosaccharides of the G protein are modified in the same manner whether the G protein is expressed from cloned cDNA or from the VSV genome (Gabel and Bergmann, 1985). Like the G protein in virus-infected cells, the G protein expressed from cloned cDNA moves from the Golgi apparatus to the plasma membrane (Rose and Bergmann, 1982, 1983). Finally, the G protein of the VSV temperature-sensitive mutant ts045 exhibits the same mutant phenotype when expressed during virus infection or when expressed from cloned cDNA (Gallione and Rose, 1985). Initial biochemical characterization of G-protein biogenesis relied heavily on *in vitro* translation of isolated mRNA and on pulse–chase protocols (brief incorporation of radioactive amino acids into G protein followed by variable time periods during which the fate of the labeled protein was followed). These studies revealed that the G protein is synthesized on membrane-bound polysomes whereas the other four VSV proteins are synthesized on free polysomes (David, 1977; Ghosh *et al.*, 1973; Grubman and Summers 1973; Grubman *et al.*, 1975; Morrison and Lodish, 1975). They also revealed that the different VSV proteins become incorporated into virions at different rates. The M and NS protein begin to enter virus particles 5–10 minutes after synthesis. In contrast, the G protein is incorporated into virus particles only after a 20-minute lag (Atkinson *et al.*, 1976; Hunt and Summers, 1976; David, 1973; Knipe *et al.*, 1977b). During these 20 minutes, it is transferred from dense membrane fractions to lighter membrane fractions—coincident with a remodeling of its asparagine-linked oligosaccharides (Knipe *et al.*, 1977a,b). In contrast, temperature-sensitive forms of the G protein expressed by the mutants M501 and ts045 fail to undergo oligosaccharide remodeling and remain associated with the dense membrane fractions at 40°C, the nonpermissive temperature (Knipe *et al.*, 1977c).

The discovery of these mutant viral proteins created a unique opportunity to synchronize the transport of the totality of an integral membrane protein, not just a subpopulation that has been radiochemically tagged. Once synchronized, this transport has proved amenable to morphological characterization.

In this chapter I will describe several experimental protocols that are illustrative of the manner in which these mutants can be used. For the sake of continuity, the required technical procedures are described in detail at the end of the chapter.

II. Synchronizing the Transport of the ts045 G Protein

A. Following Transport via Immunofluorescence Microscopy

Here I will demonstrate the synchrony achievable with ts045 using a study undertaken with Abraham Kupfer and Jon Singer (Bergmann *et al.*, 1983). A confluent monolayer of normal rat kidney (NRK) cells grown on glass coverslips was disrupted by scraping a track with a rubber policeman. This procedure "polarizes" the cytoplasm of the cells bordering the experimental wound; both the microtubule (MT)-organizing center and the Golgi apparatus become oriented toward the "wound," and the cells begin to migrate into the space. After 3 hours, the cells were infected with 50 plaque forming units (pfu)/cell, as described in Section IV,B,1 on plaque assays, using ts045 and incubated at 39.7°C for 3 hours. One such culture was fixed directly, and parallel infected cultures were shifted to an incubator at the permissive temperature (32°C) for 15, 22, and 35 minutes prior to fixation. When the cells were stained for G protein as described under technical procedures (Section IV, F), the synchrony was quite evident. In cells fixed at the nonpermissive temperature, the ts045 G protein is labeled in the nuclear envelope and in the RER (Section II, D) (Fig. 1B). It is not seen on the cell surface (Fig. 1A). However, within 15 minutes at the permissive temperature, there is a dramatic shift in the distribution of G protein. It is found concentrated in a perinuclear structure, which is oriented toward the direction of migration (Fig. 1E). We have shown by electron microscopy that this structure is the Golgi appa-

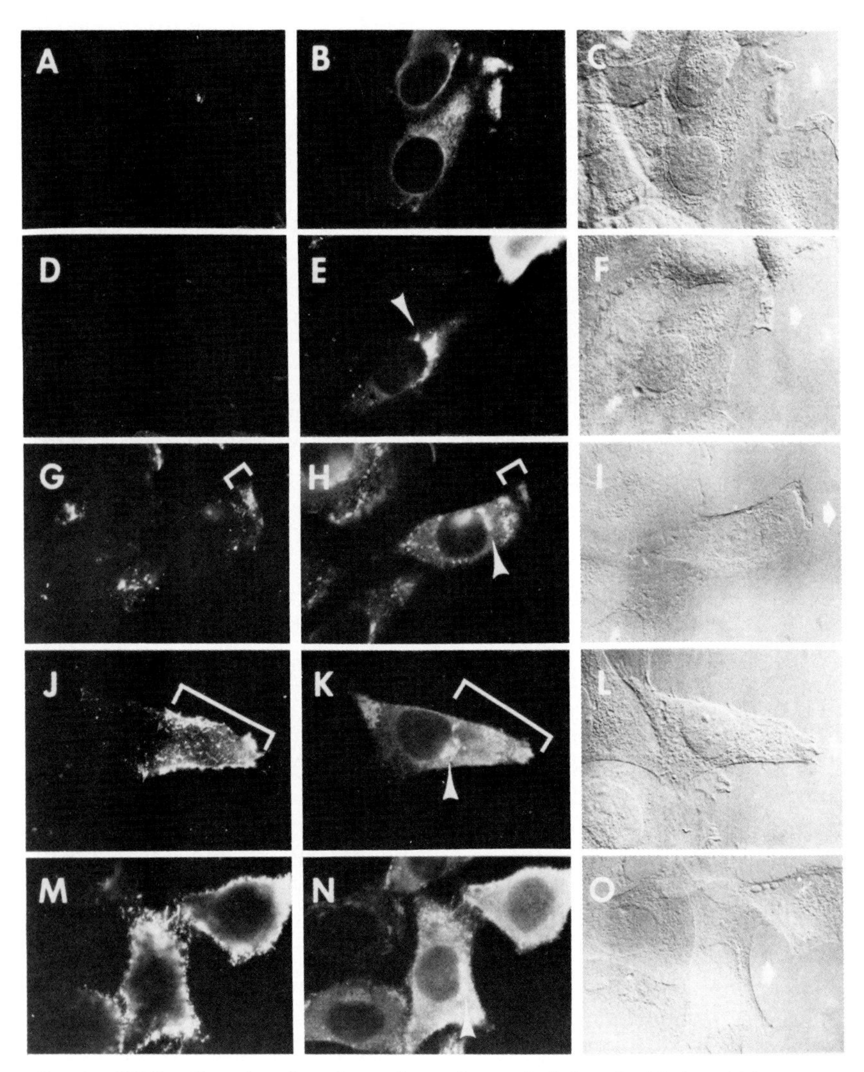

FIG. 1. NRK cells at the edge of experimental wounds, infected with the ts045 temperature-sensitive mutant of VSV. In the Nomarski-optics micrographs (C, F, I, L, and O), the leading edge of a motile cell and the direction of migration into the wound are indicated by the white arrow. The VSV G protein is immunolabeled inside the same cells (B, E, H, K, and N) and on the cell surfaces (A, D, G, J, and M). (A–C) Same group of cells at the nonpermissive temperature, just prior to the shift to the permissive temperature; (D–F) cells 15 minutes after the temperature shift; (G–I and J–L) corresponding sets of cells 22 minutes after the temperature shift; (M–O) cells 35 minutes after the temperature shift. The arrowheads in E, H, K, and N point to intracellular G protein in the Golgi apparatus that is positioned forward of the nucleus in the direction of cell migration. The small brackets in G and H designate the leading edge of the cell, and the large brackets in J and K designate the front half of the cell forward of the nucleus. Reproduced from Bergmann, Kupfer, and Singer (1983).

ratus (see Section II, D and Bergmann *et al.*, 1981; Bergmann and Singer, 1983). At this time, G protein is not yet evident on the cell surface. Using immunofluorescence microscopy, we have seen the G protein in the Golgi apparatus as early as 8 minutes after shifting such cultures to a 32°C incubator (Bergmann *et al.*, 1981). Twenty-two minutes after the temperature shift, the G protein appears on the cell surface (Fig. 1G, J). Thirty-five minutes after the temperature shift, the G protein is seen to be distributed far more evenly over the cell surface (Fig. 1M). It is evident that when fibroblasts are migrating, newly synthesized G protein is added at their leading edge. Note that the surface pattern is only evident in Fig. 1G and J; it is obscured by internal fluorescence in Fig. 1H and K. In summary, ts045 allows one to synchronize the transport of the G protein and, using immunofluorescence microscopy, at least four major steps in the biogenesis of the G protein may be resolved: blockage in the endoplasmic reticulum (ER), movement to the Golgi apparatus, appearance in a restricted region of the plasma membrane, and finally the steady-state distribution over the entire plasma membrane.

Using pharmacological agents one can study the role of one or more cellular functions in the biogenesis of the G protein. For example, we have used this protocol to investigate the role of the MT and microfilament systems in the intracellular transport of the G protein (Rogalski *et al.*, 1984). In the experiment shown in Fig. 2, NRK cells were again infected with ts045 and incubated at the nonpermissive temperature. Ninety minutes postinfection, nocodazole (30 μM, Aldrich Chemical Co., Milwaukee, WI) was added to disrupt the microtubules and the cells were incubated for an additional 90 minutes at 39.9°C. As shown in Fig. 2A and B, the intracellular distribution of ER-associated G protein was not materially altered by disruption of the MT system. As in the previous experiments, 13 minutes after the temperature shift, the G protein enters the Golgi apparatus (Fig. 2C, D). However, unlike those experiments, in nocodazole-treated cells the Golgi apparatus is not a localized perinuclear structure but is completely disrupted (compare Fig. 2C and D). Disruption of the MT and the Golgi apparatus does not affect the temperature sensitivity of the ts045-coded G protein. No G protein is seen on the cell surface during the first 13 minutes after the temperature shift in either the control or nocodazole-treated cells (Fig. 2A′,B′,C′, and D′). Thirty minutes after the temperature shift, the G protein has moved from the disrupted Golgi apparatus to the cell surface (Fig. 2E′ and F′). In the control cultures, the Golgi apparatus is sometimes polarized (arrowheads in Fig. 2E), and the site of surface appearance of the G protein is similarly polarized (arrows in Fig. 2E′). This was not seen in the nocodazole-treated cells (Fig. 2F′).

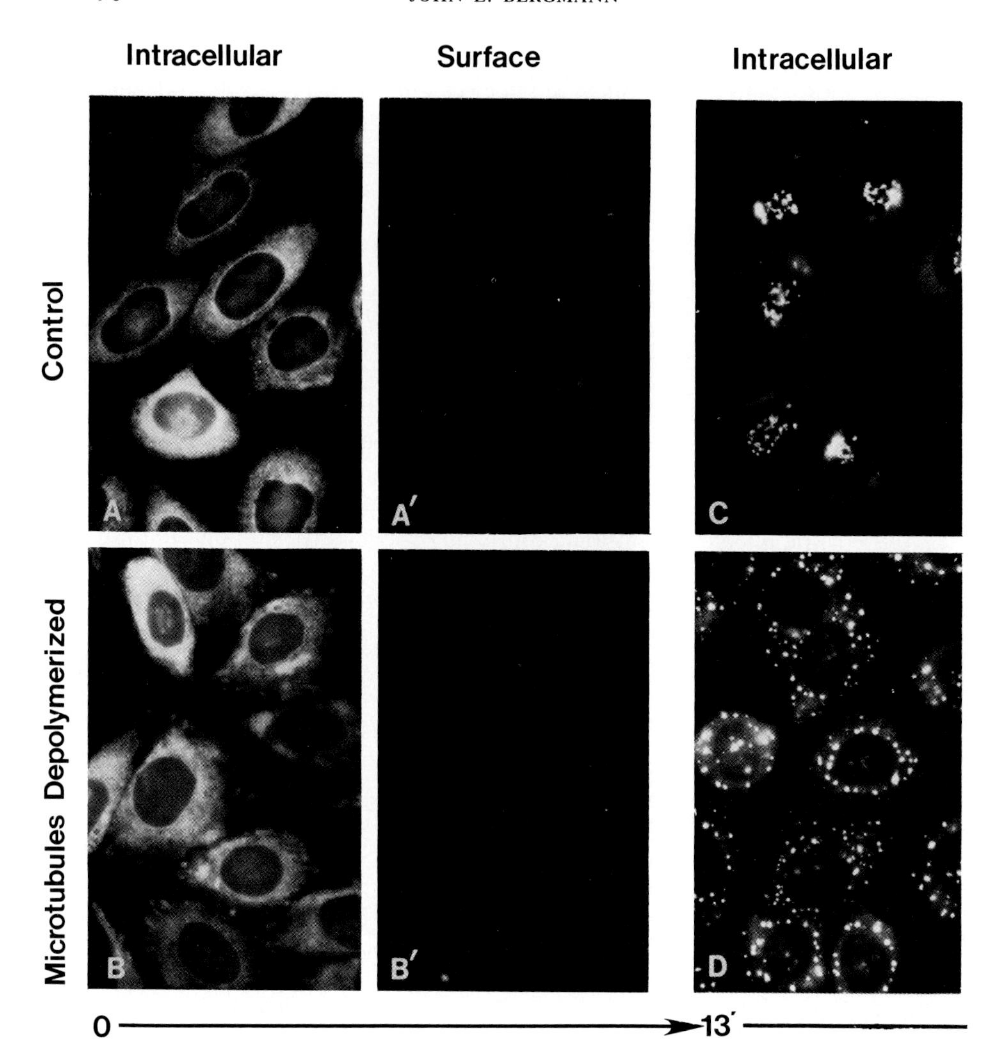

FIG. 2. Effect of MT disruption on the transport of the VSV G protein. NRK cells infected with ts045 and examined either just before the temperature shift down to 32°C (zero time, A/A′ and B/B′), 13 minutes after the shift (C/C′ and D/D′), or 30 minutes after the shift (E/E′ and F/F′). The bottom row of cells was treated with nocodazole to disassemble the cytoplasmic MT completely; the top row of cells was untreated. The cells were double-immunofluorescence-labeled for their intracellular G protein (A, B, C, D, E, and F) and for their surface-expressed G protein (A′, B′, C′, D′, E′, and F′, respectively). (E) Arrowheads point to perinuclear Golgi apparatuses labeled with G protein. On the surfaces of the corresponding cells, the first expression of the G protein is seen to be polarized (arrows), and juxtaposed to the Golgi apparatus inside the cell. No such surface G-protein polarization is seen on the nocodazole-treated cells (F′). Reproduced from Rogalski *et al.* (1984), by copyright permission of the Rockefeller University Press.

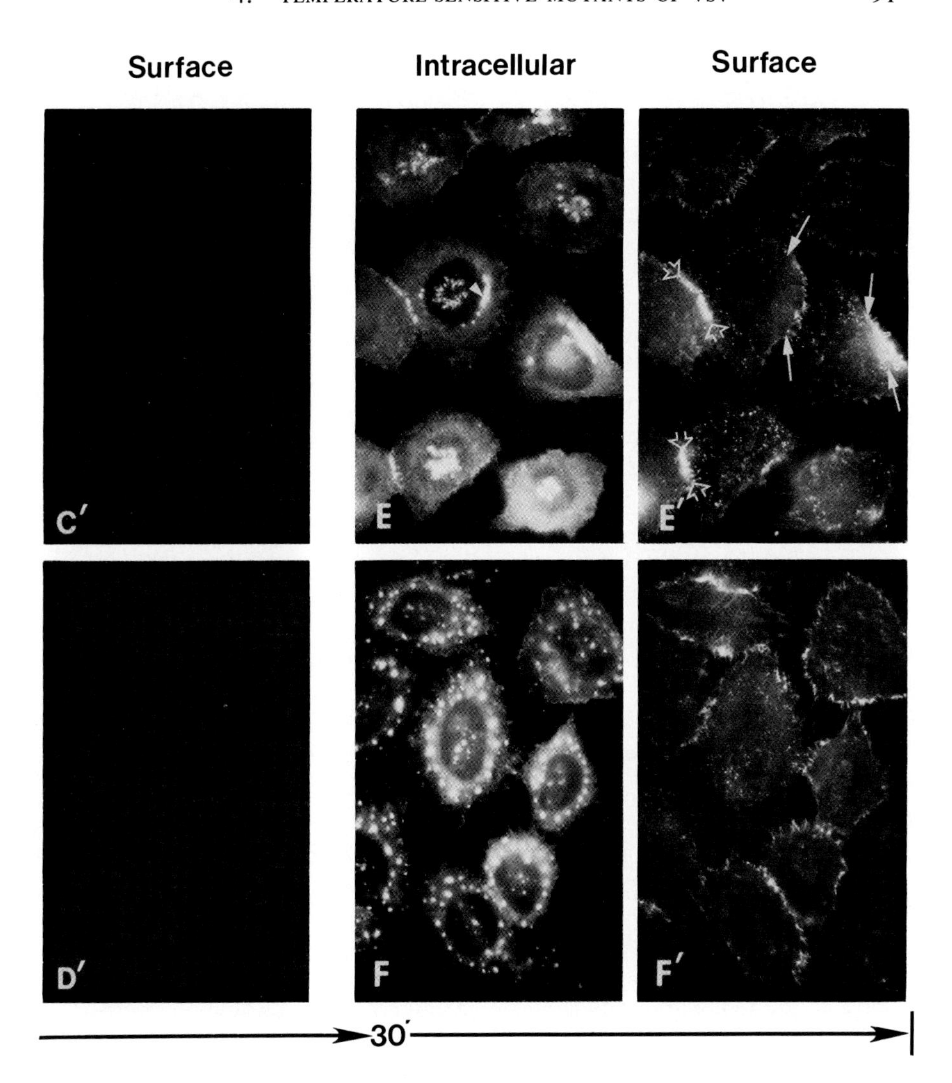

B. Biochemical Analysis of G-Protein Transport

In order to determine independently whether the G protein has been correctly transported to the Golgi apparatus and not to an inappropriate destination, one can investigate whether its asparagine-linked oligosaccharides have been appropriately remodeled. During translation, a high-mannose-type oligosaccharide containing three *N*-acetylglucosamine, nine mannose, and three glucose residues is transferred *en bloc* to each of two asparagine residues of the G protein. After traversing the Golgi apparatus, these oligosaccharides have been extensively altered. The three glucose residues and six of the mannose residues are removed and are replaced with a variable number of *N*-acetylglucosamine, galactose, and terminal sialic acid residues. To detect the addition of sialic acid residues one can take advantage of the fact that they cause a reduction in the mobility of many glycoproteins on sodium doderyl sulfate (SDS)–polyacrylamide gels. Furthermore, these residues can be removed with neuraminidase, increasing the mobility of these proteins on SDS–polyacrylamide gels. As the addition of sialic acid residues is the last modification made to the G protein in the Golgi apparatus, the presence of these residues indicates that the G protein has reached the Golgi apparatus and that at least this function of that organelle is unimpaired. For further discussion of oligosaccharide structural analysis and processing, see Chapter 5 by A. L. Tarentino *et al.* and Chapter 6 by R. D. Cummings *et al.* (this volume.)

Figure 3 shows the results of such an analysis (from Rogalski *et al.*, 1984). NRK cells were again infected with ts045. After virus adsorption, the cells were incubated for 90 minutes at 39.9°C. Nocodazole and [^{35}S]methionine (50 μCi/ml, New England Nuclear, Boston, MA) were added, and incubations were continued for another 90 minutes. Radiolabeling was terminated by the addition of 1 m*M* nonradioactive methionine, and the cells were transferred to 32°C. The mobility of the G protein decreases during the 30-minute chase period at 32°C (Fig. 3, lanes E, F, and G). Thirty minutes after the temperature shift, treatment with neuraminidase increases the mobility of all the G-protein molecules (Fig. 3, lane H). Comparable results were obtained with control cells that had not been treated with nocodazole (Fig. 3, lanes A–D). Thus, it is evident that the disruption of the MT system and the Golgi apparatus with nocodazole does not prevent the rapid and quantitative transfer of G protein from the RER to the Golgi apparatus. These ts045-coded G-protein molecules had been synthesized and held in the RER at the nonpermissive temperature for up to 90 minutes. Thus, these experiments also demonstrate that the conformational lesion that prevents ts045-coded G proteins from leaving the ER at 39.9°C is readily reversible when the temperature is dropped to 32°C. Lodish and Kong (1983) demonstrated a similar temperature reversibility in another G-protein mutant, tsM513. The reversibility of the temperature-induced block in intracellular transport suggests that ts045 may be useful in the studies aimed at better understanding subtle proofreading functions within the ER (see Section III).

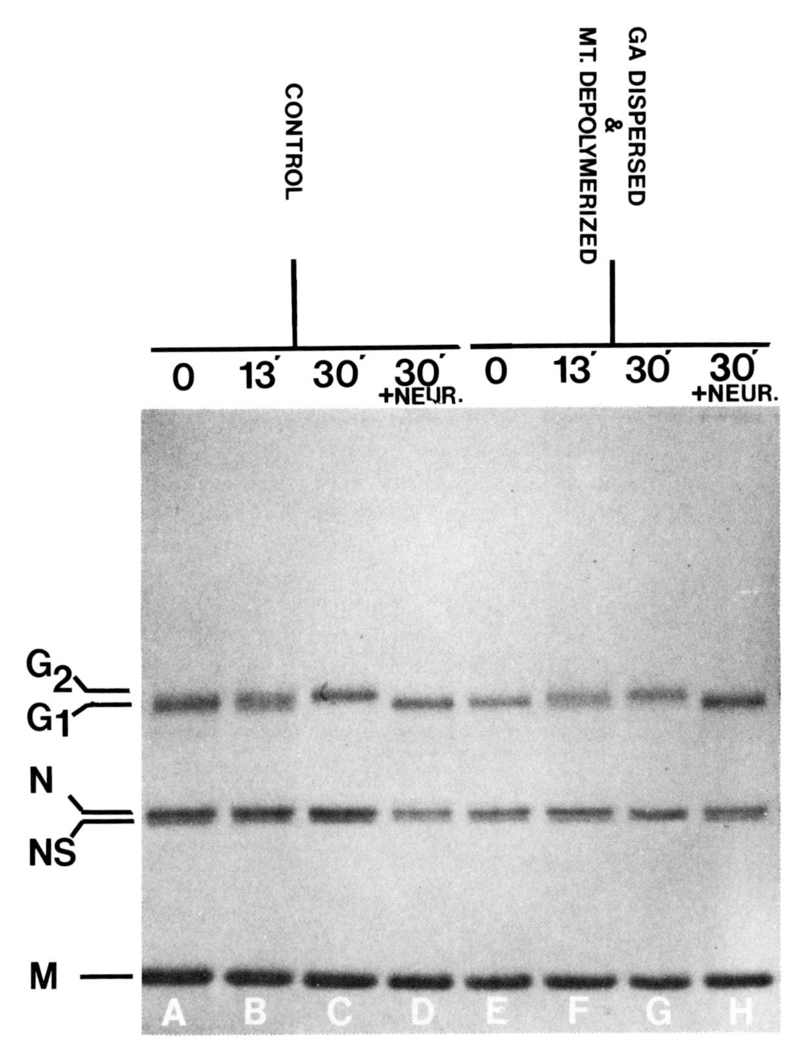

FIG. 3. Analysis of sialic acid addition to the ts045 G protein in normal and nocodazole-treated cells. SDS–PAGE of immunoprecipitates of the viral proteins from extracts of [^{35}S]methionine-labeled, ts045-infected NRK cells. Lanes A–C are from control cells 0, 13, and 30 minutes after the temperature downshift; lane D is of the 30-minute control cells after neuraminidase treatment of the extract. N, NS, and M are the other viral proteins. Lanes E–G are of similar experiments with cells that had been first treated with nocodazole before the temperature downshift, with lane H showing the corresponding neuraminidase-treated 30-minute extract. Reproduced from Rogalski *et al.* (1984), by copyright permission of the Rockefeller University Press.

C. Using Temperature-Sensitive Mutants to Uncouple the Biogenesis of Two Viral Membrane Proteins

Two of the VSV-coded proteins are membrane proteins. In addition to the G protein, the matrix (M) protein of VSV is a peripheral membrane protein that is made on free polysomes and begins to be incorporated into virus particles within 5 minutes of synthesis. Several lines of evidence indicate that both the M and G proteins play a major role in the formation of virions—a process that occurs only at the plasma membrane (summarized in Bergmann and Fusco, 1988). It is necessary to uncouple the biogenesis of these proteins in order to investigate whether this specificity of the budding site might be attributable to the intracellular distribution of M protein and, in addition, whether the M-protein distribution depends on that of the G protein. To study these issues, we used a protocol essentially identical to that used in the study shown in Fig. 1, except that we examined the distribution of M protein as well as G protein 1 hour after infection. In order to ensure antibody specificity, monoclonal antibodies to both the G and M proteins were used on samples treated in parallel. Figure 4 shows the results of such an experiment (from Bergmann and Fusco, 1988). When cells are fixed without transfer to the permissive temperature (0 minutes), the G protein has a distribution consistent with its location in the ER (Fig. 4A). In particular, the G protein is clearly present in the nuclear envelope (small arrowheads). The distribution of the M protein is strikingly different. The M protein is seen primarily on the lateral plasma membranes of the cells (arrows in Fig. 4B). No staining of the nuclear envelope is seen. Thus, it appears that the M protein is associating with the lateral plasma membrane of these cells in the absence of G protein and failing to associate with the G protein that is in the nuclear envelope. When the cells are shifted to 32°C for 10 minutes, the G protein is able to move to and concentrate within the Golgi apparatus (large arrowheads in Fig. 4C). Although the G protein can now assume its normal structure and move to the Golgi apparatus, it fails to attract M protein (Fig. 4D). When the G protein reaches the cell surface 50 minutes after the temperature shift, the M-protein distribution remains unchanged. Thus, we can conclude that the M protein associates with the plasma membrane independently of the G protein and fails to associate with the properly folded G protein on intracellular membranes. We have used this system to show that this membrane association is restricted to the basolateral plasma membrane (Bergmann and Fusco, 1988). Continued use of temperature-sensitive mutants will help to analyze further the mechanism of virion formation and to understand why it occurs only at restricted sites.

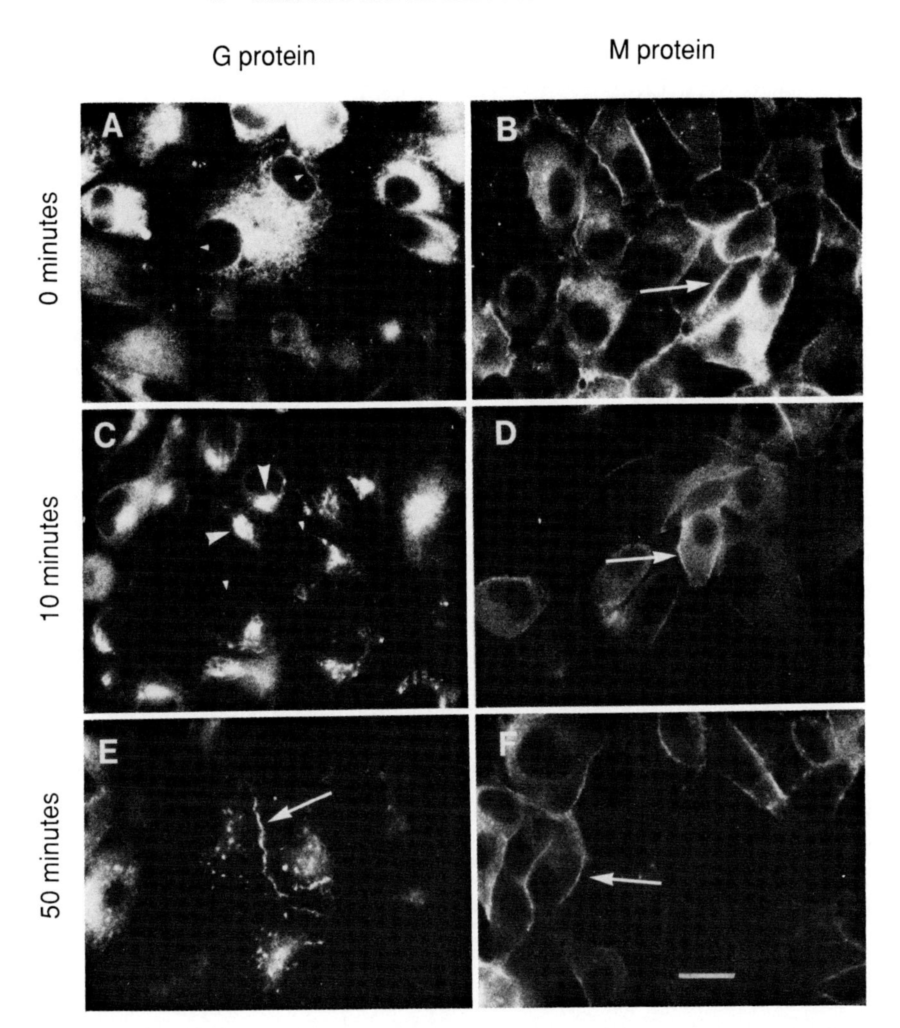

FIG. 4. Distribution of the VSV G and M proteins in ts045-infected MDCK cells. MDCK cells were infected with ts045 and incubated at 39.9°C for 1 hour. (A, B) Cells were immediately fixed. Cycloheximide was added (20 μg/ml), and the cells were shifted to 32°C for 10 minutes (C, D) or 50 minutes (E, F). (A, C, and E) Distribution of the VSV G protein. (B, D, and F) Distribution of the VSV M protein. (A, C) Small arrowheads indicate staining of the nuclear envelope. (C) Large arrowheads indicate labeling of the presumptive Golgi apparatus. Arrows indicate examples of lateral plasma membrane labeling in (B), (D), (E), and (F). Bar = 20 μm. Reproduced from Bergmann and Fusco (1988), by copyright permission of the Rockefeller University Press.

D. The Use of Electron Microscopy to Follow the Synchronized Transport of G Protein

While the synchrony achievable using mutants like ts045 is amply demonstrated in the examples given, it is even more impressive when observed using electron microscopy. Using the techniques described previously, three major steps in the biogenesis of a glycoprotein can be resolved: accumulation in the ER, transfer to the Golgi apparatus, and arrival at the plasma membrane. In contrast, using electron microscopy, a truly vast array of structures can be resolved and, in addition, there is the potential to visualize compartmentalization within each of these structures.

In order to realize this potential, it is necessary to preserve G-protein antigenicity during a far more rigorous fixation procedure. It is also necessary to maintain the temporal synchrony in a larger population of infected cells. Fortunately, the antigenicity of the G protein is maintained completely after glutaraldehyde fixation (Bergmann and Singer, 1983). This is not true of all proteins, and care should be taken to monitor the retention of antigenicity for every fixative–antibody–antigen combination. This is most easily accomplished by comparing the staining of standard and fixed "Western blots" (Bergmann and Singer, 1983). Temporal synchrony is maintained by rapid cooling of the sample and fixation at 0°C (see Section IV, H).

Figures 5–7 illustrate the types of results achievable with these procedures (Bergmann and Singer, 1983). When cells are fixed prior to transfer to the permissive temperature, the ER is lightly labeled and there is very little labeling of the Golgi apparatus (Bergmann *et al.*, 1981). Within 3 minutes after a reversal of the temperature block, G protein is seen entering the Golgi apparatus. Labeling of the entry or cis face is significantly stronger than labeling of the ER or of the bulk of the Golgi apparatus (Fig. 5). At this time, specific labeling of the Golgi apparatus is not visible using immunofluorescence microscopy. Although the G protein is concentrated in the cis face of the Golgi apparatus, when its concentration is averaged over the area resolved by light microscopy, it does not stand out above the general labeling of the ER. At this time point one never sees labeling of the entire Golgi apparatus. Four minutes after the temperature shift, one begins to see uniform labeling of the Golgi apparatus in some cells (Fig. 6B). However, many of the stacks of cisternae are still labeled only at their cis face (Fig 6A) (Bergmann and Singer, 1983). Over the next 2 minutes, G protein fills the Golgi apparatus in all the cells. Thus, the rapid cooling to and fixation at 0°C is sufficient to resolve discrete steps occurring within 3 minutes in the synchronous movement of the G protein. In

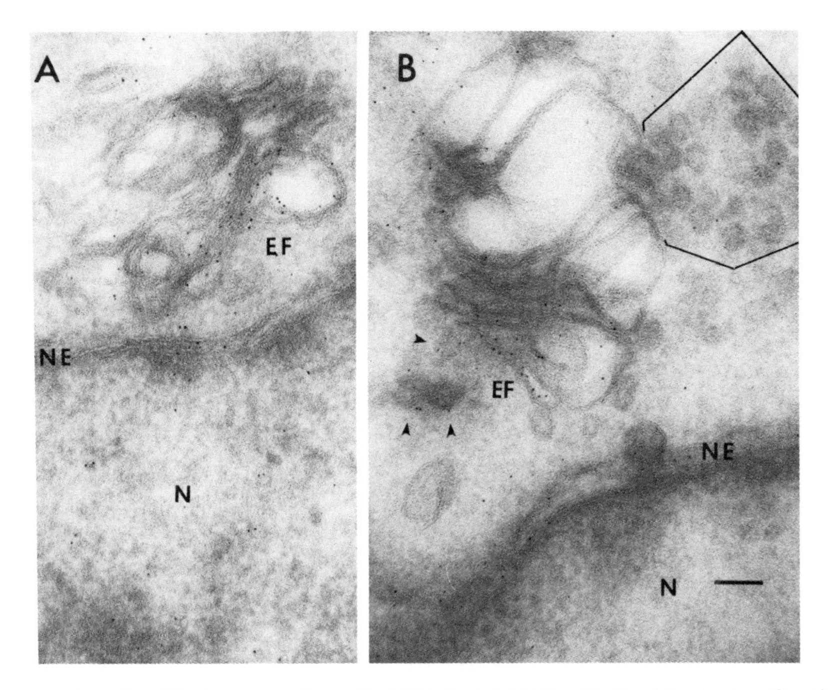

FIG. 5. Ultrathin frozen sections of ts045-infected CHO cells fixed 3 minutes after the temperature shift to 32°C. Frozen sections were indirectly immunolabeled for the G protein using colloidal gold–protein A as the secondary reagent. NE, Nuclear envelope; N, nucleus; EF, entry face of the Golgi apparatus. (A) The nucleus-proximal face of the Golgi apparatus is immunolabeled. (B) Oblique section through the Golgi apparatus. The nucleus-proximal face is immunolabeled, as are several electron-dense vesicular structures of 50–70 nm diameter (small arrowheads) near the entry face. A cluster of similar-sized and electron-dense vesicles (between brackets) is not immunolabeled. Bar = 100 nm. Reproduced from Bergmann and Singer (1983), by copyright permission of the Rockefeller University Press.

Fig. 5, G protein was visualized using protein A coupled to 5-nm gold particles; in Fig. 6, ferritin coupled to goat anti-rabbit IgG was used. The gold is more easily visualized over the stained tissue, but in our experience, ferritin staining is more sensitive.

This technology has also proved capable of resolving heterogeneity in the distribution of the G protein within the ER. Figure 7 shows a view of the nuclear envelope of infected cells 3 minutes after the temperature shift. In regions such as shown here, the nuclear envelope is dilated and the inner and outer nuclear membranes are well separated. G protein is readily labeled in both the inner and outer nuclear membranes. Though not evident from any one micrograph, labeling in these dilated regions of

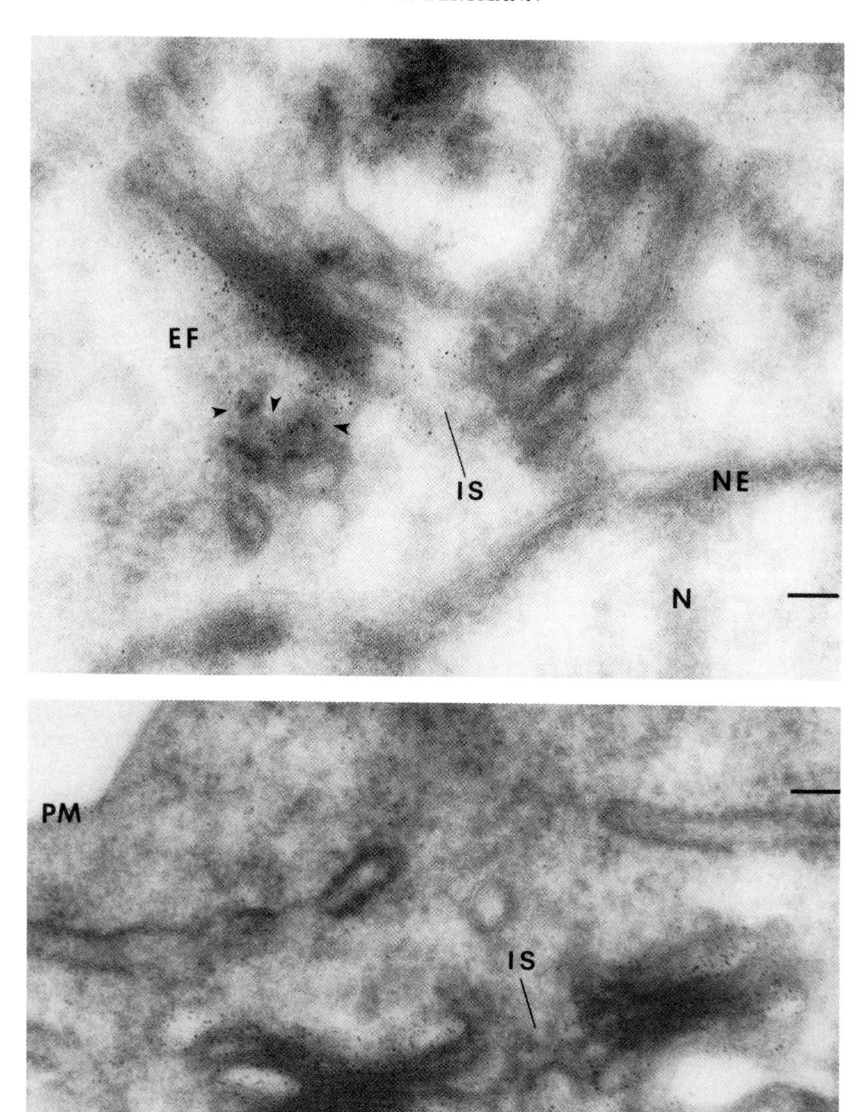

FIG. 6. Ultrathin frozen sections of ts045-infected CHO cells fixed 4 minutes after the temperature shift to 32°C. A preparation similar to that seen in Fig. 5, except that the cells were incubated 1 minute longer at the permissive temperature and that the secondary labeling reagent was ferritin antibody.The Golgi apparatus shows dense and uniform labeling across all of the saccules. In top panel the entry face (EF) is more heavily labeled. IS, intersaccular region between the Golgi stacks; N, nucleus; NE, nuclear envelope. The plasma membrane (PM) shows no labeling at this time. Bar = 100 nm. Reproduced from Bergmann and Singer (1983), by copyright permission of the Rockefeller University Press.

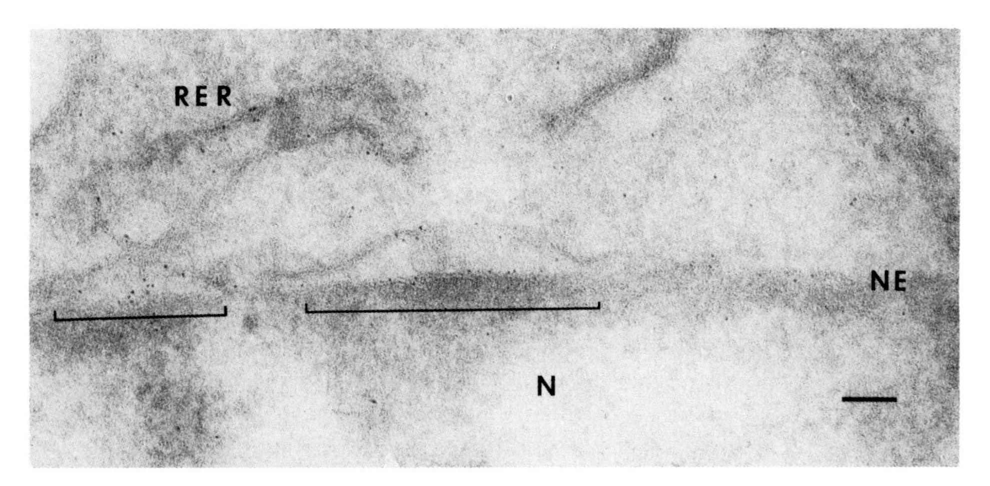

FIG. 7. Nuclear envelope 3 minutes after the temperature downshift. Blebs on the nuclear envelope are designated by the brackets, and show labeling for the G protein at both the outer and inner membrane and an overall density of labeling that is greater than in nonblebbed regions of the nuclear envelope (NE). N, nucleus. Bar = 100 nm. Reproduced from Bergmann and Singer (1983), by copyright permission of the Rockefeller University Press.

the ER is approximately twice as dense as labeling of the nondilated regions of the ER. When ts045-infected cells are chilled and fixed directly from the nonpermissive temperature, this difference in labeling densities is not seen (Bergmann and Singer, 1983). These findings suggest that when it is able to move, the G protein may be concentrated within the plane of the bilayer of the ER prior to transport to the Golgi apparatus. The differences in labeling density are so small that they were only detected because these specialized regions are morphologically distinct. It is therefore likely that as other markers become available, additional heterogeneity may be discerned.

III. Prospects for Future Uses of Temperature-Sensitive Mutants

The usefulness of viral mutants to synchronize the transport of selected membrane glycoproteins is now well established. In addition to ts045, mutants of other viruses have been used in such studies (e.g., Rindler *et al.*, 1985; Saraste and Hedman, 1983). The use of these and other mutants in a variety of cell lines and in conjunction with a number of pharmacological interventions should provide a richer understanding of the pathways followed by membrane proteins.

In addition, these mutants may prove quite useful in determining the mechanisms underlying the retention of proteins in the ER or at other specific steps along the "secretory" pathway. The rapid reversibility of the temperature-induced transport block indicates that the mechanism of retention of these mutants may involve subtle interactions with specific cellular machinery. The ability to manipulate these interactions using temperature shifts may help to elucidate those features that allow a protein to move beyond a particular "checkpoint" or, conversely, those that cause a protein to be retained. For example, Kreis and Lodish (1986) used ts045 to demonstrate that the G protein must form oligomers prior to exit from the ER. Doms and co-workers (1987) extended these studies to show that the oligomers are trimers and that ts045 forms an aggregate at the nonpermissive temperature. They further showed that disaggregation of ts045 and its subsequent trimerization require cellular ATP. Further work is sure to elucidate the components of the aggregate, the mechanism of this aggregation, and whether the wild-type G protein forms such an aggregate as a normal step in its maturation.

IV. Technical Procedures

A. Control of Incubation Temperature

1. General Considerations

As mentioned before, the discovery that the G protein in some mutant VSV strains is temperature-sensitive for exit from the ER allows one to synchronize its transport between the ER and the cell surface. Before describing specific protocols, I will discuss some overall characteristics of the VSV system and the mutant ts045.

By definition, a temperature-sensitive mutant fails to replicate at a temperature that is permissive to the growth of wild-type virus. However, at the restrictive temperature for the VSV mutants, even the wild-type virus is hampered in its replication. At 39.9°C, wild-type VSV synthesizes greatly reduced quantities of the viral proteins and fails to shut off host cell protein synthesis. As a result, if cells are incubated at 39.9°C from the time of infection, the viral proteins cannot be detected above the background of host proteins on SDS–polyacrylamide gels. They can, however, be easily detected on polyacrylamide gels following immunoprecipitation or by immunolabeling of fixed tissue. When studying the synchronized transport of the G protein using morphological techniques it is necessary to incubate cells at the restrictive temperature immediately following the adsorption period. When these studies are correlated with biochemical assays, it is often necessary to follow the identical incubation protocols for both procedures. However, there are many experiments in which pulse–chase protocols can be utilized, and a strict comparison with morphological studies is not required. It is then preferable to incubate the infected cells at 37°C for 2–3 hours prior to incubation at the restrictive temperature. This will result in a far greater incorporation of label into the viral proteins and may circumvent the immunoprecipitation step.

2. Methods of Temperature Control

Cells are generally incubated either in a CO_2 incubator or in a temperature-regulated water bath. Although most incubators may be successfully used for experiments with temperature-sensitive mutants, we have often found it difficult to adjust the temperature of such incubators within a fraction of a degree. The most convenient incubators are those using electronic temperature regulators rather than bimetallic strip thermostats. Electrical control circuits often use a potentiometer to set the desired temperature. Replacing the standard potentiometer with one that covers the same resistance range with 10 turns of the knob allows one easily to control the set temperature to within 0.1°C. We use a VWR 1820 incubator (VWR Scientific, Piscataway, NJ) that has been modified in this manner.

There are many water bath circulators on the market that can be adjusted within 0.1°C. However, it is important to remember that the temperature of the liquid within a vessel in the water bath may not be the same as the temperature of the circulating water. This is because of the discrepancy between the bath temperature and the air temperature above the bath. When preparing an experiment in which infected cells are to be incubated at the nonpermissive temperature, one must adjust the water bath so that a comparable volume of water in the culturing vessel to be used reaches the desired incubation temperature; this may require setting the temperature of the bath as much as 1–2°C above the desired incubation temperature. If the air temperature in the room is constant within 1–2°C the required temperature offset of the water bath will remain the same from experiment to experiment.

B. Propagation of Wild-Type and Temperature-Sensitive Mutants of VSV

1. The Plaque Assay

When a confluent monolayer of cells is infected and subsequently cultured under agarose, virus is transmitted only from an infected cell to its immediate neighbors. After a few rounds of infection, the virus has cleared a plaque of dead cells in the living monolayer. If the original infection was carried out at an extremely low multiplicity (e.g., 1 infectious unit per 10,000 cells), these plaques are easily resolved by eye. Each plaque is derived from a single infected cell, and thus, if the original virus was sufficiently disaggregated prior to infection, each plaque contains virus particles derived from a single virion.

Counting plaques reveals the original titer of a virus inoculum expressed in plaque-forming units (pfu). Virus in a single plaque may be harvested (picked) and further propagated, resulting in a stock that is genetically homogeneous. In the following pages I describe in detail each of these procedures for use with VSV.

a. Titering a Virus Stock. Six 35-mm dishes are used in a standard plaque assay. A six-pack such as Corning no. 25810 (Corning Glass Works, Corning, NY) is convenient, but not necessary. Indicator cells are plated at 10^6 cells per 35-mm dish and incubated at 37°C overnight in growth medium: a 1 : 1 mixture of Dulbecco's modified Eagle's medium (DMEM) and Ham's F-12 medium (pH 7.2), containing 5% fetal bovine serum (FBS). We have successfully used both mouse L cells and VERO cells as indicators. A virus stock of unknown concentration is diluted in growth medium to a final concentration of between 50 and 500 pfu/ml. Usually dilutions of 10^5, 10^6, and 10^7 are sufficient to guarantee an appropriate concentration. Duplicate dishes are infected with each virus dilution using 200 μl of inoculum per 35-mm dish. The dishes are "rocked" in such a way that the inoculum is spread evenly over the entire surface of the cell monolayer and the dishes are placed in an incubator at 37°C in an atmosphere of 5% CO_2 to allow for virus adsorption to the cells. (Note: for temperature-sensitive mutants such as ts045, adsorption of virus should proceed at or below the permissive temperature of 32°C rather than 37°C. This is also true during experiments in which the cells will be incubated at the nonpermissive temperature following the adsorption period.) Adsorption is continued for 30 minutes, with rocking every 10 minutes to redistribute the inoculum.

During this adsorption period, 10.5 ml growth medium are warmed to

37°C in a sterile 17 – 100-mm culture tube (e.g., no. 14-956-1H, Fisher Scientific, Pittsburgh, PA) and a 4% agarose solution in water is prepared using a boiling-water bath or microwave oven. At the end of the adsorption period, the inoculum is removed. A 10-ml pipet is warmed by rapidly filling and discharging the hot agarose several times, and then is used to transfer 2 ml of the 4% agarose solution to the 10.5 ml of prewarmed growth medium. The tube of medium is immediately capped, inverted two or three times, and, using the same pipet, the medium is distributed over the infected cultures (2 ml per dish). The mixture should be applied with care so that the cells are not damaged by the stream of medium containing agar. The dishes are allowed to stand until the agarose mixture has completely gelled (~10 minutes). The dishes are returned to the CO_2 incubator until the plaques become evident (usually 1 day at 37°C or 1.5–2 days at 32°C). Plaques usually appear as "rough spots," visible by eye. When observed by phase-contrast microscopy, plaques appear as areas in which the cells have rounded up and dead cells have lysed and left debris. Though visible, the plaques can be more easily visualized if the monolayer is stained with neutral red. Growth medium lacking FBS (6 ml) is mixed with 0.9 ml of 0.1% neutral red. This solution is carefully layered over the agar (1 ml per 35-mm dish). One must allow at least 4 hours for the neutral red to diffuse into the agarose and become concentrated in the living cells. Plaques will now appear as clear areas in a red background. Although this liquid overlay may be removed by aspiration, some investigators prefer to use an agar overlay prepared as described before except that 5 ml of medium lacking serum at 37°C are mixed with 900 μl neutral red and 1 ml of melted 4% agarose. This agar overlay is preferable when several independent plaques are to be picked from the same plate (see part *b* of this section).

Counting the plaques gives a direct quantitation of the infectious titer of a virus stock. Since plaques become more difficult to count as they grow and begin to overlap, care should be taken to count the plaques as early as possible. On dishes with ≤10 plaques, one may wait longer for larger, more obvious plaques to grow. The number of these plaques can be compared to that determined earlier from the plates infected with less diluted virus.

b. Plaque Purification of Virus. It is important to repurify virus periodically to ensure that stocks remain free of defective interfering (DI) particles and that stocks of mutant virus remain free of revertants (Steinhauer and Holland, 1987). We repurify stocks every three to four passages. To purify a virus, one identifies and circles a plaque that is well separated from all neighboring plaques. A sterile 12 × 75-mm culture tube (e.g., no. 14-956-1D, Fisher-Scientific) is prepared with 1 ml growth medium. The inside walls of a sterile Pasteur pipet are wetted by filling and discharging the growth medium. One should choose a pipet with a flat, undamaged end. While applying negative pressure with a 1–2 ml pipet bulb, one carefully inserts the pipet into the agar directly over the plaque until the pipet comes into contact with the plastic dish; this creates an agar plug containing the virus at the tip of the pipet. The pipet is moved up and down so as to free the agarose plug from the surrounding agar and to pull it up into the pipet. The plug is rinsed out of the pipet using the 1 ml of prepared growth medium. The tube of medium with the agar plug is vortexed vigorously to extract virus from the agarose plug. If there is reason to believe the original virus stock was highly heterogeneous, it is a good practice to repurify two more times. This is accomplished by spinning out the agarose debris (2000 rpm for 5 minutes) and beginning another plaque assay. As a plaque contains ~10,000 pfus, dilutions of 10- to 100-fold are appropriate. These subsequent plaque purifications are employed to minimize the probability that a plaque derived from a cell infected with two different viruses will be used to establish a virus stock.

2. Preparation of Infectious Stocks

Stocks of both wild-type and mutant strains of VSV are extremely easy to grow and maintain. Our usual procedure is to plate 5×10^6 COS-1 cells in 20 ml growth medium (see Section IV,B,1,a) on a 150-mm dish, and culture them at 37°C overnight. (CHO, BHK, L, and VERO cells can also be used as hosts, but in our hands they give only one-tenth the yield of virus.) The next morning, the medium is replaced with 15 ml of fresh growth medium to which are added $2–5 \times 10^6$ pfu of virus. The plate is then returned to incubate in 5% CO_2 at 37°C (32°C for temperature-sensitive mutants). When 75–90% of the cells have died and detached from the dish (~1 day), the virus-containing supernatant is harvested, the cellular debris is removed by centrifugation at 2000 rpm for 5 minutes, and the clear supernatant is divided into 40 aliquots in sterile 1.5-ml screw-cap microcentrifuge tubes (no. 72.692.005, Sarstedt, Nümbrecht, Federal Republic of Germany). The virus is frozen and stored at −80°C. Virus stocks are quite stable at −80°C, but should not be frozen and thawed more than one time. Typical yields from COS-1 cells are $2–5 \times 10^9$ pfu/ml.

As with the serial propagation of any organism, genetic variants begin to accumulate. A particular problem with virus stocks is the creation of DI particles. As these particles cannot replicate independently of a wild-type virus, passage of VSV at a low multiplicity inhibits the rate at which they accumulate in successive serially passaged stocks. However, even when propagating virus at low multiplicity, it is a good practice to plaque-purify stocks every three to four passages (see Section IV,B,1,b).

3. LARGE-SCALE PREPARATIONS AND PURIFICATION OF VSV

In order to prepare antibodies, it is necessary physically to purify milligram quantities of the virus. We have used procedures that are very similar to those worked out by McSharry and Wagner (1971) with great success. Twenty 150-mm dishes of COS-1 cells plated with 5×10^6 cells per dish should yield ~20 mg of virus. After plating, the cells are incubated overnight at 37°C. Next morning, 15 ml of growth medium are prepared per dish and 1.5×10^7 pfu per dish of VSV are added. The old medium is replaced with 15 ml of virus-containing growth medium and incubated at 37°C for wild-type virus or 32°C for temperature-sensitive mutants. When the virus is ready to harvest, the 300 ml of supernatant are pooled and clarified as described before. The virus is then pelleted by centrifugation at 19,000 rpm for 2 hours in a 19 rotor or for 1 hour at 30,000 rpm in a 60Ti rotor (Beckman Instruments). The pellet is resuspended in TE buffer (10 m*M* Tris, pH 7.5, 1 m*M* EDTA) and sonicated for 10 seconds at a setting of 5 in a closed tube using a cup horn sonicator (Heat systems W225). The disaggregated virus is applied to a 15–60% linear sucrose gradient in TE and centrifuged for 20 minutes at 27,000 rpm in an SW28 rotor (Beckman). The visible virus band is removed with a Pasteur pipet. The virus fraction is diluted 4-fold and harvested by centrifugation for 1 hour at 30,00 rpm in a 60Ti rotor. Virus is quantitated by measuring the OD_{280} in 50% acetic acid. (A solution containing 1 mg/ml VSV has an absorbance of ~1 OD_{280}.) The virus may be analyzed for purity by sodium dodecyl sulfate–polyacrylamide gel electrophoresis (SDS–PAGE) (Laemmli, 1970). Three major bands should be visible (G, 65 kDa; N, 47 kDa; M, 26 kDa). The minor structural proteins, L (241 kDa) and NS (45 kDa apparent, but has variable mobility) can be seen if sufficient material is loaded on the gel. A good preparation of VSV will yield ~15 mg of virus and has a titer of $\sim 2.0 \times 10^{12}$ pfu.

C. Inactivation of the Infectivity of VSV Stocks

For antibody or biochemical work, the virus may be inactivated by 30 minutes of ultraviolet illumination in an open Petri dish. The germicidal lamps in a tissue culture hood work well. The dish is turned every 10 minutes to guarantee uniform illumination. Before injection into animals, one must make sure the virus is completely inactivated by putting 10 μg (formerly 10^9 pfu) of virus into a dish of freshly fed COS-1 cells. If after 2 days the cells remain healthy, no infectivity remains in the stock.

D. Immunization of Rabbits

In order to achieve a timed release of immunogen in the rabbit, one prepares an emulsion with 250 μg of purified VSV in 500 μl of PBS (10 m*M* NaH_2PO_4, 150 m*M* NaCl) and 500 μl complete Freund's adjuvant. Rabbits are initially immunized by injecting ~200 μl of this emulsion into each of the popliteal lymph nodes. The remainder is divided equally between intradermal, intramuscular, and subcutaneous injections.

For booster injections, one prepares an emulsion with 250 μg of purified VSV in 500 μl of PBS and 500 μl incomplete Freund's adjuvant. After 3 and 6 weeks, the rabbit is boosted using intramuscular and subcutaneous injections. One may test the response 1–2 weeks after the second boost by comparing the staining of infected and uninfected tissue culture cells (e.g., COS-1, VERO, CHO, L, or MDCK). A good serum will give specific staining at a dilution of 1 : 250 or greater. A regular schedule of bleeding can then be initiated under the supervision of a veterinarian.

E. Preparation of a G-Protein Affinity Column

Rabbits immunized as just described make antibodies to all the major structural proteins of VSV. Antibodies specific for the G protein may be rapidly purified using a column chromatography matrix to which the G protein has been attached.

First the G protein is liberated from the purified virions by addition of Triton X-100 (Sigma) to a final concentration of 1%. As the virus is in a low ionic strength buffer, the nucleocapsids and associated M protein remain particulate and the G protein is quantitatively solubilized (Kelley *et al.*, 1972). The particulate matter is removed by centrifugation at 40,000 rpm for 1 hour in a 60Ti rotor. The supernatant containing the G protein is adjusted to 10 m*M* NaH_2PO_4, 1 m*M* $MgCl_2$, 1 m*M* $Cacl_2$, 150 m*M* NaCl, and 0.5 mg/ml sodium azide. Concanavalin A (Con A)–Sepharose (Pharmacia) is added (2 ml of a 50% slurry) and tumbled slowly for 48 hours. The supernatant is removed by filtration on a sinterred glass funnel and the Con A–Sepharose is rinsed with and then resuspended in PBS containing 1 m*M* $MgCl_2$, 1 m*M* $CaCl_2$. The G protein is then fixed to the matrix by adjusting the slurry to a final concentration of 2% glutaraldehyde and tumbling for 1 hour at 4°C. The fixation solution is removed by filtration. The matrix is rinsed with PBS, and blocked with PBS containing ovalbumin (Sigma) at 10 mg/ml and 10 m*M* glycine. This matrix can be used in standard affinity chromatography procedures (e.g., Ternynck and Avrameas, 1976).

F. Immunofluorescence Staining

There are several protocols that will discriminate between G protein present at the cell surface and intracellular G protein. Here I describe one such procedure, which requires only rabbit antibody to the G protein or a monoclonal antibody to G and commercially available secondary antibodies. Infected cells grown on glass coverslips (Corning no. 1.5) are fixed for 20 minutes in a freshly prepared solution of 3% formaldehyde in PBS. They are rinsed twice for 10 minutes in PBS containing 10 m*M* glycine. G protein on the cell surface is stained by exposure to affinity-purified rabbit anti-G protein for 10 minutes. To conserve reagents, each of the staining steps is performed by inverting the coverslip with attached cells onto a 100 μl drop of PBS containing the desired staining reagent. Inverting the coverslip brings the cells into direct contact with the reagent and minimizes drying. After staining, the coverslip is returned to the 35-mm dish with the cells again on the top side. After two 5-minute rinses in 2 ml PBS, the cells are exposed to 10 μg/ml of affinity-purified fluorescein-conjugated goat anti-rabbit IgG (Boehringer Mannheim) for 10 minutes and again rinsed twice with PBS. The internal G protein next is exposed by treating the cells for 5 minutes with 1% Triton X-100 or Emulphogene BC-720 (General Aniline and Film, Wayne, NJ) in PBS. The cells are rinsed, stained with primary antibodies, and again rinsed as described previously, except that the staining and rinse times are all doubled. The cells are next exposed to biotin-conjugated goat anti-rabbit IgG for 20 minutes (Vector Labs, Burlingame, CA), rinsed, treated for 20 minutes with Texas red conjugated to streptavidin (Bethesda Research Labs), and rinsed again. The reagents used for surface and internal labeling can be reversed. Using this protocol, the "surface" label is quite specific. In contrast, the "internal" pattern always includes surface features. However, when the majority of the G protein is internal, these surface features are only noticeable in the "surface" label (Bergmann *et al.*, 1983; and compare Fig. 1G and H, or Fig. 1J and K). Equally specific staining results are obtained using purified mouse monoclocal anti-G (10 μg/ml in PBS) as the primary antibody. Biotin-conjugated horse anti-mouse IgG (Vector Labs) and Texas red conjugated to streptavidin are then sequentially applied as described earlier.

G. Immunoprecipitation of the VSV G Protein

During infection, substantial amounts of the VSV proteins are synthesized, and the G protein is easily resolved from the other VSV proteins on SDS–polyacrylamide gels. Thus, the G protein may be immunoprecipitated with either crude rabbit anti-VSV serum or with affinity-purified rabbit anti-G protein. It is most convenient to harvest the immune complexes with fixed *Staphylococcus aureus*. We have prepared fixed *S. aureus* ourselves and have also found material supplied by Beohringer Mannheim Biochemicals and Bethesda Research Laboratories to work quite well. However, when using the standard immunoprecipitation protocols supplied with these products, one will find that one always sees substantial amounts of the N protein and some M protein, even when using affinity-purified antibody to the G protein. These proteins form nucleocapsids in the cell and appear in the immunoprecipitates. When necessary, these contaminants may be removed by centrifugation in a Sorvall SS34 rotor at 19,000 rpm for 20 minutes prior to immunoprecipitation.

H. Infection, Temperature Shift, and Fixation Protocols for Electron Microscopy

In the experiments illustrated in Figs. 5–7, we have used CHO cells. These cells are readily grown in suspension culture using Joklik's modified Eagle's medium (Gibco) supplemented with 10% FBS. They should be diluted daily to a concentration of 2×10^5 cells/ml and grown to a density of $\sim 4 \times 10^5$ cells/ml. On the day of the experiment, 4×10^7 cells are harvested by centrifugation at 1000 rpm for 5 minutes in a clinical centrifuge. The supernatant is decanted and discarded, the cell pellet is resuspended in 1 ml of medium containing 4×10^8 pfu of ts045. The virus is allowed to adsorb to the cells at 20°C for 30 minutes, with 5 seconds of gentle mixing every 10 minutes. The cells are again diluted to a concentration of 4×10^5 cells/ml and transferred to a spinner flask (no. 1969-00100, Belco Glass, Vineland, NJ), which has been gassed with 90% air–10% CO_2 and incubated for 3.5 hours at 39.9°C. (See Section IV,A for details of adjusting the incubation temperature.)

To prepare for the temperature shift, magnetic stir bars (no. R-25060 Markson Science, AZ) are placed in 20 × 150 mm glass test tubes. The tubes are gassed with 10% CO_2–90% air, tightly stoppered, and maintained at 20°C. A second water bath with a magnetic stirring motor and a nonmetallic test tube rack is warmed to 32°C. Also, 5 ml of medium are added to 50-ml conical screw-cap tubes (e.g., Falcon no. 2070, Beckton-Dickinson, Lincoln Park, NJ) and frozen by immersion in a dry ice–ethanol bath while being vigorously shaken. (The caps are tightly secured to ensure that the CO_2 liberated from the dry ice does not lower the pH of the medium.) This procedure causes the medium to freeze in a thin shell along the sides and bottom of the conical tubes. The tubes are placed in an ice bucket until needed. Care should be taken to ensure that the medium does not thaw before it is needed.

To initiate the temperature shift, 10 ml of the infected cells are transferred to each of the prepared tubes and the tubes are transferred to the 32°C bath. This procedure will cool the cells to the permissive temperature within 30 seconds. If a distance of 3 cm between adjacent tubes is maintained and care is taken properly to adjust the stirring motor speed, all the small magnetic stir bars can be kept rotating simultaneously. At each time point, a 10-ml aliquot of infected cells is poured into one of the tubes containing the shell-frozen medium while it is continuously vortexing. The vortexing is continued until nearly all the frozen medium has thawed. The tube is recapped and returned to the ice bucket. The cells are harvested from the ice-cold medium by centrifugation at 1000 rpm for 3 minutes at 0°–4°C. The tubes are returned to the ice bucket; the supernatant is then gently removed and discarded. Care must be taken not to disrupt the cell pellet. The cell pellet is carefully overlaid with fixative (3% formaldehyde and 2% glutaraldehyde in PBS precooled to 0°C). Any medium that has not been removed will float on the denser fixative. Fixation is continued for 1 hour on ice. The pellet is dislodged and poured out into a 16-ml snap-cap vial (e.g., no. 03-335-20A Fisher Scientific). The fixative is replaced with PBS and the samples are stored at 4°C. This fixation procedure results in a block of tissue that will remain in one piece if handled gently enough.

In preparation for cryoultramicrotomy, the samples are placed in a 50-mm glass Petri dish and set on the stage of a dissecting microscope. They are carefully cut into blocks ~0.5 mm on a side using a pair of scalpels. The tissue blocks are pulled into a Pasteur pipet and transferred back to the Wheaton vial. Most of the PBS in the Wheaton vial is now withdrawn while care is taken not to remove the tissue blocks. Freezing medium [0.6 *M* sucrose and 10% dimethyl sulfoxide (DMSO) in PBS] is carefully layered under the PBS in the Wheaton vial. The tissue blocks initially float, but as they become infused with the freezing medium, they sink to the bottom of the vial. The PBS–freezing medium is carefully removed by aspiration and replaced with fresh freezing medium. The samples should be allowed to stand for at least 15 additional minutes. The tissue can now be quickly frozen and ultrathin frozen sections can be cut, immunolabeled, and stained as previously described (Keller *et al.*, 1984; Tokuyasu, 1978; Griffiths *et al.*, 1984).

References

Atkinson P. H., Moyer, S. A., and Summers, D. F. (1976). *J. Mol. Biol.* **102,** 613–631.
Bergmann, J. E., and Fusco, P. J. (1988). *J. Cell Biol.* **107,** 1707–1715.
Bergmann, J. E., and Fusco, (1988). *J. Cell Biol.* (in press).
Bergmann, J. E., and Singer, S. J. (1983). *J. Cell Biol.* **97,** 1777–1787.
Bergmann, J. E., Tokuyasu, K. T., and Singer, S. J. (1981). *Proc. Natl. Acad. Sci. U.S.A.* **78,** 1746–1750.
Bergmann, J. E., Kupfer, A., and Singer, S. J. (1983). *Proc. Natl. Acad. Sci. U.S.A.* **80,** 1367–1371.
David, A. E. (1973). *J. Mol. Biol.* **76,** 135–148.
David, A. E. (1977). *Virology* **76,** 98–108.
Doms, R. W., Keller, D. S., Helenius, A., and Balch, W. E. (1987). *J. Cell Biol.* **105,** 1957–1969.
Gabel, C. A., and Bergmann, J. E. (1985). *J. Cell Biol.* **101,** 460–469.
Gallione, C. J., and Rose, J. K. (1985). *J. Virol.* **54,** 374–382.
Ghosh, H. P., Toneguzzo, F., and Wells, S. (1973). *Biochem. Biophys. Res. Commun.* **54,** 228–233.
Griffiths, G., McDowall, A., Back, R., and Dubochet, J. (1984). *J. Ultrastruct. Res.* **89,** 65–78.
Grubman, M. J., and Summers, D. F. (1973). *J. Virol.* **12,** 265–274.
Grubman, M. J., Moyer, S. A., Banerjee, A. K., and Ehrenfeld, E. (1975). *Biochem. Biophys. Res. Commun.* **62,** 531–538.
Hunt, L. A., and Summers, D. F. (1976). *J. Virol.* **20,** 637–645.
Keller, G. A., Tokuyasu, K. T., Dutton, A. H., and Singer, S. H. (1984). *Proc. Natl. Acad. Sci. U.S.A.* **81,** 5744–5747.
Kelley, J. M., Emerson, S. U., and Wagner, R. R. (1972). *J. Virol.* **10,** 1231–1235.
Knipe, D. M., Lodish, H. F., and Baltimore, D. (1977a). *J. Virol.* **21,** 1121–1127.
Knipe, D. M., Baltimore, D., and Lodish, H. F. (1977b). *J. Virol.* **21,** 1128–1139.
Knipe, D. M., Baltimore, D., and Lodish, H. F. (1977c). *J. Virol.* **21,** 1149–1158.
Kreis, T. E., and Lodish, H. F. (1986). *Cell (Cambridge, Mass.)* **46,**(6), 929–937.
Laemmli, U. K. (1970). *Nature (London)* **227,** 680–685.
Lodish, H. F., and Kong, N. (1983). *Virology* **125,** 335–348.
McSharry, J. J., and Wagner, R. R. (1971). *J. Virol.* **7,** 59–70.
Morrison, T. G., and Lodish, H. F. (1975). *J. Biol. Chem.* **250,** 6955–6962.
Rindler, M. J., Ivanov, I. E., Plesken, H., and Sabatini, D. D. (1985). *J. Cell Biol.* **100,** 136–151.
Rogalski, A. A., Bergmann, J. E., and Singer, S. J. (1984). *J. Cell Biol.* **99,** 1101–1109.
Rose, J. K., and Bergmann, J. E. (1982). *Cell (Cambridge, Mass.)* **30,** 753–762.
Rose, J. K., and Bergmann, J. E. (1983). *Cell (Cambridge, Mass.)* **34,** 513–524.
Saraste, J., and Hedman, K. (1983). EMBO *J.* **2,** 2001–2006.
Steinhauer, D. A., and Holland, J. J. (1987). *Annu. Rev. Microbiol.* **41,** 409–433.
Ternynck, T., and Avrameas, S. (1976). *Scand. J. Immunol., Suppl.* **3,** 29–35.
Tokuyasu, K. T. (1978). *J. Ultrastruct. Res.* **63,** 287–307.

Chapter 5

Enzymatic Approaches for Studying the Structure, Synthesis, and Processing of Glycoproteins

ANTHONY L. TARENTINO, ROBERT B. TRIMBLE, AND
THOMAS H. PLUMMER, JR.

Wadsworth Center for Laboratories and Research
New York State Department of Health
Albany, New York 12201

I. Introduction

A. Purpose and Scope

This chapter's intent is to provide the researcher with a practical guide for the successful application of oligosaccharide-cleaving enzymes to topics of current interest in the cell biology of glycoproteins. Such studies

Copyright © 1989 by Academic Press, Inc.
All rights of reproduction in any form reserved.

may focus on the structure–function relationships of the covalently attached carbohydrate present on isolated, purified glycoproteins or may follow the biosynthesis and modification of glycoprotein glycans as probes of cellular processes, including targeting, transport, and turnover. To this end, each enzyme that has found utility in studying cellular processes will be described in sufficient detail to provide a clear perspective on its mode of action and known substrate specificity, its commercial availability, its handling characteristics, and its application in specific experimental protocols.

The information presented here represents an amalgamation of published and anecdotal data from the many laboratories already using endoglycosidases and glycoprotein–glycopeptide amidases with that from our own, both published and unpublished. This chapter does not include an extensive review of the literature; rather, citations were chosen to provide a cross section of studies that have employed these enzymes in useful ways to answer novel questions. We have endeavored to include in the references the most recent reviews on appropriate topics relating to glycoprotein biosynthesis and secretion in order to expand the accessible information base. Omissions, which are always a hazard in summarizing a research field, are regrettably a result of unintentional oversight.

B. Endoglycosidase Utility—a Historical Perspective

Our understanding of asparagine-linked glycoprotein glycan structure, biosynthesis, and processing has grown remarkably during the past 15 years. We now know that N-linked glycosylation is a rough endoplasmic reticulum (RER) function in all eukaryotes studied to date. In most organisms, glycosylation proceeds via transfer of a preformed $Glc_3Man_9GlcNAc_2$ oligosaccharide from a dolichyl phosphate carrier molecule to some, but not all asparagine residues in the tripeptide consensus sequence, -Asn-X-Thr/Ser-, where X may be any amino acid except proline. Subsequent to en bloc transfer of $Glc_3Man_9GlcNAc_2$, trimming reactions in the ER remove the glucose and, depending on the system, a specific mannose residue. Once glycoproteins move from the ER to the Golgi, additional mannose may be trimmed, followed by outer-chain addition of GlcNAc, Gal, NeuNAc, PO_4, and SO_4. Several excellent reviews have appeared that summarize our current understanding of asparagine–oligosaccharide synthesis and processing (Hirshberg and Snider, 1987; Kornfeld and Kornfeld, 1985; Kukuruzinska *et al.,* 1987; Tanner and Lehle, 1987).

Traditional glycoprotein studies of the early 1970s were heavily focused on determining the structures of the associated N-linked glycans. Extensive proteolysis with pronase and carboxypeptidases provided glycopeptides that could be separated into compounds by size and/or charge fractionation. Glycopeptides were subjected to compositional analyses, acetolysis fragmentation, and methylation analysis to assign branching patterns. Mild acid hydrolysis to remove peripheral sialic acids and sequential exoglycosidase digestions were also employed. Chemical methods for deglycosylation were developed, and technical advances in analytical methods constantly have been introduced, including fast atom bombardment–mass spectrometry (FAB–MS) and ^{1}H-NMR spectroscopy. Together, these approaches have yielded a wealth of oligosaccharide structural information. Nevertheless, no method was available to release intact Asn-linked glycans from their host proteins without destruction of the peptide itself.

The discovery of the endoglycosidases H (Tarentino and Maley, 1974) and D (Muramatsu, 1971) provided this ability. Initially they were employed to release oligosaccharides from isolated glycoproteins such as RNase B, IgM, yeast invertase, and hen ovalbumin (Tarentino *et al.*, 1974), but were quickly applied to the physicochemical characterization of the nearly carbohydrate-free deglycosylated proteins. Thus, the endoglycosidases furnished a new approach to assess the role of the glycan moieties in glycoprotein structure and function (Trimble and Maley, 1977a,b).

As more oligosaccharides were characterized, their branching patterns and compositions revealed structural interrelationships that began to define a generalized scheme of N-linked glycan processing in eukaryotes (Kornfeld and Kornfeld, 1985). It soon became clear that because of their substrate specificities, to be described in detail in subsequent sections, endoglycosidases H and D had a limited potential for generating oligosaccharides for structural work. These enzymes could release oligosaccharides only of the high-mannose type, which are now recognized as the early intermediates in the oligosaccharide-processing pathway.

It thus became desirable to identify additional oligosaccharide-cleaving enzymes with broader, or at least different substrate specificities, which would include the more highly processed complex-oligosaccharide types. Two additional enzymes were discovered in the early 1980s and have now been characterized: Endo F hydrolyzes high-mannose and the simplest form of complex oligosaccharides (Elder and Alexander, 1982), and PNGase F, a glycopeptide amidase that releases essentially all Asn-linked glycans from glycoproteins (Plummer *et al.*, 1984). Their mechanisms of action are described in Section I,C.

The limited substrate range of Endo H has been exploited in metabolic studies to probe aspects of protein glycosylation and oligosaccharide processing (Robbins *et al.*, 1977). The products of $(Glc)_3(Man)_9(GlcNAc)_2$ trimming remain sensitive to Endo H until the action of Golgi mannosidase II forms a resistant $GlcNAc(Man)_3(GlcNAc)_2$ structure (Kornfeld and Kornfeld, 1985), which can be used as an event marker in the transport of a given glycoprotein. Thus, the initial role of Endo H as an analytical tool for biochemical and structural studies on isolated glycoprotein components has been broadened by the cell biologist to include temporal and spatial aspects of glycoprotein processing, targeting, and transit (Strous and Lodish, 1980; Lodish *et al.*, 1983). Coupled with the successively broader substrate specificities of Endo F and PNGase F, these enzymes have become powerful tools in the dissection of the cell biology of glycoprotein metabolism.

C. General Features of Enzymatic Deglycosylation

1. Reaction Mechanisms

a. Endoglycosidases. All Asn-linked glycans contain an invariant inner-core pentasaccharide, (Manα1–6)Manα1–3Manβ1-4GlcNAcβ1-4GlcNAcβAsn, which is attached to the polypeptide chain via a glycosylamine linkage between the *N*-acetylglucosamine terminus and the Asn amide nitrogen. Structural diversity of these glycoconjugates is developed by the peripheral sugars attached to the core pentasaccharide during processing and maturation (Kornfeld and Kornfeld, 1985). Enzymes that hydrolyze specifically at the inner-core di-*N*-acetylchitobiose moiety comprise a broad group of glycosidic enzymes known as endo-β-*N*-acetylglucosaminidases, or more commonly, endoglycosidases. Among these, the bacterial enzymes Endo H (Tarentino and Maley, 1974), Endo D (Koide and Muramatsu, 1974), and Endo F (Elder and Alexander, 1982) are the most widely used because of their different substrate specificities, and their commercial availability. Cleavage of a susceptible Asn-linked glycan by Endo H, D, or F redistributes the core *N*-acetylglucosamine residues equally among the reaction products such that one residue remains attached to the Asn-polypeptide, while the other becomes the reducing terminus of the liberated oligosaccharide (Fig. 1). The structural determinants for endoglycosidase activity are represented by the peripheral attachments R_1 and R_2 to the trimannosyl branches of the inner core, and very subtle structural differences in this region can be differentiated by Endo H, D, or F (Section II,A–C).

The protein moiety of an Asn-linked glycan is not a determinant per se for the bacterial endoglycosidases. However, protein conformational

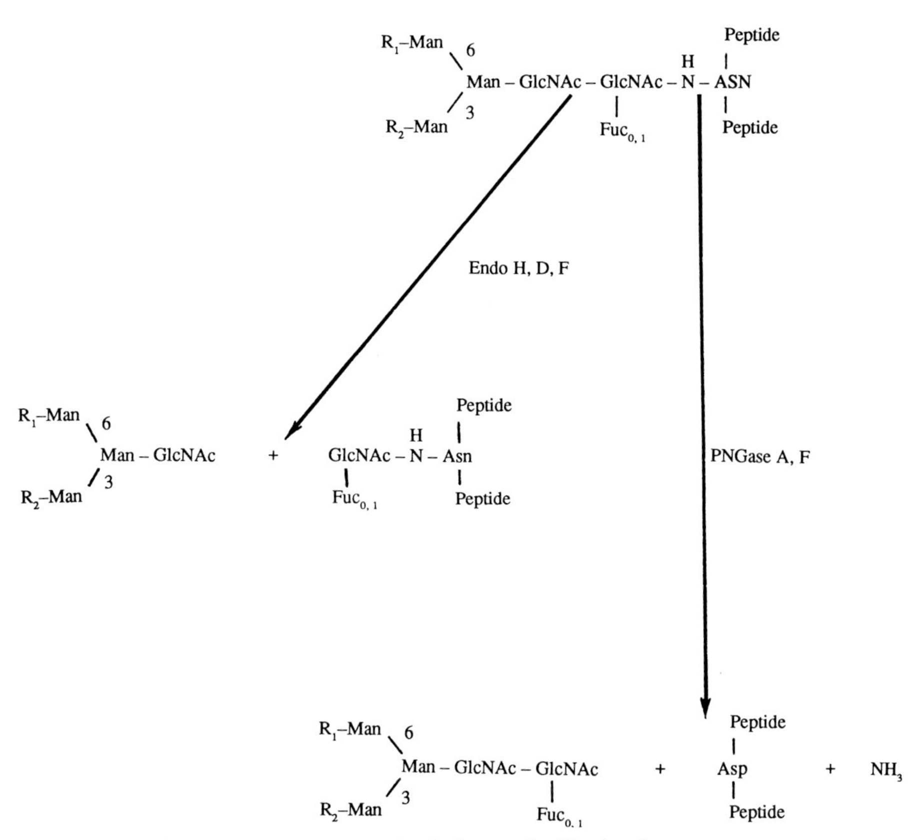

FIG. 1. Sites of hydrolysis of oligosaccharide-cleaving enzymes.

effects can influence greatly the course and extent of deglycosylation by rendering the susceptible oligosaccharides at certain glycosylation sites partially or completely "inaccessible" to endoglycosidase action (Trimble *et al.*, 1983; Ziegler *et al.*, 1988). Thus first-time users of endoglycosidases need to be aware not only of enzyme–substrate specificities, but also that the native and denatured state of a glycoprotein can present a very different oligosaccharide susceptibility picture because of the "inaccessibility" phenomena.

b. Glycoprotein–Asn Amidases. Deglycosylation of Asn-linked glycans can also be achieved by a different mechanism with another group of enzymes classified as glycoprotein/peptide-N^4-(N-acetyl-β-glucosaminyl)Asn amidases (PNGases). In this case, the reaction products are different from those generated by endoglycosidase action, since an aspartic acid residue is formed at the site of hydrolysis with the release of a 1-amino oligosaccharide (Fig. 1). The latter is converted nonenzymatically to ammonia and an oligosaccharide containing di-*N*-acetylchitobiose at the reducing end.

PNGase A (almonds) and PNGase F (*Flavobacterium meningosepticum*) are the only commercially available preparations at this time, although PNGase-type enzymes are ubiquitously distributed in plants (Plummer *et al.*, 1987). Unlike the bacterial endoglycosidases, which are restricted in oligosaccharide specificity, the PNGase-type enzymes are very broad and will cleave most Asn-linked glycans regardless of oligosaccharide class. The primary structural determinants involve recognition of the polypeptide chain in close proximity to the glycosylation site and the di-*N*-acetylchitobiose region of the oligosaccharide (Chu, 1986). As with the endoglycosidases, the same considerations regarding oligosaccharide accessibility apply to the PNGase-type enzymes.

2. Techniques for Promoting Oligosaccharide Accessibility

Various denaturants have been employed to expose otherwise inaccessible oligosaccharide cores, including ionic, nonionic, and zwitterionic detergents as well as chaotropic salts and thiols. Of these, heating in sodium dodecyl sulfate (SDS) appears to be the simplest and most effective method for unfolding glycoproteins. Tritons, CHAPS, and octylglucoside have been reported to enhance the ability of oligosaccharide-cleaving enzymes to deglycosylate certain glycoproteins in metabolically labeled cell extracts. However, this may be more an effect of solubilizing an otherwise susceptible but cryptic glycoprotein product from a membranous matrix rather than specifically unfolding the protein of interest to expose the oligosaccharides. For an initial trial with any of the oligosaccharide-cleaving enzymes, treatment of paired samples, one of which is boiled in SDS, is the recommended approach. Addition of thiols may be required to promote complete accessibility by reducing disulfide bridges.

Boiling in SDS radically alters tertiary and quaternary structure resulting in the loss of biological activity in many glycoproteins, only a few of which have been reversed by removal of the SDS. Thus, under certain circumstances it might be possible to perturb a glycoprotein's structure sufficiently with chaotropic salts to allow access to oligosaccharide-cleaving enzymes without loss of biological activity. This has been accomplished by treating bovine thyroglobulin with Endo H in the presence of 0.5 *M* NaSCN. All of the high-mannose oligosaccharides were rapidly removed (Trimble and Maley, 1984), and following desalting, the thyroglobulin was shown to retain full activity by bioassay (unpublished experiments). Other chaotropic salts such as guanidinium-HCl, sodium perchlorate, or urea may be helpful in this regard. It should be noted that a 50% decrease in the viscosity of F-actin requires 0.26 *M* KSCN or 3.4 *M* urea, which serves to relate their relative denaturing potential (Nagy and Jencks, 1965).

Finally, substrate glycoproteins may be denatured by reduction and either carboxymethylation or—to provide a more soluble product—aminoethylation. However, many glycoproteins so treated become insoluble as their oligosaccharides are progressively removed by the oligosaccharide-cleaving enzymes, which may prevent the reaction from going to completion. A low level of SDS can be included with reduced alkylated glycoproteins to maintain their solubility.

The time course for enzymatic deglycosylation of a large-scale preparative reaction can be monitored by assaying small aliquots for the release of soluble carbohydrate after precipitating the substrate glycoprotein. The phenol sulfuric acid assay for neutral hexose (Dubois *et al.*, 1956) and the thiobarbituric acid chromophore assay for sialic acid (Skoza and Mohas, 1976) are sensitive methods to be considered. Precipitation of the substrate glycoprotein with cold 10% (w/v) trichloroacetic acid (TCA) is usually successful (Trimble and Maley, 1984), unless the glycoproteins have been denatured by previous boiling in SDS. Precipitation of proteins complexed with SDS may be accomplished by addition of potassium salts to 0.2 *M* in the presence of 10% (w/v) TCA. Alternatively, addition of two volumes of cold methanol or acetone followed by incubation at −20°C for 1 hour prior to centrifugation in the cold may be useful methods. Phosphotungstic–TCA mixtures are also potent protein precipitants. Any method chosen should be verified to precipitate all measurable carbohydrate in a test aliquot before adding the carbohydrate-cleaving enzyme.

3. Applications

This section will outline briefly some of the ways the oligosaccharide-cleaving enzymes can be applied to problems in cell biology. Typically, for these enzymes to be effective they must be used in conjunction with highly sensitive analytical techniques for characterizing the carbohydrate and/or protein moieties. The experimental approach is generally dictated by the system to be studied, the information desired, and the quantity of sample.

a. General Considerations. A glycoprotein of unknown carbohydrate class (N- versus O-linked) and type (high-mannose versus complex, etc.) can be tested for sensitivity to one of the endoglycosidases, H, D, or F, or PNGase A or F, either separately or sequentially. The simplest method of determining whether deglycosylation has occurred is by sodium dodecyl sulfate–polyacrylamide gel electrophoresis (SDS–PAGE) of control and enzyme-treated material. A significant decrease in molecular weight (i.e., increase in electrophoretic mobility) is a good indication that Asn-linked oligosaccharides have been enzymatically released. However, confirmation should be obtained whenever possible by lectin-binding experiments performed directly on a Coomassie blue-stained gel (Chu *et al.,* 1981), or after Western blotting (Glass *et al.,* 1981), because changes in electrophoretic mobility are relative to initial molecular weights and number and types of oligosaccharides, and results sometimes can be deceptive. For instance, removal of the single high-mannose chain from ribonuclease B (MW 15,000) with Endo H or PNGase F results in deglycosylated proteins that are separated from each other and the parent protein on a standard 12.5% gel. Deglycosylation of the high-mannose oligosaccharide of egg albumin (MW 46,000) with Endo H or PNGase F, however, results in a very small decrease in molecular weight, and the deglycosylated products are not separable under these conditions. Indeed, for glycoproteins of very high molecular weight (>100,000) and few Asn-linked oligosaccharides, complete deglycosylation may not substantially alter the molecular weight, and lectin-binding experiments are essential to detect glycosylation.

The observed decrease in molecular weight following an enzyme digestion is sometimes used as a first approximation of the number of oligosaccharide chains in a glycoprotein. Such estimations should be regarded with caution unless it can be established that no resistant Asn-linked glycans remain after enzyme treatment. PNGase F, which has the broadest oligosaccharide specificity known, would be the most likely choice for such studies, but some unidentified carbohydrate structures resistant even to this enzyme have been reported. Furthermore, glycoproteins with many oligosaccharide chains may undergo a net increase in negative charge following PNGase F digestion (i.e., invertase) resulting in anomalous electrophoretic behavior (Chu, 1986).

The presence of O-linked sugars must be excluded, since these would adversely affect estimations of the number of oligosaccharides in a glycoprotein. A further decrease in molecular weight following sequential digestion of a PNGase F-treated glycoprotein with neuraminidase and O-glycanase would strongly imply the presence of O-linked sugars. Alternatively, a limit molecular weight for carbohydrate-free material can be obtained by chemical deglycosylation with trifluoromethanesulfonic acid (TFMS) (Edge *et al.*, 1981).

Sequential enzyme digestion is an effective method for determining whether more than one type of Asn-linked glycan is present in the same protein. Using a Waldenströms immunoglobulin M, one can show on SDS–PAGE a progressive decrease in molecular weight and corresponding loss of lectin binding [concanavalin A (Con A), wheat germ agglutinin] upon sequential digestion with Endo H, Endo F, and PNGase F (Fig. 2). The purified oligosaccharides isolated at each step from scaled-up reactions were used to verify by direct analytical characterization that Endo H, Endo F, and PNGase F released the $C_4 + C_5$ high-mannose chains, the C-1 complex biantennary chain, and the $C_2 + C_3$ complex triantennary chains, respectively, from immunoglobulin M.

Inferences regarding the kinds of oligosaccharides in Asn-linked glycans should be made cautiously when based only on known enzyme specificities. Thus, while susceptibility to Endo H is likely to indicate the presence of high-mannose structures, such oligosaccharides might also be hybrids, one of several "bisected" ovalbumin types, or contain secondary posttranslational modifications. Similarly, PNGase F can release a variety of oligosaccharide types (high-mannose, hybrids, polylactosamine, polysialic acid, etc.) including sulfated and phosphorylated glycans. Several techniques are available for qualitative characterization and structural analysis of oligosaccharides, and these should be used whenever possible to characterize the enzymatically cleaved products.

It is always a good policy to check the activity of commercial enzyme preparations by including a positive control in an experimental protocol,

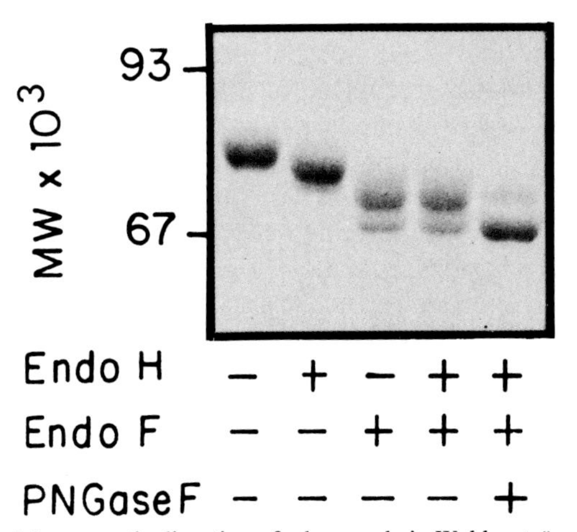

FIG. 2. Sequential enzymatic digestion of a heavy-chain Waldenströms immunoglobulin M (Ga). Reduced aminoethylated heavy chain (30 μg) was digested sequentially (37°C, 18 hours) with Endo H, Endo F, and PNGase F (20 mU/ml) as indicated in the figure key. Aliquots (3μg) were subjected to 10% SDS–PAGE.

especially if a negative result was obtained in preliminary studies. RNase B is an excellent test substrate for the enzymes listed in Section II, because only 0.6 m U enzyme/ml is needed for deglycosylation. It is commercially available (Sigma) but has to be purified on Con A–Sepharose (Baynes and Wold, 1976). Direct fluorimetric assay for oligosaccharide-cleaving activity at the start and end of an incubation can also be done using Resorufin-*N*-glycopeptide (Boehringer Mannheim). This is a derivatized $(Man)_6(GlcNAc)_2Asn$, which is suitable for endoglycosidases H and F or PNGase-type enzymes.

b. Identification and Structural Analysis of Oligosaccharides. Structural analysis can be conducted on micro amounts of starting material by derivatizing purified, enzymatically released oligosaccharides with a fluorescent label and characterizing the components by high-performance liquid chromatography (HLPC) (Takahashi *et al.*, 1987). Alternatively, oligosaccharides can be radiolabeled with NaB^3H_4 and fractionated by size, charge, and linkage into homogeneous species followed by structural analysis using lectin affinity techniques (Green *et al.*, 1987b). When larger amounts of oligosaccharides can be obtained, lectin affinity studies can be supplemented with direct analytical characterization including chemical analysis (methylation, acetolysis), high-field ^1H-NMR, and FAB–MS. The utility of this approach is illustrated by structural analysis of the Endo H-release oligosaccharides of invertase, which provided a clear processing pathway of single-step mannose additions from $Man_8GlcNAc_2$ to $Man_{13}GlcNAc_2$ (Trimble and Atkinson, 1986).

For proteins containing multiple Asn-linked oligosaccharides, chemical and/or proteolytic fragmentation followed by HPLC is recommended for production of individual glycosylation sites (Frutiger *et al.*, 1988; Ziegler *et al.*, 1988). This approach allows one to study normal site-specific oligosaccharide processing, and potential changes during differentiation and development, viral transformation, and host cell-dependent glycosylation (Sweidler *et al.*, 1985). In addition, the amount of enzyme required for removing oligosaccharides from defined glycopeptides is usually substantially less than that needed for intact proteins, and a significant cost savings is achieved.

c. Identification of Glycosylation Sites. In addition to providing oligosaccharides for structural analysis, enzymatic deglycosylation can be used to determine the exact location of all glycosylation sites in a protein, as well as the fidelity of glycosylation at each site (Carr and Roberts, 1986; Reddy *et al.*, 1988). The principle is based on conversion of the glycosylation-site Asn(s) in a protein to either an aspartic acid residue(s) by PNGase F or an asparaginyl-*N*- acetylglucosamine (Asn-GlcNAc) residue(s) by Endo H or F. Depending on the oligosaccharide-cleaving enzyme used, proteolytic fragmentation gives the corresponding Asp-peptide(s) or AsnGlcNAc-peptide(s), which can be resolved from the digest by HPLC. These peptides are 1 and 203 Da greater, respectively, in molecular mass than the nonglycosylated Asn-peptide(s) deduced from the DNA sequence, and each can be identified by FAB–MS. Peptide microsequencing techniques can then be used to establish the sequon attachment site(s) from the purified Asp-peptide(s) or Asn-GlcNAc-peptide(s).

High-resolution HPLC analysis of proteolytic digests of glycosylated and deglycosylated protein amount to a "difference spectrum" that has many applications in cell and molecular biology. For example, using this approach it has been shown that the acquisition of virulence by the chicken/Pennsylvania 83 influenza virus was associated with a deletion mutation in the hemagglutinin of the avirulent strain which eliminated a glycosylation site necessary for normal processing and transport of this viral surface glycoprotein (Deshpande *et al.*, 1987).

d. Biosynthetic/Molecular Studies. Endoglycosidases can be used to study the biosynthesis, processing, and secretion of Asn-linked glycans as well as the physiological role of the carbohydrate moiety. Such experiments rely on high-density isotopic labeling with [^{35}S]methionine or 2-[^{3}H]mannose, and purification of the radiolabeled glycoprotein from cell lysates or medium by immunoprecipitation or affinity chromatography. Enzyme digests are analyzed by SDS–PAGE and fluorography, and susceptibility to deglycosylation is determined by an increase in electrophoretic mobility for a [^{35}S]methionine-labeled glycoprotein and disappearance of a radioactive band when the carbohydrate moiety is labeled with 2-[^{3}H]mannose.

During glycoprotein biosynthesis the oligosaccharide moiety undergoes extensive processing at specific intracellular locations and the deglycosylating enzymes can be used in a pulse–chase format to distinguish these site-specific structural modifications. Endo H is a specific probe for cotranslational glycosylation in the RER where attachment of enzyme-sensitive, Asn-linked high-mannose oligosaccharides occurs (Kornfeld and Kornfeld, 1985). The acquisition of resistance to Endo H corresponds to the intracellular transport time required for these glycosylated polypeptides to move from the RER to elements of the Golgi, where processing reactions convert them to Endo H-resistant, complex chains. Endo H is being used as a probe for studying the role of the carbohydrate moiety in directing the proper folding of the polypeptide chain, and in forming the appropriate disulfide bonds and subunit interactions prior to secretion (Machamer and Rose, 1988; Copeland *et al.*, 1988). Such experiments have employed native glycoproteins and those whose glycosylation sites have been altered through molecular techniques.

Complex Endo H-resistant Asn-linked oligosaccharides are sensitive to PNGase F throughout the entire maturation process, as are those glycans containing specialized peripheral additions that lead to more diversified structures (polylactosamine, polysialic acid, sulfation, phosphorylation, etc.). Endo F is similar in specificity to Endo H, except it also hydrolyzes complex biantennary oligosaccharides. This enzyme should become a useful probe for defining the parameters involved in differential processing of complex biantennary to triantennary and tetraantennary oligosaccharides.

II. Oligosaccharide-Cleaving Enzymes

A. Endo H

1. Substrate Specificity

Endo H was initially thought to be specific for only high-mannose oligosaccharides and that addition of α-1,6-linked fucose to the Asn-proximal GlcNAc during formation of complex oligosaccharides was the structural determinant that eliminated these compounds as substrates (Tarentino and Maley, 1975). Later studies show, however, that hybrid oligosaccharides formed in the presence of the mannosidase II inhibitor, swainsonine (Elbein *et al.,* 1981; Elbein, 1987), which have a high-mannose upper arm (Fig. 1, R_1), an acidic lower arm (R_2), and a fucosylated core are sensitive to Endo H, while nonfucosylated biantennary structures are not. In combination with earlier work on a series of structurally defined oligosaccharides generated from ovalbumin $Man_6GlcNAc_2Asn$ by partial digestion with jack bean α-mannosidase (Trimble *et al.,* 1978), these results suggest that the structural determinant conferring sensitivity to Endo H is the α-1,3-linked mannose in the linear pentasaccharide Manα1,3Man-α1,6Manβ1,4$GlcNAc_2$Asn. By contrast, the trimannosyl core found in complex oligosaccharides, (Manα1,6)Manα1,3Manβ1,4$GlcNAc_2$Asn, is poor substrate for Endo H (Table I). Addition of fucose to the branched trimannosyl core eliminates it as an Endo H substrate (Tarentino and Maley, 1975). Therefore, all high-mannose or hybrid oligosaccharide intermediates from the time $Glc_3Man_9GlcNAc_2$ is added to proteins in the RER until processed by mannosidase II in the Golgi compartment to remove

TABLE I

COMPARATIVE RATES OF HYDROLYSIS FOR ENDO F AND ENDO H

R_1-Man, R_2-Man >Man-GlcNAc-GlcNAc-Asn[^{3}H]dns		Hydrolysis rate (U/mg)[a]	
R_1	R_2 (substituent)	Endo F	Endo H
(Manα1-6)Manα1-3-	Manα1–3	43	42
H	H	0.04	10^{-3}
SA-Gal-GlcNAcβ1-2-	SA-Gal-GlcNAcβ1-2-	0.15	10^{-5}
SA-Gal-GlcNAcβ1-2-	(SA-Gal-GlcNAcβ1-4)SA-Gal-GlcNAcβ1-2-	0	0

[a]Hydrolysis was performed at 37°C at a substrate concentration of 0.5 m*M*.

the α-3-linked mannose structural determinant should be good substrates for Endo H.

2. Commercial Availability

Endo H (EC 3.2.1.96) activity originally was discovered in a commercial chitinase preparation produced by a bacterium identified as *Streptomyces griseus* (Tarentino and Maley, 1974). Later, however, the culture was more correctly identified as *Streptomyces plicatus* and is on deposit with the American Type Culture Collection (ATCC no. 27800). The utility of Endo H in glycoprotein research justified attempts to increase the availability of this enzyme beyond the 2–3 U/liter secreted by *S. plicatus* into cultural filtrates. Accordingly, the gene was isolated, introduced into pKC30, an *Escherichia coli* plasmid with a *ts* λ repressor, and the 200-fold-amplified gene product was isolated from induced *E. coli* cells (Trumbly *et al.*, 1985). The cloning of Endo H into *E. coli* not only provided a higher level of enzyme, but also allowed a purification scheme that virtually eliminated protease contamination. This product is commercially available (Table II). The Endo H gene has also been cloned into *Streptomyces lividans* and the purified enzyme is commercially available as well (Table II). It is recommended that a supplier be chosen who provides enzyme (cloned or not) that has been certified as to the contaminant level of protease. This can be an important factor in evaluating the results from experiments in which mobility changes in the molecule of interest serves as a measure of endoglycosidase sensitivity.

TABLE II

COMMERCIAL SOURCES OF OLIGOSACCHARIDE-CLEAVING ENZYMES[a]

Suppliers	Oligosaccharide-cleaving enzymes						
	PHGase A	PNGase F	Endo F	Endo F PNGase F	Endo D	Endo H	Endo H cloned
Genzyme Inc.		10 u[b]	15 mU				0.2 U
		50 u					2.0 U
Boehringer		20 u[b]	3 u[c]	6 u[c]	0.1 U	0.1 U	0.1 U
Mannheim		100 u					0.5 U
Calbiochem					0.1 U		0.1 U
					0.5 U		0.5 U
Sigma		10 u[b]	3 u[c]	6 u[c]	0.1 U	0.1 U	
		20 u					0.5 U
ICN Biomedicals	1 mU				0.1 U		0.2 U
					0.5 U		2.0 U
New England Nuclear (NEN)				10 u[c]		6 μg	

[a] Enzyme activities are listed as milliunits (mU) or units (U) representing cleavage of substrate to product of 1 nmol or 1 μmol, respectively, per minute at 37°C. The term unit (u), is *not* an international unit and is defined for each enzyme.

[b] Units of PNGase F supplied by all companies are in fact milliunits.

[c] Unit designations are per hour and not per minute and should be divided by 60 to correspond to international units.

3. General Properties

Endo H is the most stable of the oligosaccharide-cleaving enzymes studied to date. It is extremely resistant to proteolysis (Tarentino and Maley, 1974) and even retains some activity after boiling (Trumbly *et al.*, 1985), which probably reflects the absence of cysteine in its amino acid composition. Endo H can be stored either as a lyophilized powder or in solution and is stable frozen or at 0–4°C indefinitely. Precautions should be taken to prevent microbial growth in enzyme stocks, by freezing, inclusion of 5 m*M* sodium azide, or sterile filtration. Many suppliers provide sterile enzyme, which is to be preferred. The enzyme may be freeze-thawed repeatedly with no loss of activity and need not be stored with protective agents such as albumin, sucrose, or glycerol. In fact it is advisable to avoid glycerol, because during oligosaccharide hydrolysis, Endo H as well as Endo F can add glycerol to the reducing end of the released oligosaccharide. Depending on the experimental protocol, this can be problematic, because the presence of the additional three-carbon alcohol increases the apparent size of the oligosaccharides on Bio-Gel P4 by an amount equal to a GlcNAc or two neutral hexose residues and eliminates reduction by $NaBH_4$ (Trimble *et al.*, 1986).

Endo H at concentrations below 5–10 μg/ml have a tendency to adsorb to glass walls, so silanized-glass or plastic vials are preferred for storage. The enzyme is stable above pH 5 and exhibits a broad pH optimum at ~5.5 with 50% activity retained at pH 7.5. The enzyme's mass is ~30 kDa and specific activity depends on the substrate. With $(Man)_5(GlcNAc)_2Asn$, 42–45 U/mg at 37°C is commonly obtained with homogeneous cloned Endo H.

Endo H retains full activity in 0.5 *M* NaSCN, guanidinium-HCl, $NaClO_4$, and 4*M* urea, which may be sufficient to expose inaccessible oligosaccharide substrates as discussed earlier. We have not systematically evaluated Endo H stability at higher salt concentrations, but have verified that activity is retained for several hours in 1 *M* NaSCN and 6 *M* urea. Therefore, should higher levels of these salts be used in an effort structurally to perturb substrate glycoproteins, it would be prudent to verify that Endo H retains sufficient activity for the hydrolysis reactions to reach end point. As a final note on chaotropic salts, guanidinium-HCl has the undesirable property of forming an insoluble precipitate (guanidinium dodecyl sulfate) in the presence of SDS, so its use in deglycosylation protocols that subsequently employ SDS–PAGE is limited.

As already indicated (see Section I,C,2), SDS has been found to be the most effective and facile glycoprotein denaturant to potentiate oligosaccharide-cleaving enzymes. Endo H is quite stable to SDS, despite some controversy generated by our studies that suggested otherwise (Trimble and Maley, 1984). The SDS used in that study has since been found to be a mixture containing a high proportion of longer chain fatty acid sulfates, and it is these compounds that appear to inactivate the Endo H. Therefore a certified "electrophoresis grade" of SDS should be used for glycoprotein denaturation.

As a general rule, SDS is most effective at about a 1.2-fold weight excess over the protein content in the sample of interest. For very low protein concentrations, addition of SDS at 0.05% is sufficient. Though optional for Endo H, excess SDS can be complexed if desired by addition of bovine serum albumin (BSA), sequestered in mixed micells with a 2- to 3-fold weight excess of Triton X-100 or Nonidet P-40 (NP-40), precipitated at 0°C overnight, or partially removed by dialysis against a buffer such as 50 m*M* sodium phosphate, pH 6.5. In time course work it may be desirable to inactivate Endo H. This is somewhat problematic because of its stability, but boiling in SDS (e.g., Laemmli sample buffer: Laemmli, 1970) followed by storage below −20°C appears effective.

Finally, although Endo H is extremely active, there are occasions where a threshold level of enzyme must be added before deglycosylation is seen, despite the fact that the full activity added to a hydrolysis reaction can be assayed with a test substrate. This appears to be due to sequestration of enzyme by components present in certain experimental protocols and generally is overcome by adding of Endo H at or above 20 mU/ml.

4. Deglycosylation Protocol

The actual protocol employed will depend on the researcher's goal. Experimental applications can be diverse, ranging from hydrolysis of bulk, purified glycoproteins for isolation of susceptible oligosaccharides to determining when during secretion a glycoprotein's oligosaccharides become Endo H-resistant. In general, such types of experiments can be performed by simply scaling a "generic" reaction mixture up to suit the conditions.

The first question to be asked is whether the experiment will differentiate between accessible and inaccessible Endo H-sensitive oligosaccharides. If this is desired, then a pair of samples should be prepared, one of which is denatured. If maintenance of biological activity is not a concern, SDS is the denaturant of choice; otherwise, chaotropic salts may be useful (Section II,A,3.).

For denaturation with SDS, the sample containing a suitable amount of label (^{35}S, ^{32}P, ^{3}H, ^{14}C), protein, or carbohydrate for the detection method to be employed (e.g., autoradiography, Coomassie stain or Western blotting of gels, direct measurement of released carbohydrate by scintillation counting, phenol sulfuric acid or fluorescent derivative assays) is heated at 100°C for 3–5 minutes in the presence of 0.1 *M* β-mercaptoethanol and SDS at 0.05% or a 1.2-fold weight excess over protein, whichever is higher. The sample is then treated with Endo H by assembling reactions in the following order:

Addition (in microliters)	1	2
Sample (prepared as before)	10	10
0.5 *M* Sodium citrate, pH 5.5	5	5
H_2O	33	31
10% Phenylmethylsulfonyl fluoride (PMSF) in isopropanol	2	2
Endo H (0.5 U/ml)	—	2
	50	50

Incubation is at 30°C or 37°C overnight. Longer incubations should contain a thymol crystal or small drop of toluene to prevent microbial growth. After hydrolysis, the sample can be worked up for the method chosen to evaluate the extent of deglycosylation. In the generalized protocol just given, any amount of the water can be replaced with additional protease inhibitors, or a chaotropic salt if the sample was not initially denatured in SDS.

B. Endo D

1. Substrate Specificity

Endo D has the most restricted substrate specificity of all the endoglycosidases. The key structural determinant for enzyme activity is the lower Manα1–3 branch of the core pentasaccharide (Fig. 1). When this residue is unsubstituted, or contains an R_2 substitution in the 4-position, the oligosaccharide is Endo D-sensitive (Mizuochi *et al.*, 1984). R_2 substitutions in the 2-position render all Asn-linked glycans resistant to Endo D. Thus $(Man)_3(GlcNAc)_2(Fuc)_{0,1}$Asn and $(Man)_5(GlcNAc)_2$Asn are substrates for Endo D, but not $(Man)_6(GlcNAc)_2$Asn or higher Asn-oligomannosides. Since most complex oligosaccharides contain an *N*-acetylglucosamine residue attached β1–2 to the key determinant, they, too, are Endo D-resistant.

Endo D can be used in conjunction with exoglycosidases (neuraminidase, β-galactosidase, β-*N*-acetylglucosaminidase) to release complex oligosaccharide cores from intact glycoproteins or glycopeptides (Muramatsu, 1971). However, since the introduction of Endo F and PNGase F, its main use now is limited to discriminating subtle structural differences in oligosaccharides.

2. Commercial Availability and General Properties

Endo D is a high molecular weight protein (280,000) purified from the cultural filtrate of *Diplococcus pneumoniae* (Koide and Muramatsu, 1974). The enzyme has a pH optimum of 6.5 and is stable at 37°C in the presence of 0.025% BSA. The enzyme is available from numerous sources. It is supplied as a lyophilized powder prestabilized with BSA, and is certified to be free of protease. All commercial Endo D preparations have <1% exo-β-*N*-acetylglucosaminidase activity.

3. Protocol for Oligosaccharide Susceptibility

Oligosaccharides released from proteins by chemical methods (Takasaki *et al.*, 1982) or enzymatically with PNGase F can be radiolabeled and purified as described in Section I,C,3,b. For enzyme digestions, reduced oligosaccharides (1–5 $\times$ 10^5 cpm) are incubated at 37°C with Endo D (10–100 mU) in 0.2 *M* citrate-phosphate buffer, pH 6.0 (reaction volume, 20–40 μl). Susceptibility to Endo D is determined by monitoring the release of *N*-acetyl-[^{3}H]glucosaminitol by chromatography on Bio-Gel P4.

C. Endo F

1. Substrate Specificity

Susceptibility to Endo F is determined by R_1 and R_2 attached to the pentasaccharide core (Fig. 1). The enzyme hydrolyzes high-mannose and some complex Asn-linked glycans (Elder and Alexander, 1982), but at greatly different rates (Table I). High-mannose glycans are the preferred substrates for Endo F, and structures ranging from linear $(Man)_3(GlcNAc)_2$Asn to $(Man)_{>25}(GlcNAc)_2$Asn are hydrolyzed. A Manα1-3 substituent at the R_1 position, as in $(Man)_5(GlcNAc)_2$Asn (Table I) confers susceptibility to Endo F as well as to Endo H. This is an absolute requirement for Endo H, and oligosaccharides that lack this determinant, such as the branched trimannosyl pentasaccharide or a complex biantennary chain, are not hydrolyzed at an appreciable rate. This determinant is somewhat more relaxed for Endo F, and these oligosaccharides are hydrolyzed, but at a rate 300- to 400-fold slower than $(Man)_5(GlcNAc)_2$Asn. Sulfated biantennary oligosaccharides are also hydrolyzed by Endo F (Green *et al.*, 1987a), as are some bisected ovalbumin hybrids. The degree of substitution by R_2 is also a determinant for Endo F activity. Bisected ovalbumin hybrids, ovomucoid hybrids, and triantennary and tetraantennary oligosaccharides, which are all di-substituted on R_2 with *N*-acetylglucosamine, are not hydrolyzed by Endo F. The determinant that appears to eliminate oligosaccharides as substrates for Endo F is a β-1,4 substitution on R_2 (Tarentino *et al.*, 1985; Tarentino and Plummer, 1987).

Endo F, like Endo H, is active on intact glycoproteins, glycopeptides, Asn-oligosaccharides, and free oligosaccharides with di-*N*-acetylchitobiose on the reducing end.

2. Commercial Availability

a. Endo F. Endo F is isolated from the cultural filtrate of *Flavobacterium meningosepticum* (ATCC no. 33958). The pure enzyme is available from several sources (Table II), and depending on the supplier, is sold as a ready-for-use solution (Boehringer Mannheim; 3 units, equals 50 mU) or a lyophilized powder (Genzyme; 15 mU and Sigma; 3 units, equals 50 mU). All preparations of pure Endo F perform equally well, and can be stored for months at 4°C where the enzyme is very stable, or aliquoted and stored at −70°C. The enzyme should *not* be subjected to freeze–thaw cycles.

Endo F from all suppliers contains a trace metalloprotease. To prevent proteolysis of glycoprotein substrates, some suppliers include 50 m*M* EDTA in their enzyme preparations, while others recommend addition of 10–20 m*M* 1,10-*o*-phenanthroline, if necessary, directly to the enzyme digest.

b. Endo F–PNGase F. The crude mixture was the first commercially available preparation and was called Endo F before the more potent PNGase F activity was discovered. Since the introduction of pure Endo F and PNGase F (Section II,D), it is not as widely used. The mixed-enzyme preparation is sold in quantities of 6 units (100 mU) or 10 units (167 mU) under the ambiguous name Endo F (grade II) or worse, Endo F. The crude mixture can contain up to twice as much PNGase F as Endo F, which, at pH 6.2 (the pH suggested by the supplier) exhibits ~65% of its maximal activity seen at pH 8.6. Under these conditions, and in view of the specificity differences of the two enzymes, use of this preparation gives undefined deglycosylation products and therefore a low informational return. It is recommended by suppliers for "those who wish to get rid of the carbohydrate," but compared to the cost of pure Endo F or pure PNGase F, and the value of a clearly defined system, it is not cost effective. One Endo F–PNGase preparation (NEM) is supplied in 50%

glycerol, which can add to the reducing end of released oligosaccharides (Trimble *et al.*, 1986), further complicating structural analysis of a heterogeneous mixture (see Section II,A,3).

3. General Properties

Endo F (M_r 32,000) has a broad pH optimum between pH 4 and 6. It is 60–70% active at pH 7.0. The enzyme is used at pH 6, where it is stable at 37°C for 24 hours at moderate enzyme concentrations (0.5 mU/ml) without detergents (SDS–NP-40) present. In very dilute solutions (0.01–0.05 mU/ml), 0.1% BSA must be present to prevent rapid loss of activity.

Endo F is inhibited by SDS, and reactions with denatured proteins must contain a 6- to 10-fold excess (w/w) of NP-40 to protect the enzyme. The enzyme is very stable in β-mercaptoethanol–NP-40 and the detergents CHAPS and octylglucoside. These reagents have been used to promote oligosaccharide accessibility in some proteins.

Endo F is completely stable in 10–20 m*M* 1,10-*o*-phenanthroline or 50–75 m*M* EDTA. It is not affected by protease inhibitors such as 10 m*M* diisopropylfluorophosphate (DFP) or PMSF, 100 μ*M* leupeptin, and 625 KI units Trasylol/ml, and these may be used if trace proteolytic activity is present in the glycoprotein substrate.

4. Protocol for Deglycosylation

The following protocol can be used to release Endo F-susceptible oligosaccharides from glycoprotein samples. Because of the high cost of oligosaccharide-cleaving enzymes, small reactions are important for increasing enzyme concentrations to the threshold levels necessary for complete deglycosylation. The protein is boiled for 3–5 minutes in 0.5% SDS–0.1 *M* β-mercaptoethanol to expose fully all glycosylation sites. If the glycoprotein is unlabeled, a concentration of 2 mg/ml is a convenient working stock if samples are analyzed by SDS–PAGE and visualized by Coomassie blue staining. For purified, metabolically labeled proteins, sufficient radioactivity should be used per well (^{35}S, 200–300 dpm; ^{3}H, 1000–2000 dpm) for clear identification by SDS–PAGE and fluorography. Reagents (in microliters) are added in the order indicated:

Sample	10
0.5 *M* Sodium acetate, pH 6.0	10
0.1 *M* 1,10-*o*-Phenanthroline in methanol	3
10% NP-40	5
Endo F (15–50 mU/ml)	2
	30

The reactions are performed at 30°C instead of 37°C because some loss of Endo F activity occurs at the higher temperature even with a 10 : 1 weight excess of NP-40 to SDS. Reactions at 30°C do not adversely affect the performance of the enzyme.

The amount of enzyme needed for complete deglycosylation and the time of incubation must be empirically determined for each protein. If a protein ladder is generated in an 18-hour reaction, it is a good indication of partial deglycosylation due to insufficient Endo F. Typically, 1–5 mU/ml should be sufficient to remove all Endo F-susceptible oligosaccharides and convert the starting material to a single sharp band on SDS–PAGE. If more Endo F is needed, increase the protocol buffer to 1 *M* and add 5 μl to the reaction. This allows for as much as an additional 5 μl of enzyme. The considerations in Section I,C,1,a should be taken into account when interpreting results.

D. PNGase F

1. Substrate Specificity

The major determinants for cleavage by PNGase F [EC 3.5.1.52, peptide-N^4-(*N*-acetyl-β-glucosaminyl)asparagine amidase F; other names, glycopeptidase F, *N*-Glycanase (Genzyme Inc.); trivial name, peptide : *N*-glycosidase F] are the polypeptide chain in close proximity to the glycosylation site and, minimally, the di-*N*-acetylchitobiose moiety of the oligosaccharide chain. Both the Asn amino and carboxyl functions must be in peptide linkage for efficient hydrolysis to occur. The oligosaccharide type has little effect on hydrolysis except for small differences in rate with triantennary > biantennary > high-mannose > tetraantennary. More highly processed glycans are also good substrates for PNGase F. These include Asn-linked polysialyl complex oligosaccharides (McCoy and Troy, 1987), as well as direct additions to the trimannosyl pentasaccharide (Fig. 1) of polylactosamine chains (R = Galβ1,4GlcNAcβ1,3)$_n$ (Morrison *et al.*, 1986) or sulfate (R = SO_4-4-GalNAcβ1,4GlcNAcβ1,2) (Green and Baenziger, 1988).

The effects of posttranslational modifications of amino acids are unknown at this time, but the presence of phosphorylated derivatives of Tyr or Ser/Thr in the vicinity of the Asn-oligosaccharide may influence binding and hydrolysis by PNGase F.

Glycopeptides as well as glycoproteins are hydrolyzed by PNGase F. An investigator must take precautions in glycopeptide production to avoid misinterpretation of data. Pronase, which was commonly used with endoglycosidase digestion in the 1970s, is not recommended with

PNGase-type enzymes because of its ability to produce undesirable free amino or carboxyl groups on the glycosylated Asn. Trypsin, chymotrypsin, or thermolysin produce better glycopeptides for PNGase F than pronase. Even with these enzymes, and possibly the newer endoproteinase-Lys, -Arg, or -Glu, the mode of binding of the protease must be considered if a resistant glycopeptide is observed. For example, the theoretical glycopeptide R-Lys-Asn(CHO)-R would be cleaved by trypsin to R-LysCOOH + NH_2Asn(CHO)-R. The glycopeptide R-Tyr-Asn(CHO)-Leu-R would be cleaved by thermolysin to R-COOH + NH_2Tyr-Asn-(CHO)COOH + NH_2Leu-R or by chymotrypsin to R-TyrCOOH + NH_2Asn(CHO)-LeuCOOH + NH_2R. The hypothetical tryptic cleavage has been observed in IgM, and the chymotryptic and thermolytic cleavages in ovalbumin. All three of these glycopeptides are nearly resistant to cleavage by PNGase F, even though they contain susceptible oligosaccharides. Therefore, any glycopeptide found resistant to PNGase F should be tested with a second protease of different specificity before conclusions about the oligosaccharide are made.

2. Commercial Availability

PNGase F is available from three sources as a homogeneous preparation supplied in solution at pH 7.0–7.2 (Table II). *N*-Glycanase is prepared in 50% glycerol at 250 mU/ml versus 200 mU/ml from the other suppliers. Note that all suppliers quantitate by dansyl-fetuin glycopeptide assays but use an improper designation of units instead of mU. The enzyme performs equally well from all commercial sources. The PNGase F preparations are essentially free of Endo F activity, and no deleterious effects are caused by the use of glycerol-containing enzyme. All commercial preparations contain a trace protease that in rare instances will degrade glycoproteins unless inhibited by 10 m*M* 1,10-*o*-phenanthroline (Tarentino and Plummer, 1987).

3. General Properties

PNGase F has been purified to homogeneity, and has a molecular weight of 35,500. The pH optimum is 8.6, but the enzyme retains 60% of its activity at pH 6.5 or 9.5. Furthermore, the enzyme retains 10% of its activity at pH 5.0 and 10.0, and is stable at these pH extremes. Since the enzyme is expensive, it is preferable to do digestions near the pH optimum (8.6), but any reactions between pH 6.0 and 9.5 should be successful.

Commercial preparations of PNGase F can be stored at 4°C for at least

6 months or can be frozen at −70°C indefinitely. The enzyme should not be subjected to repeated freeze–thaw cycles. Stock dilutions of ≤0.1 mU/ml should be made in 0.1% BSA to reduce nonspecific adsorption on glass and plastic surfaces.

PNGase F activity is stable to a wide variety of reagents and conditions. It is not inhibited by stabilizing reagents such as glycerol and sodium azide, by 100 m*M* β-mercaptoethanol, by chelators such as EDTA (50–75 m*M*) or 1,10-*o*-phenanthroline (10–20 m*M*), or by peptide inhibitors or serine inhibitors such as DFP or PMSF.

PNGase F is completely stable in 2.5 *M* urea at 37°C for 24 hours and loses only 40% activity in 5 *M* urea. We have noted that some native monomeric glycoproteins can be completely deglycosylated by PNGase F using just urea, especially those that show partial accessibility to the enzyme (i.e., 10–20% cleavage on SDS–PAGE) (unpublished observations). The benefits of using urea whenever it is effective are 2-fold: (1) it allows deglycosylation to go to completion at a *minimum* PNGase F level, and (2) it can be dialyzed after the reaction with the possibility of restoring the native conformation (i.e., biological activity). NaSCN can also be used in conjunction with PNGase F to promote deglycosylation without detergents, but the enzyme is more unstable in NaSCN than in urea, and concentrations should not exceed 0.25 *M*, nor the temperature, 30°C.

4. Deglycosylation Protocol

For complete deglycosylation using PNGase F, the protocol in the preceding section (Endo F) can be applied. As with Endo F, the SDS–NP-40 format should be conducted at 30°C for maximum PNGase F stability.

For initial determination of cleavage, the glycoprotein is boiled for 3–5 minutes in 0.5% SDS–0.1 *M* β-mercaptoethanol to expose all sites. Radiolabeled glycoprotein substrates solubilized from membranes or immunoprecipitates with SDS–β-mercaptoethanol can be used directly in this protocol. Reagents (in microliters) are added in the order indicated:

Reagent	μl
Sample	10
0.5 *M* Potassium phosphate, pH 8.6	10
0.1 *M* 1,10-*o*-Phenanthroline in methanol	3
10% NP-40	5
PNGase F (200–250 mU/ml)	2
	30

Use of the commercially supplied enzyme in the protocol will give 13.3 or 16.6 mU PNGase F/ml, respectively. This concentration is sufficient for the release of carbohydrate from RNase B (0.6 mU/ml), native fetuin or transferrin (5.0 mU/ml), but will not completely deglycosylate α_1-glycoprotein (60 mU/ml). The amount of enzyme can be adjusted depending on the results observed. If more PNGase F is needed, increase the buffer concentration and decrease the volume as described for Endo F (Section II,C,4).

PNGase F can deglycosylate proteins in the absence of denaturants, but higher amounts of enzyme are needed. It may not be feasible for investigators using commercial enzyme to deglycosylate some proteins. Human carboxypeptidase N, a multimeric protein containing 20% carbohydrate, can be deglycosylated at 2000 mU PNGase F per milliliter in 1 hour; in the usual 25–200 mU/ml range, some chains are removed, but the reaction will not go to completion regardless of the length of incubation. This threshold effect is typical for PNGase F and other oligosaccharide-cleaving enzymes (Section II,A,3).

III. Related Matters

A. Other Enzymes

1. PNGase A

PNGase A [EC 3.5.1.52, peptide-N^4-(N-acetyl-β-glucosaminyl)asparagine amidase A; other names, glycopeptidase A; trivial name, peptide : *N*-glycosidase A], isolated from almond emulsin, was the first glycopeptide amidase to be detected and characterized (Takahashi and Nishibe, 1981). The substrate specificity of PNGase A on glycopeptides is similar to PNGase F, and deglycosylation rates increase only slightly (20%) with prior removal of sialic acid (Plummer *et al.,* 1987). PNGase A is much less efficient that PNGase F on intact glycoproteins, possibly because of its higher molecular weight (79,500); with few exceptions it is used primarily with glycopeptides. Unlike PNGase F, PNGase A will recognize and cleave the GlcNAc "stub" generated by endoglycosidases, but large amounts of enzyme are required (Tarentino and Plummer, 1982).

PNGase A is available commercially (ICN Biomedicals Inc.), but the 1 mU vial costs more than the 10 or 20 mU vials of PNGase F. The enzyme has remarkable stability to sodium thiocyanate (100% active in 0.75 *M*), but has no other real advantage. If the enzyme is used, the protocol for PNGase F can be followed with the substitution of 0.5 *M* sodium acetate (pH 5.1) for the buffer.

2. ENDO-β-GALACTOSIDASE

Endo-β-galactosidase (EC 3.2.1.103, keratan-sulfate endo-1,4-β-D-galactanohydrolase) is commonly used to demonstrate the presence of poly-*N*-acetyllactosamine sequences in glycoproteins. The enzyme will hydrolyze either glycolipid or glycoprotein structures having the sequence GlcNAcβ1,3-Galβ1,4GlcNAc at the β-1,4 linkage. Branching on the Gal residue by β1,6-linked GlcNAc, occurrence of fucose near the cleavage site, or the disaccharides and trisaccharides GlcNAcβ1,3Gal and Galβ1,4GlcNAcβ1,3Gal inhibit enzyme action (Scudder *et al.*, 1983).

Endo-β-Galactosidase is sold as a protease-free, lyophilized solid (0.1 or 0.2 U per vial). The enzyme is a monomer of M_r = 32,000; it has a pH optimum of 5.7–5.8 and a temperature optimum of 55°C. To maintain full activity, glycoproteins should be digested in the presence of 0.2 mg BSA per milliliter whereas glycolipids require an additional 2 mg sodium taurodeoxycholate per milliliter (Scudder *et al.*, 1984). Glycoproteins at 0.4–1.0 mg/ml can be digested with endo-β-galactosidase (50–500 mU/ml) in 50 m*M* sodium acetate, pH 5.8 (Cairns *et al.*, 1984). Susceptibility to hydrolysis has been monitored by a change in migration on SDS–PAGE, by loss in immunoreactivity on nitrocellulose, or by Bio-Gel filtration.

3. ENDO-α-*N*-ACETYLGALACTOSAMINIDASE

Endo-α-*N*-acetylgalactosaminidase (EC 3.2.1.97) hydrolyzes the Galβ1, 3GalNAc core disaccharide from Ser/Thr residues in glycoproteins or glycopeptides. The enzyme has a strict specificity for this disaccharide, and prior removal of sialic acid or other residues (GlcNAc or Fuc) is essential for activity. This enzyme provides an activity complementary to PNGase F. If neuraminidase treatment after PNGase F cleavage results in a further change in migration on SDS–PAGE, the possibility of O-linked oligosaccharides can be suspected. A further change in migration after endo-α-acetylgalactosaminidase treatment can indicate the presence of O-linked sugars. As noted, removal of these oligosaccharides may be incomplete because core disaccharides containing additional residues of Fuc and/or GlcNAc are common, and no enzyme is currently available to remove these structures.

Endo-α-*N*-acetylgalactosaminidase is available from two suppliers (*O*-Glycanase, Genzyme Inc.; Boehringer Mannheim) in 25- or 125-mU vials as a 30% glycerol solution in 20 m*M* Tris-maleate buffer, pH 6.0. The enzyme is stable for at least 6 months at 4°C and for at least one freeze–thaw cycle. Commercial preparations are free of protease and exoglycosidases with the exception of very low β-galactosidase activity.

Enzyme activity is quantitated by release of disaccharide from asialofetuin.

Endo-α-*N*- acetylgalactosaminidase is isolated from filtrates of *Diplococcus pneumoniae* cultures (Umemoto *et al.*, 1977). It has a molecular weight of 160,000 and a pH optimum of 6.0, but it retains at least 60% activity between pH 5.0 and 8.0. Little information is available as to the effects of chaotropic reagents or denaturants on enzyme activity. If such conditions are used, a positive control such as PNGase F-treated fetuin should be included.

Reaction mixtures require 1–4 mU of enzyme for 20 μg of glycoprotein in 20 m*M* Tris-maleate, pH 6.0. Neuraminidase may be added to produce the asialoderivative, and D-galactono-γ-lactone (40 mg/ml) may be added to suppress β-galactosidase activity (2 μl per 40-μl reaction). Released disaccharide can be measured with the Morgan–Elson method for *N*-acetylamino sugars, since the disaccharide gives 110% the color yield of free GalNAc (Glascow *et al.*, 1977). The cleavage of glycoproteins can be demonstrated by a change in migration on SDS–PAGE.

4. Endo-*N*-Acyl-Neuraminidase (Endo N)

Endo-*N*-acyl-neuraminidase is a soluble enzyme from KIF phage lysates. The enzyme is specific for hydrolyzing oligo- or poly-α-2,8-linked sialosyl units in glycoproteins. These polysialyl units have recently been shown to be present in embryonic neural cell adhesion molecules (N-CAM) and to have a function in cell–cell interaction and neural development. Endo N requires a minimum of five sialyl residues for activity. The enzyme is not yet commercially available but may be in the near future. It can be purified and assayed by published methods (Hallenbeck *et al.*, 1987).

B. Alternatives to Commercial Suppliers

Endo H, Endo F, and PNGase F provide the greatest versatility for enzymatic deglycosylation. Whether to purchase these from a commercial supplier or undertake purification is a decision that depends on cost, quantity of enzyme required, time, and personnel. The cloned commercial Endo H is a cost-effective preparation that is relatively inexpensive for the large 2-Unit quantity ($500). Considering that 0.5 U is enough Endo H to release all the accessible high-mannose oligosaccharides from 37 mg of yeast invertase or 128 mg of IgM, it is doubtful that many laboratories would need to purify this enzyme. Endo F and PNGase F, however, are 10 and 15 times more expensive, respectively, then Endo H, and purification is an attractive alternative for routine users, especially since both enzymes are isolated from the same culture medium using one protocol. The organism, *Flavobacterium meningosepticum,* can be obtained from the ATCC (no. 33958). The growth of this organism and the separation and purification of Endo F and PNGase F are detailed elsewhere (Tarentino and Plummer, 1987).

1. Methods of Assay

Radiolabeled substrates specific for Endo F ([^{3}H]dansyl-Asn(GlcNAc)$_2$-(Man)$_5$) and PNGase F ([^{3}H]dansyl-Leu-Ala-Asn(Oligo)-([^{3}H]dansylAE)-Cys-Ser) were originally used for detection and quantitation. However, since the preparation of these substrates requires a large input of time and energy, it would be desirable to have inexpensive substrates that are commercially available and a simple detection system for each enzyme.

a. Principle. The enzymatic removal of oligosaccharides from glycoprotein substrates is monitored by SDS–PAGE. Fetuin (Gibco) is used as a specific substrate for PNGase F, since it contains three complex triantennary oligosaccharide chains that are resistant to Endo F. RNase B (see Section II,C) is used to detect Endo F; PNGase F also cleaves RNase B, but is separated from Endo F on TSK HW-55(S). The substrates are heat-denatured in SDS to increase their susceptibility to deglycosylation, and 1,10-*o*-Phenanthroline is included to prevent proteolysis.

b. Procedure

	Endo F	PNGase F
2 mg RNase B/ml boiled in 0.5% SDS	10	—
2 mg Fetuin/ml boiled in 0.5% SDS	—	10
0.5 *M* Sodium acetate, pH 5.1	10	—
0.5 *M* Sodium phosphate, pH 8.6	—	10
0.2 *M* 1,10-*o*-Phenanthroline in methanol	5	5
10% NP-40	5	5
Enzyme + H_2O	20	20

The reactions are incubated at 30°C for varying lengths of time, depending on the stage of purification, and 2–4 μg of glycoprotein are subjected to electrophoresis on a 12.5% acrylamide gel. Negative controls (no enzyme) are included with each run. A positive control using commercial Endo F or PNGase F is desirable.

c. Comments. Since fetuin has three PNGase F-susceptible oligosaccharides, a product ladder is to be expected early in the purification when the enzyme concentration is low. Endo F is the first activity peak off the TSK HW-55 column; it elutes very sharply and tests positive with RNase B and negative with fetuin. PNGase F elutes later as a very broad peak that cleaves both RNase B and fetuin. The purification procedure was originally developed for 8 liters of culture medium, but 2 liters of broth can provide roughly 2–3 U (μmol/min) of both enzymes in <1 month.

References

Baynes, J. W., and Wold, F. (1976). *J. Biol. Chem.* **251,** 6016–6024.

Cairns, M. T., Elliot, D. A., Scudder, P. R., and Baldwin, S. A. (1984). *Biochem. J.* **221,** 179–188.

Carr, S. A., and Roberts, G. D. (1986). *Anal. Biochem.* **157,** 396–406.

Chu, F. K. (1986). *J. Biol. Chem.* **261,** 172–177.

Chu, F. K., Maley, F., and Tarentino, A. L. (1981). *Anal. Biochem.* **116,** 152–160.

Copeland, C. S., Zimmer, K.-P., Wagner, K. R., Healey, G. A., Mellman, I., and Helenius, A. (1988). *Cell (Cambridge, Mass.)* **53,** 197–209.

Deshpande, K. L., Fried, V. A., Ando, M., and Webster, R. G. (1987). *Proc. Natl. Acad. Sci. U.S.A.* **84,** 36–40.

Dubois, M., Gilles, K. S., Hamilton, J. K., Rebers, P. A., and Smith, F. (1956). *Anal. Chem.* **28,** 350–356.

Edge, A. S. B., Faltyner, C. R., Hof, L., Reichert, L. E., and Weber, P. (1981). *Anal. Biochem.* **118,** 131–137.

Elbein, A. D. (1987). *Annu. Rev. Biochem.* **56,** 497–534.

Elbein, A. D., Sof, R., Dorling, P. R., and Vosbeck, K. (1981). *Proc. Natl. Acad. Sci. U.S.A.* **78,** 7393–7397.

Elder, J. H., and Alexander, S. (1982). *Proc. Natl. Acad. Sci. U.S.A.* **79,** 4540–4544.

Frutiger, S., Hughes, G. J., Hanly, W. C., and Jaton, J.-C. (1988). *J. Biol. Chem.* **263,** 8120–8125.

Glascow, L. R., Paulson, J.-C., and Hill, R. L. (1977). *J. Biol. Chem.* **252,** 8615–8623.

Glass, W. F., III, Briggs, R. C., and Hnilica, L. S. (1981). *Anal. Biochem.* **115,** 219–224.

Green, E. D., and Baenziger, J. U. (1988). *J. Biol. Chem.* **263,** 25–35.

Green, E. D., Van Halbeck, H., Boime, I., and Baenziger, J. U. (1987a). *J. Biol. Chem.* **260,** 15623–15630.

Green, E. D., Brodbeck, R. M., and Baenziger, J. U. (1987b). *Anal. Biochem.* **167,** 62–75.

Hallenbeck, P. C., Vimr, E. R., Yu, F., Bassler, B., and Troy, F. A. (1987). *J. Biol. Chem.* **262,** 3553–3561.

Hirschberg, C. B., and Snider, M. D. (1987). *Annu. Rev. Biochem.* **56,** 63–87.

Koide, N., and Muramatsu, T. (1974). *J. Biol. Chem.* **249,** 4897–4904.

Kornfeld, R., and Kornfeld, S. (1985). *Annu. Rev. Biochem.* **54,** 631–664.

Kukuruzinska, M. A., Bergh, M. L. E., and Jackson, B. J. (1987). *Annu. Rev. Biochem.* **56,** 915–944.

Laemmli, U. K. (1970). *Nature (London)* **227,** 680–685.

Lodish, H. F., Kong, N., Snider, M., and Strous, G. J. A. M. (1983). *Nature (London)* **304,** 80–83.

Machamer, C. E., and Rose, J. K. (1988). *J. Biol. Chem.* **263,** 5955–5960.

McCoy, R. D., and Troy, F. A. (1987). *In* "Methods in Enzymology" (V. Ginsburg, ed.), Vol. 138, pp. 627–637. Academic Press, Orlando, Florida.

Mizuochi, T., Amano, J., and Kobata, A. (1984). *J. Biochem. (Tokyo)* **95,** 1209–1213.

Morrison, M. H., Lynch, R. A., and Esselman, W. J. (1986). *Mol. Immunol.* **23,** 63–72.
Muramatsu, T. (1971). *J. Biol. Chem.* **246,** 5535–5539.
Nagy, B., and Jencks, W. P. (1965). *J. Am. Chem. Soc.* **87,** 2480–2488.
Plummer, T. H., Jr., Elder, J. H., Alexander, S., Phelan, A. W., and Tarentino, A. L. (1984). *J. Biol. Chem.* **259,** 10700–10704.
Plummer, T. H., Jr., Phelan, A. W., and Tarentino, A. L. (1987). *Eur. J. Biochem.* **163,** 167–173.
Reddy, A. V., Johnson, R. S., Biemann, K., Williams, R. S., Ziegler, F. D., Trimble R. B., and Maley, F. (1988). *J. Biol. Chem.* **263,** 6978–6985.
Robbins, P. W., Hubbard, S. C., Turco, S. J., and Wirth, D. P. (1977). *Cell (Cambridge, Mass.)* **12,** 893–900.
Scudder, P., Uemura, K.-I., Dobly, J., Fukuda, M. N., and Feizi, T. (1983). *Biochem. J.* **213,** 485–494.
Scudder, P., Hanfland, P., Uemura, K.-I., and Feizi, T. (1984). *J. Biol. Chem.* **259,** 6586–6592.
Skoza, L., and Mohas, S. (1976). *Biochem. J.* **159,** 457–462.
Strous, G. J. A. M., and Lodish, H. F. (1980). *Cell (Cambridge, Mass.)* **22,** 709–717.
Sweidler, S. J., Freed, J. H., Tarentino, A. L., Plummer, T. H., Jr., and Hart, G. W. (1985). *J. Biol. Chem.* **260,** 4046–4054.
Takahashi, N., Ishii, I., Ishihara, H., Mori, M., Tejima, S., Jefferis, R., Endo S., and Arata, Y. (1987). *Biochemistry* **26,** 1137–1144.
Taskaski, S., Mizuochi, T., and Kobata, A. (1982). *In* "Methods in Enzymology" (V. Ginsburg, ed.), Vol. 83, pp. 263–268. Academic Press, New York.
Tanner, W., and Lehle, L. (1987). *Biochim. Biophys. Acta* **906,** 81–99.
Tarentino, A. L., and Maley, F. (1974). *J. Biol. Chem.* **249,** 811–817.
Tarentino, A. L., and Maley, F. (1975). *Biochem. Biophys. Res. Commun.* **67,** 455–462.
Tarentino, A. L., and Plummer, T. H., Jr. (1982). *J. Biol. Chem.* **257,** 10776–10780.
Tarentino, A. L., and Plummer, T. H., Jr. (1987). *In* "Methods in Enzymology" (V. Ginsburg, ed.), Vol. 138, pp. 770–778. Academic Press, Orlando, Florida.
Tarentino, A. L., Plummer, T. H., Jr., and Maley, F. (1974). *J. Biol. Chem.* **249,** 818–824.
Tarentino, A. L., Gomez, C. M., and Plummer, T. H., Jr. (1985). *Biochemistry* **24,** 4665–4671.
Trimble, R. B., and Atkinson, P. H. (1986). *J. Biol. Chem.* **261,** 9815–9824.
Trimble, R. B., and Maley, F. (1977a). *Biochem. Biophys. Res. Commun.* **78,** 935–944.
Trimble, R. B., and Maley, F. (1977b), *J. Biol. Chem.* **252,** 4409–4412.
Trimble, R. B., and Maley, F. (1984). *Anal. Biochem.* **141,** 515–522.
Trimble, R. B., Tarentino, A. L., Plummer, T. H., Jr., and Maley, F. (1978). *J. Biol. Chem.* **253,** 4508–4511.
Trimble, R. B., Maley, F., and Chu, F. K. (1983). *J. Biol. Chem.* **258,** 2562–2567.
Trimble, R. B., Atkinson, P. H., Tarentino, A. L., Plummer, T. H., Jr., Maley, F., and Tomer, K. B. (1986). *J. Biol. Chem.* 12000–12005.
Trumbly, R. J., Robbins, P. W., Belfort, M., Ziegler, F. D., Maley, F., and Trimble, R. B. (1985). *J. Biol. Chem.* **260,** 5683–5690.
Umemoto, J., Bhavanandan, V. P., and Davidson, E. A. (1977). *J. Biol. Chem.* **252,** 8609–8614.
Ziegler, F. D., Maley, F., and Trimble, R. B. (1988). *J. Biol. Chem.* **263,** 6986–6992.

Chapter 6

Separation and Analysis of Glycoprotein Oligosaccharides

RICHARD D. CUMMINGS, ROBERTA K. MERKLE, AND NANCY L. STULTS

Department of Biochemistry
The University of Georgia
Athens, Georgia 30602

METHODS IN CELL BIOLOGY, VOL. 32

Copyright © 1989 by Academic Press, Inc.
All rights of reproduction in any form reserved.

I. Introduction

Many recent studies have demonstrated that a wide variety of intracellular and cell surface receptors as well as other membrane-associated and secreted proteins contain covalently bound carbohydrate residues. These studies have shown that the functionality, antigenicity, and biosynthesis of these glycoproteins is closely related to the nature and extent of glycosylation. The oligosaccharide moieties of glycoproteins appear to represent the most diverse type of posttranslational modifications of proteins; surprisingly little, however, is known about the detailed structures of the sugar chains of most cellular glycoproteins. Many cell surface receptors have been identified as glycoproteins, and several examples are listed in Table I. In only a few cases [e.g., asialoglycoprotein receptor, epidermal growth factor (EGF) receptor, low-density lipoprotein (LDL) receptor, and mannose 6-phosphate (Man6P) receptor] have the oligosaccharide structures on receptor glycoproteins been elucidated in detail.

Information about the carbohydrate portions of animal cell glycoproteins is limited because the determination of oligosaccharide structures is complicated by a number of critical factors:

1. The hundreds of known and possible structures present a formidable problem in attempts to obtain purified samples for structural analysis. Even many of the known oligosaccharide structures exist as positional isomers, and their separation by currently available methods is often either tedious or impossible.
2. Relatively minor structural species, which may be the most important in terms of their contribution to the fate of a glycoprotein, may go undetected or unrecognized using most conventional purification approaches. An illustration of this problem is found in the targeting of acid hydrolases to lysosomes. Although most lysosomal enzymes possess numerous sugar side chains, it appears that it is necessary for only one chain to carry the Man6P recognition signal to allow the glycoprotein to interact successfully with the Man6P receptor (Fischer *et al.,* 1982; Varki and Kornfeld, 1983).
3. Many glycoproteins of interest are not abundant, and purification of amounts sufficient for detailed chemical structural characterization is often impractical. For example, the receptor for LDL occurs in many mammalian cells on the order of 10,000–100,000 molecules per cell which represents less than 0.01% of the total cell protein.

Given these problems it is clear that special techniques and approaches may be required for the analysis of glycoproteins present in trace quanti-

TABLE I

EXAMPLES OF GLYCOSYLATED RECEPTORS

Receptors for	References
Acetylcholine	Herron and Schimerlik (1983); Rauh *et al.* (1986)
Asialoglycoproteins	Lowe and Nilsson (1983); Halberg *et al.* (1987)
Adenosine A1	Stiles (1986)
β-Adrenergic ligands	Stiles *et al.* (1984); Cervantes-Olivier *et al.* (1985); Dohlman *et al.* (1987)
Calcitonin	Moseley *et al.* (1983)
Class III collagen	Carter and Wayner (1988)
Complement C3b/C4b	Lublin *et al.* (1986)
Cholecystokinin	Rosenzweig *et al.* (1984)
Dopamine D_2	Grigoriadis *et al.* (1988)
EGF	Soderquist and Carpenter (1984); Cummings *et al.* (1985)
LDL	Cummings *et al.* (1983)
Fc	Green *et al.* (1985)
Formyl chemotactic peptide	Malech *et al.* (1985)
Glucocorticoids	Blanchardie *et al.* (1986)
Growth hormone	Asakawa *et al.* (1986); Yamada *et al.* (1987)
IgE	Keegan and Conrad (1987)
Insulin	Hedo *et al.* (1983); McElduff *et al.* (1986)
Man6P	Goldberg *et al.* (1983); Gasa and Kornfeld (1987)
Nerve growth factor	Grob *et al.* (1983)
Parathyroid hormone	Karpf *et al.* (1987); Shigeno *et al.* (1988)
Platelet-derived growth factor	Claesson-Welsh *et al.* (1987); Daniel *et al.* (1987)
Somatostatin	Zeggari *et al.* (1987)
Transferrin	Van Driel and Goding (1985)
Transforming growth factor-β	Segarini and Seyedin (1988)
Tumor necrosis factor	Tsujimoto *et al.* (1986)
Vasoactive intestinal peptide	Nyugen *et al.* (1986); El Battari *et al.* (1987)

ties. Using the methodology outlined in this article, it should be possible to define the structures of the oligosaccharide chains of any cellular glycoprotein, including receptor molecules. Our discussion will focus largely on those techniques which we have found useful in our laboratory for analyzing the structures of sugar chains of metabolically radiolabeled animal cell-derived glycoproteins. The advantages of the techniques we describe are that they are inexpensive, easy to set up, require no elaborate instrumentation, and are applicable to extremely small amounts of material. These methods complement rather than replace the older and more traditional chemical methods of structural analysis described by other investigators (Lindberg and Lönngren, 1978; McNeil *et al.*, 1982; Vliegenthart *et al.*, 1983; Reinhold, 1987).

II. Diversity of Sugar Chains in Animal Cell Glycoproteins

There are many ways in which carbohydrate chains have been found linked to peptide in animal cell glycoproteins. Each of the forms represents structurally different sugar chains that appear to be synthesized by unique pathways. The common linkage groups known to date are listed in Table II. Some glycoproteins may have oligosaccharides in one or more of these types of linkages. The following discussion is related to the two most common linkage groups.

A. GlcNAc–Asn Sugar Chains

Most, if not all, animal cell surface glycoproteins and a wide variety of secreted and intracellular glycoproteins contain sugar chains in N-glycosidic linkage to Asn. These glycosylated Asn residues occur in the sequence-Asn-X-Ser/Thr-, where X represents any amino acid except perhaps Pro or Asp (Marshall, 1974). These Asn-linked sugar chains are characterized by a common core pentasaccharide sequence containing mannose (Man) and *N*-acetylglucosamine (GlcNAc). This core sequence is usually substituted in a number of ways to generate additional types of chains generally classified as high-mannose, hybrid, or complex type. Complex-type chains can be further classified based on their branching pattern as biantennary, triantennary, or tetraantennary. Examples of these types are shown in Fig. 1. These oligosaccharides may also possess

TABLE II

EXAMPLES OF THE DIVERSITY OF SUGAR CHAINS IN ANIMAL CELL GLYCOPROTEINS

Linkage form	References
Mannose-containing chains N-linked through R-GlcNAc-Asn	Kornfeld and Kornfeld (1985)
Mucin-type chains O-linked through R-GalNAc-Ser/Thr	Sadler (1984)
GlcNAc O-linked to Ser/Thr	Torres and Hart (1984); Holt and Hart (1986)
Mannose O-linked to Ser/Thr	Krusius *et al.* (1986)
Glycosaminoglycan-type chains O-linked through Xyl-Ser/Thr or GalNAc-Ser/Thr	Rodén (1980); Hascall (1981)
Galactose O-linked to hydroxylysine (collagen chains)	Butler and Cunningham (1966); Spiro (1967)
Mannose-containing glycolipids linked to proteins	Tse *et al.* (1985); Fatemi and Tartakoff (1986); Homans *et al.* (1988)

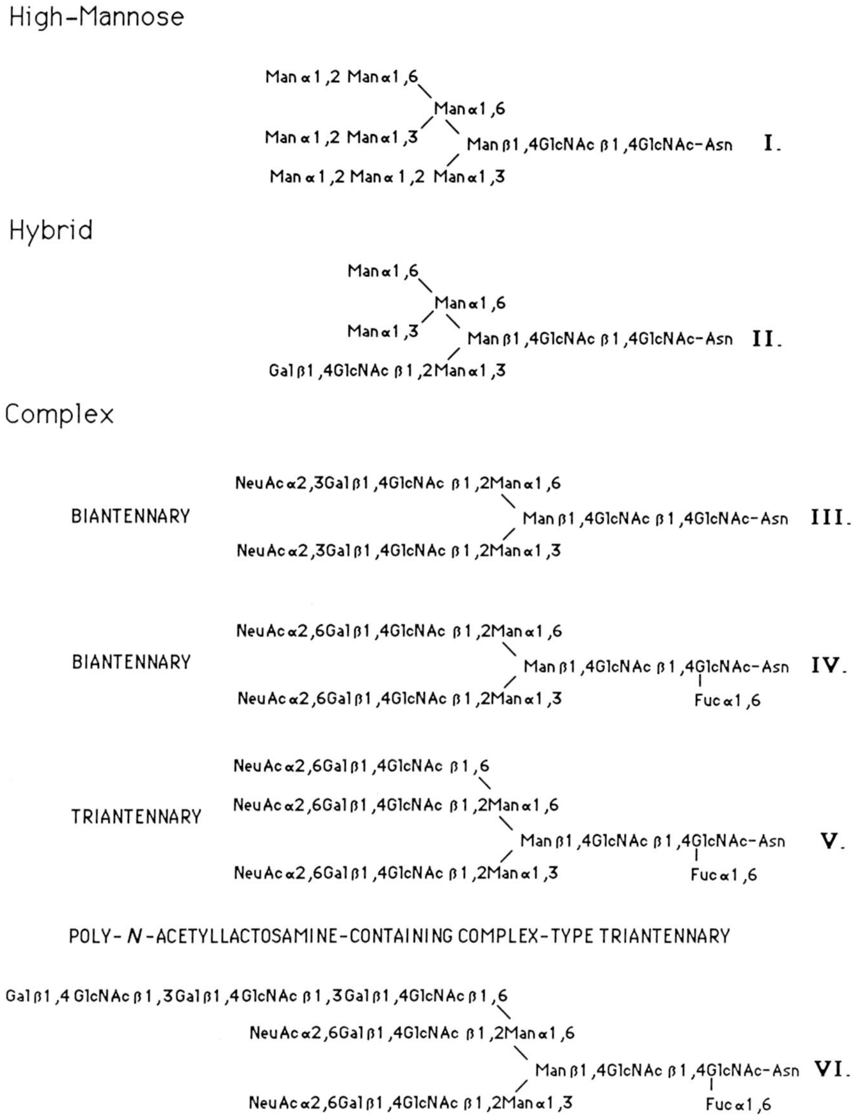

High-Mannose

```
Man α1,2 Manα1,6
                \
                 Manα1,6
                /       \
Man α1,2 Manα1,3         Man β1,4GlcNAc β1,4GlcNAc-Asn   I.
                        /
Man α1,2 Manα 1,2 Manα1,3
```

Hybrid

```
                 Manα1,6
                        \
                         Manα1,6
                        /       \
                 Manα1,3         Man β1,4GlcNAc β1,4GlcNAc-Asn   II.
                                /
Galβ1,4GlcNAc β1,2Manα1,3
```

Complex

```
             NeuAcα2,3Galβ1,4GlcNAc β1,2Manα1,6
                                               \
BIANTENNARY                                     Man β1,4GlcNAc β1,4GlcNAc-Asn   III.
                                               /
             NeuAcα2,3Galβ1,4GlcNAc β1,2Manα1,3


             NeuAcα2,6Galβ1,4GlcNAc β1,2Manα1,6
                                               \
BIANTENNARY                                     Man β1,4GlcNAc β1,4GlcNAc-Asn   IV.
                                               /                  |
             NeuAcα2,6Galβ1,4GlcNAc β1,2Manα1,3                 Fucα1,6


             NeuAcα2,6Galβ1,4GlcNAc β1,6
                                        \
TRIANTENNARY NeuAcα2,6Galβ1,4GlcNAc β1,2Manα1,6
                                               \
                                                Man β1,4GlcNAc β1,4GlcNAc-Asn   V.
                                               /                  |
             NeuAcα2,6Galβ1,4GlcNAc β1,2Manα1,3                 Fucα1,6
```

POLY-*N*-ACETYLLACTOSAMINE-CONTAINING COMPLEX-TYPE TRIANTENNARY

```
Galβ1,4 GlcNAcβ1,3Galβ1,4GlcNAc β1,3Galβ1,4GlcNAcβ1,6
                                                    \
             NeuAcα2,6Galβ1,4GlcNAc β1,2Manα1,6
                                               \
                                                Man β1,4GlcNAc β1,4GlcNAc-Asn   VI.
                                               /                  |
             NeuAcα2,6Galβ1,4GlcNAc β1,2Manα1,3                 Fucα1,6
```

FIG. 1. Examples of Asn-linked sugar chains. Each of the six glycopeptides is numbered as indicated on the right-hand side.

a bisecting GlcNAc residue linked β-1,4 to the β-linked core mannose. The high-mannose-type chains are precursors to hybrid- and complex-type chains (Kornfeld and Kornfeld, 1985). Although most mature surface glycoproteins commonly contain hybrid- and complex-type Asn-linked chains, some glycoproteins also contain high-mannose-type chains. An example of a surface glycoprotein with both complex- and high-mannose-type chains is the receptor for EGF (Cummings *et al.*, 1985). An interesting structural feature of the complex-type Asn-linked sugar chains of many surface glycoproteins is the repeating disaccharide sequence [3Galβ1,4GlcNAcβ1]$_n$, or poly-*N*-acetyllactosamine (Järnefelt *et al.*, 1978; Li *et al.*, 1980; Fukuda and Fukuda, 1984; Yamashita *et al.*, 1984; Merkle and Cummings, 1987b, 1988). These poly-*N*-acetyllactosamine chains are seldom found in serum glycoproteins.

Available evidence indicates that the structures of Asn-linked sugar chains in cellular glycoproteins are determined by a number of factors, including the array of glycosyltransferases and glycosidases present in cells, the position of the glycosylated Asn residue within the polypeptide, and the subcellular localization of the glycoprotein. The oligosaccharide structure may even differ at a given glycosylation site on individual molecules of a single glycoprotein. For example, ovalbumin, which has only a single glycosylation site, has been found to possess at least nine different oligosaccharide structures (Yamashita *et al.*, 1983b). In addition, for some virus glycoproteins and some native surface receptor glycoproteins, the oligosaccharides on the glycoproteins are structurally different depending on the cell type of origin (Hsieh *et al.*, 1983; Hubbard, 1987; Yet *et al.*, 1988a,b).

B. GalNAc–Ser/Thr Sugar Chains

A second and common form of attached sugar chains in membrane and soluble glycoproteins is the mucin-type chain. These chains are characterized by a common core sugar, *N*-acetylgalactosamine (GalNAc) in O-glycosidic linkage and α configuration to the Ser and Thr residues. A variety of structures can be elaborated on the GalNAc by substitutions of that sugar with GlcNAc, galactose (Gal), and sialic acid (NeuAc). Some mature surface glycoproteins contain these mucin-type chains, such as erythrocyte glycophorin (Tomita and Marchesi, 1975) and the LDL receptor (Cummings *et al.*, 1983). However, other receptors such as the receptor for EGF (Childs *et al.*, 1984; Cummings *et al.*, 1985) lack these chains.

C. Other Types of Glycosylation

Although the GlcNAc–Asn- and GalNAc–Ser/Thr-linked oligosaccharides represent major types of sugar additions to animal cell glycoproteins, they are by no means the only kinds of posttranslational sugar modifications known. Table II lists several other glycosylation forms of proteins including a number of other sugars attached to Ser/Thr residues.

These other forms of glycosylation are listed both to inform and to caution investigators that demonstration that a protein does or does not contain classical Asn- and Ser/Thr-linked chains does not necesarily exclude the possibility that a protein may also be glycosylated in these other ways.

Many aspects of the biosynthesis of Asn-linked and Ser/Thr-linked sugar chains are understood and have been reviewed (Hubbard and Ivatt, 1981; Kornfeld and Kornfeld, 1985; Sadler, 1984; Snider, 1984). Readers interested in the pathways of biosynthesis and other details about alternative forms of glycosylation should consult these reviews and the articles cited in Table II.

III. Evidence Indicating that a Protein Is Glycosylated

If the actual or deduced amino acid sequence of the glycoprotein of interest is known, the presence of the N-linked glycosylation consensus sequence Asn-X-Ser/Thr is highly suggestive of sugar modification. The simplest test to determine whether a protein is glycosylated is to determine its reactivity with sugar reagents. Provided that sufficient quantity (nanomole range of sugar) of glycoprotein is available, the presence of sugar can be detected by the phenol sulfuric acid assay (Dubois *et al.*, 1956) or by a number of chemical assays for reducing sugar (Park and Johnson, 1949; Dygert *et al.*, 1965) or for sialic acid (Warren, 1959; Powell and Hart, 1986). In addition, chemical detection of glycoproteins after electrophoresis in acrylamide gels can be accomplished by periodic acid oxidation followed by reaction with Schiff's reagent to yield a red aldehyde addition product (Kapitany and Zebrowski, 1973).

The glycoprotein nature of a protein can be further confirmed or determined by demonstrating the properties listed in Table III, which include (1) metabolic incorporation of radiolabeled sugars ([^{14}C]- or [^{3}H]-labeled

TABLE III

METHODOLOGY TO DETERMINE THAT A PROTEIN IS GLYCOSYLATED

Compositional analysis	Metabolic radiolabeling
Chemical staining	External radiolabeling
Binding to lectins	Inhibition of protein glycosylation
Sensitivity to exoglycosidases	Inhibition of oligosaccharide processing
Sensitivity to endoglycosidases	Chemical deglycosylation
Binding to carbohydrate-directed antibodies	

mannose, galactose, or glucosamine); (2) radiolabeling of sugar chains on the glycoprotein by external labeling techniques; (3) identification of the sugar composition; (4) susceptibility to enzymatic or chemical deglycosylation; (5) ability to bind lectins or antibodies directed against carbohydrate determinants; and (6) sensitivity of the glycoprotein to inhibitors of oligosaccharide biosynthesis or processing.

Numerous studies have shown that living cells can take up radioactive sugar precursors from growth medium and incorporate the radioactive derivatives into the sugar chains of newly synthesized glycoproteins, glycolipids, and glycosaminoglycans. An excellent review by Yurchenco *et al.* (1978) discusses many aspects of metabolic radiolabeling of animal cells with radioactive monosaccharides. This technique, as it has been applied to the structural analysis of glycoconjugates, will be discussed in detail in a following section. In the case in which a glycoprotein cannot be metabolically radiolabeled, it may be possible to radiolabel oligosaccharide moieties by external radiolabeling procedures as outlined later. For example, NeuAc residues are sensitive to periodate oxidation yielding a C-7 aldehyde derivative which can reduced with NaB^3H_4 (Van Lenten and Ashwell, 1972). Likewise, exposed Gal residues can be converted to NaB^3H_4-reducible aldehydes enzymatically using galactose oxidase (Morell and Ashwell, 1972).

Incorporation of radiolabeled sugar or chemical detection of sugar per se is not convincing enough evidence that a protein is a glycoprotein. The incorporated radioactivity should be identified to be in a sugar molecule, and compositional analysis of labeled and unlabeled glycoproteins should, in most cases, demonstrate the presence of specific sugars in ratios consistent with known or anticipated oligosaccharide structures. Many methods are available for the determination of the component monosaccharides at the nanomole level in glycoconjugates including anion exchange chromatography of carbohydrate–borate complexes (Lee, 1972), gas–liquid chromatography of alditol acetates or trimethylsilyl derivatives (Laine *et al.*, 1972), high-performance liquid chromatography (HPLC) involving precolumn or postcolumn derivatization (reviewed by Honda, 1984), and more recently, anion exchange chromatography followed by pulsed amperometric detection (Hardy *et al.*, 1988). As will be discussed in detail later, if the glycoprotein has been metabolically radiolabeled, the sugar composition can be assessed after strong acid hydrolysis by descending paper chromatography.

Following glycosidase digestion, an apparent reduction in molecular size of a protein, usually determined by sodium dodecyl sulfate–polyacrylamide gel electrophoresis (SDS–PAGE), is often considered as evidence for the presence of oligosaccharide chains. For example, glycoproteins possessing typical complex-type N-linked chains might be sensitive to sequential treatment with neuraminidase, β-galactosidase, and β-hexosaminidase. It is important in such preliminary experiments to use enzymes with a broad substrate specificity such that fine-structural details do not preclude a positive result. For example, a good selection of exoglycosidases for use in this type of experiment would include *Arthrobacter ureafaciens* neuraminidase, bovine testicular β-galactosidase, and jack bean β-*N*-acetyl-D-glucosaminidase. Endoglycosidases that cleave between the chitobiosyl core of *N*-linked oligosaccharide chains—including Endo F and Endo H, and now more recently, the peptide glycosidases including *N*- and *O*-Glycanase—are commonly used to assess protein glycosylation, once again by monitoring an increase in SDS–PAGE mobility. Based on the substrate specificities of these different enzymes, it is possible to draw some general conclusions about the types of oligosaccharide chains present; however, this approach is seldom definitive.

A major problem with using glycosidases to assess the glycosylated nature of a protein is that some of these enzymes may be contaminated with proteases. It is important to perform controls to ensure that changes in the size of a protein are due, in fact, to release of sugar. This may be accomplished in several ways. First, in the case of exoglycosidases, specific inhibitors should eliminate their effect on the apparent size or properties of a glycoprotein. For example, D-galactono-1,4-lactone inhibits many β-galactosidases (Meisler, 1972; Sloan, 1972). Second, a protein lacking any sugar moieties, such as bovine serum albumin (BSA), should also be treated under similar conditions and its apparent size determined before and after the treatment to rule out protease contamination.

It is also possible to deglycosylate glycoproteins chemically using anhydrous trifluoromethanesulfonic acid (TFMS) or hydrogen fluoride (HF) (Sojar and Bahl, 1987) with complete destruction of the carbohydrate resulting in a reduction in apparent molecular weight of the protein in question. There are several advantages to TFMS over HF, including greater potency and easier handling. Since in some cases these reagents minimally perturb the quaternary structure of glycoproteins and remove both N- and O-linked oligosaccharide chains, they have been used for studies on the structure of the polypeptide as well as on the function of the carbohydrate (Kaylan and Bahl, 1983).

Carbohydrate-binding proteins, such as plant lectins, are not only useful for the detection of glycoproteins, but as will be discussed in detail later, are powerful tools for the isolation and purification of oligosaccharide structures in complex mixtures (Merkle and Cummings, 1987a,b; Osawa and Tsuji, 1987). Wheat germ agglutinin (WGA) and concanavalin A (Con A) have been used most frequently to screen for the presence of carbohydrate (reviewed by Lis and Sharon, 1984, 1986a,b). The ability of a putative glycoprotein to bind to carbohydrate-binding proteins can be ascertained by passage over a column of immobilized lectin and subsequent elution of any bound material with low molecular weight haptens. This approach may, however, be complicated by the possibility that the protein in question is itself unglycosylated, but is associated with a glycoprotein that can bind to the immobilized lectin. Alternatively, using radioactively labeled lectins, glycoproteins can be identified after gel electrophoresis in SDS gels (Burridge, 1978) or following transfer to nitrocellulose (Bartles *et al.*, 1985). Though less generally available, animal lectins (reviewed in Olden and Parent, 1987) and antibodies directed against carbohydrate (Gooi *et al.*, 1983; Childs *et al.*, 1984; Feizi, 1985; Pendu *et al.*, 1985; Magnani, 1987; Zopf *et al.*, 1987) can be used in a similar fashion for the detection of glycoconjugates.

Another approach for identifying glycoproteins has been to demonstrate an effect of inhibitors of oligosaccharide biosynthesis and processing on their apparent molecular weight (Schwarz and Datema, 1982; Elbein, 1987). Tunicamycin is a nucleoside antibiotic from *Streptomyces lysosuperificus* that prevents N glycosylation of proteins by blocking the synthesis of the precursor of the dolichol pyrophosphate oligosaccharide. Consequently, tunicamycin inhibits biosynthesis of Asn-linked sugar chains with usually no effect on that of other glycosylation forms, such as Ser/Thr-linked chains. Inhibitors of Asn-linked oligosaccharide processing (reviewed by Fuhrmann *et al.*, 1985), including castanospermine, deoxynojirimycin, deoxymannojirimycin, and swainsonine, may not necessarily result in significant alterations in the apparent molecular weight of a newly synthesized glycoprotein. However, concomitant use of other techniques, such as serial lectin affinity chromatography, which will be discussed later, would confirm that the inhibitors actually effected a change in oligosaccharide structure. Studies with these inhibitors, in addition to tunicamycin, have provided useful information regarding the role of oligosaccharide chains in the biological activity of a glycoprotein. For example, the effects of these drugs on receptor function have been addressed in the case of the acetylcholine receptor (Smith *et al.*, 1986), β_2-adrenergic receptor (Boege *et al.*, 1988), asialoglycoprotein receptor (Breitfeld *et al.*, 1984), EGF receptor (Soderquist and Carpenter, 1984; Slieker *et al.*, 1986), and insulin receptor (Ronnett *et al.*, 1984; Duronio *et al.*, 1986; Arakaki *et al.*, 1987).

IV. Radiolabeling of Sugar Chains

There are several methods that can be used to radiolabel the glycoprotein oligosaccharides whose structures are to be analyzed. The three major techniques to be discussed are metabolic radiolabeling, end-labeling using reduction with tritiated borohydride or acetylation with [^{14}C]-labeled acetic anhydride, and external labeling techniques employing enzymatic addition of radiolabeled sugar to the existing oligosaccharide.

A. Metabolic-Radiolabeling Techniques

If a glycoprotein in question is derived from cultured cells, it is often advantageous to use metabolic radiolabeling to allow a study at the micro level of the sugar chains on the newly synthesized glycoproteins. This method circumvents many of the complications encountered in the analysis of glycoproteins. Metabolic radiolabeling allows the study of glycoproteins that are available in only very small quantities (from 10^9 cells the glycoproteins usually occur in the picogram to microgram range). Additionally, it is often possible to distinguish between heterogeneous oligosaccharides that may occur even at a given glycosylation site. For example, by deriving the glycoprotein of interest from cells grown in media containing different precursor sugars, then comparing their fractionation patterns on immobilized lectins, and finally subjecting them to compositional analysis, it is possible to discern very slight structural differences. Another advantage of metabolic radiolabeling is that it makes possible the analysis of even unusual types of oligosaccharide chains or sugar residues that may occur in the glycoprotein. Metabolic radiolabeling ensures that only biosynthetic products are analyzed and thus eliminates the worry that a sample is contaminated with glycoconjugates from other sources.

The basic technique for metabolic labeling of animal cell glycoproteins using radioactive precursor monosaccharides was reviewed by Yurchenko *et al.* (1978). The recent use of radioactive sugars in long-term radiolabeling of animal cells has been described by Cummings and Kornfeld (1982b) and Cummings *et al.* (1983). Cells in culture are labeled while in log phase in normal medium with [2-^{3}H]mannose, [6-^{3}H]glucosamine, or [6-^{3}H]galactose at a concentration of 50–500 μCi/ml of medium. For long-term radiolabeling (12–36 hours) it is important not to reduce the level of glucose or alter the concentrations of other components in the normal tissue culture medium. Since most cultured animal cells rapidly deplete the medium of glucose, decreasing the glucose concentration will often result in lack of growth after a few hours. If the radiolabeled sugar is supplied in ethanol, it should be dried under N_2 and reconstituted in ~1.0 ml of normal, complete medium. The medium containing the radiolabeled sugar can then be filter-sterilized and added to the culture. The specific monosaccharides listed earlier are employed because they are known to be metabolized to a limited number of sugar derivatives. Animal cells metabolize [2-^{3}H]mannose to radiolabeled GDP-Man and GDP-Fuc without loss of the isotope at position C-2, whereas conversion to any other sugar results in the loss of the radiolabel. [6-^{3}H]Glucosamine can be used to radiolabel the amino sugars. Under normal culture conditions this precursor is restricted in metabolism to GlcNAc, GalNAc, and NeuAc (Cummings *et al.*, 1983, 1985; Nyame *et al.*, 1987). The [6-^{3}H]glucosamine precursor may cause radiolabeling of some other sugars such as glucose if exogenous glucose levels are low. The galactose and glucose residues in glycoproteins can be metabolically radiolabeled by using [6-^{3}H]galactose as the precursor sugar.

After radiolabeling, the glycoprotein of interest can be isolated from a cell extract by immunoprecipitation and then subjected to SDS–PAGE and identified by fluorography. Alternatively, it may be possible to isolate the protein of interest using affinity chromatography. The amount of radioactivity that is incorporated into carbohydrate chains by this technique depends on the turnover rate of the glycoprotein, the pool sizes of sugar nucleotides within the cells, the efficiency of transport of precursor sugars within the cells, the efficiency of recovery of glycoproteins, and the degree of heterogeneity of the oligosaccharide chains. Thus, it is unlikely that all radiolabeled sugars will be labeled to the identical specific radioactivity, although the difference in specific radioactivity between different sugars is found not to be greater than 5- to 10-fold. In our experience in analyzing oligosaccharides the method just described is satisfactory for radiolabeling the carbohydrate chains of glycoproteins derived from a variety of cell types. Typically, 10,000–100,000 cpm of radiolabeled glycoprotein are needed to allow the detailed analysis of the sugar chains as described in this review.

It should be noted that there are limitations in analyzing metabolically radiolabeled glycoproteins. It is not always possible to identify unequivocally a complex sugar chain. As described later, for example, not all partially methylated sugar standards are available for using the technique of methylation for structural analysis. Another problem relates to the specific sugars labeled by this method. If an oligosaccharide contains an unusual sugar, such as sulfated glucuronic acid, it might be difficult to define the structure. It must be emphasized, though, that clues to the presence of an unexpected sugar or unusual modification would probably be apparent.

B. "End-Labeling" of Glycoprotein Oligosaccharides

Glycoproteins can also be "end-labeled" by sequential oxidation and reduction using galactose oxidase followed by reduction with tritiated borohydride (Morell and Ashwell, 1972) if the oligosaccharide has galactose as a terminal nonreducing sugar residue. This method also allows one to label intact cells (Gahmberg, 1978). In this case, only glycoconjugates located on the cell surface are radiolabeled. If the glycoprotein oligosaccharide terminates in sialic acid, this residue may be removed to expose the penultimate galactose and then labeled. An alternative method of reductive labeling using NaB^3H_4 can be employed with N-linked oligosaccharides or with O-linked oligosaccharides that are released, respectively, from protein by hydrazinolysis or β elimination (Takasaki *et al.*, 1982; Tsuji *et al.*, 1981). The sugar residue that had been linked to the peptide is radiolabeled using this procedure. However, this method is complicated by the fact that it requires milligram quantities of material, it is inefficient, and it requires radioisotopes with very high specific activity.

Alternatively, terminal sialic acid may be radiolabeled by exposure to periodic acid followed by reductive labeling with tritiated borohydride (Van Lenten and Ashwell, 1972). Another method of end-labeling that can be used to label glycopeptides is N acetylation of the peptide portion with radiolabeled acetic anhydride (Finne and Krusius, 1982). This method has the advantage of being able to label all glycopeptides regardless of their nonreducing sugar termini.

While end-labeling methods are useful for labeling glycoproteins for their initial fractionation, they may have limited utility for the fine-structural analyses to be described later. For example, if one was to use an exoglycosidase approach for analysis of a glycoprotein labeled only at the peptide, little, if any, structural information could be derived. If the oligosaccharide being studied contained the repeating poly-*N*-acetyllactosamine sequence Galβ1,4GlcNAc, end-labeling would not permit the sequence determination that is possible using the same glycoprotein that had been metabolically labeled with galactose (Merkle and Cummings, 1987b; Cummings and Kornfeld, 1984). The use of end-labeling methods also limits the identity of the sugar residue that can be labeled and thus does not allow the comparison of glycoproteins as can be done with those metabolically labeled in parallel with galactose, glucosamine, and mannose.

C. Radiolabeling by External Glycosylation

Another way to radiolabel the sugar chains of glycoproteins is to add enzymatically a radiolabeled monosaccharide to the sugar chains of an unlabeled glycoprotein. A variation of this approach has been used to confirm the existence of the unusual terminal O-linked GlcNAc. In this case, the O-linked GlcNAc had been metabolically radiolabeled, and unlabeled Gal was added to it by the enzyme UDP-Gal : GlcNAcβ1,4-galactosyltransferase using UDP-Gal as a substrate (Nyame *et al.*, 1987). If radiolabeled UDP-Gal is used instead, one can add a radiolabeled galactosyl residue to the O-linked GlcNAc in the previously unlabeled glycopeptide (Torres and Hart, 1984; Holt and Hart, 1986; Abeijon and Hirschberg, 1988). It is also possible to radiolabel terminal galactosyl residues using sialyltransferases and radiolabeled CMP-sialic acid (Passaniti and Hart, 1988).

V. Enzymatic and Chemical Methods for Releasing Sugar Chains from Proteins

Oligosaccharide structures can be analyzed as part of a glycopeptide, or as a free oligosaccharide following enzymatic or chemical release from the glycoprotein or glycopeptides. In the case of metabolically labeled cells, the glycoprotein of interest must first be isolated from the cell extract, preferably by an affinity method using ligand or antibody coupled to Sepharose, or by immunoprecipitation. If necessary, the glycoprotein can be further purified by SDS–PAGE, identified by fluorography if radiolabeled, and excised from the gel for analysis. In some cases the glycoprotein of interest may possess a phosphatidylinositol glycolipid anchor (Low and Kincade, 1985; Reiser *et al.*, 1986; Homans *et al.*, 1988). To simplify the isolation and subsequent analyses of such a glycoprotein, the lipid portion of the molecule can be cleaved using the phosphatidylinositol-specific phospholipase C from *Staphylococcus aureus* (Low, 1981).

A. Enzymatic Release of Glycopeptides and Oligosaccharides

The most frequently used method for preparation of glycopeptides is exhaustive proteolysis using pronase, a commercial preparation of nonspecific proteases, which generates glycopeptides containing one to three amino acid residues (Finne and Krusius, 1982). The digestion is routinely carried out in a conical glass test tube (15 ml) containing the dried glycoprotein in 0.1 *M* Tris-HCl (pH 8.0) containing 1 m*M* $CaCl_2$ and 10 mg/ml pronase from *Streptomyces griseus* (Calbiochem) at 60°C for 16–24 hours. Prior to use, pronase is preincubated at 37°C for 15 minutes to destroy possible contaminant glycosidase activities. A 1-ml quantity of 10 mg/ml pronase is usually sufficient for milligram or less quantities of glycoproteins. The digestion should be conducted in a toluene atmosphere to inhibit bacterial growth. Toluene may be added directly to the incubation mixture (one drop per milliliter of buffer) or placed in a small glass tube (6 × 50 mm) and suspended by a string (waxed dental floss works well) in the test tube. The incubation tubes are then capped (rubber stopper) and sealed on the outside with parafilm. To terminate the digestion, the reaction mixture is heated for at least 5 minutes at 100°C to inactivate the enzyme. The pronase glycopeptides are then desalted by gel filtration using Sephadex G-25 (1 × 45 cm) in a volatile buffer [e.g., 0.1 *M* pyridine-acetate buffer (pH 5.4) or 7% *n*-propyl alcohol], which can be removed by lyophilization or vacuum evaporation.

To release glycopeptides from an SDS–polyacrylamide gel, the gel slice of interest is excised from the dried gel and incubated with pronase (Cummings *et al.*, 1983, 1985). Each 1-cm^2 piece of gel is incubated with 1 ml of 10 mg/ml pronase as described earlier. Any residual fluor in the gel does not interfere with the enzyme digestion. After incubation, 10 ml of water are added to the 1-ml digest and the mixture is boiled for 10 minutes. After reserving the supernatant liquid, this step is repeated with 5 ml of water. The supernatant solutions are combined, evaporated or lyophilized, and desalted by gel filtration prior to further analysis. The gel pieces may be counted directly in scintillation fluor to determine the remaining radioactivity in the gel.

Oligosaccharides can be released intact from glycoproteins by a number of bacterial endoglycosidases including the endo-β-D-*N*-acetylglucosaminidases and endoglycopeptidases listed in Table IV (Thotakura and Bahl, 1987). In most cases, highly purified preparations of these enzymes are now currently available from Boehringer Mannheim, Genzyme, and ICN Biomedical. Although some general guidelines for the use of these enzymes are outlined later, it is recommended that the incubation conditions suggested by the supplier be followed (a useful reference is *Biochemica Information,* 1987, J. Keesey (ed.), available from Boehringer Mannheim Biochemicals, Indianapolis, IN). However, depending on the glycoprotein or glycopeptide substrate, it may be necessary to determine empirically optimal conditions for cleavage (i.e., time and enzyme concentration). In general, for every nanomole of oligosaccharide–glycopeptide, 1–2 mU of enzyme should be included in the digestion. As described

TABLE IV

ENZYMATIC RELEASE OF OLIGOSACCHARIDE CHAINS FROM GLYCOPROTEINS

Enzyme	Source	Specificity	References
Endo-β-*N*-acetylglucosaminidases			
Endo C_I	*Clostridium perfringens*	$Man_5GlcNAc_2Asn$	Kobata (1978)
Endo C_{II}	*C. perfringens*	High-mannose, hybrid	Kobata (1978)
Endo D	*Diplococcus pneumoniae*	$Man_5GlcNAc_2Asn$	Muramatsu (1978)
Endo F	*Flavobacterium meningosepticum*	High-mannose, biantennary hybrid, biantennary complex	Tarentino *et al.* (1985)
Endo H	*Streptomyces plicatus*	High-mannose, hybrid	Tarentino *et al.* (1978); Trimble *et al.* (1987)
Endoglycopeptidases			
N-Glycanase	*F. meningosepticum*	High-mannose, hybrid, complex	Tarentino *et al.* (1985); Tarentino and Plummer (1987)
O-Glycanase	*D. pneumoniae*	Galβ1,3GalNAc–Ser/Thr	Umemoto *et al.* (1977); Kobata and Takasaki (1978)

earlier, for the use of pronase, overnight enzyme incubations are routinely carried out in a toluene atmosphere to inhibit bacterial growth, and the digestion is terminated by addition of 1 ml of H_2O and then heating the reaction mixture at 100°C for at least 5 minutes. In addition, in the case of a several-day incubation, a fresh portion of enzyme should be added after 18–24 hours at 37°C, since most endoglycosidases will have lost all activity by this time. Since some enzyme preparations contain trace amounts of contaminating glycosidases and/or proteases, excessive amounts of enzyme and prolonged incubation periods should be avoided.

The endoglycosidases discussed here are more active toward denatured substrates. Denaturation (0.5–1% SDS with 1% β-mercaptoethanol at 100°C for 3–5 minutes) facilitates accessibility of the enzyme to all susceptible glycosylation sites and therefore significantly reduces the amount of enzyme required for complete hydrolysis. In order to protect the enzymes from denaturation, a nonionic detergent such as Triton X-100, Nonidet P-40, octyl glucoside, or CHAPS must be included in large excess over SDS in the digestion mixture. However, if necessary, use of high enough enzyme concentrations in most cases will allow the complete hydrolysis of susceptible chains from native glycoproteins. Release of oligosaccharide chains from glycoproteins by any of these enzymatic methods can be assessed by comparing their electrophoretic mobility using SDS–PAGE or their elution behavior from Sephadex G-50 in the presence of SDS before and after enzyme treatment (e.g., Varki and Kornfeld, 1983; Freeze and Wolgast, 1986). The purification of the released chains from the peptide backbone is simultaneously accomplished by the gel filtration approach.

The endo-β-D-glucosaminidases including Endo D, Endo F, Endo H, and Endo C_I and C_{II} cleave between the di-*N*-acetylchitobiose core of Asn-linked oligosaccharides. Endoglycosidase F (EC 3.2.1.96) from *Flavobacterium meningosepticum* cleaves most high-mannose and biantennary hybrid oligosaccharide chains (Tarentino *et al.*, 1985). Hybrid structures possessing bisecting *N*-acetyglucosamine residues linked β-1,4 to the core mannose are resistant to cleavage. Triantennary and tetraantennary complex-type structures are not hydrolyzed by Endo F; however, biantennary complex-type structures are cleaved slowly. The digestion is routinely carried out in 0.25 *M* sodium acetate (pH 5–7) containing 10 m*M* EDTA and 10 m*M* β-mercaptoethanol. Prior to addition of Endo F, SDS-denatured glycoproteins should be diluted (to 0.1% SDS) and mixed with a 6-fold excess of Nonidet P-40.

Endoglycosidase H (EC 3.2.1.96) from *Streptomyces plicatus* (Tarentino *et al.*, 1978; Trimble *et al.*, 1987), like Endo F, cleaves N-linked high-mannose and certain hybrid oligosaccharides and does not act on complex-type chains. Since Endo H requires that the α-1,6-linked Man be substituted with one or more Man residues, a loss in sensitivity to Endo H treatment is diagnostic of oligosaccharide processing to complex-type chains. In contrast to EndoF, Endo H will hydrolyze hybrid chains having a bisecting *N*-acetylglucosamine residue. Digestion with Endo H is routinely carried out in 10–50 m*M* sodium phosphate, citrate, or acetate buffer, pH 5.5. If SDS denaturation is required, the SDS concentration should not exceed 1 mg/ml in the reaction mixture and BSA (1 mg/ml) should be added to stabilize the enzyme.

Endoglycosidase D (EC 3.2.1.96) from *Diplococcus pneumoniae* (Muramatsu, 1978) has a more narrow specificity than the other endoglycosidases, cleaving high-mannose chains that lack substitution at C-2 of the α-1,3-linked core Man residue. As a result, Endo D will not cleave high-mannose structures larger than $Man_5(GlcNAc)_2$, nor will it cleave hybrid or complex-type chains unless sialic acid, galactose, and *N*-acetylglucosamine are removed from the α-1,3-linked arm by sequential treatment with the appropriate exoglycosidases. Although Endo D has somewhat limited utility for release of oligosaccharide chains for glycoproteins, it has proved useful for discriminating between similar structures (Taniguchi *et al.*, 1986). Two endo- β-*N*-acetylglucosaminidases, C_I and C_{II} (not yet commercially available) have been purified from the culture filtrate of *Clostridium perfringens* (Kobata, 1978). Endo C_I has the same specificity as *Diplococcus* Endo D, while Endo C_{II} is more similar to *Streptomyces* Endo H.

Use of endoglycopeptidases—including *N*-Glycanase and *O*-Glycanase, which cleave the *N*-acetylglucosaminyl-asparagine and *N*-acetylgalactosaminyl-serine and threonine linkages, respectively, eliminates any size or charge interference by the peptide in subsequent analyses. *N*-Glycanase (peptide *N*-glycosidase F, EC 3.2.2.18) from *Flavobacterium meningosepticum* will release high-mannose, hybrid, biantennary, triantennary, or tetraantennary complex oligosaccharide chains with the reducing GlcNAc residue intact (Tarentino *et al.*, 1985; Tarentino and Plummer, 1987). Oligosaccharides located at the *N*- or *C*-terminal residue are resistant to *N*-Glycanase; therefore, if glycopeptides are prepared prior to *N*-Glycanase treatment, a protease with limited specificity such as trypsin should be employed. The digestion is routinely carried out in 0.2 *M* phosphate buffer (pH 8.6) containing 10 m*M* 1,10-*o*-phenanthroline and 10 m*M* β-mercaptoethanol. Sodium dodecyl sulfate-denatured glycoproteins should be diluted and treated with excess nonionic detergent as described for Endo F before addition of *N*-Glycanase.

A peptide N-glycosidase from almond emulsin has also been purified, which exhibits similar specificity to that from *F. meningosepticum,* but because of its larger molecular size, it is often unable to remove completely all susceptible oligosaccharide chains from glycoproteins, even under denaturing conditions (Taga *et al.*, 1984). Although this enzyme is useful for the deglycosylation of glycopeptides, it is not yet commercially available.

O-Glycanase (endo-α-*N*-acetyl-D-galactosaminidase) from *Diplococcus pneumoniae* hydrolyzes the O-glycosidic linkage between α-*N*-acetyl-D-galactosamine and serine or threonine of the disaccharide Galβ1,3GalNAc (Umemoto *et al.*, 1977; Kobata and Takasaki, 1978). The enzyme exhibits the specific requirement that both the Gal and GalNAc residues be unsubstituted. Therefore, it may be necessary to desialylate the glycoprotein or glycopeptide substrate by mild acid hydrolysis or neuraminidase treatment prior to *O*-Glycanase digestion. Digestion with *O*-Glycanase is carried out at pH 6 using phosphate or Tris-maleate buffers (e.g., Daniel *et al.*, 1987; Yoshimura *et al.*, 1987). The application of *Diplococcus O*-Glycanase to the analysis of O-linked oligosaccharides is limited because of its restricted substrate specificity. However, a recent report describes another endo-β-*N*-acetylgalactosaminidase in the culture filtrate of *Streptomyces* sp. that releases, in addition to Galβ1,3GalNAc, larger O-linked oligosaccharides (Iwase *et al.*, 1988). This enzyme may prove very useful once it is purified and further characterized.

B. Chemical Release of Oligosaccharides

Chemical release of intact N- and O-linked oligosaccharide chains is generally accomplished by hydrazinolysis and β elimination, respectively. These methods have the advantage of being less selective than the enzymatic methods. However, the potential for chemical degradation of the oligosaccharide exists and, as such, prolonged reaction times must be avoided. Hydrazinolysis cleaves the GlcNAc–Asn linkage, yielding a glycosyl amine which is immediately *N*-acetylated to reduce the danger of isomerization and decomposition. The peptide and any O-linked chains are degraded during the reaction. Takasaki *et al.* (1982) have optimized the conditions for the hydrazinolysis reaction, which should be carried out under a N_2 atmosphere. Briefly, the glycopeptide or glycoprotein (≤ 1 mg) is suspended in a small volume (≤ 1 ml) of freshly distilled hydrazine and heated at 100°C for 8–12 hours in a sealed tube. After drying the reaction mixture under reduced pressure over H_2SO_4, hydrazine is removed by repeated evaporation with toluene. *N* acetylation is then carried out by adding 0.2 ml of acetic anhydride with an equal volume of saturated $NaHCO_3$ and vortexing continuously for 10 minutes. After passing the reaction mixture over a 3-ml Dowex 50W-X8(H^+) column in water, the released glycopeptides can be purified by gel filtration (e.g., Sephadex G-25). When using trace amounts of radiolabeled material, carrier protein or glycopeptides should be included in the hydrazinolysis reaction to minimize loss of the sample.

Treatment of glycopeptides or glycoproteins with mild alkaline borohydride results in the complete release of O-linked oligosaccharides via β elimination, yielding stable sugar alditols with destruction of the peptide (Iyer and Carlson, 1971; Baenziger and Kornfeld, 1974). The dried sample is suspended in 1 ml of freshly prepared 1 *M* $NaBH_4$ in 50 m*M* NaOH for 16 hours at 45°C, after which the unreacted borohydride is destroyed by adding 4 *N* acetic acid dropwise until bubbling of the sample stops. To remove sodium borate, the reaction mixture is applied to a 3-ml column of Dowex 50W-X8(H^+) and eluted with 10–20 ml water. Residual borate ion can be removed by repeated evaporation (two to four times) with methanol containing 0.1 *M* acetic acid. To assess release of O-linked oligosaccharides and to effect their purification, the β-eliminated radiolabeled glycopeptides are applied to a Bio-Gel P-10 column in 0.1 *M* NH_4-HCO_3 or 0.1 *M* pyridine-acetate buffer (pH 5.4) and their elution position compared to that prior to β elimination (e.g., Cummings *et al.*, 1983). The released oligosaccharides, which elute in the included volume, are pooled and concentrated for further analysis.

To assess the efficacy of the enzymatic and chemical treatments for removing sugar chains from glycoproteins under study, it is important to perform parallel control studies with authentic glycoproteins and nonglycosylated proteins. For this purpose the glycoproteins fetal bovine serum (FBS) fetuin and hen ovalbumin and the nonglycosylated protein BSA are ideally suited. Fetuin contains three complex-type Asn-linked sugar chains and three mucin-type Ser/Thr-linked sugar chains (Spiro and Bhoyroo, 1974; Baenziger and Fiete, 1979a). Ovalbumin contains a single Asn-linked sugar chain that occurs usually as a variety of hybrid-type structures (Tai *et al.*, 1975, 1977). These three proteins are commercially available (Sigma) and provide suitable substrates for most known or useful exoglycosidases and endoglycosidases. Predictable behavior of these glycoproteins following endoglycosidase and/or exoglycosidase treatment reduces the possibility that the observed alterations in protein size are artifactual due to contaminant glycosidases or proteolytic enzymes.

VI. Separation of Sugar Chains by Serial Lectin Affinity Chromatography

Most animal cell glycoproteins contain a variety of oligosaccharides. Some of these oligosaccharides are related in structure because they represent intermediates in a common biosynthetic pathway. This is often seen in the case of Asn-linked sugar chains. For example, from hen egg white ovalbumin, which contains a single Asn-linked sugar chain, it is possible to derive over a dozen structurally related, but different, oligosaccharides (Tai *et al.*, 1975, 1977). Some glycoproteins, such as the LDL receptor, contain both Asn- and Ser/Thr-linked oligosaccharides (Cummings *et al.*, 1983). Thus, if one prepares oligosaccharides or glycopeptides from animal cell glycoproteins, it is to be expected that a variety of oligosaccharide species will have to be separated from one another before attempts are made to characterize the chains structurally.

For this reason it is important to use analytical procedures that are specific and sensitive and that give reliable, rapid, and inexpensive separations of complicated mixtures of oligosaccharides. A method of separation that meets these requirements is lectin affinity chromatography, which involves the fractionation of oligosaccharides on the basis of their interactions with immobilized plant and animal lectins. When multiple lectins are used the technique is called serial lectin affinity chromatography or SLAC. The technique is effective in a wide range of analyses, since separation is not based simply on differences between oligosaccharides in terms of size, charge, or hydrophobicity.

Affinity chromatography, in conjunction with other separation techniques, is particularly effective for separating and analyzing oligosaccharides. Other separation techniques, which we will not discuss in detail, include size exclusion column chromatography (Yamashita *et al.*, 1982); paper chromatography (Yoshima *et al.*, 1980); paper electrophoresis (Narasimhan *et al.*, 1980) and HPLC (reviewed by Honda, 1984) including the use of amine-adsorption (Mellis and Baenziger, 1983; Blanken *et al.*, 1985; Green and Baenziger, 1986); reversed-phase (Tomiya *et al.*, 1987), ion exchange (Baenziger and Natowicz, 1981; Hardy and Townsend, 1988; Townsend *et al.*, 1988), and amide-adsorption (Tomiya *et al.*, 1988) techniques. These methods combined with affinity chromatography can facilitate the complete purification of isomers and chemically similar oligosaccharide species.

Lis and Sharon (1984; 1986a,b), Osawa and Tsuji (1987), and Merkle and Cummings (1987a) have reviewed the biochemistry of lectins and their utilization in a variety of studies. Lectins have been isolated from a variety of sources, including plants, microorganisms, and animals. A number of these lectins have been studied with regard to the determinants within oligosaccharides to which they interact with high affinity. In addition, many lectins have been immobilized (usually on Sepharose or agarose) and utilized in the separation of complex mixtures of oligosaccharides.

Since different immobilized lectins interact with different oligosaccharide determinants with high affinity, these differences are exploited as the basis for SLAC. The immobilized lectins are used in tandem to facilitate purification of oligosaccharides. Table V contains a list of some of these lectins, the major sugar determinants to which they bind with high affinity, and appropriate references describing their use.

An example of SLAC of a mixture of glycopeptides containing Asn-linked sugar chains from a hypothetical glycoprotein is shown in Fig. 2. The glycoprotein contains six different sugar chains represented by the six different structures shown in the upper part of Fig. 1. Pronase treatment of the purified glycoprotein generates glycopeptides **(I–VI)**, each of which contains a single type of sugar chain. This mixture of glycopeptides is first applied to a column of Con A–Sepharose. As indicated in Table V, this immobilized lectin binds to the high-mannose-type, hybrid-type, and biantennary complex-type Asn-linked chains represented by compounds **I, II, III,** and **IV.** The biantennary complex-type chains bind with lower affinity to the lectin than the high-mannose- and hybrid-type chains and are efficiently separated by this procedure. In contrast, the triantennary complex-type Asn-linked chains represented by glycopeptides **V** and **VI** are not bound by Con A–Sepharose. These three pools of glycopep-

TABLE V

LECTINS USEFUL FOR SERIAL LECTIN AFFINITY CHROMATOGRAPHY OF OLIGOSACCHARIDES AND THE CARBOHYDRATE DETERMINANTS RECOGNIZED WITH HIGH AFFINITY

Lectin	Carbohydrate determinant recognized	References
Con A *(Canavalia ensiformis)*	High-mannose-, hybrid-, and biantennary complex-type Asn-linked chains	Baenziger and Fiete (1979b); Krusius *et al.* (1976); Ogata *et al.* (1975)
Pea lectin *(Pisum sativum)*; Lentil lectin *(Lens culinaris)*	± [R-GlcNAcβ1,6] \\ ± [R′-GlcNAcβ1,2]Manα1,6 \\ Manβ1,4GlcNAcβ1,4GlcNAcβ1-Asn / | ± [R″-GlcNAcβ1,2]Manα1,3 Fucα1,6	Kornfeld *et al.* (1981); Yamamoto *et al.* (1982)
L_4-Phytohemagglutinin (PHA) *(Phaseolus vulgaris)*	Galβ1,4GlcNAcβ1,6 \\ Galβ1,4GlcNAcβ1,2Manα-R	Cummings and Kornfeld (1982a); Hammarström *et al.* (1982)
E_4-PHA *(P. vulgaris)*	Galβ1,4GlcNAcβ1,2Manα1,6 \\ GlcNAcβ1,4Manβ1-R / Galβ1,4GlcNAcβ1,2Manα1,3	Cummings and Kornfeld (1982a); Irimura *et al.* (1981); Yamashita *et al.* (1983a)
Datura stramonium agglutinin (DSA)	[3Galβ1,4GlcNAcβ1]$_n$ and Galβ1,4GlcNAcβ1,6 \\ Galβ1,4GlcNAcβ1,2Manα-R	Crowley *et al.* (1984); Cummings and Kornfeld (1984); Yamashita *et al.* (1987)
Tomato lectin *(Lycopersicon esculentum)*	[3Galβ1,4GlcNAcβ1]$_n$	Merkle and Cummings (1987b)
Maackia amurensis leukoagglutinin (MAL)	NeuAcα2,3Galβ1,4GlcNAc-R	Wang and Cummings (1988)
Sambucus nigra agglutinin (SNA) (elderberry bark)	NeuAcα2,6Gal/GalNAc-R	Shibuya *et al.* (1987a,b)
Helix pomatia agglutinin	GalNAcα1,3-R	Hammarström and Kabat (1969); Torres *et al.* (1988)

TABLE V (*continued*)

Lectin	Carbohydrate determinant recognized	References
Ricinus communis agglutinin (RCA I and RCA II)	Galβ1,4Glc/GlcNAc-R	Baenziger and Fiete (1979c); Narasimhan *et al.* (1985); Kornfeld *et al.* (1981)
Pokeweed mitogen (*Phytolacca americana*)	Galβ1,4GlcNAcβ1,6 \ [3Galβ1,4GlcNAcβ1]$_n$	Katagari *et al.* (1983); Irimura and Nicolson (1983)
Griffonia simplicifolia I	GalNAcα-R or Galα-R	Blake and Goldstein (1980; 1982); Wang *et al.* (1988)
Calf heart agglutinin	[3Galβ1,4GlcNAcβ1]$_n$	Merkle and Cummings (1988)
Bovine Man6P receptor	PO_4^{2-}-6-Man-R	Fischer *et al.* (1982); Varki and Kornfeld (1983)

tides can be further purified by chromatography on other immobilized lectins. For example, glycopeptides **V** and **VI** can be separated from each other by chromatography on tomato lectin–Sepharose, which interacts with high affinity with the poly-*N*-acetyllactosamine chains on compound **VI** and does not bind with high affinity to glycopeptide **V.** Glycopeptides **III** and **IV** can be separated on pea lectin–Sepharose, since the lectin, which binds with high affinity to glycopeptides containing α-1,6-linked fucose residues in the chitobiosyl core, binds to **IV** and not to **III.** The hybrid- and high-mannose-type chains of glycopeptides **I** and **II** can be separated by affinity chromatography on immobilized *Ricinus communis* agglutinin (RCA-I), which binds to the terminal β-1,4-linked galactose residue in glycopeptide **II.** Finally, glycopeptides **III** and **IV** could be further differentiated on the basis of sialic acid linkages by chromatography on immobilized *Maackia amurensis* leukoagglutinin (MAL). This lectin binds with high affinity to the α-2,3-linked sialic acid in glycopeptide **III,** but does not bind to the isomeric glycopeptide **IV,** in which sialic acid is linked α-2,6.

The scheme of SLAC is organized in this way because, both practically and empirically, it has been found to minimize the number of steps necessary for the analysis in most cases. All of the lectins that appear to recognize the core sugars and their substitution pattern (e.g., Con A, E_4-PHA, and pea lectin) are utilized early in the fractionation. Lectins that recognize more peripheral determinants (e.g., RCA-I and MAL) are employed later in the analysis. In this way glycopeptides may be separated based

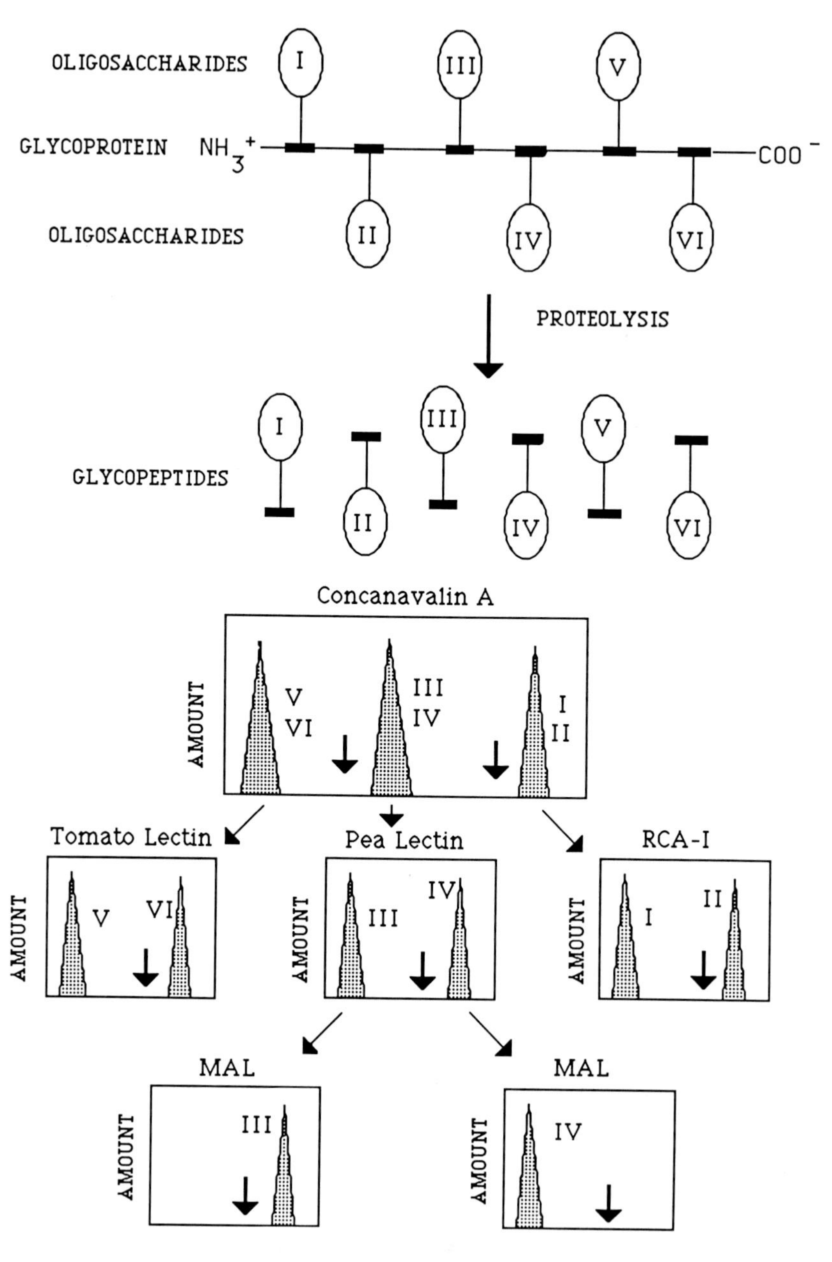
OLIGOSACCHARIDES
I
III
V
GLYCOPROTEIN
NH_3^+
COO^-
OLIGOSACCHARIDES
II
IV
VI
PROTEOLYSIS
GLYCOPEPTIDES
I
II
III
IV
V
VI
Concanavalin A
AMOUNT
V
VI
III
IV
I
II
Tomato Lectin
Pea Lectin
RCA-I
AMOUNT
V
VI
III
IV
I
II
MAL
MAL
AMOUNT
III
IV
FRACTION NUMBER

on their class distinctions, such as complex-type or high-mannose-type or complex-type containing poly-*N*-acetyllactosamine chains. For this reason, Con A–Sepharose is often used in the early steps of the separation, since it interacts with different affinities with the three major classes of Asn-linked sugar chains.

The fractionation of glycopeptides or oligosaccharides by even these few immobilized lectins is impressive, but it certainly does not allow purification of all known Asn-linked oligosaccharides. However, extensive purification of many sugar chains can be achieved by the inclusion of more lectins with different carbohydrate-binding specificities coupled with use of physical methods of separation such as HPLC and size exclusion chromatography. For example, Mellis and Baenziger (1983) and Green and Baenziger (1986) demonstrated that ion-suppression amine adsorption HPLC on Micropak AX-10 and AX-5 can resolve sialylated or phosphorylated sugar chains linked to small peptides based on size. The peptide and other potentially charged moieties contribute little to the retention times, allowing the analysis of glycopeptides as well as oligosaccharides. Alternatively, ion exchange HPLC on the same columns can be performed to separate oligosaccharides on the basis of charged residues (Green and Baenziger, 1986). Recently, Hardy and Townsend (1988) have reported the separation of positional isomers of oligosaccharides and glycopeptides using high performance anion exchange chromatography on Dionex CarboPac columns with pulsed amperometric detection.

As with any fractionation technique, it is important to use well characterized authentic standards of high purity when using HPLC to purify and identify oligosaccharide structures. For HPLC separations based on size, we routinely standardize our columns with ^{14}C-labeled starch oligosaccharides, prepared as described by Nishigaki *et al.* (1978) and with [^{3}H]mannose-labeled high mannose type oligosaccharides derived from the mouse lymphoma cell line BW5147 (Nyame *et al.*, 1988) and from Chinese hamster ovary (CHO) cells (Li *et al.*, 1978).

One should always be cautious in interpreting the observed fraction-

---

FIG. 2. Generalized scheme for serial lectin affinity chromatography (SLAC) in the separation of glycopeptides. In the procedure shown, glycopeptides are generated by proteolytic cleavage of a glycoprotein. In the top illustration the glycoprotein contains six different Asn-linked chains designated by numbers that refer to the structures shown in Fig. 1. The mixture of glycopeptides is applied to a column of Con A–Sepharose. Some glycopeptides are not bound, and some are bound and differentially eluted with haptenic sugars as indicated by the arrows. Pooled glycopeptides are subsequently applied to other immobilized lectins as shown to facilitate purification. Although the scheme shown utilizes glycopeptides, SLAC is also effective using oligosaccharides enzymatically released from either the original glycoprotein or the derived glycopeptides.

ation pattern of glycopeptides on immobilized lectins. The strength of interactions of glycopeptides with the lectins depends greatly on the coupling density of the lectin, the percentage of active lectin conjugated, and flow rates and geometry of the columns (as discussed by Merkle and Cummings, 1987a). Thus, not all batches of lectin conjugates may behave identically. It is of course wise, as in any technique, to standardize immobilized lectin columns with chemically defined structures. Sources of these standards have been described previously (Merkle and Cummings, 1987a).

In general, it is best if the lectins are conjugated at a density of at least 10 mg protein per milliliter of gel. This is often far above that commercially available. The reason for the high coupling density is that the binding affinity (Kd) of the lectins for sugar determinants on a glycopeptide are usually on the order of 10^{-5} to $10^{-6}M$. It can thus be calculated that to bind 99% of a ligand that binds to a lectin with an affinity in this range, the coupling density must be at least 10 mg/ml. This conclusion was essentially reached by Baenziger and Fiete (1979b,c) and has been experimentally verified in a number of laboratories. Commercial sources of immobilized lectins are available from E·Y Laboratories, Sigma, Pharmacia-LKB Biotechnology, and Vector Laboratories; however, the coupling density of many of the lectins sold by these companies is often too low to be useful. To achieve higher coupling densities of immobilized lectins, it is advisable to purchase or otherwise obtain the purified lectin and perform the coupling in the laboratory. The lectin is coupled directly to Affi-Gel-10 or Affi-Gel-15 (from Bio-Rad), following the directions from the supplier. The coupling reaction should be done under conditions where the lectin is at or just slightly above the desired coupling density (i.e., ~10–12 mg/ml. Affigel-coupled lectins are highly stable, require little prehandling of the gel, and give reproducible and high coupling efficiencies. Lectins may also be covalently coupled to other supports such as CNBr-activated Sepharose 4B. Although SLAC usually involves simple column-chromatographic procedures, Green *et al.* (1987a,b) have shown the utility of HPLC methods for lectin chromatography. In this approach lectins are covalently attached to activated diol silica and the samples are eluted essentially as described for more traditional SLAC methods.

In most cases, analyses involving immobilized lectins are performed with extremely small amounts of radiolabeled glycopeptides. Thus, saturation of the columns is usually of minimal concern. To test this possibility, however, it is advisable to recover the glycopeptides apparently not bound by an immobilized lectin and pass them back over the column to ensure the complete removal of all glycopeptides that can possibly bind. Because the sample sizes are so small, it is usually not necessary to use columns containing more than 1–2 ml of immobilized lectin on the gel. Such columns with small amounts of gel often have high flow rates that do not allow sufficient time for the glycopeptides to interact with the lectins. Thus it is important to pack columns so that the flow rates are no faster than 1–2 ml/minute. This can be accomplished by packing them in 1-ml disposable plastic pipets plugged tightly at the bottom with glass wool. Most lectins are stable at room temperature and can be used repeatedly for up to 50 or 100 column runs before detectable loss of activity or capacity. In between column runs it is advisable to store the columns at 4°C. All of the chromatographic steps may be performed at 4°C if so desired; low temperature enhances the affinity of most lectins for glycopeptides. In any case, columns should be tested routinely to ensure their activity.

Because so many immobilized lectins are now used for SLAC, it is not possible to discuss each one and the conditions for their use in this space. The reader is urged to refer to the appropriate recent references listed in Table V for the precise steps in the procedures. In addition, Merkle and Cummings (1987a) have discussed many of these lectins and the exact manner in which the columns are used.

Lectin affinity chromatography at best affords relatively purified samples, which should then be analyzed for structure and purity by other methods when possible. Although the immobilized lectins interact with glycopeptides as indicated in Table V and Fig. 2, new and previously untested glycopeptides may behave somewhat differently. Another concern in SLAC is the nature of the sample. Some immobilized lectins, such as pea and lentil lectins, are known to have different binding affinities for glycopeptides versus reduced or unreduced oligosaccharides derived from those same glycopeptides. It has been shown that oligosaccharides released by *N*-Glycanase and subsequently reduced behave differently from glycopeptides when subjected to lectin affinity chromatography (Green and Baenziger, 1987). It is important to be aware of these differences for each lectin used, and one should consult the primary literature for further information.

The SLAC method is especially useful in those studies in which there is a need to compare the types of glycoconjugates isolated from a receptor or from total cell glycoproteins to those from other receptors or other cells. It is difficult using most traditional methods of carbohydrate analysis to analyze in a comparative way an array of glycoconjugate structures. For example, there is often a need to analyze a number of glycoconjugates from a variety of cells or to analyze glycoconjugates before and after cellular differentiation, neoplastic transformation, recycling of receptors, and so forth. In this regard, it is often sufficient for investigators to determine that glycosylation differences exist between glycoprotein species; detailed structural studies may not always be required. The SLAC technique is rapid and reliable, and is especially suited to allow comparative analyses of oligosaccharides.

VII. Analysis of Metabolically Radiolabeled Sugar Chains

The analysis of the structures of oligosaccharides derived from glycoproteins cannot be accomplished by a single method, and usually a combination of techniques is required. The particular combination necessary to define a structure is dictated by that structure and therefore it is not possible to list a precise series of steps. In the following section we will consider the fractionation and purification of glycoprotein-derived oligosaccharides. In many cases the manner of purification and the affinity techniques employed provide important clues about certain structural features of an oligosaccharide. We will consider here the structural analysis of an oligosaccharide that is metabolically radiolabeled with a radioactive precursor sugar.

A. Compositional Analysis

The initial strategy for determination of the structure of a glycopeptide or oligosaccharide is to subject it to acid hydrolysis and determine the sugar composition of the oligosaccharide. Using samples that have been metabolically radiolabeled using [^{3}H]mannose, [^{3}H]galactose, and [^{3}H]glucosamine and fractionated in an identical manner, it is possible to determine whether the oligosaccharide in question contains radiolabeled mannose, fucose, *N*-acetylglucosamine, *N*-acetylgalactosamine, galactose, glucose, and/or sialic acid. Besides determining the actual composition, these data are useful for approximating the ratios of radioactive sugars and comparing these ratios from one glycopeptide to another.

[^{3}H]Mannose-and [^{3}H]galactose-labeled glycopeptides or oligosaccharides are hydrolyzed in conical-bottom glass tubes by treating with 2 *N* HCl in a final volume of 0.2 ml for 4 hours at 100°C. Marbles are placed on top of the tubes to reduce evaporation rates. Alternatively, 1-ml capped microreaction vessels may be used. Following the hydrolysis the acid is removed by evaporation under reduced pressure, the sample is resuspended in 20 μl water, and the samples are subjected to descending paper chromatography for 24 ([^{3}H] mannose) or 48 ([^{3}H] galactose) hours on

Whatman No. 1 filter paper in the solvent system ethyl acetate–pyridine–water (8 : 2 : 1). Six lanes (1.5 in. wide) are penciled on sheets and marked into 1-cm sections (starting 3 in. from the top, 48 cm from orgin to end) and the bottom is cut in a zig-zig pattern. The sample is spotted 5 μl at a time, drying completely between spottings (drying time may be reduced by using a blower gun set on cool). After chromatography the paper is allowed to dry, then lanes are separted and the 1-cm sections are cut for liquid scintillation counting in 0.4 ml water and 4 ml scintillation fluid. The labeled sugar residues are identified by cochromatography with standards. Unlabeled standards (100 nmol each, mannose, fucose, and galactose) are run in parallel lanes and are located by the silver nitrate dip assay (Trevelyan *et al.*, 1950).

[^{3}H]Glucosamine-labeled glycopeptides are hydrolyzed in a similar manner except that 4 *N* HCl is used. After hydrolysis and removal of the acid, it is necessary to reacetylate the sugars in 0.3 ml of H_2O with 0.03 ml of 4 *N* Na_2CO_3 and 0.045 ml of freshly prepared ice-cold 12.5% acetic anhydride solution in H_2O. After 10 minutes at room temperature followed by 5 minutes at 100°C, the mixture is desalted over a 3-ml column of Amberlite MB-3 (Sigma) mixed-bed resin. The column is washed with 5 ml of H_2O and the eluate containing the reacetylated sugars is dried by evaporation under reduced pressure. The samples are subjected to descending paper chromatography on borate-impregnated paper in the solvent system 1-butanol–pyridine–water (6 : 4 : 3) for 48–60 hours (Cardini and Leloir, 1957). Standards are prepared as described by Nyame *et al.* (1987).

Since strong acid hydrolysis destroys sialic acid, it is advisable to desialylate a portion of the [^{3}H]glucosamine-labeled glycopeptides with mild acid or neuraminidase in order to identify and quantify sialic acid. For acid desialylation the sample is dissolved in 2 *N* acetic acid in a final volume of 0.2 ml and incubated for 1 hour at 100°C. The acid is removed by evaporation under reduced pressure or by lyophilization. It should be noted that this mild acid hydrolysis can also remove most terminal fucose residues. Sialic acid can also be removed by neuraminidase treatment as described later. Released and radiolabeled sialic acid may be detected by descending paper chromatography using the solvent system ethyl acetate–pyridine–acetic acid–water (5 : 5 : 1 : 3) (Cummings and Kornfeld, 1984).

B. Ion Exchange Chromatography

Oligosaccharides that contain sialic acid, phosphate, sulfate, or other charged moieties will interact with anion exchange resins. A decrease in the net charge of an oligosaccharide by treatments with neuraminidase, mild acid, alkaline phosphatase (Varki and Kornfeld, 1983), or solvolysis (Freeze *et al.*, 1983), or combinations of these treatments provide indications as to the nature of the charged group and often the number of such moieties on an oligosaccharide. The efficacy of these treatments is monitored by ion exchange chromatography on QAE–Sephadex or DEAE–Sephadex before and after treatment. For example, a biantennary glycopeptide having two sialic acid residues as the nonreducing termini would behave as a species with two negative charges upon chromatography on QAE–Sephadex and would become a neutral species after removal of sialic acid with neuraminidase.

To perform the anion exchange chromatography on QAE–Sephadex (Sigma), a 2-ml column is equilibrated in 2 m*M* Tris base, pH 9.0. Glycopeptides are applied to the column, and six 2-ml fractions are collected at room temperature. Bound glycopeptides are eluted by a step gradient of 12 ml each of 2 m*M* Tris (pH 9.0) containing 0, 20, 70, 140, 200, 250, or 1000 m*M* NaCl. Glycopeptides containing one, two, or three negative charges are bound by QAE–Sephadex and eluted with 20, 70, and 140 m*M* NaCl, respectively. A complication in this analysis is that the peptide moiety may contribute to the net charge of the glycopeptide. To circumvent this problem, the negatively charged oligosaccharides can be absorbed to QAE–Sephadex columns and eluted using a step gradient or linear gradient of pyridine-acetate (2–500 m*M* pyridine-acetate, pH 5.4) (Goldberg and Kornfeld, 1981; Gabel and Kornfeld, 1982). Under these conditions the contribution of peptide to the net charge is minimized.

Sialylated oligosaccharides can also be fractionated on DEAE–cellulose by suspending the sample in 2 m*M* pyridine-acetate buffer (pH 5.4) and applying it to a column of DE-52 equilibrated in the same buffer. Neutral oligosaccharides are found in the flowthrough when the column is washed with the same buffer. The sialyl oligosaccharides are eluted with 12 m*M* and 60 m*M* pyridine-acetate buffer, pH. 5.4 (Smith *et al.*, 1978). Another method of fractionation of charged oligosaccharides is anion exchange HPLC using a TSK–DEAE column. The columns are equilibrated in 25 m*M* potassium phosphate buffer (pH 5.0), and elution is accomplished using a linear gradient to 0.4 *M* buffer (Sasaki *et al.*, 1987).

C. Methylation

Methylation analysis is employed to define the glycosyl linkage positions of mannosyl, fucosyl, and galactosyl residues. When this approach is combined with sequential exoglycosidase treatment (see later), a great deal of information about the structure of the oligosaccharide is obtained.

[^{3}H]Galactose-labeled glycopeptides are methylated by the procedure of Hakomori (1964) and then hydrolyzed in 2 *N* H_2SO_4 for 4 hours at 100°C. The methylated galactose species are separated by thin-layer chromatography (TLC) on Silica-gel G in the solvent system acetone–water–ammonium hydroxide (250 : 3 : 1.5) (Stoffyn *et al.*, 1971). Methylated [^{3}H]mannose and fucose species are separated by TLC in the solvent system acetone–benzene–water–ammonium hydroxide (80 : 20 : 1.2 : 0.6). Sample lanes are scraped in 0.5-cm sections and radioactivity is determined by scintillation counting. Alternatively, the methylated mannose and fucose residues may be reduced with $NaBH_4$ and then separated by reverse-phase HPLC using a Zorbax ODS column (250 × 4.6 mm) at 45°C (Szilagyi *et al.*, 1985; Wang and Cummings, 1988).

D. Endoglycosidase Treatment

The sensitivity of glycopeptides or oligosaccharides to treatments with endoglycosidases can provide specific structural information. As described earlier, for example, susceptibility of the glycopeptide to digestion with endoglycosidase H indicates the probability of either a high-mannose or hybrid structure. However, the digestion of complex structures with an endoglycosidase such as endo-β-galactosidase can provide specific structural data. Treatment with the enzyme endo-β-galactosidase from *Escherichia freundii* can indicate the presence of poly-*N*-acetyllactosamine sequences. This enzyme cleaves oligosaccharides contained in the sequence R-GlcNAcβ1,3Galβ1,4GlcNAc-R′ at internal galactose residues and can cleave at multiple sites to release low molecular weight oligosaccharides (Nakagawa *et al.*, 1980; Fukuda *et al.*, 1978). For example, the sequence:

Galβ1,4GlcNAcβ1,3Galβ1,˅4GlcNAcβ1,3Galβ1,˅4GlcNAc-Man-R

on glycopeptide **VI** (Fig. 1) can be cleaved at the sites indicated by the arrowheads to generate 1 mol of trisaccharide Galβ1,4GlcNAcβ1,3Gal and 1 mol of disaccharide GlcNAcβ1,3Gal. These products can be separated, identified, and quantified using descending paper chromatography in the solvent system ethyl acetate–pyridine–acetic acid–water (5 : 5 : 1 : 3) (Cummings and Kornfeld, 1984; Merkle and Cummings, 1987b). Performance of this analysis requires metabolic labeling with [^{3}H]galactose. Treatment with this endoglycosidase, especially in conjunction with fractionation or analysis using immobilized tomato lectin (see earlier), can be used to deduce the approximate length of a poly-*N*-acetyllactosamine chain.

For treatment with *E. freundii* endo-β-galactosidase (V-Labs and ICN Biomedical), dried glycopeptides are treated in a 15-ml conical tube with

2.4 mU enzyme in 0.1 *M* sodium acetate (pH 5.6), containing 20 μg BSA in a final volume of 0.02 ml at 37°C for 24 hours in a toluene atmosphere (as described earlier for pronase digestion). The reaction is terminated by addition of 0.4 ml 7% *n*-propyl alcohol or H_2O and boiling for 5 minutes. To separate released oligosaccharides from residual glycopeptides, the material is applied either to Sephadex G-25 (1 × 45 cm) or adjusted to 0.1 *M* NH_4HCO_3 and applied to a column (1.5 × 90 cm) of Bio-Gel P-10. Fractions (1 ml) are collected and portions are subjected to liquid scintillation counting. Alternatively, the reaction mixture may be subjected directly to descending paper chromatography (Cummings and Kornfeld, 1984; Merkle and Cummings, 1987b).

Another endoglycosidase useful for structural analysis is the *Escherichia coli* bacteriophage-derived endo-*N*-acetylneuraminidase (Endo N), which specifically hydrolyzes α-2,8-sialyl linkages. This enzyme (not commercially available, however, at this time) has proved useful for studies of glycoproteins with a high sialic acid content such as those found in neuronal tissue (Troy *et al.*, 1987).

Other endoglycosidases are useful in the identification of glycosaminoglycan (GAG) or GAG-like oligosaccharides. For example, susceptibility to keratanase, an enzyme that cleaves β-galactoside linkages of nonsulfated galactosyl residues in keratan sulfate (Nakazawa and Suzuki, 1975; Pierce and Arango, 1986), indicates the presence of a polymer of alternating $(Gal\beta1,4GlcNAc\beta1,3)_n$ containing sulfate which may be covalently linked to both Gal and GlcNAc residues.

E. Sequential Exoglycosidase Treatment

Glycopeptides or oligosaccharides can be structurally analyzed by using sequential digestion with exoglycosidases such as β-*N*-acetylhexosaminidase, α-*N*-acetylgalactosaminidase, β-galactosidase, α-galactosidase, neuraminidase, α-mannosidase, α-fucosidase, and β-mannosidase. After each digestion, an aliquot of the oligosaccharides can be methylated to provide the complete linkage description. The oligosaccharides can be subjected to size exclusion chromatography using conventional resins or HPLC (Wang and Cummings, 1988) to compare their sizes before and after treatment as removal of sugar residues should reduce the size of the oligosaccharides. Alternatively, the digestion mixture may be subjected directly to descending paper chromatography to determine the identity and amount of released sugar residues.

Although specific conditions for exoglycosidase treatments are shown in Table VI, this should not be considered to be a complete list. This table may be used as a starting point, but we suggest that the investigator utilize

TABLE VI

CONDITIONS FOR SELECTED EXOGLYCOSIDASE TREATMENTS

Enzyme	Source	Amount required (U/ml)	Incubation conditions	References
α-Galactosidase	Green coffee bean	1	0.1 *M* Phosphate, pH 6.4, 24 hours, 37°C	Cummings and Mattox (1988)
β-Galactosidase	*Escherichia coli*	—	0.05–0.5 *M* Potassium phosphate, pH 7	
	Jack bean	0.4	0.05 *M* Sodium citrate, pH 4.6, 48 hours, 37°C	Kornfeld *et al.* (1981)
	Diplococcus	0.05–0.2	0.05 *M* Cacodylate, pH 6.0, 4–6 hours, 37°C	Baenziger and Fiete (1979a)
	Bovine testes	0.01–0.04	0.1 *M* Citrate-phosphate, pH 4.3, 12 hours, 37°C	Distler and Jourdian (1978)
β-*N*-Acetylhexosaminidase	Jack bean	10	0.05 *M* Sodium citrate, pH 5.0, 18 hours, 37°C	Yamashita *et al.* (1979)
	Diplococcus	0.04–0.2	0.05 *M* Sodium citrate, pH 6.0, 18 hours 37°C	Yamashita *et al.* (1981)
α-*N*-Acetylgalactosaminidase	*Charonia lampas*	0.25	0.1 *M* Citrate, pH 4.6, 48 hours, 37°C	Cummings *et al.* (1985)
α-Mannosidase	Jack bean	2–20	0.05 *M* Sodium citrate, pH 4.6, 48 hours, 37°C	Li and Li (1972)
β-Mannosidase	*Aspergillus niger*	—	0.05 *M* Glycine, pH 3.5, 12 hours, 37°C	Baenziger (1979)
α-Fucosidase	Bovine epididymus	0.3	0.025 *M* Sodium citrate, pH 4.5, 48 hours, 37°C	Kornfeld *et al.* (1981)
	Almond	0.003	0.05 *M* Sodium acetate, pH 4.0	Yamashita *et al.* (1979)
Neuraminidase	*Vibrio cholerae*	250	0.05 *M* Acetate, pH 4.6, 0.15 *M* NaCl, 10 m*M* $CaCl_2$, 72 hours, 37°C	Kornfeld *et al.* (1981)
	Arthrobacter ureafaciens	0.2	0.1 *M* Sodium acetate, pH 4.8, 12 hours, 37°C	Wang and Cummings (1988)

the technical information obtained with the enzymes from the supplier in order to optimize reaction conditions and to avoid inclusion of inhibitors of enzyme activity. The various exoglycosidases are available from a number of suppliers including Boehringer Mannheim, V-Labs, Sigma, Calbiochem, and ICN Biomedical. When more than one source of a glycosidase is available, an important consideration is the choice of the exoglycosidase. Often the specificity of cleavage in terms of the linkage or substitution of the susceptible residues should be taken into account. For example, the neuraminidase from *Arthrobacter ureafaciens* has broad specificity, capable of cleaving *N*-acetylneuraminic acid in α-2,3, α-2,6, and α-2,8 linkages as well as being able to cleave the α-ketosidic linkage of *N*-glycolylneuraminic acid (Uchida *et al.*, 1979). On the other hand, the choice of limited-specificity glycosidases may help to distinguish a certain linkage. For example, jack bean β-galactosidase cleaves Galβ-1,4 linkages in preference to Galβ-1,3 linkages (Li and Li, 1972), whereas bovine testicular β-galactosidase will readily cleave Galβ-1,3 as well as the β-1,4 linkage from the nonreducing termini of disaccharides (Distler and Jourdian, 1978). The lack of cleavage does not always indicate the absence of the residue from a glycoprotein, as the residue may be substituted in a way that sterically interferes with the glycosidase activity. For example, although diplococcal β-*N*-acetylglucosaminidase readily hydrolyzes the GlcNAcβ1,2 linkage in GlcNAcβ1,4(GlcNAcβ1,2)Man, it does not readily cleave the linkage in GlcNAcβ1,6(GlcNAcβ1,2)Man (Yamashita *et al.*, 1981).

Prior to treatment the glycopeptides or oligosaccharides are dried under vacuum. The enzyme digestions are terminated by dilution with H_2O (0.5 ml), and the reaction tubes are heated in a boiling-water bath for 2 minutes. It is important that the exoglycosidase be tested with a known substrate. A further precaution is that some of these enzymes contain contaminating activities that may be apparent when larger amounts of enzyme than recommended are used or when incubation conditions are altered.

The combination of exoglycosidase treatment with the analytical procedures described earlier provides the methylation pattern, the identity of sugars released, and the decrease in apparent size of the oligosaccharides after treatment. Such data should provide clear information as to the structures of the oligosaccharides being studied. It should be emphasized that the use of these methods also depends on the availability of standards for size comparison and structure comparison by cochromatography.

F. Example of the Separation and Analysis of Metabolically Radiolabeled Sugar Chains from a Cellular Glycoprotein

It is important to conclude this chapter by discussing an example in which these techniques can be applied and to illustrate the rationale for the analytical approach. Consider that one has a mixture of the following two complex-type Asn-linked sugar chains, represented by glycopeptides **III** and **IV**, shown in Fig. 1. Consider also that these glycopeptides could be derived from cells grown in media containing [2-^{3}H]mannose, [6-^{3}H]galactose, or [6-^{3}H]glucosamine. Initial fractionation is accomplished by SLAC. Both glycopeptides would bind to Con A–Sepharose. Glycopeptide **IV** would bind to immobilized pea lectin whereas glycopeptide III would not. Glycopeptide **III,** but not **IV,** would bind to immobilized MAL.

After fractionation the sugar composition is determined. A portion of each of the separated glycopeptides would be hydrolyzed in strong acid and the radiolabeled monosaccharides identified by cochromatography with standards. Both glycopeptides **III** and **IV,** derived from [^{3}H]mannose-labeled cells, would contain radioactive mannose, but only glycopeptide **IV** would contain radiolabeled fucose. When isolated from cells grown in media containing [6-^{3}H]galactose, both **III** and **IV** would be radiolabeled and all the radioactivity would be recovered as galactose following strong acid hydrolysis. When isolated from cells grown in media containing [6-^{3}H]glucosamine, both **III** and **IV** would be found to contain radioactive *N*-acetylglucosamine and sialic acid (the latter would be identified after mild acid hydrolysis).

More detailed information about the structures of these glycopeptides could be obtained. For simplicity, let us just consider the structural analysis of glycopeptide **IV.** Upon ion exchange column chromatography, the oligosaccharide would behave as a species with two negative charges and would become a neutral species following removal of sialic acid with neuraminidase, indicating that the chain is disialylated. To determine if sialic acid residues are in α linkage to the galactose residues and in what positions, the [^{3}H]galactose-labeled oligosaccharides could be methylated before and after neuraminidase treatment (which cleaves α-linked sialic acids). The partially methylated and radiolabeled galactose residues could be recovered directly by strong acid hydrolysis, separated by either TLC or HPLC, and compared to a number of standards. Methylation before treatment would generate the species 2,3,4-tri-*O*-methylgalactose, whereas methylation after neuraminidase treatment would generate the species 2,3,4,6-tetra-*O*-methylgalactose. These results would indicate that sialic acid is present in α linkage to position C-6 of penultimate galactose residues.

Methylation analysis of the [^{3}H]mannose-labeled oligosaccharide could also be performed before and after treatment with specific exoglycosidases. For example, methylation of glycopeptide **IV** would generate 3,4,6-tri-*O*-methylmannose and 2,4-di-*O*-methylmannose in the approximate ratio of 2 : 1. Now, suppose the glycopeptide were treated with a combination of neuraminidase, β-galactosidase, and β-*N*-acetylhexosaminidase before this methylation. Methylation of the residual glycopeptide would generate 2,3,4,6-tetra-*O*- methylmannose and 2,4-di-*O*-methylmannose in the approximate ratio of 2 : 1. If this treated glycopeptide were subsequently treated with α-mannosidase, two-thirds of the mannose would be released, since two of the three mannose residues are α-linked. Methylation of the α-mannosidase-treated sample would now generate only 2,3,4,6-tetra-*O*-methylmannose. This would indicate that the two mannose residues are in α linkage to positions C-3 and C-6 of a β-linked mannosyl residue.

Additional information about this structure could be obtained depending on the amount of radiolabeled sample available. However, the type of analysis previously described should illustrate the detailed structural information that can be obtained by SLAC and metabolic radiolabeling.

VIII. Conclusions

In this chapter we have attempted to describe some of the procedures and experimental approaches to identify and partially characterize the sugar chains of less abundant animal cell glycoproteins. It is relevant to ask why such detailed analysis is necessary. How far should one carry out the analysis, and how much detail should one attempt to obtain? The complete description of a receptor or other glycoprotein under study requires a knowledge not only of its amino acid sequence but also of all cotranslational and posttranslational modifications of the protein, such as phosphorylation, proteolytic cleavage, and sugar addition. Often the lack of understanding of these modifications in chemical detail can be a severe hindrance to understanding the biosynthesis and function(s) of proteins, especially in cases in which the glycoprotein is altered in some way, for example, either through mutation or through the administration of exogenous agents. With the approaches and techniques described in this chapter, it should be possible for the nonspecialist in carbohydrate research to investigate the glycosylated state of a protein and partially define the composition and structures of extremely small amounts of material.

Acknowledgments

The authors gratefully acknowledge the support of NIH grants RO1 CA 37626 and RO1 EY 05971 to R. D. C., National Research Service Award 1F32 CA 07795 to R. K. M., and a Bioresources and Biotechnology grant from the University of Georgia Research Foundation to N. L. S. We thank Dr. Carl Bergmann for critically reviewing this manuscript.

REFERENCES

Abeijon, C., and Hirschberg, C. (1988). *Proc. Natl. Acad. Sci. U.S.A.* **85,** 1010–1014.

Arakaki, R. F., Hedo, J. A., Collier, E., and Gorden, P. (1987). *J. Biol. Chem.* **262,** 11886–11892.

Asakawa, K., Hedo, J. A., McElduff, A., Rouiller, D. G., Waters, M. J., and Gorden, P. (1986). *Biochem. J.* **238,** 379–386.

Baenziger, J., and Kornfeld, S. (1974). *J. Biol. Chem.* **249,** 7270–7281.

Baenziger, J. U. (1979). *J. Biol. Chem.* **254,** 4063–4071.

Baenziger, J. U., and Fiete, D. (1979a). *J. Biol. Chem.* **254,** 789–795.

Baenziger, J. U., and Fiete, D. (1979b). *J. Biol. Chem.* **254,** 2400–2407.

Baenziger, J. U., and Fiete, D. (1979c). *J. Biol. Chem.* **254,** 9795–9799.

Baenziger, J. U., and Natowicz, M. (1981). *Anal. Biochem.* **112,** 357–361.

Bartles, J. R., Braiterman, L. T., and Hubbard, A. L. (1985). *J. Biol. Chem.* **260,** 12792–12802.

Blake, D. A., and Goldstein, I. J. (1980). *Anal. Biochem.* **102,** 103–109.

Blake, D. A., and Goldstein, I. J. (1982). *In* "Methods in Enzymology" (V. Ginsburg, ed.), Vol. 83, pp. 127–132. Academic Press, New York.

Blanchardie, P., Lustenberger, P., Denis, M., Orsonneau, J. L., and Bernard, S. (1986). *J. Steroid Biochem.* **24,** 263–267.

Blanken, W. M., Bergh, M. L. E., Koppen, P. L., and van den Eijnden, D. H. (1985). *Anal. Biochem.* **145,** 322–330.

Boege, F., Ward, M., Jürss, R., Hekman, M., and Helmreich, E. J. M. (1988). *J. Biol. Chem.* **263,** 9040–9049.

Breitfeld, P., Rup, D., and Schwartz, A. L. (1984). *J. Biol. Chem.* **259,** 10414–10421.

Burridge, K. (1978). *In* "Methods in Enzymology" (V. Ginsburg, ed.), Vol. 50, pp. 54–64. Academic Press, New York.

Butler, W. T., and Cunningham, L. W. (1966). *J. Biol. Chem.* **241,** 3882–3888.

Cardini, C. E., and Leloir, L. F. (1957). *J. Biol. Chem.* **249,** 7270–7281.

Carter, W. G., and Wayner, E. A. (1988). *J. Biol. Chem.* **263,** 4193–4201.

Cervantes-Olivier, P., Durieu-Trautmann, O., Delavier-Klutchko, C., and Strosberg, A. D. (1985). *Biochemistry* **24,** 3765–3770.

Childs, R. A., Gregoriou, M., Scudder, P., Thorpe, S. J., Rees, A. R., and Feizi, T. (1984). *EMBO J.* **3,** 2227–2233.

Claesson-Welsh, L., Rönnstrand, L. K., and Heldin, C. H. (1987). *Proc. Natl. Acad. Sci. U.S.A.* **84,** 8796–8800.

Crowley, J. F., Goldstein, I. J., Arnarp, J., and Lönngren, J. (1984). *Arch. Biochem. Biophys.* **231,** 524–533.

Cummings, R. D., and Kornfeld, S. (1982a). *J. Biol. Chem.* **257,** 11230–11234.

Cummings, R. D., and Kornfeld, S. (1982b). *J. Biol. Chem.* **257,** 11235–11240.

Cummings, R. D., and Kornfeld, S. (1984). *J. Biol. Chem.* **259,** 6253–6260.

Cummings, R. D., and Mattox, S. A. (1988). *J. Biol. Chem.* **263,** 511–519.

Cummings, R. D., Kornfeld, S., Schneider, W. J., Hobgood, K. K., Tolleshaug, H., Brown., M. S., and Goldstein, J. L. (1983). *J. Biol. Chem.* **248,** 15261–15273.

Cummings, R. D., Soderquist, A. M., and Carpenter, G. (1985). *J. Biol. Chem.* **260,** 11944–11952.

Daniel, T. O., Milfay, D. F., Escobedo, J., and Williams, L. T. (1987). *J. Biol. Chem.* **262,** 9778–9784.

Distler, J. J., and Jourdian, G. W. (1978). *In* "Methods in Enzymology" (V. Ginsburg, ed.), Vol. 50, 514–520. Academic Press, New York.

Dohlman, H. G., Bouvier, M., Benović, J. L., Caron, M. G., and Lefkowitz, R. J. (1987). *J. Biol. Chem.* **262,** 14282–14288.

Dubois, M., Giles, K. H., Hamilton, J. K., Rebers, P. A., and Smith, F. (1956). *Anal. Chem.* **28,** 350–356.

Duronio, V., Jacobs, S., and Cuatrecasas, P. (1986). *J. Biol. Chem.* **261,** 970–975.

Dygert, S., Li, L. H., Florida, D., and Thoma, J. A. (1965). *Anal. Biochem.* **13,** 367–374.

El Battari, A., Luis, J., Martin, J. M., Fantini, J., Muller, J. M., Marvaldi, J., and Pichon, J. (1987). *Biochem. J.* **242,** 185–191.

Elbein, A. D. (1987). *In* "Methods in Enzymology" (V. Ginsburg, ed.), Vol. 138, pp. 661–709. Academic Press, Orlando, Florida.

Fatemi, S. H., and Tartakoff, A. M. (1986). *Cell (Cambridge, Mass.)* **46,** 653–657.

Feizi, T. (1985). *Nature (London)* **314,** 53–57.

Finne, J., and Krusius, T. (1982). *In* "Methods in Enzymology" (V. Ginsburg, ed.). Vol. 83, pp. 269–277. Academic Press, New York.

Fischer, H. D., Creek, K. E., and Sly, W. S. (1982). *J. Biol. Chem.* **257,** 9938–9943.

Freeze, H. H., and Wolgast, D. (1986). *J. Biol. Chem.* **261,** 135–141.

Freeze, H. H., Yeh, R., Miller, A. L., and Kornfeld, S. (1983). *J. Biol. Chem.* **258,** 14874–14879.

Fuhrmann, U., Bause, E., and Ploegh, H. (1985). *Biochim. Biophys. Acta* **825,** 95–110.

Fukuda, M., and Fukuda, M. N. (1984). *In* "Biology of Glycoproteins" (R. J. Ivatt, ed.), pp. 183–234. Plenum, New York.

Fukuda, M. N., Watanabe, K., and Hakomori, S. (1978). *J. Biol. Chem.* **253,** 6814–6819.

Gabel, C. A., and Kornfeld, S. (1982). *J. Biol. Chem.* **257,** 10605–10612.

Gahmberg, C. G. (1978). *In* "Methods in Enzymology" (V. Ginsburg, ed.), Vol. 50, pp. 204–206. Academic Press, New York.

Gasa, S., and Kornfeld, S. (1987). *Arch. Biochem. Biophys.* **257,** 170–176.

Goldberg, D. E., and Kornfeld, S. (1981). *J. Biol. Chem.* **256,** 13060–13076.

Goldberg, D. E., Gabel, C. A., and Kornfeld, S. (1983). *J. Cell Biol.* **97,** 1700–1706.

Gooi, H. C., Schlessinger, J., Lax, I., Yarden, Y., Libermann, T. A., and Feizi, T. (1983). *Biosci. Rep.* **3,** 1045–1053.

Green, E. D., and Baenziger, J. U. (1986). *Anal. Biochem.* **158,** 42–49.

Green, E. D., and Baenziger, J. U. (1987). *J. Biol. Chem.* **262,** 12018–12029.

Green, E. D., Brodbeck R. M., and Baenziger, J. U. (1987a). *J. Biol. Chem.* **262,** 12030–12039.

Green, E. D., Brodbeck, R. M., and Baenziger, J. U. (1987b). *Anal. Biochem.* **167,** 62–75.

Green, S. A., Plutner, H., and Mellman, I. (1985). *J. Biol. Chem.* **260,** 9867–9874.

Grigoriadis, D. E., Niznik, H. B., Jarvie, K. R., and Seeman, P. (1988). *FEBS Lett.* **227,** 220–224.

Grob, P. M., Berlot, C. H., and Bothwell, M. A. (1983). *Proc. Natl. Acad. Sci. U.S.A.* **80,** 6819–6823.

Hakomori, S. (1964). *J. Biochem. (Tokyo)* **55,** 205–208.

Halberg, D. F., Wager, R. E., Farrell, D. C., Hildreth, J., Quesenberry, M. S., Loeb, J. A., Holland, E. C., and Drickamer, K. (1987). *J. Biol. Chem.* **262,** 9828–9838.

Hammarström, S., and Kabat, E. A. (1969). *Biochemistry* **8,** 2696–2700.

Hammarström, S., Hammarström, M.-L., Sundblad, G., Arnarp, J., and Lönngren, J. (1982). *Proc. Natl. Acad. Sci. U.S.A.* **79,** 1611–1615.

Hardy, M. R., and Townsend, R. R. (1988). *Proc. Natl. Acad. Sci. U.S.A.* **85,** 3289–3293.

Hardy, M. R., Townsend, R. R., and Lee, Y. C. (1988). *Anal. Biochem.* **170,** 54–62.

Hascall, V. C. (1981). *In* "Biology of Carbohydrates" (V. Ginsburg, ed.), Vol. 1, pp. 1–49. Wiley, New York.

Hedo, J. A., Kahn, C. R., Hayashi, M., Yamada, K. M., and Kasuga, M. (1983). *J. Biol. Chem.* **258,** 10020–10026.
Herron, G. S., and Schimerlik, M. I. (1983). *J. Neurochem.* **41,** 1414–1420.
Holt, G. D., and Hart, G. W. (1986). *J. Biol. Chem.* **261,** 8049–8057.
Homans, S. W., Gerguson, M. A. J., Dwek, R. A., Rademacher, T. W., Anand, R., and Williams, A. F. (1988). *Nature (London)* **333,** 269–272.
Honda, S. (1984). *Anal. Biochem.* **140,** 1–47.
Hsieh, P., Rosner, M. R., and Robbins, P. W. (1983). *J. Biol. Chem.* **258,** 2548–2554.
Hubbard, S. C. (1987). *J. Biol. Chem.* **262,** 16403–16411.
Hubbard, S. C., and Ivatt, R. J. (1981). *Annu. Rev. Biochem.* **50,** 555–583.
Irimura, T., and Nicolson, G. L. (1983). *Carbohydr. Res.* **120,** 187–195.
Irimura, T., Tsuji, T., Tagami, S., Yamamoto, K., and Osawa, T. (1981). *Biochemistry* **20,** 560–566.
Iwase, H., Ishii, I., Ishihara, K., Tanaka, Y., Omura, S., and Hotta, K. (1988). *Biochem. Biophys. Res. Commun.* **151,** 422–428.
Iyer, R. N., and Carlson, S. M. (1971). *Arch. Biochem. Biophys.* **142,** 101–105.
Järnefelt, J., Rush, J., Li, Y.-T., and Laine, R. A. (1978). *J. Biol. Chem.* **253,** 8006–8009.
Kapitany, R. A., and Zebrowski, E. J. (1973). *Anal. Biochem.* **56,** 361–369.
Karpf, D. B., Arnaud, C. D., King, K., Bambino, T., Winer, J., Nyiredy, K., and Nissenson, R. A. (1987). *Biochemistry* **26,** 7825–7833.
Katagari, Y., Yamamoto, K., Tsuji, T., and Osawa, T. (1983). *Carbohydr. Res.* **120,** 283–292.
Kaylan, N. K., and Bahl, O. P. (1983). *J. Biol. Chem.* **258,** 67–74.
Keegan, A. D., and Conrad, D. H. (1987). *J. Immunol.* **139,** 1199–1205.
Kobata, A. (1978). *In* "Methods in Enzymology" (V. Ginsburg, ed.), Vol. 50, pp. 567–574. Academic Press, New York.
Kobata, A., and Takasaki, S. (1978). *In* "Methods in Enzymology" (V. Ginsburg, ed.), Vol. 50, pp. 560–567. Academic Press, New York.
Kornfeld, R., and Kornfeld, S. (1985). *Annu. Rev. Biochem.* **54,** 631–664.
Kornfeld, S., Reitman, M. L., and Kornfeld, R. (1981). *J. Biol. Chem.* **256,** 6633–6640.
Krusius, T., Finne, J., and Rauvala, H. (1976). *FEBS Lett.* **71,** 117–120.
Krusius, T., Finne, J., Margolis, R. K., and Margolis, R. U. (1986). *J. Biol. Chem.* **261,** 8237–8242.
Laine, R. A., Esselman, W. J., and Sweeley, C. C. (1972). *In* "Methods in Enzymology" (V. Ginsburg, ed.), Vol. 28, pp. 159–167. Academic Press, New York.
Lee, Y. C. (1972). *In* "Methods in Enzymology" (V. Ginsburg, ed.), Vol. 28, pp. 63–73. Academic Press, New York.
Li, E., Tabas, I., and Kornfeld, S. (1978). *J. Biol. Chem.* **253,** 7762–7770.
Li, E., Gibson, R., and Kornfeld, S. (1980). *Arch. Biochem. Biophys.* **199,** 393–399.
Li, Y.-T., and Li, S.-C. (1972). *In* "Methods in Enzymology (V. Ginsburg, ed.), Vol. 28, pp. 702–713. Academic Press, New York.
Lindberg, B., and Lönngren, J. (1978). *In* "Methods in Enzymology" (V. Ginsburg, ed.), Vol. 50, pp. 3–32. Academic Press, New York.
Lis, H., and Sharon, N. (1984). *In* "Biology of Carbohydrates" (V. Ginsburg and P. W. Robbins, eds.), Vol. 2, pp. 1–85. Wiley, New York.
Lis, H. and Sharon, N. (1986a). *In* "The Lectins: Properties, Functions, and Applications in Biology and Medicine" (I. E. Liener, N. Sharon, and I. J. Goldstein, eds.), pp. 293–370. Academic Press, Orlando, Florida.
Lis, H., and Sharon, N. (1986b). *Annu. Rev. Biochem.* **55,** 35–68.

Low, M. G. (1981). *In* "Methods in Enzymology" (J. M. Lowenstein, ed.), Vol. 71, pp. 741–746. Academic Press, New York.
Low, M. G., and Kincade, P. W. (1985). *Nature (London)* **318,** 62–64.
Lowe, M., and Nilsson, B. (1983). *J. Biol. Chem* **258,** 1885–1887.
Lublin, D. M., Griffith, R. C., and Atkinson, J. P. (1986). *J. Biol Chem.* **261,** 5736–5744.
Magnani, J. L. (1987). *In* "Methods in Enzymology" (V. Ginsburg, ed.), Vol. 138, pp. 195–207. Academic Press, Orlando Florida.
Malech, H. L., Gardner, J. P., Heiman, D. F., and Rosenzweig, S. A. (1985). *J. Biol Chem.* **260,** 2509–2514.
Marshall, R. D. (1974). *Biochem. Soc. Symp.* **40,** 17–46.
McElduff, A., Watkinson, A., Hedo, J. A., and Gorden, P. (1986). *Biochem. J.* **239,** 679–683.
McNeil, M., Darvill, A. G., Aman, P., Franzen, L.-E., and Albersheim, P. (1982). *In* "Methods in Enzymology" (V. Ginsburg, ed.), Vol. 83, pp. 3–45. Academic Press, New York.
Meisler, M. (1972). *In* "Methods in Enzymology" (C. H. W. Hirs and S. N. Timasheff, eds.), Vol. 27, pp. 820–824. Academic Press, New York.
Mellis, S. J., and Baenziger, J. U. (1983). *Anal. Biochem.* **134,** 442–449.
Merkle, R. K., and Cummings, R. D. (1987a). *In* "Methods in Enzymology" (V. Ginsburg, ed.), Vol. 138, pp. 232–259. Academic Press, Orlando, Florida.
Merkle, R. K., and Cummings, R. D. (1987b). *J. Biol. Chem.* **262,** 8179–8189.
Merkle, R. K., and Cummings, R. D. (1988). *J. Biol. Chem.* **263,** 16143–16149.
Morell, A. G., and Ashwell, G. (1972). *In* Methods in Enzymology" (V. Ginsburg, ed.), Vol. 28, pp. 205–208. Academic Press, New York.
Moseley, J. M., Findlay, D. M., Gorman, J. J., Michelangeli, V. P., and Martin, T. J. (1983). *Biochem. J.* **212,** 609–616.
Muramatsu, T. (1978). *In* "Methods in Enzymology" (V. Ginsburg, ed.), Vol. 50, pp. 555–559. Academic Press, New York.
Nakagawa, J. H., Yamada, T., Chien, J.-L., Gardas, A., Kitamikado, M., Li, S.-C., and Li, Y. T. (1980). *J. Biol. Chem.* **255,** 5955–5959.
Nakazawa, K., and Suzuki, S. (1975). *J. Biol. Chem.* **250,** 912–917.
Narasimhan, S., Harpaz, N., Longmore, G., Carver, J. P., Grey, A. A., and Schachter, H. (1980). *J. Biol. Chem.* **255,** 4876–4884.
Narasimhan, S., Freed, J. C., and Schachter, H. (1985). *Biochemistry* **24,** 1694–1700.
Nishigaki, M., Yamashita, K., Matsuda, I., Arashima, S., and Kobata, A. (1978). *J. Biochem. (Tokyo)* **84,** 823–834.
Nyame, K., Cummings, R. D., and Damian, R. (1987). *J. Biol. Chem.* **262,** 7990–7995.
Nyame, K., Cummings, R. D., and Damian, R. (1988). *Mol. Biochem. Parasitol.* **28,** 265–274.
Nyugen, T. D., Williams, J. A., and Gray, G. M. (1986). *Biochemistry* **25,** 361–368.
Ogata, S., Muramatsu, T., and Kobata, A. (1975). *J. Biochem. (Tokyo)* **78,** 687–696.
Olden, K., and Parent, J. B. (1987). "Vertebrate Lectins." Van Nostrand-Reinhold, Co., New York.
Osawa, T., and Tsuji, T. 1987). *Annu. Rev. Biochem.* **56,** 21–42.
Park, J. T., and Johnson, M. J. (1949). *J. Biol. Chem.* **181,** 149–151.
Passaniti, A., and Hart, G. W. (1988) *J. Biol. Chem.* **263,** 7591–7603.
Pendu, J. L., Fredman, P., Richter, N. D., Magnani, J. L., Willingham, M. C., Pastan, I., Oriol, R., and Ginsburg, V. (1985). *Carbohydr. Res.* **141,** 347–349.
Pierce, M., and Arango, M. P. (1986) *J. Biol. Chem.* **261,** 10772–10777.
Powell, L. D., and Hart, G. W. (1986). *Anal. Biochem.* **157,** 179–185.

Rauh, J. J., Lambert, M. P., Cho, N. J., Chin, H., and Klein, W. L. (1986). *J. Neurochem.* **46,** 23–32.
Reinhold, V. N. (1987). *In* "Methods in Enzymology (V. Ginsburg, ed.), Vol. 138, pp. 59–83. Academic Press, Orlando, Florida.
Reiser, H., Oettgen, H., Yeh, E. T. H., Terhorst, C., Low, M. G., Benacerraf, B., and Rock, K. L. (1986). *Cell (Cambridge, Mass.)* **47,** 365–370.
Rodén, L. (1980). *In* "The Biochemistry of Glycoproteins and Proteoglycans" (W. J. Lennarz, ed.), pp. 267–372. Plenum, New York.
Ronnett, G. V., Knutson, V. P., Kohanski R. A., Simpson, T. L., and Lane, M. D. (1984). *J. Biol. Chem.* **259,** 4566–4575.
Rosenzweig, S. A., Madison, L. D., and Jamieson, J. D. (1984). *J. Cell Biol.* **99,** 1110–1116.
Sadler, J. E. (1984). *In* "Biology of Carbohydrates" (V. Ginsburg and P. W. Robbins, eds). Vol. 2, pp. 199–288. Wiley, New York.
Sasaki, H., Bothner, B., Dell, A., and Fukuda, M. (1987). *J. Biol. Chem.* **362,** 12059–12076.
Schwarz, R. T., and Datema, R. (1982). *In* "Methods in Enzymology" (V. Ginsburg, ed.), Vol. 83, pp. 432–443. Academic Press, New York.
Segarini, P. R., and Seyedin, S. M. (1988). *J. Biol. Chem.* **263,** 8366–8370.
Shibuya, N., Goldstein, I. J., Broekaert, W. F., Nsimba-Lubaki, M., Peeters, B., and Peumans, W. J. (1987a). *Arch. Biochem. Biophys.* **254,** 1–8.
Shibuya, N., Goldstein, I. J., Broekaert, W. F., Nsimba-Lubaki, M., Peeters, B., and Peumans, W. J. (1987b). *J. Biol. Chem.* **262,** 1596–1601.
Shigeno, C., Hiraki, Y., Westerberg, D. P., Potts, J. T., and Segré, G. V. (1988). *J. Biol. Chem.* **263,** 3872–3878.
Slieker, L. J., Martensen, T. M., and Lane, M. D. (1986). *J. Biol. Chem.* **260,** 15233–15241.
Sloan, H. R. (1972). *In* "Methods in Enzymology" (C. H. W. Hirs and S. N. Timasheff, eds.), Vol. 27, pp. 868–874. Academic Press, New York.
Smith, D., Zopf, D. A., and Ginsburg, V. (1978). *In* "Methods in Enzymology" (V. Ginsburg, ed.), Vol. 50, pp. 221–226. Academic Press, New York.
Smith, M. M., Schlesinger, S., Lindstrom, J., and Merlie, J. P. (1986). *J. Biol. Chem.* **261,** 14825–14832.
Snider, M. (1984). *In* "Biology of Carbohydrates" (V. Ginsburg and P. W. Robbins, eds.), Vol. 2, pp. 163–198. Wiley, New York.
Soderquist, A. M., and Carpenter, G. (1984). *J. Biol. Chem. 259,* 12586–12594.
Sojar, H. T., and Bahl, O. P. (1987) *In* "Methods in Enzymology" (V. Ginsburg, ed.), Vol. 138, pp. 341–350. Academic Press, Orlando, Florida.
Spiro, R. G. (1967). *J. Biol. Chem.* **242,** 4813–4823.
Spiro, R. G., and Bhoyroo, V. D. (1974). *J. Biol. Chem.* **249,** 5704–5717.
Stiles, G. L. (1986). *J. Biol. Chem.* **261,** 10839–10843.
Stiles, G. L., Benović, J. L., Caron, M. C., and Lefkowitz, R. J. (1984). *J. Biol. Chem.* **259,** 8655–8663.
Stoffyn, P., Stoffyn, A., and Hauser, G. (1971). *J. Lipid Res.* **12,** 318–323.
Szilagyi, P. J., Arango, J., and Pierce, M. (1985). *Anal. Biochem.* **148,** 260–267.
Taga, E., Waheed, A., and Van Etten, R. L. (1984). *Biochemistry* **23,** 815–822.
Tai, T., Yamashita, K., Ogata-Arakawa, M., Koide, N., Muramatsu, T., Iwashita, S., Inoue, Y., and Kobata, A. (1975). *J. Biol. Chem.* **250,** 8569–8575.
Tai, T., Yamashita, K., Ito, S., and Kobata, A. (1977). *J. Biol. Chem.* **252,** 6687–6694.
Takasaki, S., Mizuochi, T., and Kobata, A. (1982). *In* "Methods in Enzymology" (V. Ginsburg, ed.), Vol. 83, pp. 263–268. Academic Press, New York.
Taniguchi, T., Adler, A. J., Mizuochi, T., Kochibe, N., and Kobata, A. (1986). *J. Biol Chem.* **261,** 1730–1736.

Tarentino, A. L., and Plummer, T. H., Jr. (1987). *In* "Methods in Enzymology" (V. Ginsburg, ed.), Vol. 138, pp. 770–778. Academic Press, Orlando, Florida.
Tarentino, A. L., Trimble, R. B., and Maley, F. (1978). *In* "Methods in Enzymology" (V. Ginsburg, ed.), Vol. 50, pp. 574–580. Academic Press, New York.
Tarentino, A. L., Gomez, C. M., and Plummer, T. H., Jr. (1985). *Biochemistry* **24,** 4665–4671.
Thotakura, N. T., and Bahl, O. P. (1987). *In* "Methods in Enzymology" (V. Ginsburg, ed.), Vol. 138, pp. 350–359. Academic Press, Orlando, Florida.
Tomita, M., and Marchesi, V. T. (1975). *Proc. Natl. Acad. Sci. U.S.A.* **72,** 2964–2968.
Tomiya, N., Kurano, M., Ishihara, H., Tejima, S., Endo, S., Arata, Y., and Takahashi, N. (1987). *Anal. Biochem.* **163,** 489–499.
Tomiya, N., Awaya, J., Kurano, M., Endo, S., Arata, Y., and Takahashi, N. (1988). *Anal. Biochem.* **171,** 73–90.
Torres, B. V., McCrumb, D. K., and Smith, D. F. (1988). *Arch. Biochem. Biophys.* **262,** 1–11.
Torres, C.-R., and Hart, G. W. (1984). *J. Biol. Chem.* **259,** 3308–3317.
Townsend, R. R., Hardy, M. R., Olechno, J. D., and Carter, S. R. (1988). *Nature (London)* **335,** 379–380.
Trevelyan, W. E., Proctor, D. P., and Harrison, J. (1950). *Nature (London)* **166,** 444–445.
Trimble, R. B., Trumbly, R. J., and Maley, F. (1987). *In* "Methods in Enzymology" (V. Ginsburg, ed.), Vol. 138, pp. 763–770. Academic Press, Orlando, Florida.
Troy, F. A., Hallenbeck, P. C., McCoy, R. D., and Vimr, E. R. (1987). *In* "Methods in Enzymology" (V. Ginsburg, ed.), Vol. 138, pp. 169–185. Academic Press, Orlando, Florida.
Tse, A. G. D., Barclay, A. N., Watts, A., and Williams, A. F. (1985). *Science* **230,** 1003–1008.
Tsuji, T., Irimura, T., and Osawa, T. (1981). *J. Biol. Chem.* **256,** 10497–10502.
Tsujimoto, M., Feinman, R., Kohase, M., and Vilček, J. (1986). *Arch. Biochem. Biophys.* **249,** 563–568.
Uchida, Y., Tsukada, Y., and Sugimori, T. (1979). *J. Biochem (Tokyo)* **86,** 1573–1585.
Umemoto, J., Bhavanandan, V. P., and Davidson, E. A. (1977). *J. Biol. Chem.* **252,** 8609–8614.
Van Driel, I. R., and Goding, J. W. (1985). *Eur. J. Biochem.* **149,** 543–548.
Van Lenten, L., and Ashwell, G. (1972). *In* "Methods in Enzymology" (V. Ginsburg, ed.), Vol. 28, pp. 209–211. Academic Press, New York.
Varki, A., and Kornfeld, S. (1983). *J. Biol. Chem.* **258,** 2808–2818.
Vliegenthart, J. F. G., Dorland, L., and van Halbeek, H. (1983). *Adv. Carbohydr. Chem. Biochem.* **41,** 209–374.
Wang, W.-C., and Cummings, R. D. (1988). *J. Biol Chem.* **263,** 4576–4585.
Wang, W.-C., Clark, G. F., Smith, D. F., and Cummings, R. D. (1988). *Anal. Biochem.* **175,** 390–396.
Warren, L. (1959). *J. Biol Chem.* **234,** 1971–1975.
Yamada, K., Lipson, K. E., and Donner, D. B. (1987). *Biochemistry* **26,** 4438–4443.
Yamamoto, K., Tsuji, T., and Osawa, T. (1982). *Carbohydr. Res.* **110,** 283–289.
Yamashita, K., Tachibana, Y., Takada, S., Matsuda, I., Arashima, S., and Kobata, A. (1979). *J. Biol Chem.* **254,** 4820–4827.
Yamashita, K., Ohkura, T., Yoshima, H., and Kobata, A. (1981). *Biochem. Biophys. Res. Commun.* **100,** 226–232.
Yamashita, K., Mizuochi, T., and Kobata, A. (1982). *In* "Methods in Enzymology (V. Ginsburg, ed.), Vol. 83, pp. 105–126. Academic Press, New York.

Yamashita, K., Hitoi, A., and Kobata, A. (1983a). *J. Biol. Chem.* **258,** 14753–14755.
Yamashita, K., Ueda, I., and Kobata, A. (1983b). *J. Biol. Chem.* **258,** 14144–14147.
Yamashita, K., Ohkura, T., Tachibana, Y., Takasai, S., and Kobata, A. (1984). *J. Biol. Chem.* **259,** 10834–10840.
Yamashita, K., Totani, K., Ohkura, T., Takasaki, S., Goldstein, I. J., and Kobata, A. (1987). *J. Biol. Chem.* **262,** 1602–1607.
Yet, M. G., Chin, C. C. Q., and Wold, F. (1988a). *J. Biol. Chem.* **263,** 111–117.
Yet, M. G., Shao, M. C., and Wold, F. (1988b). *FASEB J.* **2,** 22–31.
Yoshima, H., Takasaki, S., and Kobata, A. (1980). *J. Biol. Chem.* **255,** 10793–10804.
Yoshimura, A., Yoshida, T., Seguchi, T., Waki, M., Ono, M., and Kuwano, M. (1987). *J. Biol. Chem.* **262,** 13299–13308.
Yurchenco, P. D., Ceccarini, C., and Atkinson, P. H. (1978). *In* "Methods in Enzymology" (V. Ginsburg, ed.), Vol. 50, pp. 175–204. Academic Press, New York.
Zeggari, M., Viguerie, N., Susini, C., Granier, M., Esteve, J. P., and Ribet, A. (1987). *Eur. J. Biochem.* **164,** 667–673.
Zopf, D., Schroer, K., Hannson, G., Dakour, J., and Lundblad, A. (1987). *In* "Methods in Enzymology" (V. Ginsburg, ed.), Vol. 138, pp. 307–312. Academic Press, Orlando, Florida.

Chapter 7

Protein Folding and Intracellular Transport: Evaluation of Conformational Changes in Nascent Exocytotic Proteins

MARY-JANE GETHING

Howard Hughes Medical Institute and
Department of Biochemistry
University of Texas Southwestern Medical Center
Dallas, Texas 75235

KAREN McCAMMON

Howard Hughes Medical Institute
University of Texas Southwestern Medical Center
Dallas, Texas 75235

JOE SAMBROOK

Department of Biochemistry
University of Texas Southwestern Medical Center
Dallas, Texas 75235

Copyright © 1989 by Academic Press, Inc.
All rights of reproduction in any form reserved.

I. Introduction

It has been known for some time that different secretory and membrane proteins traverse the exocytotic pathway at different speeds (reviewed by Lodish, 1988), with most of the variation in the rate of transport occurring prior to the arrival of the nascent proteins in the medial cisternae of the Golgi apparatus. With the advent of recombinant DNA technology and the ability to construct and express wild-type and mutant forms of membrane and secretory proteins in eukaryotic cells, it has become clear that many mutant proteins display defects in exocytosis and that with few exceptions, the most affected step is transport out of the endoplasmic reticulum (ER) to the Golgi apparatus (reviewed by Gething, 1985; Garoff, 1986). It was initially proposed that exocytosis may be facilitated by the recognition of specific "transport epitope(s)" on newly synthesized polypeptides by one or more cellular receptors; the rate of exit of each nascent molecule from the ER would then be proportional to its affinity for its receptor (Fitting and Kabat, 1982). However, it has been difficult to reconcile this model of receptor-mediated transport with observations that mutations in membrane proteins causing impaired or blocked transport are not clustered within any single location in the primary or tertiary structures of the molecules (reviewed by Garoff, 1985). In the best-studied example, the hemagglutinin (HA) glycoprotein of influenza virus, mutations leading to transport defects are dispersed throughout the luminal, cytoplasmic, and membrane-spanning domains of the molecule (reviewed by Gething, 1985). By utilizing a number of different assays designed to monitor the state of folding and assembly of nascent HA, evidence has been obtained that attainment of a "correct" trimeric structure is a primary requirement for entry of HA into the exocytotic pathway and that defective transport from the ER correlates with defective folding of the HA polypeptide (Gething *et al.*, 1986; Copeland *et al.*, 1986, 1988). Similarly, it has been shown that correct folding and trimerization is also required for efficient intracellular transport of the G glycoprotein of vesicular stomatitis virus (VSV) (Kreis and Lodish, 1986; Doms *et al.*, 1987, 1988), and that misfolded or unassembled forms of a number of cellular membrane and secretory proteins are retained in the ER (Owen *et al.*, 1980; Weiss and Stobo, 1984; Peters *et al.*, 1985; Bole *et al.*, 1986; Kishimoto *et al.*, 1987; reviewed by Carlin and Merlie, 1986). Although the cellular processes that facilitate and regulate the folding, assembly, and transport of nascent membrane and secretory proteins are not yet understood, evidence has been obtained for the involvement of resident proteins of the ER such as BiP/GRP78 (Bole *et al.*, 1986; Gething *et al.*, 1986;

Munro and Pelham, 1986) and protein disulfide isomerase (PDI/GRP58, Freedman, 1984; Roth and Pierce, 1987). Observations of transient or permanent associations between these ER proteins and newly synthesized wild-type or mutant proteins led to the suggestion that they may play a role in either facilitating or monitoring the folding and assembly of the nascent polypeptide chains (Pelham, 1986; Gething *et al.*, 1986; Hendershot *et al.*, 1987).

A full understanding of the role of protein structure in intracellular transport will undoubtedly be greatly facilitated by further studies on the biosynthesis, folding, and transport of a wide range of different membrane and secretory proteins. This chapter describes a number of assays that have been developed to analyze the *in vivo* process of protein folding and/or oligomerization within the ER. These assays fall into two groups: those that probe conformational changes in the nascent polypeptide chain and those that follow the assembly of the polypeptides into oligomers. The full range of these assays are appropriate for multimeric proteins, whether they consist of identical or heterologous subunits.

II. Methods for Analysis of Protein Conformation

A. Pulse–Chase Labeling of Nascent Polypeptides

To determine the time course *in vivo* of folding and assembly of membrane or secretory proteins by any of the methods described in this article, it is necessary to label the nascent polypeptides with a radioactive amino acid, usually [^{35}S]methionine and/or [^{35}S]cysteine. Optimally, the period of incorporation of isotope should be short relative to the time course of folding so that only the completely unfolded polypeptide will be labeled at the start of the chase period. In practice this goal may be difficult to achieve, since initial folding of the polypeptide may commence even before translation and translocation of the chain is completed. Moreover, a compromise must be made between the shortest possible labeling period and the expense of the amount of isotope required to provide sufficient signal for autoradiography. A 3-minute pulse-labeling period has been used successfully to study the folding and oligomerization of influenza HA when it trimerizes rapidly ($t_{1/2}$ = 7–10 minutes) following expression at high levels (~10^8 molecules/cell) from a Simian Virus 40 (SV40)-based vector (Gething *et al.*, 1986; Copeland *et al.*, 1986). Following expression from a bovine papilloma virus-based vector at a level closer to those found for cellular membrane proteins (2 × 10^6 molecules/

cell), HA trimerizes somewhat more slowly ($t_{1/2}$ = 30 minutes), and a 5-minute labeling period was found to be satisfactory (Hearing *et al.*, 1989).

The choice of time intervals for chase with nonradioactive amino acids may be guided by initial studies of the rate of intracellular transport of the protein of interest. For glycoproteins, this is most conveniently measured as the rate of acquisition of resistance to endoglycosidase H (see Chapter 5).

For rapidly transported proteins ($t_{1/2}$ Endo H^r = 15–30 minutes), the following time intervals are suggested: pulse **3 minutes;** chase **0 minutes,** 2 minutes, **7 minutes,** 12 minutes, **17 minutes,** 27 minutes.

For more slowly transported proteins ($t_{1/2}$ Endo H^r = 1–2 hours), longer time intervals are suggested: pulse **5 minutes;** chase **0 minutes,** 5 minutes, **15 minutes,** 25 minutes, **55 minutes,** 115 minutes.

Detailed studies will require analyses of the folding process at all the time points. The time points shown in **bold type** should suffice for preliminary experiments or for surveillance of a large number of mutant proteins.

Labeling Protocol

1. Take an almost-confluent monolayer of cells that express the protein of interest, and aspirate the growth medium.

2. Wash the cells once with prewarmed, methionine-free (or methionine- and cysteine-free) growth medium (labeling medium), and then incubate in 3 ml of this medium for 20 minutes at 37°C to starve the cells for the nonradioactive amino acid(s).

3. Aspirate the medium and add prewarmed labeling medium containing [^{35}S]methionine (500–1000 Ci/mmol, 50–250 μCi per 10^6 cells in 250 μl); incubate for 3 or 5 minutes at 37°C. Note: the least expensive source of [^{35}S]methionine available at present is Trans^{35}S-label (ICN Radiochemicals), which is produced from a cellular hydrolysate of *Escherichia coli* grown in the presence of carrier-free $^{35}SO_4$. Because the product contains ~15% [^{35}S]cysteine in addition to the specified activity of [^{35}S]methionine, additional incorporation of isotope can be achieved if the cells are starved for cysteine as well as methionine.

4. Remove the labeling solution for discard to radioactive waste.

5a. For the 0-minute chase sample: wash monolayer once with 4 ml ice-cold phosphate-buffered saline (PBS) or Tris-buffered saline, and then add 1 ml of ice-cold lysis buffer (see recipes later). Leave culture dish on ice until ready to proceed.

5b. For chase samples: wash monolayers twice with 4 ml prewarmed growth medium supplemented with 2 m*M* nonradioactive methionine, and incubate at 37°C until the end of the desired chase period. Then continue as described in 5a.

6. Scrape the lysed cells and debris from the culture plate, and transfer to a microfuge tube on ice. Vortex gently and then spin for 1 minute in a microfuge (12,000 *g*) to pellet nuclei and cell debris. Remove supernatant to a fresh tube, taking care not to disturb or transfer any of the pellet. Store on ice for the minimum time before proceeding to perform folding assays.

Note a. A number of alternative cell lysis buffers have been used in folding studies.

Nonidet P-40 Lysis Buffer

1% Nonidet P-40 (Sigma) in 50 m*M* Tris-HCl (pH 8.0), containing 100 m*M* NaCl

Triton X-100 Lysis Buffer

1% Triton X-100 (Sigma) in 20 m*M* methylethane sulfonate (MES), 30 m*M* Tris (pH 7.4), containing 100 m*M* NaCl

Octyl Glucoside Lysis Buffer

40 m*M* Octyl-β-D-glucopyranoside (Calbiochem) in 50 m*M* Tris-HCl (pH 7.4), containing 100 m*M* NaCl

Lauryl Maltoside Lysis Buffer

6 m*M* Dodecyl-β-D-maltoside (Calbiochem) in 50 m*M* Tris-HCl (pH 7.4), containing 100 m*M* NaCl

Nonidet P-40 and Triton X-100 are commonly used nonionic detergents for the solubilization of lipid bilayers and membrane proteins (Helenius and Simons, 1975), and have been successfully employed in folding studies (Gething *et al.*, 1986; Copeland *et al.*, 1986; Doms *et al.*, 1987, 1988). Octyl glucoside, a small molecule of simple defined structure, is also an effective membrane-solubilizing agent (e.g., Baron and Thompson, 1975). The high critical micelle concentration of this detergent, which permits easy removal by dialysis, has led to its frequent use in reconstitution studies (e.g., Felgner *et al.*, 1979), and we have found it effective in studies of the folding and trimerization of the HA derived from the A/Japan strain of influenza virus (Gething *et al.*, 1986). Although these detergents do not usually function as protein denaturants, they have been reported to cause loss of biological activity of some membrane proteins (reviewed by Helenius and Simons, 1975). The sensitivity of cytochrome oxidase to its detergent environment led Rosevear *et al.* (1980) to investigate the use of different alkyl glycoside detergents and to report that lauryl maltoside was superior to octyl glucoside and other alkyl glycosides, and to other commercially available nonionic detergents including Triton X-100, in its physical properties (small uniform micelles) and its ability to stabilize the active dimeric form of the enzyme. We have found that lauryl maltoside appears also to stabilize the quaternary structure of HA derived from the A/X31 strain of influenza, since chemical crosslinking of the trimeric form of X31 HA can be observed following solubilization in lauryl maltoside, but not following solubilization in octyl glucoside or Nonidet P-40 (our unpublished results). Before embarking on extensive folding studies, preliminary experiments should be performed to compare the effect of different detergents on the structure of the protein of interest.

Note b. Except when protease sensitivity studies are to be performed, the lysis buffer should contain protease inhibitors [aprotinin (0.1 TIU/ml), soybean trypsin inhibitor (SBTI, 200 μg/ml), and phenylmethylsulfonyl fluoride (PMSF), 1 m*M*), all from Sigma or Boehringer Mannheim] to prevent nonspecific degradation of the nascent polypeptides.

Note c. Both the pulse-labeling and chase incubations may be carried out at temperatures other than 37°C for determinations of the effect of temperature on folding.

B. Analysis of Protein Folding Using Antibodies Specific for Different Conformational States

These assays depend on the availability of antibodies that recognize different conformational forms of the protein of interest, and in particular antibodies that discriminate between unfolded and fully folded forms of the polypeptide chain. The methods used to develop such antibodies are beyond the scope of this article, but are discussed fully in Harlow and Lane (1988). Polyclonal or monoclonal antibodies raised against reduced and alkylated polypeptides are likely to recognize the fully unfolded conformations and may not recognize the folded molecule. Monoclonal antibodies (mAb) raised against the native protein should include those that recognize epitopes only on the fully folded mature protein. In the case of multimeric proteins, mAb that recognize epitopes that are present only following formation of the quaternary structure are particularly useful. It is preferable to have at least one antibody that can quantitatively precipitate from a cell extract all molecules of the polypeptide of interest regardless of their conformational state. However, in practice this is not always possible. Polyclonal sera raised against the native protein will usually contain antibodies that recognize both native and partially unfolded forms of the molecule, but antibodies that recognize the fully unfolded, non-disulfide-bonded conformation of the polypeptide may not be present. Figure 1 illustrates the use of polyclonal and monoclonal antibodies that recognize different conformational forms of the A/Japan HA to characterize wild-type and mutant forms of the protein. The published studies on

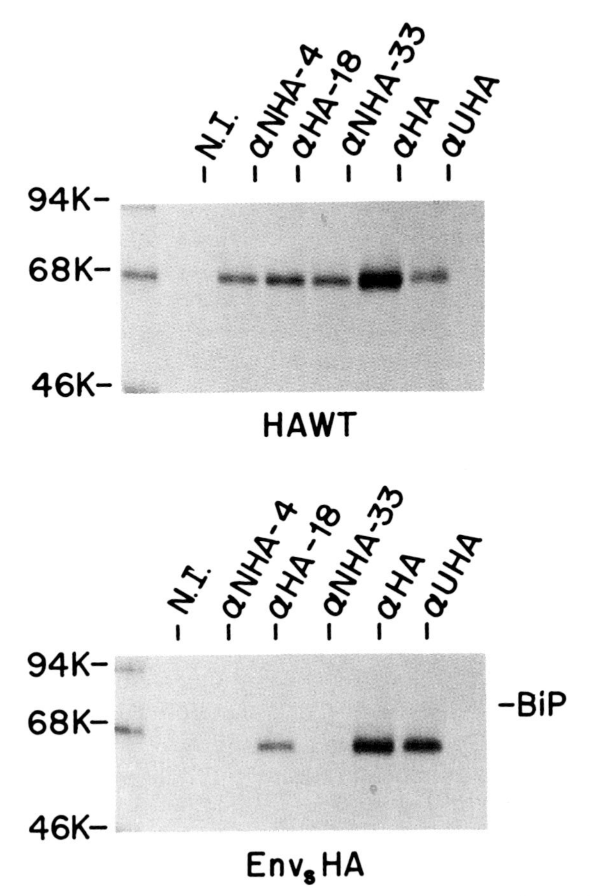

FIG. 1. Characterization of the recognition of wild-type and mutant forms of HA by anti-HA antibodies. CV-1 cells expressing wild-type HA (top) or the mutant env$_s$HA (bottom) were labeled for 15 minutes with [^{35}S]methionine before cell extracts were prepared in Nonidet P-40 lysis buffer and immunoprecipitated either with nonimmune serum (N.I.), or with one of three monoclonal anti-HA antibodies (αNHA-4 or αNHA-33, which recognize only correctly folded trimeric HA, or αHA-18, which recognizes all forms of HA), or with rabbit polyclonal antisera directed against all forms of HA (αHA) or unfolded HA (αUHA). The immunoprecipitated proteins were separated by SDS–PAGE and visualized by autoradiography. From Gething *et al.* (1986).

influenza HA (Gething *et al.*, 1986; Copeland *et al.*, 1986, 1988; Yewdell *et al.*, 1988) and VSV G (Doms *et al.*, 1987, 1988) provide additional examples of the use of conformation-specific antibodies to follow the folding process.

It is important to characterize as fully as possible antibodies that are to be used for folding assays. For the studies to be quantitative, it is necessary to ensure by serial precipitation of labeled cell extracts that sufficient antibody and immunoadsorbant are used to precipitate all of the target protein present in the extract. Serial precipitation of extracts can also be used to define whether different antibodies recognize distinct or overlapping populations of molecules. Immunoprecipitation of extracts of cells labeled using pulse–chase protocols such as those outlined earlier will provide information on the timing of appearance and/or disappearance of epitopes recognized by the various antibodies. Finally, it is important to note that for glycoproteins the sequential processing of oligosaccharides in the ER and the Golgi apparatus may result in the appearance or disappearance of antigenic epitopes consisting of either protein or carbohydrate elements. Such epitopes may or may not be useful determinants of the overall process of polypeptide folding and assembly. The initial characterization of the antibodies will undoubtedly be augmented during folding studies; for example, a particular antibody known to bind to the mature protein might recognize only oligomeric, protease-resistant molecules (Gething *et al.*, 1986; Copeland *et al.*, 1986). Similarly, analysis of different mutant forms of the protein of interest using a panel of antibodies may not only help to define the conformation of the mutant polypeptide but also serve to refine the current understanding of the structures recognized by the antibodies. There is therefore an element of circularity in the process of characterization; nevertheless a constant review of the internal consistency of the data arising from the different assay systems can reveal nuances in the patterns of folding of the polypeptide chain under different circumstances and between different mutant forms of the protein.

When analyzing the results of immunoprecipitation studies, careful attention should be paid to additional labeled protein species, often of minor intensity, that may represent cellular proteins that coprecipitate with nascent polypeptides. A 78-kDa protein (BiP) that is now known to correspond to the previously identified Ig heavy-chain-binding protein (Morrison and Scharff, 1975; Haas and Wabl, 1983; Bole *et al.*, 1986; Gething *et al.*, 1986) and to the glucose-regulated protein GRP78 (Shiu *et al.*, 1977; Munro and Pelham, 1986) is frequently observed to coprecipitate with unfolded or unassembled polypeptides in the ER (see, for example, Fig. 1, also Bole *et al.*, 1986; Gething *et al.*, 1986; Dorner *et al.*, 1987). The expression of BiP is induced under conditions that lead to the accumulation of unfolded proteins in the ER (Kozutsumi *et al.*, 1988). Other ER proteins such as the 58-kDa PDI (Freedman, 1984; Edman *et al.*, 1985) and GRP94 [Lee *et al.*, 1981; also named ERP99 (Lewis *et al.*, 1985; Mazzarella and Green, 1987) or endoplasmin (Koch *et al.*, 1987)] have also been observed to associate with nascent proteins in the ER (Roth and Pierce, 1987; our unpublished results). Because these proteins are very abundant constituents of the ER (Freedman, 1984; Lewis *et al.*, 1985; Munro and Pelham, 1986; Koch *et al.*, 1987), their specific activities tend to be low following pulse–chase labeling protocols; we have found them difficult to observe after labeling for periods <15 minutes.

Immunoprecipitation Protocols

A multitude of immunoprecipitation procedures exist, and most laboratories will have evolved a protocol that is satisfactory for the particular protein under study. The following procedures have been used successfully for a range of different proteins; they are presented in order of increasing stringency of the incubation or washing steps. A fuller discussion of immunoprecipitation techniques is available in Harlow and Lane (1988).

For folding studies where the aim is to define conformational differences, care should be taken to minimize denaturation of the polypeptides. Cell extracts should be kept on ice and analyzed as soon as possible. Freezing and thawing of samples should be avoided. All steps should be carried out at low temperatures (on ice or in a cold room), and short incubation and washing periods should be used. Although more stringent conditions are sometimes required to reduce nonspecific precipitation, preliminary studies should be undertaken to ensure that the type and concentration of detergent or salt used does not lead to unfolding of the polypeptide after preparation of the extract.

a. Protocol 1

1. Dilute a sample of the extract obtained from $\sim 10^5$ cells to 1.0 ml with NET–gel buffer: 50 m*M* Tris-HCl, pH 7.5, 150 m*M* NaCl, 1 m*M* EDTA, 0.1% Nonidet P-40, 0.02% sodium azide, 0.25% gelatin (diluted from an autoclaved 2.5% stock solution in H_2O). It is convenient when working with multiple samples to include the appropriate amount of antiserum with the NET–gel buffer. Alternatively, the antiserum can be added separately to the diluted extracts.
2. Incubate for 30 minutes to 2 hours at 4°C.
3. Add protein A–Sepharose (Pharmacia; 100 μl of a 10% suspension in NET–gel buffer) or prewashed, fixed *Staphylococcus aureus* (available as Pansorbin from Calbiochem; 100 μl of a 10% suspension), and incubate for 1–2 hours at 4°C.
4. Collect the immune complexes by brief centrifugation in a microfuge (1 minute, 12,000 *g*) or bench centrifuge (5 minute, 1000 *g*) and wash twice

with NET–gel buffer. When protein A–Sepharose is used as the immunoadsorbant, 10- to 15-minute washes should be used to allow equilibration of the buffer with the solution in the interior of the beads.

5. Aspirate the last wash solution carefully, and add 25 μl of gel sample buffer (20 m*M* Tris-HCl, pH 6.8, 2% sodium dodecyl sulfate (SDS), 0.02% bromophenol blue). For analysis of the immunoprecipitated proteins under reducing conditions, the sample buffer should contain 0.1 *M* dithiothreitol (DTT, Sigma or Calbiochem; added freshly before use from a 1.0 *M* stock solution in 1 m*M* HCl, stored frozen). Release the immunoprecipitated proteins from the immunoadsorbant by boiling for 3 minutes or heating to 70°C for 10 minutes.

6. Analyze the immunoprecipitated proteins by sodium dodecyl sulfate–polyacrylamide gel electrophoresis (SDS–PAGE) (Laemmli, 1970) and autoradiography.

b. Protocol 2. Follow protocol 1 except that the two washes with NET–gel should be replaced by one 20-minute wash with each of the following three solutions:

(i) NET–gel buffer supplemented with 0.5 *M* NaCl
(ii) NET–gel buffer supplemented with 1% Nonidet P-40 and 0.1% SDS
(iii) 10 m*M* Tris-HCl (pH 7.5), containing 0.1% Nonidet P-40

c. Protocol 3. Follow protocol 1 except that the NET–gel buffer should be replaced with RIPA buffer: 50 m*M* Tris (pH 8.0), 150 m*M* NaCl, 1% Nonidet P-40, 1% sodium deoxycholate, 0.1% SDS.

C. Protease Sensitivity As an Assay for Conformational Changes in Nascent Proteins

The majority of plasma membrane and secreted proteins have evolved so that their mature structures are resistant to degradation by proteases in the external milieu. However, protease-sensitive sites are frequently exposed on the unfolded or partially assembled polypeptides. Thus the acquisition of protease resistance by a protein during a pulse–chase labeling protocol provides a convenient measure of attainment of tertiary and quaternary structure. Trypsin and proteinase K are commonly used in such assays, but many other proteases can be employed. Preliminary experiments should be undertaken to establish the protease concentration and temperature and duration of incubation that yield complete degradation of unfolded molecules while allowing quantitative recovery of fully folded molecules. The reaction should be terminated by the addition of protease inhibitor(s) and/or an excess of unlabeled protein [serum, bovine serum albumin (BSA), or gelatin]. Conditions that have proved successful for assaying the protease sensitivity of influenza HA are as follows:

1. To an aliquot of cell extract or sucrose gradient fraction add TPCK–trypsin (L-1-tosylamide-2-phenylethyl-chloromethylketone–trypsin, Sigma.) to a final concentration of 100 μg/ml.
2. Incubate on ice for 30 minutes.
3. Add SBTI, aprotinin, and PMSF to final concentrations of 200 μg/ml, 0.1 TIU, and 1 m*M*, respectively. If all of the sample is to be immunoprecipitated, these inhibitors can be added in 0.5 ml NET–gel buffer.
4. Immunoprecipitate all or part of the sample as described before.

Note: Trypsin digestions using a concentration of 25 μg/ml for 30 minutes at a range of temperatures (e.g., 0°C, 30°C, 37°C, 45°C) can reveal differences in the stability of different mutant forms of HA (P. Gallagher, unpublished results).

D. Analysis of Disulfide Bond Formation in Nascent Proteins

Disulfide bond formation is an early event in the folding of membrane and secretory proteins. For some proteins, the majority of the intramolecular disulfide bonds form cotranslationally, with the remainder forming within 1–2 minutes of chain completion (Bergman and Kuehl, 1979; Peters and Davidson, 1982; Ruddon *et al.*, 1987). In other cases, for example the β subunit of human chorionic gonadotropin (Ruddon *et al.*, 1987), disulfide bond formation in the ER can occur relatively slowly ($t_{1/2}$ = 30 minutes) and intermediates containing reduced sulfhydryl groups can be isolated. The formation of correct disulfide bonds is thought to occur via formation of intermediate mismatched bonds (reviewed by Creighton, 1988), which are then rearranged by protein disulfide isomerase (Freedman, 1984). However, such intermediates have not as yet been identified or characterized during *in vivo* studies.

Experimentally, the formation of disulfide bonds can be monitored very simply by observing differences in the mobility of a nascent protein on SDS–polyacrylamide gels run under reducing and nonreducing conditions. In many cases it has been observed that the electrophoretic mobilities of polypeptides containing intramolecular disulfide bonds are significantly altered from those of the fully reduced proteins (see, for example, Ruddon, *et al.*, 1987). A method for the quantitative analysis of the proportion of free sulfhydryl groups and intramolecular disulfide pairs in different molecular forms of a nascent protein has been described by Ruddon *et al.* (1987). This technique relies on use of sulfhydryl reagents such as vinylpyridine or iodoacetic acid to derivatize free-thiol groups in polypeptides labeled with [^{35}S]cysteine in a pulse–chase protocol. Following immunoprecipitation the labeled polypeptides are separated on SDS–PAGE run under nonreducing conditions, and the gel bands containing the species of interest are identified by autoradiography and excised. The labeled polypeptides are then eluted, lyophilized, and hydrolyzed in 6 *M* HCl under reducing conditions before determining the ratio of free [^{35}S]cysteine (from disulfide-bonded cystines) and derivatized [^{35}S]cysteine (from free thiols in the protein) following separation on high-performance liquid chromatography (HPLC).

III. Methods for Analysis of the Assembly of Nascent Proteins into Oligomers

The methods just discussed can be used to follow the conformational changes that take place as a nascent polypeptide folds in the lumen of the ER: they are applicable to both monomeric and oligomeric proteins. Many membrane and secreted proteins are composed of several polypeptide subunits, and their assembly into their mature homo- and hetero-oligomeric forms appears to be a primary requirement for transport along the secretory pathway (Bole *et al.*, 1986; Gething *et al.*, 1986; Copeland *et al.*, 1986). In this section are described three procedures for analyzing the extent and time course of assembly of newly synthesized polypeptides into oligomeric structures.

A. Analysis of Oligomerization Using Chemical Crosslinking

There is an element of serendipity in the use of chemical crosslinking to study protein oligomerization, since the method requires that reactive amino acids (usually those containing amino or sulfhydryl groups) be present on two different subunits of the oligomer at an appropriate distance apart. Nevertheless, a wide range of bifunctional reagents are available that contain different-length alkyl spacers. These crosslinkers and general considerations for their use are described in detail in the Pierce Chemical Company catalog.

Homobifunctional imidoesters have been widely used as reagents for chemical crosslinking of lysine residues in protein structural studies. Dimethyl adipimidate dihydrochloride (DMA), dimethyl pimelimidate dihydrochloride (DMP), and dimethyl suberimidate dihydrochloride (DMS) vary in the length of their alkyl spacers, which consist of four, five, and six methyl groups, respectively. Dimethyl-3,3′-dithiobispropionimidate

(DBTP) is a disulfide-containing crosslinker that can be cleaved by the addition of thiol reagents, allowing analysis of the composition of crosslinked complexes by two-dimensional SDS–PAGE. Two general considerations arise from the reactivity of these reagents with primary amines: (1) basic conditions are recommended for coupling in order to maintain the nonprotonated state of the amines; (2) buffers containing primary amine functions, such as Tris or glycine, should generally be avoided, since the crosslinkers may react preferentially with them. Compatible buffers include those that are acetate-, borate-, citrate-, or phosphate-based as well as triethanolamine hydrochloride (see later). Nevertheless, we have found in practice that aliquots of cell extracts prepared in buffers containing Tris (see earlier) can be successfully crosslinked using protocol 1 (this section).

The homobifunctional *N*-hydroxysuccinimide (NHS) esters comprise a second series of crosslinkers that differ from the imidoesters in their reactivity at physiological pH and their long half-lives in aqueous media. Commonly used NHS crosslinkers include disuccinimidyl suberate (DSS) and the thiol-cleavable reagents dithiobis(succinimidylpropionate) (DSP) and 3,3′-dithiobis(sulfosuccinimidylpropionate) (DTSSP).

Crosslinking reactions are most frequently carried out *in vitro* on purified proteins or cell extracts (see protocol 1, this section; Davies and Stark, 1970; Gething *et al.*, 1986). However, to visualize transient or unstable protein–protein interactions that may be lost upon detergent extraction and/or dilution, crosslinking reactions may be performed by adding the reagents either to the intact cells prior to lysis (protocol 2, this section; Kreis and Lodish, 1986; Roth and Pierce, 1987) or with the lysis buffer (protocol 3, this section; P. Gallagher and K. McCammon, unpublished results). These protocols can be taken as a general guide for performing crosslinking reactions, but the concentrations of protein and crosslinker, buffer conditions, and reaction temperature should be optimized in each case. Analysis of the crosslinked proteins frequently requires a gel system that can resolve high molecular weight complexes (>200,000) that are not well separated by the discontinuous SDS–PAGE systems developed by Laemmli (1970). Protocol 4 (this section) describes the use of electrophoresis on 3.5% polyacrylamide gels employing a borate–acetate buffer to resolve proteins in the molecular weight range ~35,000–500,000. Finally, protocol 5 (this section) outlines the preparation of a molecular-weight standard by crosslinking the hexameric protein glutamate dehydrogenase to provide a ladder comprising monomers, dimers, trimers, tetramers, pentamers, and hexamers of MW 52,000 subunits.

It should be noted that protein crosslinking is usually performed under conditions that provide partial crosslinking, since nonspecific crosslinking and aggregation become problems when the reaction is driven to completion. Separation of a crosslinked oligomer under denaturing conditions will reveal the monomer, the fully crosslinked oligomer, and all the intermediate multimers. For example, analysis of crosslinked samples of pulse-labeled HA reveals monomer, dimer, and trimer forms of the protein (see Fig. 2), even at time points when sucrose density gradient analysis shows that the protein is entirely present as the mature trimer (Gething *et al.*, 1986). Although crosslinking can therefore only provide a qualitative measure of the presence of oligomeric forms of a protein, the technique can be used effectively to define the time course of their assembly (Fig. 2) and to analyze the extent of oligomerization of mutant polypeptides (Gething *et al.*, 1986).

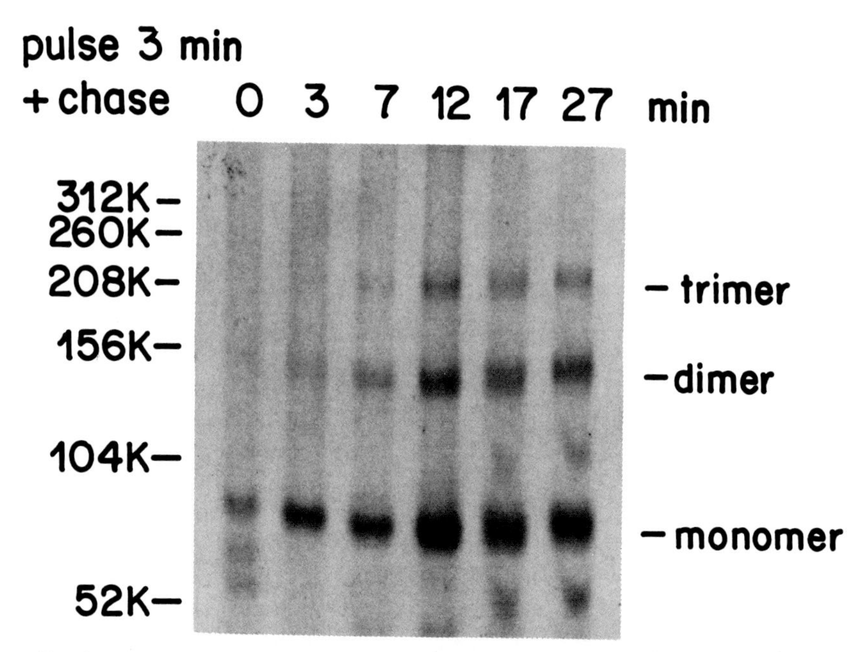

FIG. 2. Analysis by crosslinking with DMS of the time course of trimerization of HA in CV-1 cells. CV-1 cells expressing the wild-type A/Japan HA were pulse-labeled for 3 minutes with [^{35}S]methionine and then chased with an excess of nonradioactive methionine for the times indicated before cell extracts were prepared in octyl glucoside lysis buffer. Aliquots were crosslinked using DMS, and HA polypeptides were isolated by immunoprecipitation, separated by gel electrophoresis, and visualized by autoradiography. From Gething *et al.* (1986).

PROTOCOL 1. CROSSLINKING OF PROTEINS IN LABELED CELL EXTRACTS

1. Radiolabel cells as described in Section II,A either by a continuous pulse (15 minutes to several hours) or using a pulse–chase protocol.
2. Prepare cell extracts as described in Section II,A using 1 ml of lysis buffer per 10^6 cells. It may be preferable to use a PBS-based lysis buffer, although extracts prepared in the Tris-containing buffers listed on in Section II,A (note a) can be crosslinked successfully (Gething *et al.*, 1986).
3. Take 100 μl of cell extract and add the crosslinking reagent.
   (a) Members of the imidoester series (e.g., DMS) should be prepared immediately before use in 0.2 *M* triethanolamine-HCl (pH 8.5) at a concentration of 1 mg/ml (range 0.25–5.0 mg/ml). Add 400 μl to the 100 μl aliquot of extract, and incubate at room temperature for 30–120 minutes.
   (b) Members of the NHS-ester series (e.g., the thiol-cleavable reagent DSP) should be added at a concentration of 300 μg/ml in PBS (range 50–1000 μg/ml). Incubate for 30 minutes on ice.
4. Add 500 μl NET–gel buffer and immunoprecipitate samples as described in Section II,B.
5. Analyze the immunoprecipitated proteins by SDS–PAGE (protocol 4, this section). When a thiol-cleavable crosslinker is used the proteins should be analyzed both under nonreducing conditions on 3.5% polyacrylamide gels as described in protocol 4 [to determine the size(s) of the cross-linked complexes] and under reducing conditions on 10% polyacrylamide gels [to determine the size(s) of the proteins in the crosslinked complexes].

PROTOCOL 2. CROSSLINKING OF PROTEINS IN CELLS *in Vivo*

1. Radiolabel cells as described in Section II,A either by a continuous pulse (15 minutes to several hours) or using a pulse–chase protocol.
2. Remove the labeling medium and wash the cells twice with ice-cold PBS.
3. Add crosslinking reagent (e.g., the thiol-cleavable reagent DSP or another member of the NHS-ester series) at a concentration of 150 μg/ml in PBS. Incubate for 30 minutes on ice.
4. Remove crosslinker and wash the cells twice with ice-cold PBS.
5. Lyse the cells using one of the buffers described in Section II,A (note a), and utilize the supernatants for immunoprecipitations (see Section II,B).

6. Analyze the immunoprecipitated proteins as just outlined in protocol 1.

PROTOCOL 3. CROSSLINKING OF PROTEINS AT THE TIME OF CELL LYSIS

1. Radiolabel cells as described in Section II,A, either by a continuous pulse (15 minutes to several hours) or using a pulse–chase protocol.
2. Remove the labeling medium and wash the cells twice with ice-cold PBS.
3. Add 1 ml of lysis buffer containing the crosslinking reagent.
   (a) Members of the NHS-ester series (e.g., the thiol-cleavable reagent DSP), should be added at a concentration of 150 μg/ml in PBS containing the desired detergent. Incubate for 30 minutes on ice.
   (b) Members of the imidoester series (e.g., DMS), should be added at a concentration of 1 mg/ml in 0.1 *M* triethanolamine-HCl (pH 8.5), containing the desired detergent. Incubate at room temperature for 30 minutes.
4. Collect the lysates by scraping into microfuge tubes, and remove nuclei and debris by centrifugation (see Section II,A). Utilize the supernatants for immunoprecipitations (see Section II,B).
5. Analyze the immunoprecipitated proteins as just outlined in protocol 1.

PROTOCOL 4. ACRYLAMIDE GEL SEPARATION OF CROSSLINKED PROTEINS

1. Preparation of 3.5% acrylamide gel (for 42 ml gel solution):

5 ml Acrylamide stock solution (30% acrylamide, 0.8% bisacrylamide)
37.5 ml Gel buffer–
  0.1 *M* boric acid (6.18 g/liter)
  0.1 *M* Sodium acetate (13.6g/liter)
  0.1% SDS
  adjusted to pH 8.5
  (can be prepared and stored as a 5× stock)
420 μl 20% ammonium persulfate
21 μl *N,N,N′-N′*-tetramethylethylenediamene (TEMED)

Mix well and pour quickly, and then immediately insert a well former with the desired number of slots. This recipe is suitable for the preparation of one standard-sized acrylamide gel or for four minigels, such as those run using the Hoeffer Mighty Small electrophoresis apparatus, on which very good resolution can be achieved. These gels can be prepared using Gel-Bond film (FMC Co.) attached to one glass plate, which greatly improves the handling of the sloppy gels during staining and destaining procedures.

2. Assemble gel in apparatus, and fill upper and lower reservoirs with borate–acetate gel buffer (see recipe just given).

3. Add samples in a minimum volume. Place molecular-weight standard (crosslinked GDH, see protocol 5) in outside wells. Since there is no stacking gel, the broadness of the protein bands will be increased by large sample volumes. Carry out electrophoresis at 5–10 V/cm.

4. Dismantle the gel apparatus, and fix and stain gel for 30 minutes in 40% methanol, 7% acetic acid containing 0.2% Coomassie brilliant blue. Destain gel in 40% methanol, 7% acetic acid to reveal the positions of the crosslinked GDH subunits. Gels attached to gel-bond film should be dried under a heating lamp before autoradiography. Free gels should be carefully picked up on a sheet of 3-mm paper cut to approximately the dimensions of the gel, covered with plastic film and dried under vacuum before autoradiography.

Protocol 5. Preparation of Crosslinked Bovine Glutamate Dehydrogenase (GDH) As a Molecular-Weight Standard

1. Take 250 μl (5 mg) of bovine L-glutamate dehydrogenase (supplied as a 20 mg/ml solution in 2 *M* $(NH_4)_2SO_4$ by Boehringer Mannheim).
2. Add 750 μl of 0.2 *M* triethanolamine-HCl, pH 8.5 (TE-HCl), and dialyze for several hours against the same buffer (100 ml, two changes). This dialysis is necessary to remove the primary amine, which will react with the crosslinker.
3. To the dialyzed solution (5mg of GDH in ~1 ml), add 1 ml of a solution of DMS (5 mg/ml in TE-HCl buffer; prepared immediately before use). Mix and incubate for 2 hours at room temperature.
4. Add 100 μl 20% SDS and 50 μl of β-mercaptoethanol, and heat for 15 minutes at 70°C. Add 400 μl of glycerol and 25 μl of 1% bromophenol blue dye. Use 5 μl/well of the minigel for a molecular-weight standard.

B. Formation of SDS-Resistant Oligomers As an Assay for Assembly of Protein Multimers

Oligomeric proteins such as influenza HA have been observed to be partially resistant to denaturation by heating in SDS, particularly in the absence of reducing agents (Doms and Helenius, 1986; Boulay *et al.*, 1988). This property provides a qualitative but simple assay for the presence of oligomeric forms of a protein because species corresponding to the monomer, the full-sized oligomer, and the intermediate forms can be visualized on acrylamide gels. The protocol requires only that immunoprecipitated samples be resuspended and heated in sample buffer (20 m*M* Tris-HCl, pH 6.8, 20% glycerol, 1% SDS, 0.02% bromophenol blue) lacking reducing agents before separation by SDS–PAGE.

C. Sucrose Density Gradient Centrifugation As an Assay for Oligomerization

Centrifugation of sucrose density gradients can be used to separate monomeric and oligomeric forms of a protein and, unlike chemical crosslinking or analysis by SDS–PAGE under nonreducing conditions, provides a quantitative measure of the presence of the different forms. Figure 3 illustrates the use of this technique to characterize the extent of trimerization of HA synthesized in Chinese hamster ovary (CHO) cells (Hearing *et al.*, 1989). Combined with a detailed pulse–chase labeling protocol, sucrose gradient centrifugation can yield an accurate time course for the assembly of the multimeric protein from its newly synthesized subunits (Gething *et al.*, 1986; Copeland *et al.*, 1986; Doms *et al.*, 1987). The procedure for the preparation and separation of proteins on sucrose density gradients is described in the following paragraphs.

A drawback of this technique is that some oligomers are not stable during centrifugation on sucrose gradients. Core-glycosylated forms of X-31 HA run as monomers on sucrose gradients, even though they can be shown by other folding assays to have assembled into trimers (Copeland *et al.*, 1986). Similarly, trimers of VSV G protein are unstable on sucrose gradients under neutral-pH conditions; acidification is required to maintain the trimeric state upon centrifugation (Doms *et al.*, 1987). It is inter-

---

FIG. 3. Analysis of HA trimer formation in HA-CHO cells using sucrose density gradient centrifugation. CHO cells that express HA from a bovine papilloma virus vector were pulse-labeled for 5 minutes with [^{35}]methionine and then incubated for 0 or 60 minutes with medium containing an excess of nonradioactive methionine before extracts were prepared in octyl glucoside lysis buffer. Aliquots of the extracts were subjected to velocity sedimentation (40,000 rpm, 16 hours in a Beckman SW41 rotor) on 5–20% sucrose gradients prepared in the lysis buffer. Following fractionation of the gradients into 21 aliquots, the HA molecules were immunoprecipitated and analyzed by SDS–PAGE and fluorography. The densest fractions are shown at the left of the photograph.

After a 5-minute pulse, the core-glycosylated HA sediments as monomers nearer to the top of the gradient. Following a 60-minute chase, much of the labeled HA sediments to a denser portion of the gradient. These trimeric HA molecules include both core-glycosylated and terminally glycosylated forms, indicating that trimer formation precedes transport to the medial cisternae of the Golgi apparatus. From Hearing *et al.* (1989).

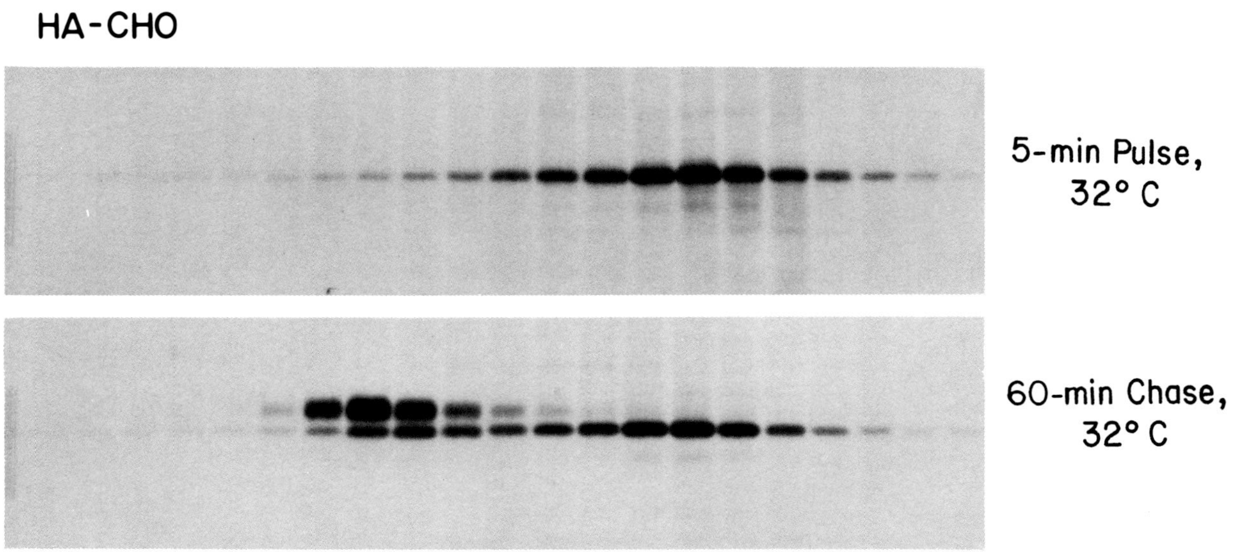

HA-CHO

5-min Pulse,
32° C

60-min Chase,
32° C

esting to note that it appears to be the physical process of centrifugation rather than the presence of sucrose that leads to disassembly of the oligomers, since X-31 HA trimers are quite stable in sucrose solutions (our unpublished results). Thus the results obtained using this technique should be interpreted in conjunction with other folding assays and should not be used to provide evidence against the presence of protein multimers.

Finally, it should be noted that both crosslinking and sucrose density gradient centrifugation provide evidence for the presence of protein oligomers, but do not necessarily discriminate between fully folded, native molecules and oligomers that do not complete the folding process. For example, many mutant forms of HA that are retained in the ER assemble into trimeric structures with the same time course as the wild-type protein. However, analysis with conformation-specific antibodies and protease sensitivity assays reveal that these mutant trimers do not achieve a wild-type conformation (Gething *et al.*, 1986; our unpublished results). It is therefore important to use a combination of folding assays to characterize each protein fully. In practice this is best achieved by combining the procedures outlined in this chapter. The protein of interest is labeled in a pulse–chase protocol, and then sucrose density gradient centrifugation is used to separate the monomeric and oligomeric forms of the protein, which can then be analyzed independently using other folding assays including crosslinking, conformation-specific antibodies, and protease sensitivity assays (Gething *et al.*, 1986).

Protocol

1. Prepare solutions of 5 and 20% sucrose in 50 m*M* Tris-HCl, pH 7.4, containing 100 m*M* NaCl. For analysis of membrane proteins the buffer should contain detergent (6 mM lauryl maltoside, or 40 m*M* octyl glucoside, or 0.1% Nonidet P-40, or 0.1% Triton X-100; see note a in Section II,A). Addition of bromophenol blue dye (to ~0.02%) to the denser solution can provide a visual check of the quality of the gradients. Use a standard apparatus to prepare 5-ml or 10-ml continuous 5–20% sucrose gradients in the appropriate centrifuge tubes. If high molecular weight aggregates are likely to be present, the gradients can be layered on a small cushion (0.5–1.0 ml) of 60% sucrose.

2. Layer an aliquot (100–500 μl) of labeled cell extract onto the top of each gradient, and centrifuge for 16 hours at 4°C either at 40,000 rpm (10-ml gradient in a Beckman SW40 or a Sorvall TH641 rotor) or at 47,000 rpm (5-ml gradient in a Beckman SW50.1 or a Sorvall AH650 rotor).

3. Collect each gradient into 15–20 fractions, and aliquot samples for analysis by immunoprecipitation or other folding assays (protease sensitivity, crosslinking, etc.).

IV. Conclusion

We have at present only a limited understanding of the cellular processes and mechanisms that facilitate and regulate the folding, assembly, and intracellular transport of membrane and secretory proteins in eukaryotic cells. Analyses of the biosynthesis of a few such proteins has provided evidence that efficient transport from the ER to the Golgi apparatus requires that the nascent proteins attain their "correct" tertiary structure. For oligomeric proteins this requirement includes assembly into their correct quaternary structure. Preliminary evidence has also been obtained for the involvement of cellular ER proteins in these processes. Elucidation of the mechanisms governing the early stages of exocytosis will undoubtedly be facilitated by further studies on the folding and transport of a wide variety of exocytotic proteins using techniques such as those outlined in this chapter.

REFERENCES

Baron, C., and Thompson, T. E. (1975). *Biochim. Biophys. Acta* **382,** 276–284.

Bergmann, L. W., and Kuehl, W. M. (1979). *J. Biol. Chem.* **254,** 5690–5694.

Bole, D. G., Hendershot, L. M., and Kearney, J. F. (1986). *J. Cell Biol.* **102,** 1558–1566.

Boulay, F., Doms, R. W., Webster, R. G., and Helenius, A. (1988). *J. Cell Biol.* **106,** 629–639.

Carlin, B. E., and Merlie, J. P. (1986). *In* "Protein Compartmentalization" (A. N. Straus, G. Kreil, and I. Boime, eds.), pp. 71–86. Springer-Verlag, New York.

Copeland, C. S., Doms, R. W., Bolzau, E. M., Webster, R. G., and Helenius, A. (1986). *J. Cell Biol.* **103,** 1179–1191.

Copeland, C. S., Zimmer, K.-P., Wagner, K. R., Healey, G. A., Mellman, I., and Helenius, A. (1988). *Cell (Cambridge, Mass.)* **53,** 197–209.

Creighton, T. E. (1988). *BioEssays* **8,** 57–63.

Davies, G. E., and Stark, G. R. (1970). *Proc. Natl. Acad. Sci. U.S.A.* **66,** 651–656.

Doms, R. W., and Helenius, A. (1986). *J. Virol.* **60,** 833–839.

Doms, R. W., Keller, D. S., Helenius, A., and Balch, W. E. (1987). *J. Cell Biol.* **105,** 1957–1969.

Doms, R. W., Ruusala, A., Machamer, C., Helenius, J., Helenius, A., and Rose, J. K. (1988). *J. Cell Biol.* **107,** 89–99.

Dorner, A. J., Bole, D. G., and Kaufman, R. J. (1987). *J. Cell Biol.* **105,** 2665–2674.

Edman, J. C., Ellis, L., Blacher, R. W., Roth, R. A., and Rutter, W. J. (1985). *Nature (London)* **317,** 267–270.

Felgner, P. L., Messer, J. L., and Wilson, J. E. (1979). *J. Biol. Chem.* **254,** 4946.
Fitting, T., and Kabat, D. (1982). *J. Biol. Chem.* **257,** 14011–14017.
Freedman, R. B. (1984). *Trends Biochem. Sci.* **9,** 438–441.
Garoff, H. (1985). *Annu. Rev. Cell. Biol.* **1,** 403–445.
Gething, M.-J. (1985). *In* "Protein Transport and Secretion" (M.-J. Gething, ed.), pp. 1–20. Cold Spring Harbor Lab., Cold Spring Harbor, New York.
Gething, M.-J., McCammon, K., and Sambrook, J. (1986). *Cell (Cambridge, Mass.)* **46,** 939–950.
Haas, I. G., and Wabl, M. (1983). *Nature (London)* **306,** 387–389.
Harlow, E., and Lane, D. P. (1988). "Antibodies." Cold Spring Harbor Lab., Cold Spring Harbor, New York (in press).
Hearing, J., Gething, M.-J., and Sambrook, J. (1989). *J. Cell Biol.* **108,** 355–365.
Helenius, A., and Simons, K. (1975). *Biochim. Biophys. Acta* **415,** 29–79.
Hendershot, L., Bole, D., and Kearney, J. F. (1987). *Immunol. Today* **8,** 111–114.
Kishimoto, T. K., Hollander, N., Roberts, T. M., Anderson, D. C., and Springer, T. A. (1987). *Cell (Cambridge, Mass.)* **50,** 193–202.
Koch, G., Smith, M., Macer, D., Webster, P., and Mortara, R. (1987). *J. Cell Sci.* **86,** 217–232.
Kozutsumi, Y., Segal, M., Normington, K., Gething, M.-J., and Sambrook, J. (1988). *Nature (London)* **332,** 462–464.
Kreis, T. E., and Lodish, H. (1986). *Cell (Cambridge, Mass.)* **46,** 929–937.
Laemmli, U. K. (1970). *Nature (London)* **227,** 680–685.
Lee, A. S., Delegeane, A., and Scharff, D. (1981). *Proc. Natl. Acad. Sci. U.S.A.* **78,** 4922–4925.
Lewis, M. J., Mazzarella, R. A., and Green, M. (1985). *J. Biol. Chem.* **260,** 3050–3057.
Lodish, H. F. (1988). *J. Biol. Chem.* **263,** 2107–2110.
Mazzarella, R. A., and Green, M. (1987). *J. Biol. Chem.* **262,** 8875–8883.
Morrison, S. L., and Scharff, M. D. (1975). *J. Immunol.* **114,** 655–659.
Munro, S., and Pelham, H. R. B. (1986). *Cell (Cambridge, Mass.)* **46,** 291–300.
Owen, M. J., Kissonerghis, A.-M., and Lodish, H. F. (1980). *J. Biol. Chem.* **225,** 9678–9684.
Pelham, H. R. B. (1986). *Cell (Cambridge, Mass.)* **46,** 959–961.
Peters, T., Jr., and Davidson, L. K. (1982). *J. Biol. Chem.* **257,** 8847–8853.
Peters, B. P., Hartle, R. J., Krzesicki, R. F., Kroll, T. G., Perini, F., Balun, J. E., Goldstein, J., and Ruddon, R. W. (1985). *J. Biol. Chem.* **260,** 14732–14742.
Rosevear, P., VanAken, T., Baxter, J., and Ferguson-Miller, S. (1980). *Biochemistry* **19,** 4108–4115.
Roth, R. A., and Pierce, S. B. (1987). *Biochemistry* **28,** 4179–4182.
Ruddon, R. W., Krzesicki, R. F., Norton, S. E., Saccuzzo Beebe, J., Peters, B. P., and Perini, F. (1987). *J. Biol. Chem.* **262,** 12533–12540.
Shiu, R. P. C., Pouyssegur, R., and Pastan, I. (1977). *Proc. Natl. Acad. Sci. U.S.A.* **74,** 3840–3844.
Weiss, A., and Stobo, J. D. (1984). *J. Exp. Med.* **160,** 1284–1299.
Yewdell, J. W., Yellen, A., and Bachi, T. (1988). *Cell (Cambridge, Mass.)* **52,** 843–852.

Chapter 8

Glycosaminoglycan Modifications of Membrane Proteins

KAREN S. GIACOLETTO, TARA RUMBARGER, AND
BENJAMIN D. SCHWARTZ

Howard Hughes Medical Institute and
Division of Rheumatology
Department of Medicine
Washington University School of Medicine
St. Louis, Missouri 63110

I. Introduction

Several membrane molecules previously thought to exist on the cell membrane only as proteins and glycoproteins, have also been recently found on the cell membrane as proteoglycans (PG), that is with the addi-

Copyright © 1989 by Academic Press, Inc.
All rights of reproduction in any form reserved.

tion of a glycosaminoglycan (GAG) (Bumol and Reisfeld, 1982; Sant *et al.*, 1985; Frannson *et al.*, 1984; Giacoletto *et al.*, 1986). Some of these molecules, such as transferrin receptor, have been shown to subserve the same function in both the glycoprotein and proteoglycan forms (Frannson *et al.*, 1984; Cresswell, 1985). For others, such as the major histocompatibility complex (MHC) class II-associated invariant chain (I_i), the function of neither form is known (Rosamond *et al.*, 1987). These observations suggest that other membrane proteins and glycoproteins may also exist in a PG form. This chapter provides the methodology with which to seek and prove the existence of this alternative form of a given protein.

Our own interest in characterizing PG began when a M_r 40,000–70,000 sulfated protein-containing molecule was immunoprecipitated from lymphocytes by antibodies specific for MHC class II molecules (Sant *et al.*, 1983, 1984). Class II molecules of the MHC are crucial to the initiation of an immune response (Rosenthal and Shevach, 1973; Unanue, 1984; Streicher *et al.*, 1984). They are heterodimers composed of an α-glycoprotein chain of M_r ~34,000 and a β-glycoprotein chain of M_r ~29,000 (Barnstable *et al.*, 1978). At the time we discovered the M_r 40,000–70,000 component, the class II heterodimer was already known to be associated with several non-MHC-encoded molecules during its biosynthesis and/or expression. Most prominent among these molecules was a nonpolymorphic glycoprotein known as the invariant chain (I_i) (Jones *et al.*, 1979). The discovery that the class II molecules were also associated with a large heterogeneous sulfated molecule prompted the characterization of this molecule. This characterization eventually led to the identification of this molecule as a PG with a core protein of invariant chain, and a GAG component of chondroitin sulfate (CS-I_i) (Sant *et al.*, 1985; Giacoletto *et al.*, 1986). This system will be used as the model system throughout this chapter.

II. Biochemical Characterization of a PG

Biochemical characterization of a PG depends on several features unique to this molecule (Rodén, 1980). Proteoglycans are extremely heterogeneous in apparent molecular weight, are highly negatively charged, bear sulfate on a high molecular weight protease-resistant structure, and are sensitive to specific glycosaminoglycanases. Identification of a particular molecule as a PG requires the demonstration that the molecule possesses these unique features.

A. Radiolabeling and Isolation of the PG

Most helpful in biochemical characterization is the availability of a cell line that can be utilized to radiolabel the protein and putative PG biosynthetically, and of an antibody to the protein of interest. The molecules of interest can then be radiolabeled, immunoprecipitated, and analyzed by a number of appropriate methods.

In our experiments, cultured lymphoblastoid cell lines were biosynthetically labeled with $^{35}SO_4$ to identify the PG moiety or with [^{35}S]methionine or [^{3}H]leucine to label specifically the core protein. However, depending on the amino acid composition of the protein under investigation, other amino acids may be more appropriate for biosynthetic labeling.

Prior to labeling, the cells were washed three times in phosphate-buffered saline (PBS) containing 1% dialyzed serum, then preincubated in precursor-deficient medium for 30 minutes at 37°C. The cells were pelleted and resuspended at a density of 2×10^6 cells/ml in medium containing the radioactive precursor as the sole source of that precursor. For $^{35}SO_4$, the medium used was Earle's balanced salt solution, with $MgCl_2$ substituted for $MgSO_4$, supplemented with amino acids and vitamins from an RPMI 1640 Selectamine Kit (Gibco, Grand Island, NY), glutamine, HEPES, 10% dialyzed serum, and $^{35}SO_4$ (New England Nuclear, Boston, MA, 42.8 Ci/mmol) at a concentration of 350 μCi/ml. For [^{35}S]methionine and [^{3}H]leucine the medium used was RPMI 1640 lacking either methionine or leucine, and supplemented with 10% dialyzed serum, glutamine, and either [^{35}S]methionine (NEN, 800 Ci/mmol) or [^{3}H]leucine (NEN, 140.2 Ci/mmol) at a concentration of 250 μCi/ml. The cells were incubated for 5 hours at 37°C in a humidified atmosphere containing 7% CO_2. After labeling, the cells were washed one time with PBS and frozen at −20°C. The cells were thawed and lysed for 15 minutes at 4°C in NP-40 cell lysis buffer: 0.5% Nonidet P-40 (NP-40, Particle Data, Inc., Elmhurst, IL) in PBS (pH 7.2), containing the protease inhibitors phenylmethylsulfonyl fluoride (PMSF, 200 μg/ml), L-1-tosylamide-2-phenylethyl-chloromethylketone (TPCK, 50 μg/ml), and *N*-α-*p*-tosyl-L-lysine chloromethylketone (TLCK, 50 μg/ml) (Sigma Chemical Co., St. Louis, MO). The detergent-solubilized cell lysates were ultracentrifuged at 100,000 *g* for 60 minutes to remove insoluble debris. The soluble supernatant was reacted with antibody for 2 hours at 4°C. The antibody–antigen complexes were pelleted using 50μl packed protein A–Sepharose per 10^7 cells (Pharmacia, Uppsala, Sweden) for 30 minutes at 4°C, and the pellets were washed three times with PBS containing 0.25% NP-40 (PSN).

In the initial stages of our work, before the core protein of our PG had been identified, we precipitated the PG using an anti-class II antibody. This procedure was successful because of the association of this PG with class II molecules, and was used extensively. After the core protein had been identified as invariant chain, we were able to precipitate the PG directly with an anti-invariant chain antibody. As noted before, in situations where a PG form of a known protein is being sought, an antibody against the protein, when available, greatly facilitates the analyses.

B. Molecular-Weight Determination by SDS–PAGE

In preparation for sodium dodecyl sulfate–polyacrylamide gel electrophoresis (SDS–PAGE) analysis, the radiolabeled antigens were eluted from the protein A pellets with SDS-reducing elution buffer [0.062 *M* Tris, 2.0% SDS, 2% 2-mercaptoethanol (2-ME), 10% glycerol, and 0.001% phenol red] and boiled for 2 minutes. Samples were electrophoresed through a 1-cm 4% stacking gel and a 11.5-cm 11% running gel using a modification of the Laemmli–Maizel system (Laemmli, 1970; Maizel, 1971). Gels were fixed in a solution of 10% glacial acetic acid, 30% methanol, and 60% deionized water for 30 minutes. They were then placed in 100 ml of Amplify (Amersham Corporation, Arlington Heights, IL) on a rocker for 30 minutes, dried on a gel dryer, and autoradiographed.

The results of our initial studies identified a sulfated moiety that migrated as a diffuse band with an apparent molecular weight of 40,000–70,000 (Fig. 1). These results suggested the possibility that this molecule was a PG.

C. Charge Determination

1. Isoelectric Focusing (IEF)

Proteoglycans are polyanionic molecules with a high average charge–mass ratio due to their high content of negatively charged sugars (glucuronic or iduronic acid) and sulfate groups, and consequently they can be distinguished from proteins and glycoproteins on the basis of net charge (Rodén, 1980). To investigate whether our sulfated molecule belonged to this class of macromolecules, two analyses were done. First, $^{35}SO_4$-labeled immunoprecipitated material was analyzed by IEF using SDS two-dimensional gels as described by O'Farrell (1975) with modifications (Karr *et al.*, 1982) in order to determine the charge of our protein. The protein A pellets were eluted with IEF elution buffer (9.5 *M* urea, 0.2% NP-40, 1.6% ampholytes pH 3.5–10, 5% 2-ME). The first-dimension

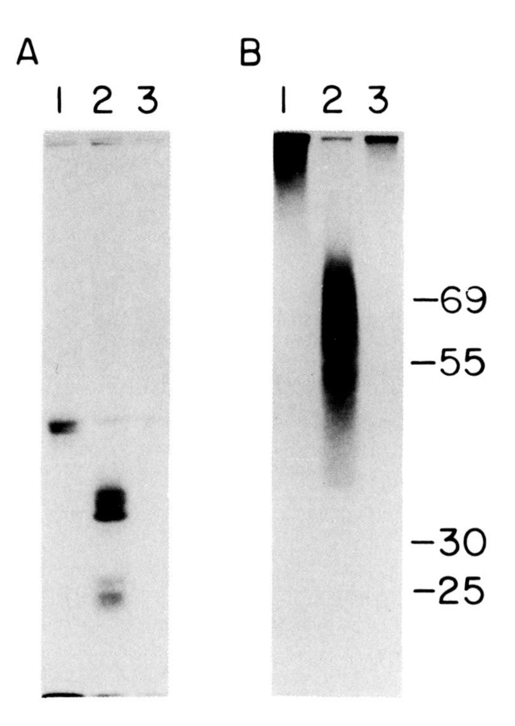

FIG. 1. Initial identification of the CS PG form of invariant chain. The lymphoblastoid B-cell line Swei was biosynthetically labeled with either [^{3}H]leucine (A) or $^{35}SO_4$(B). Detergent extracts of radiolabeled cells were immunoprecipitated with rabbit antihuman β_2-microglobulin to precipitate human class I molecules (lane 1); SG171, a human anti-HLA-DR class II monoclonal antibody (lane 2); or control ascites fluid (lane 3). Immunoprecipitated material was analyzed by SDS–PAGE (10% acrylamide). The positions of ^{14}C-labeled molecular-weight markers ($M_r \times 10^3$) run in an adjacent lane are indicated.

separation (IEF) was performed in 5 × 130-mm cylindrical gels containing 4.0% ampholytes yielding an expanded pH 3.5–10 gradient. This pH 3.5–10 gradient was made of an ampholine mixture containing 30% pH 3.5–10, 20% pH 8–9.5, and 50% pH 5–8 Ampholines (LKB Instruments, Rockville, MD). The apparatus for IEF gel electrophoresis consisted of a top chamber (negative electrode) containing 0.02 *M* NaOH and the bottom chamber (positive electrode) containing 0.01 *M* H_3PO_4. Samples were electrophoresed at 1 W/gel until the voltage reached 800 V, then under constant voltage (800 V) for 16 hours, and then at 1200 V for the final hour. The pH gradient was determined by measuring the pH of 1-cm slices of a blank gel run in parallel with the samples. The IEF gels were equilibrated in 10% (w/v) glycerol, 5% (v/v) 2-ME, 2.3% (w/v) SDS,

0.0625 *M* Tris (pH 6.8) for 4 hours with hourly changes of the equilibration buffer. The equilibrated gels were embedded in agarose on the top of a slab gel (3.5-cm 4% acrylamide SDS stacking gel and 12.5-cm 11% acrylamide SDS running gel) and electrophoresis was performed in the second dimension (Fig. 2). The results of this analysis suggested that the sulfated moiety was negatively charged.

2. DEAE–Sephacel Anion Exchange Chromatography

Further characterization of the negatively charged sulfated moiety was accomplished by ion exchange chromatography. Immunoprecipitated material from $^{35}SO_4$- and [^{3}H]leucine-labeled cells was fractionated on the anion exchange resin DEAE–Sephacel (Pharmacia) in a dissociative buffer under conditions in which PG, but not proteins or glycoproteins, would bind (Chang *et al.*, 1983).

Protein A pellets prepared as described before were eluted for 18 hours

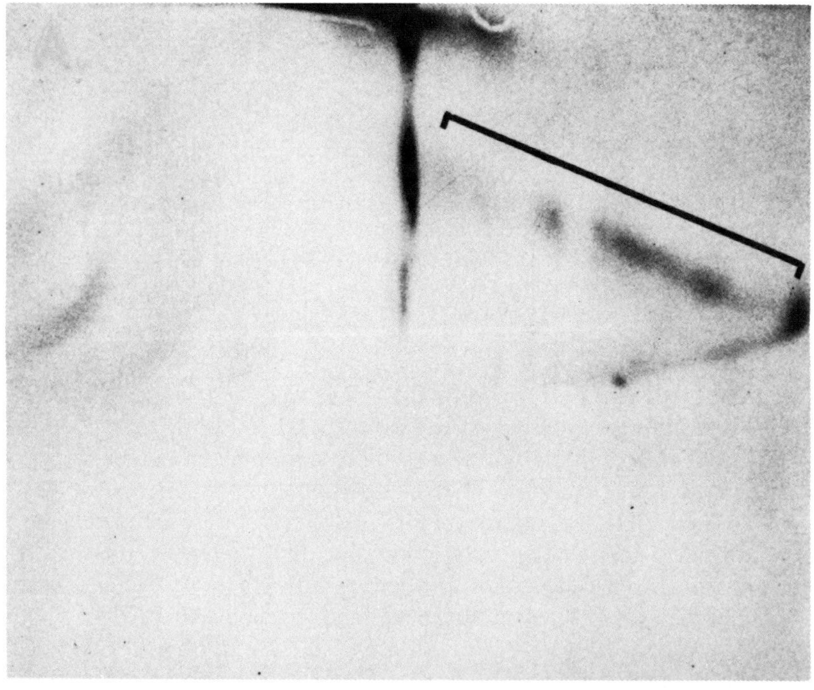

Fig. 2. Two-dimensional gel electrophoresis of anti-class II immunoprecipitates prepared from lysates of $^{35}SO_4$-labeled Swei cells. The $^{35}SO_4$-labeled component (bracket) migrates as a series of spots heterogeneous in size and charge.

at 25°C in 0.5 ml of column-starting dissociative DEAE buffer (8 *M* urea, 0.15 *M* NaCl, 0.05 *M* sodium acetate, 0.5% Triton X-100, pH 6.0). Insoluble material was pelleted by centrifugation and the supernatant was applied to a 75 × 200-mm DEAE–Sephacel column equilibrated in the same buffer. The column was washed with 5.0 ml of the starting buffer, after which bound material was eluted using a linear salt gradient (0.15–1.4 *M* NaCl). Fractions of 0.7 ml were collected, and aliquots of each fraction were assayed for radioactivity. For the [^{3}H]leucine sample, >95% of the recovered radioactivity failed to bind to the DEAE–Sephacel. This was the expected result because the majority of the [^{3}H]leucine was incorporated into the class II glycoproteins. In contrast, 80% of the $^{35}SO_4$-labeled immunoprecipitated material was bound by the resin and eluted with 0.3–0.4 *M* NaCl (Fig. 3). This result again suggested that the sulfated moiety was a PG.

D. Determination that the Sulfated Moiety Contains Protein

The demonstration that an amino acid-radiolabeled species, isolated by the same methods as the sulfated moiety, comigrated with the sulfated moiety on SDS–PAGE would provide evidence that the sulfated moiety

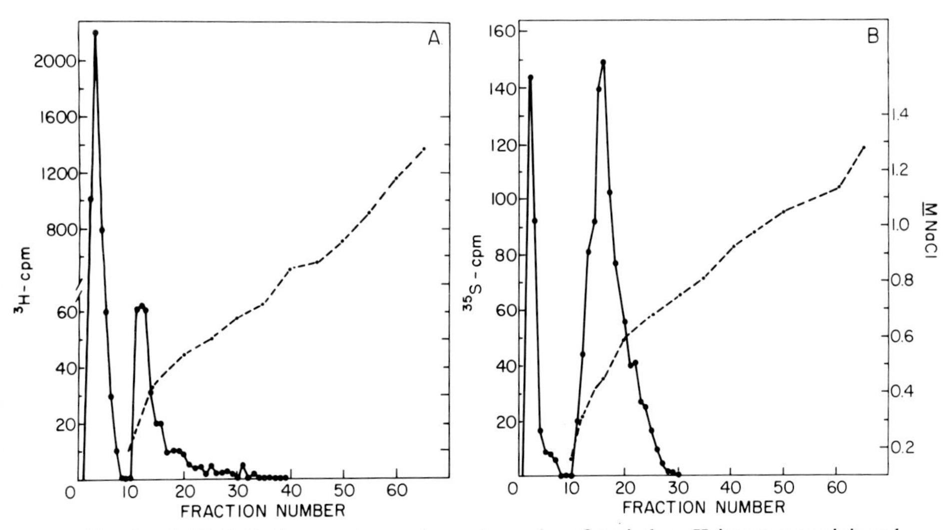

FIG. 3. DEAE–Sephacel column chromatography of anti-class II immunoprecipitated material from (A) [^{3}H]leucine-labeled (●) or (B) $^{35}SO_4$-labeled (●) Swei cells. Material was eluted with 0.3–0.4 M NaCl (----).

also contains protein. However, in our initial studies we were unable to detect any such amino acid-radiolabeled species, almost certainly because of the low representation of our putative PG compared to total cell protein. We therefore adopted other methodologies to demonstrate that the sulfated moiety contained protein.

We first tested the sensitivity of the sulfated moiety to protease digestion. Immunoprecipitated material from $^{35}SO_4$-labeled cells was subjected to degradation by staphylococcal V8 protease (*S. aureus* V8, ICN Immunobiologicals, Irvine, CA), which cleaves at aspartyl and glutamyl residues (Drapeau *et al.*, 1972). A stock solution of enzyme (1 mg/ml) was prepared in SDS nonreducing elution buffer (2% SDS, 0.062 *M* Tris, pH 6.8) and added to SDS eluates of immunoprecipitates at a final concentration of 125 μg/ml. Enzyme-treated and untreated control samples were incubated at 37°C for 2 hours and then boiled for 2 minutes to inactivate the enzyme. 2-Mercaptoethanol (BioRad Laboratories, Richmond, CA) was added to a final concentration of 2%, and samples were analyzed on SDS–PAGE slab gels. Comparison of the untreated sample with the enzyme-treated sample demonstrated that the $^{35}SO_4$-labeled molecule was reduced to a lower molecular weight species after treatment with V8 protease (Fig. 4). Sensitivity of this molecule to V8 protease thus indicated that it contains polypeptide components.

Eventually, we were able to demonstrate directly that the sulfated moiety could be radiolabeled with an amino acid precursor. We purified the amino acid-radiolabeled sulfated moiety by immunoprecipitation with an anti-invariant chain antibody followed by double passage over DEAE–Sephacel under dissociative conditions, and were able to show that the protein-labeled species comigrated with the sulfated moiety on SDS–PAGE.

E. Characterization of the Sulfated Component

1. Pronase Digestion

It was next necessary to characterize the sulfated component of this molecule. We first studied the manner in which sulfate was linked to this molecule. Proteoglycans contain sulfate groups O- or N-linked to sugar residues of large GAG side chains, which in turn are covalently attached to the core protein (Rodén, 1980). Exposure of this molecule to pronase digestion will result in the sulfate still being found on a high molecular weight species. To determine the type of moiety to which $^{35}SO_4$ groups were linked, eluates of immunoprecipitates in SDS elution buffer were made up to a final concentration of 1 mg/ml of carrier protein, and any

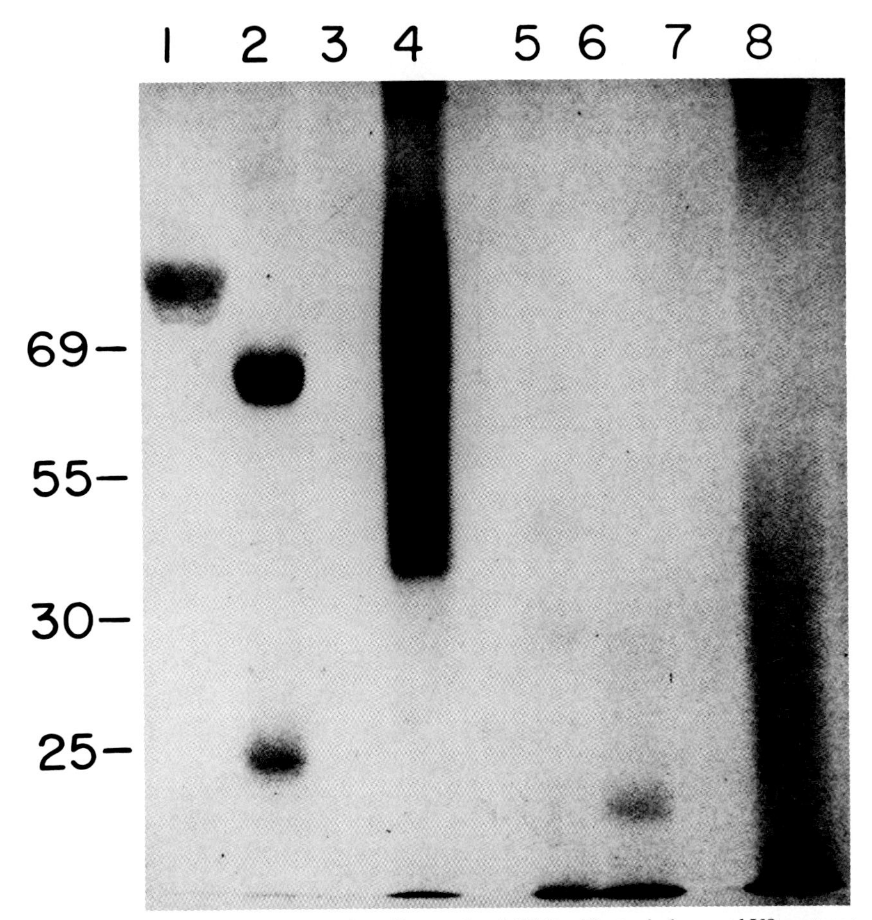

FIG. 4. Treatment of the HLA class II-associated CS-I_i with staphylococcal V8 protease. SDS–PAGE profile of mock-digested (lanes 1–4) or V8 protease-digested (lanes 5–8) samples. ^{14}C-Labeled BSA (lanes 1 and 5) and ^{14}C-labeled IgG (lanes 2 and 6) were used as positive controls for digestion. The $^{35}SO_4$-labeled CS-I_i precipitated from Swei cells (lane 4) is reduced to a lower molecular weight species after treatment with V8 protease (lane 8). Lanes 3 and 7 are control immunoprecipitates prepared from $^{35}SO_4$-labeled Swei cells. Molecular weights are indicated as $M_r \times 10^{-3}$.

protein, PG, and GAG moieties were precipitated with three volumes of 95% ethanol containing 1.3% potassium acetate overnight at −20°C (Kimata *et al.*, 1974). The pellets were washed three times in 95% ethanol containing 1.3% potassium acetate, and once in absolute ethanol, and then were dried overnight at 37°C. Ethanol precipitates were solubilized in 0.5 ml of pronase digestion buffer (10 m*M* Tris, 1 m*M* $CaCl_2$, pH 8.0).

Predigested pronase (*Streptomyces griseus* protease, Calbiochem-Behring Corp., San Diego, CA) was added to a final concentration of 5 mg/ml in pronase digestion buffer. Samples were incubated in a toluene atmosphere at 50°C for 12 hours, a second aliquot of pronase equivalent to the first was added, and digestion was allowed to continue for an additional 12 hours. A control sample was incubated in the digestion buffer in the absence of enzyme for 24 hours at 50°C.

After digestion, samples were boiled for 2 minutes to inactivate the enzyme and applied to a Bio-Gel P-10 column (BioRad Laboratories; −400 mesh, 1 × 40 cm) equilibrated in 0.1 *M* ammonium bicarbonate. The column was eluted with 0.1 *M* ammonium bicarbonate at a flow rate of 1.5 ml/hour. Fractions of 0.9 ml were collected, and the radioactivity in an aliquot from each fraction was determined. Void and included volumes were determined for each column by monitoring hemoglobin (Hb) and mannose (Man) standards, which were added to each sample before application to the column. Bio-Gel P-10 has an exclusion limit of M_r ~20,000 for globular proteins. If the sulfate were attached to an amino acid or to a carbohydrate group of a conventional glycoprotein, the radioactivity would be expected to elute in the included volume of the column. Alternatively, if the sulfate were attached to a large pronase-resistant glycan, the radioactivity would elute in the void volume. A single radioactive peak was recovered, eluting in the void volume of the column (Fig. 5). This result was suggestive evidence that the sulfate-bearing moiety was a relatively large, pronase-resistant macromolecule.

However, the possibility remained that the sulfated material was pronase-resistant, because it contained clusters of serine or threonine residues to which small O-linked sulfated carbohydrate groups were attached (Chang *et al.*, 1983; Bhavanandan *et al.*, 1977). To address this question, we treated an aliquot of the pronase-resistant P-10 pool with alkaline borohydride under conditions that would release O-linked carbohydrate groups by β elimination. An aliquot of the $^{35}SO_4$-labeled fraction peak obtained by Bio-Gel P-10 chromatography of pronase-digested material was lyophilized to dryness, resuspended in 1.0 ml of hydrolysis buffer (0.05 *M* NaOH, 50 μ*M* *N*-acetylglucosamine, 1 *M* $NaBH_4$), and incubated at 45°C for 16 hours (Baenziger and Kornfeld, 1974). Acetic acid was then added dropwise to neutralize the sample. The neutralized sample was dried, resuspended in 0.1 *M* ammonium bicarbonate containing hemoglobin and mannose, and rechromatographed on a Bio-Gel P-10 column. Again, a single radioactive peak eluted in the void volume of the column (Fig. 5). This result excluded the possibility that the $^{35}SO_4$ was attached to small O-linked oligosaccharides, and indicated that the sulfate residues were attached to a large pronase-resistant glycoconjugate, typical of PG.

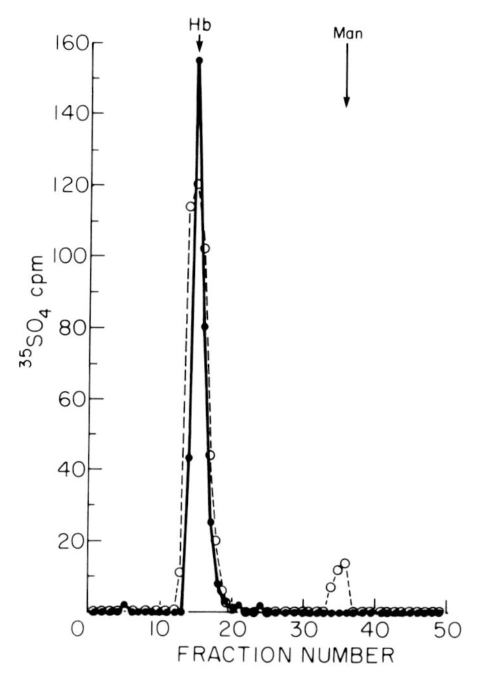

FIG. 5. (A) Bio-Gel P-10 column chromatography of pronase-digested (●—●) and alkaline borohydride-treated (○----○) anti-class II immunoprecipitates prepared from $^{35}SO_4$-labeled Swei cells. The $^{35}SO_4$-labeled material remains in the excluded volume after pronase digestion and alkaline borohydride treatment.

2. Chondroitinase Digestion to Remove Chondroitin Sulfate

To establish further that our molecule was indeed a PG, we tested its sensitivity to enzymes that are specific for the carbohydrate units of PG. The GAG components of PG consist of long repeating units of disaccharides composed of a hexosamine and, with the exception of keratan sulfate, a uronic acid. Immunoprecipitated material from $^{35}SO_4$-labeled cells was resuspended in chondroitinase buffer: 50 m*M* Tris-HCl (pH 6.0 or 8.0), containing 0.1 mg/ml bovine serum albumin (BSA). Chondroitinase AC or ABC (ICN Immunobiologicals) was added at a final concentration of 1 U/ml, and digestion was carried out for 14 hours at 37°C. Chondroitinase ABC will degrade both dermatan sulfate and chondroitin sulfate to 4,5-unsaturated disaccharides and has a pH optimum of 6.0 (Yamagata *et al.*, 1968). Chondroitinase AC specifically degrades CS and has a pH optimum of 8.0. Chondroitin sulfate (Sigma) was digested in parallel as a positive control for the efficacy of digestion. Mock-treated and enzyme-digested samples were boiled for 2 minutes and then subjected to SDS–PAGE analysis. The results of these digestions showed that our $^{35}SO_4$-bearing moiety was degraded by both chondroitinase ABC (Fig. 6) and AC (data not shown), indicating that this macromolecule is a PG with CS GAG side chains.

Paper chromatography of the digestion products was also performed to determine the position of the sulfated carbohydrate. Chondroitinase AC or ABC-digested samples were spotted on Whatman no. 1 paper (American Scientific Products, McGaw Park, IL) and analyzed by descending paper chromatography using a solvent system of 1-butanol, glacial acetic acid, and 1 *N* NH_4OH (2 : 3 : 1). Disaccharide standards run in adjacent

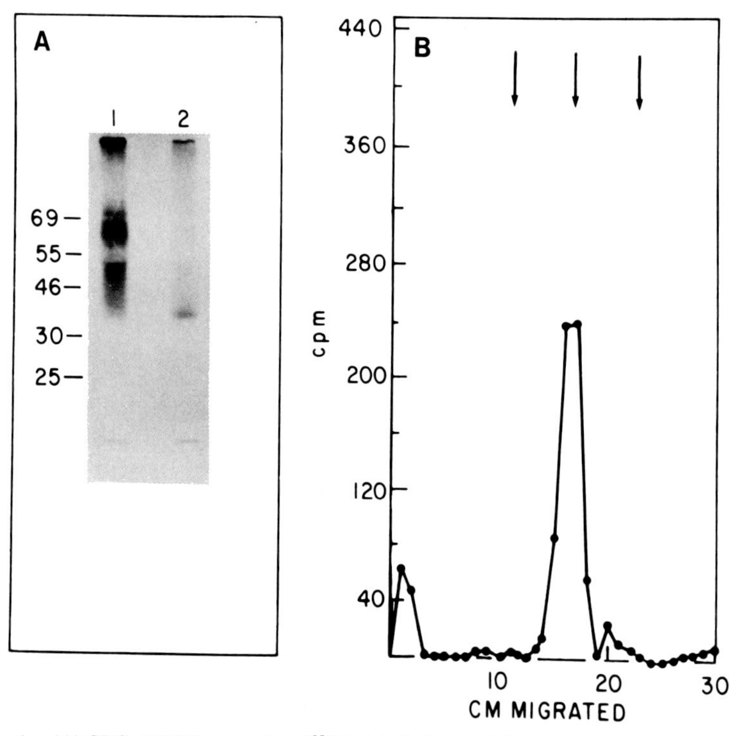

FIG. 6. (A) SDS–PAGE analysis of $^{35}SO_4$-labeled material remaining on an anti-class II immunoprecipitate after mock digestion (lane 1) or chondroitinase ABC digestion (lane 2). Molecular weights are indicated as $M_r \times 10^{-3}$. (B) Paper-chromatographic analysis of $^{35}SO_4$-labeled material released from anti-class II immunoprecipitates by mock digestion (○—○) or by chondroitinase ABC digestion (●—●). The mock-digested released material remains at the origin. Arrows indicate the positions of the standards, from left to right 6-sulfated, 4-sulfated, and nonsulfated N-acetylgalactosamine-uronic acids.

lanes included 2-acetamido-2-deoxy-3-*O*-(β-D-gluco-4-enepyranosyluronic acid)-4-*O*-D-galactose (Δ-Di-4S), 2-acetamido-2-deoxy-3-*O*-(β-D-gluco-4-enepyranosyluronic acid)-6-*O*-sulfo-D-galac-tose (Δ-Di-6S), and 2-acetamido-2-deoxy-3-*O*-(β-D-gluco-4-enepyranosyluronic acid)-D-galactose (Δ-Di-0S). Results showed that the chondroitinase-digested radiolabeled material from human sources migrated at the position of the 4-sulfated reference disaccharide, indicating that virtually all of the sulfation in this molecule occurs at the 4-position of *N*-acetylgalactosamine (Fig. 6).

Other GAG besides chondroitin sulfate and dermatan sulfate may also be present on PG. Methods for determining the presence of heparan sulfate using nitric acid, and keratan sulfate using keratan sulfate–β-endogalactosidase are given in a report by Hart (1976) and for convenience are briefly described here. It should be possible to substitute Bio-Gel P-10 for Sephadex G-50.

3. Nitrous Acid Digestion to Remove Heparan Sulfate

Peak fractions eluted in the void volume of a Sephadex G-50 fine column (Pharmacia, 1 × 200 cm) run in chromatography solvent [0.1 *M* ammonium acetate in 20% (v/v) ethanol] were pooled, lyophilized to dryness, and subjected to treatment with HNO_2 to degrade selectively N-sulfated GAG, such as heparin and heparan sulfates (Cifonelli and King, 1972, 1973). Samples were solubilized in 2 ml of glass-distilled water, mixed with 1 ml of 1 *N* HCl and 1 ml of freshly prepared 20% *n*-butyl nitrite (v/v) (Kodak, Rochester, NY) in absolute ethanol, and incubated in an open vessel for 2 hours at room temperature with gentle shaking. One sample of heparin (Wilson, Chicago, IL) (2 mg) underwent identical treatment with HNO_2 and served as the positive control. A second sample of heparin (2 mg) was mock-treated by substituting absolute ethanol for the *n*-butyl nitrite, and served as the negative control. Degradation of the heparin controls was assayed by cetylpyridinium chloride precipitation (Antonopoulos *et al.*, 1961) or by chromatography on Sephadex G-50. The mock-treated heparin produced a very turbid precipitate upon addition of 1% (w/v) cetylpyridinium chloride and was excluded from G-50. The HNO_2-treated heparin showed no turbidity, and was eluted from G-50 in the included fractions. This method selectively and completely degrades heparin and heparan sulfates after only one treatment (Conrad and Hart, 1975). The HNO_2 reaction was stopped by neutralizing the reaction with 1 ml of 1 *N* NaOH, and the solutions were lyophilized. The lyophilized, HNO_2-treated material then was dissolved in chromatography solvent and rechromatographed over Sephadex G-50.

4. Keratan Sulfate–β-Endogalactosidase Treatment to Remove Keratan Sulfate

The peak fractions remaining excluded from Sephadex G-50 after treatment with pronase, HNO_2, and chondroitinase ABC were pooled, lyophilized, redissolved in 3 ml of glass-distilled water, and split in half into an enzyme-treated and a mock-treated sample. Each half was transferred to a 1.5-ml plastic disposable conical centrifuge tube and dried at room temperature using a stream of air. Each sample was then redissolved in 0.10 ml of 0.05 *M* Tris-HCl, pH 7.2. Keratan sulfate–β-endogalactosidase (Nakazawa *et al.*, 1975; Nakazawa and Suzuki, 1975) (0.01 ml of 150 units/ml) in saturated ammonium sulfate was added. Controls received boiled enzyme or buffer alone. The samples were incubated 48 hours at 37°C, and fresh enzyme was added after 24 hours. The reaction was stopped by placing the sample in a boiling-water bath for 5 minutes. After cooling to room temperature, 1 ml of chromatography solvent was added. Degradation was assayed by chromatography on Sephadex G-50.

III. Demonstration that the PG Core Protein Is Identical to the Given Unmodified Protein

A. Biochemical Determination

1. Purification of the Core Protein

The ability of a monoclonal antibody (mAb) reactive with a given protein to immunoprecipitate a PG molecule suggests, but does not prove, that the core protein of the PG and the original protein are identical. It is possible, for example, that the PG core protein and the original protein are different but share an epitope recognized by the mAb; that is, the proteins are immunologically identical, but biochemically different. It is therefore necessary to demonstrate rigorously the biochemical identity of the core protein and the given protein.

If an antibody is available that reacts with both a protein and the putative PG form of that protein, the most straightforward method for proving identity is to purify the core protein of the PG and compare it to the unmodified protein. If such an antibody is available, then the antibody can be utilized to purify both moieties that have been differentially radiolabeled, and the proteins can be compared by double-label tryptic peptide mapping (see Section III,A,3). In our original study, it was not clear that the anti-invariant chain antibody was reacting directly with the PG or indirectly immunoprecipitating the PG because the latter was associated with class II molecules, and the class II molecules were in turn associated with invariant chain. Because a similar situation may arise in other systems, the strategy we utilized in preparing the PG core protein without the use of a specific anti-core protein antibody is given.

This strategy involved several purification steps (Fig. 7). In brief, cells

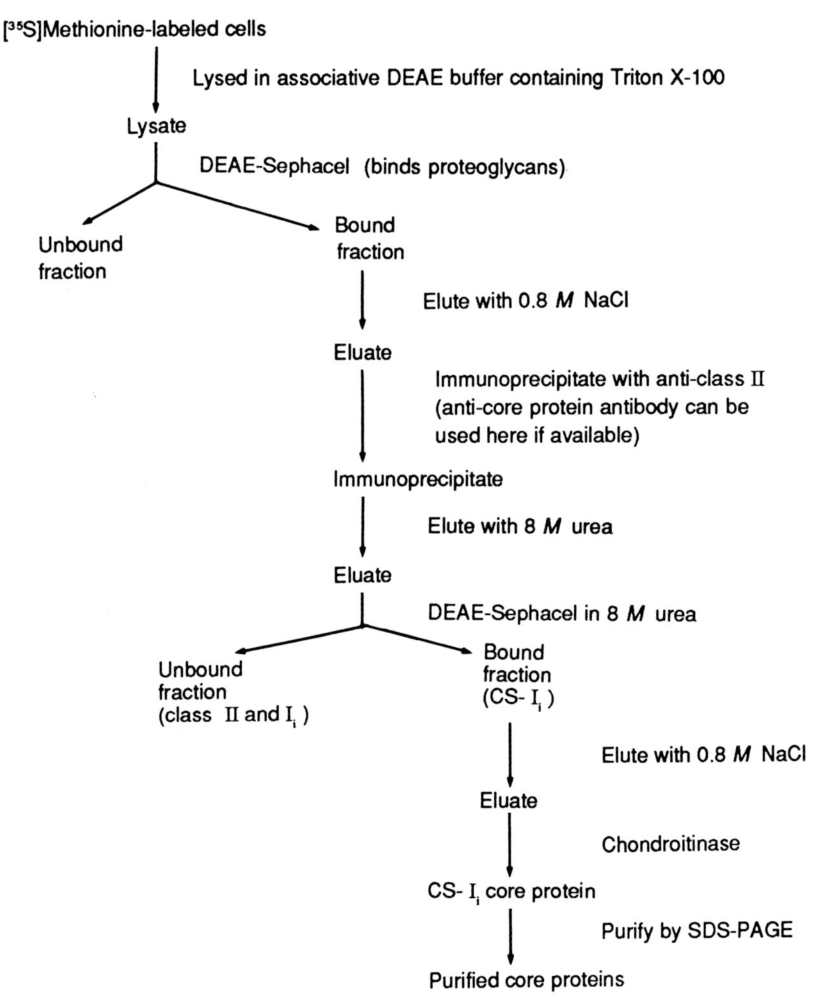

FIG. 7. Strategy for the characterization of the core protein.

were labeled with [^{35}S]methionine and solubilized in associative DEAE starting buffer (0.05 *M* sodium acetate, 0.15 *M* NaCl, 0.5% Triton X-100, pH 6.0) containing protease inhibitors. Associative conditions were used to maintain the class II-PG complex. The cell lysates were ultracentrifuged at 100,000 *g* for 60 minutes, and applied in batch under associative conditions to DEAE–Sephacel at a ratio of 0.1 ml of packed DEAE–Sephacel per 10^8 cell equivalents. Unbound material was collected, and the DEAE–Sephacel was washed with associative DEAE starting buffer. The bound material was eluted with DEAE elution buffer (0.8 *M* NaCl, 0.05 *M* sodium acetate, 0.002% Triton X-100, pH 6.0). After elution, the eluate was adjusted to 0.2 *M* NaCl (pH 7.4) and precleared using a rabbit anti-human immunoglobulin antibody (Accurate Chemical, Westbury, NY). The complex containing the class II molecules and the PG of interest was isolated from other PG in the eluate by specific immunoprecipitation with an anti-class II mAb, and the antibody–antigen complexes were pelleted using protein A–Sepharose.

Purification of the PG was accomplished by eluting the immunoprecipitates using dissociative DEAE starting buffer (see Section II,C,2). The eluates were applied batchwise to fresh DEAE–Sephacel in dissociative buffer to preclude the reassociation of the PG with the class II molecules. Unbound material was collected, and the DEAE–Sephacel was washed twice with dissociative DEAE starting buffer. The material that was bound to the DEAE–Sephacel was eluted with DEAE elution buffer.

The DEAE elution buffer containing the doubly purified PG was diluted with three volumes of chondroitinase buffer pH 6.0 and divided into two aliquots. One aliquot was treated with chondroitinase AC (1.5 U/ml) to yield to core protein of the PG, and the other aliquot was mock-digested. The samples were incubated for 2 hours at 37°C. The core protein or mock-digested PG was precipitated by the addition of trichloroacetic acid (TCA) to a final concentration of 20% (w/v). The core protein was purified by SDS–PAGE and visualized by autoradiography. Chondroitinase digestion of the PG resulted in a major band that corresponded to a core protein of ~33 kDa. This size was similar to that of the MHC class II α chain (33–34 kDa) and that of invariant chain (31 kDa).

2. V8 Protease Digestion

These size similarities suggested the possibility that either the α chain or the invariant chain was the core protein of this PG. To test this hypothesis, samples containing α, β, and invariant chains (unfractionated immunoprecipitates) and samples containing the purified core protein were electrophoresed separately using the modified Laemmli–Maizel discontinuous SDS–10% PAGE system in 5 × 120-mm cylindrical gels (Laemmli, 1970; Maizel, 1971). The gels were then embedded into the top of the second-dimension slab gel, using 1.0% low melting point agarose containing 75 μg/ml V8 protease. A modified phosphate–urea gel system (Pugsley and Schnaitman, 1979) was used for the second-dimension gels.

The separation gel contained 0.035 *M* Na_2HPO_4, 0.152 *M* NaH_2PO_4, pH 7.1, 6 *M* urea, 15% acrylamide–bisacrylamide (37.5 : 1), 0.1% SDS, 0.02% ammonium persulfate, and 0.05% *N,N,N′,N′*-tetramethylethylenediamene (TEMED). The stacking gel contained 0.035 *M* Na_2HPO_4, 0.152 *M* NaH_2PO_4, 4.0% acrylamide–bisacrylamide (37.5 : 1), 0.1% SDS, 0.02% ammonium persulfate, and 0.18% TEMED. The upper-electrode buffer (−) was 0.1 *M* cacodylic acid, 0.1% SDS, pH 6.0. The lower-electrode buffer (+) was 0.035 *M* Na_2HPO_4, 0.0152 *M* NaH_2PO_4, pH 7.1. Samples were electrophoresed using constant current, at 30 mA per slab gel through the stacking gel at 25°C, and at 80 mA per slab gel through the separating gel at 4°C. The position of the fragments was again visualized by autoradiography.

The results of this analysis indicated that the V8-generated proteolytic fragments of the core protein and the invariant chain were highly similar, and raised the possibility that the invariant chain was the core protein of the PG.

3. Tryptic Digestion

To determine if the generated core protein was indeed the invariant chain, we then compared the core protein and the invariant chain by double-label tryptic peptide analysis. The [^{35}S]methionine-labeled core protein was generated as described before. A second batch of cells was labeled with [^{3}H]methionine. The [^{3}H]methionine-labeled cells were washed and solubilized in NP-40 cell lysis buffer, and the cell lysates were ultracentrifuged. The lysate was precleared with rabbit antihuman immunoglobulin and rabbit antimouse immunoglobulin (RαMIg, Accurate Chemical), reacted with a mAb specific for the human invariant chain, and the antigen–antibody complexes were bound to protein A–Sepharose. The antigen–antibody complexes were eluted and analyzed by SDS–PAGE. The bands containing the [^{35}S]methionine-labeled core protein and the [^{3}H]methionine-labeled invariant chain were localized by autoradiography. The pieces of gel containing the bands were excised, reswollen in 30% methanol–10% acetic acid, washed in distilled water, and then placed in 0.5 ml of 0.1 *M* NH_4HCO_3 (pH 8.2), containing 0.1 ml of TPCK–trypsin (200 μg/ml; Worthington Biochemical Corp., Freehold, NJ) in 0.001 *N* HCl. The mixture was incubated for 30 minutes at 37°C, at which time 10 μl of 0.4 *M* NH_4HCO_3 (pH 8.2) and an additional 0.1 ml of TPCK–trypsin was added, and the incubation was continued overnight. The reaction was stopped by the addition of 0.1 ml of 0.8 *N* acetic acid. The gel pieces were removed by centrifugation, and the eluates were chromatographed on a C-18 reverse-phase column (Vydac, Hesperia, CA) using a gradient of 0–50% acetonitrile (J. T. Baker Chemical Co., Phillipsburg, NJ) in trifluoroacetic acid (Pierce Chemical Co., Rockford, IL) and a high-performance liquid chromatography (HPLC) system (Waters Associates, Milford, MA). The elution profiles for these proteins showed some quantitative differences, but virtually all of the [^{35}S]methionine-labeled and [^{3}H]methionine-labeled peptides coeluted, indicating the nearly complete structural identity of the core protein and the invariant chain (Fig. 8).

B. Immunochemical Determination

To determine whether these proteins were antigenically similar, reprecipitation of the PG and its core protein was performed. The mock-digested and chondroitinase-treated samples containing the intact PG and its core protein, respectively, were each pretreated by immunoprecipita-

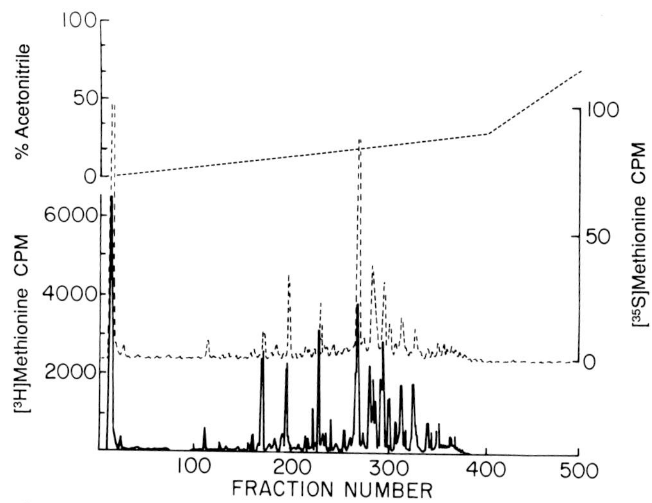

FIG. 8. Comparative tryptic peptide maps of the CS-I_i core protein and I_i. SDS–PAGE gel pieces containing the [^{35}S]methionine-labeled CS-I_i core protein (----) and the corresponding [^{3}H]methionine-labeled conventional invariant-chain species (—) were excised from a slab gel and combined, and the proteins digested with trypsin. The resulting tryptic peptides were chromatographed on a C-18 reverse-phase column using a gradient of 0–50% acetonitrile and 0.1% trifluoroacetic acid (· · ·).

tion with RαMIg and protein A–Sepharose to remove any radiolabeled material that potentially could bind nonspecifically to the immunoprecipitates. The samples were then reacted with anti-invariant antibodies as well as irrelevant antibodies. Results (Fig. 9) indicated that the intact PG and its core protein bear epitopes that are recognized by anti-invariant chain antibodies, and further confirmed our finding that the human invariant chain is the core protein of the class II-associated CS PG.

IV. Determination of a Membrane Location of the PG

The demonstration that a particular membrane protein exists in a PG form does not necessarily indicate that the PG is expressed on the cell surface. If the PG form represents a small fraction of all forms of this protein, a membrane localization may indeed be difficult to demonstrate. Three methodologies are offered to prove the membrane location of a PG.

A. Extrinsic Radiolabeling

If a PG is present in sufficient amount on the cell surface, it may be possible to radiolabel it extrinsically using a method such as lactoperoxidase-catalyzed radioiodination or chloramine-T radioiodination. After radiolabeling, the cells are simply solubilized, and the cell lysate, precleared as described earlier, is reacted with specific antibody. Provided that the antibody has been demonstrated to react directly with the PG core protein, visualization of a radiolabeled band with the mobility of the PG on SDS–PAGE analysis indicated that the PG is on the cell surface.

B. Reaction of Antibody with Biosynthetically Radiolabeled Intact Cells

In our case, neither invariant chain nor its PG form could be extrinsically radiolabeled, so we turned to alternative methodologies. Cells were biosynthetically radiolabeled with $^{35}SO_4$ or [^{35}S]methionine, and the *intact* cells were reacted with antibody for 30 minutes at 4°C. In our system, it was already known that anti-invariant chain antibody did not bind well to intact cells, and so an anti-class II antibody was used. After washing, the cells were lysed in NP-40 cell lysis buffer containing a 5- to 10-fold excess of unlabeled cells to bind to any unoccupied antibody-binding sites, and thus prevent the reaction of antibody with intracellular radiolabeled molecules. The antigen–antibody complexes were pelleted with protein

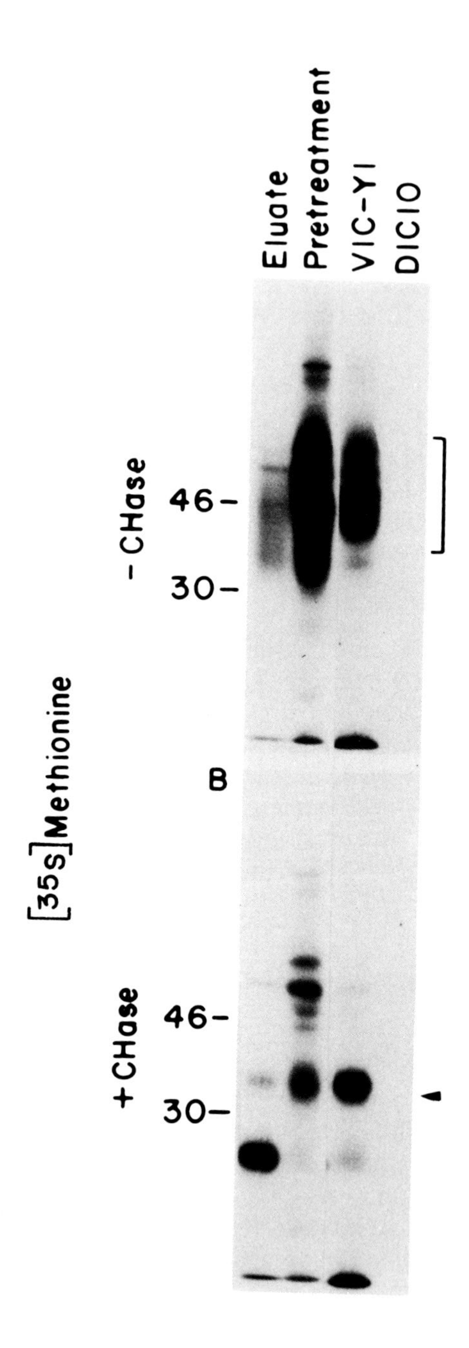
Eluate
Pretreatment
VIC-Y1
D1C10
- CHase
46-
30-
[35S]Methionine
B
+ CHase
46-
30-

A–Sepharose. Analysis by SDS–PAGE revealed that $^{35}SO_4$-labeled PG molecules were indeed associated with cell surface class II molecules. Because we had used an anti-class II antibody in this system, we could not directly conclude that the PG itself was expressed on the cell surface. However, the visualization of an $^{35}SO_4$-labeled PG component on SDS–PAGE after reaction of the intact cells with a PG core protein-specific antibody would allow the conclusion that the PG was cell surface-expressed.

C. Enzymatic Treatment of Intact Cells

Our failure to demonstrate a cell surface location for our PG by the means just described prompted us to develop additional methods. Cells were biosynthetically radiolabeled with $^{35}SO_4$ or [^{3}H]leucine, and resuspended in 0.15 *M* NaCl, 0.005 *M* Tris (pH 7.4), containing 0.01% BSA, 10 m*M* sodium azide, 10 m*M* sodium fluoride, and 1 m*M* sodium cyanide. The latter three substances were added to prevent turnover and endocytosis of the PG (Steinman *et al.*, 1974). Chondroitinase AC (1.5 U/ml) was added to half of each preparation, and the samples were incubated at 37°C for 30 minutes. The cells were washed three times. In this type of analysis, it is critical to demonstrate that the cells remain viable throughout the treatment. Our cells showed >98% viability as judged by trypan blue exclusion. The cells were solubilized using NP-40 cell lysis buffer, and the lysates were precleared as described before. Anti-invariant chain antibody was added, and the antigen–antibody complexes were pelleted with protein A–Sepharose, and analyzed by SDS–PAGE on 11% slab gels. Chondroitinase treatment of the intact cells resulted in the total absence of the PG, which was readily apparent in the mock-digested cell sample (Fig. 10). The results indicated that virtually all our PG was accessible to chondroitinase, and therefore must be expressed on the cell surface.

FIG. 9. [^{35}S]Methionine-labeled CS-I_i and its core proteins can be reprecipitated with anti-invariant chain antibody. Aliquots of a highly purified [^{35}S]methionine-labeled Cs-I_i (A) and its core proteins (B) were used to test reprecipitation with anti-invariant chain antibodies. The samples were precleared and reacted with a murine antihuman invariant chain mAb (VIC-Y1), and an irrelevant control mAb (D1C10). Immunoprecipitates were made with protein A–Sepharose alone or with protein A–Sepharose armed with rabbit antimouse immunoglobulin. (A) The bracket shows the area of the gel in which [^{35}S]methionine-labeled CS-I_i is found. (B) The arrowhead indicates the 38,000 MW core protein. The gel patterns of the elute and pretreatment immunoprecipitate are shown for comparison. Chase, Chondroitinase.

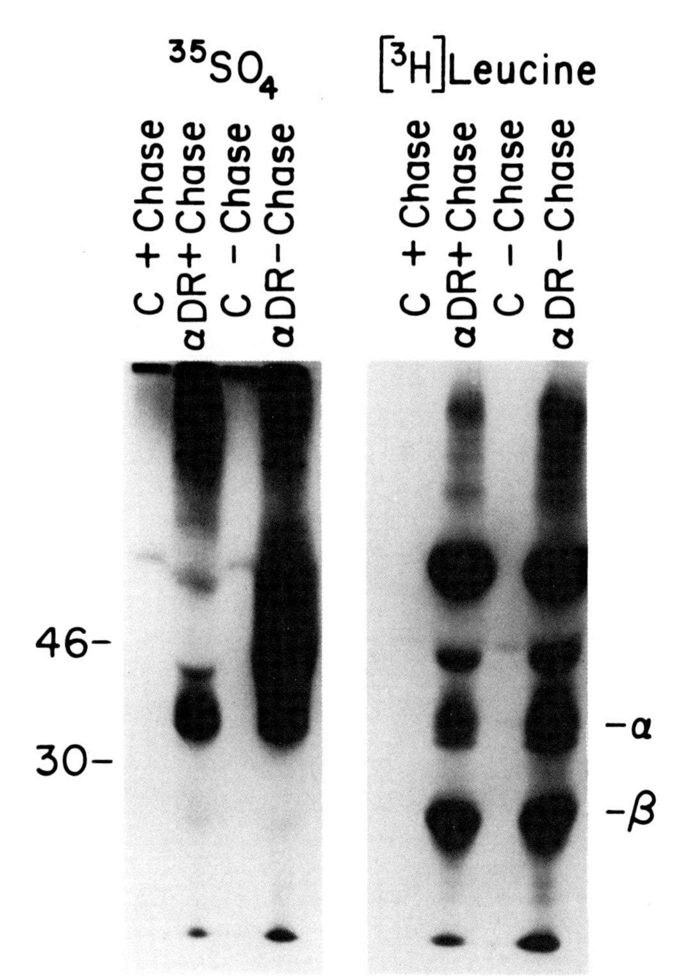

FIG. 10. The CS-I_i is on the cell surface. Swei cells were labeled with $^{35}SO_4$ (left lanes) or [^{3}H]leucine (right lanes) and washed extensively. Samples of labeled intact cells were divided in half and either digested with chondroitinase (+ Chase) or mock-digested (− Chase). The cells were again washed extensively and solubilized with NP-40 and the lysates reacted with a control or an anti-HLA-DR class II mAb. The [^{3}H]leucine-labeled α and β chains of the class II molecules are seen in both the chondroitinase-treated and mock-digested samples. In contrast, the $^{35}SO_4$-labeled CS-I_i is seen only in the mock-digested samples. The residual $^{35}SO_4$ band seen after chondroitinase treatment is the sulfated α chain.

V. Conclusion

In this chapter we have described the approach used to characterize the sulfated molecule found associated with the class II molecules of the MHC. This characterization identified this M_r 40,000–70,000 heterogeneous molecule as chondroitin sulfate PG, with the human invariant chain as its core protein. This approach should be useful in identifying PG forms of other proteins and glycoproteins, and characterizing these PG with respect to both the GAG and core proteins.

References

Antonopoulos, C. A., Borelius, E., Gardell, S., Hamnstrom, B., and Scott, J. E. (1961). *Biochim. Biophys. Acta* **54,** 213–226.

Baenziger, J., and Kornfeld, S. (1974). *J. Biol. Chem.* **249,** 7270–7281.

Barnstable, L. J., Jones, E. A., and Crumpton, M. J. (1978). *Br. Med. Bull.* **32,** 241–246.

Bhavanandan, V., Umemot, J., Banks, J. R., and Davidson, E. A. (1977). *Biochemistry* **16,** 4426–4437.

Chang, Y., Yanagishita, M., Hascall, V. C., and Wight, T. (1983). *J. Biol. Chem.* **258,** 5679–5688.

Cifonelli, J. A., and King, J. (1972). *Carbohydr. Res.* **21,** 173–186.

Cifonelli, J. A., and King, J. (1973). *Biochim. Biophys. Acta* **320,**

Conrad, G. W., and Hart, G. W. (1975). *Dev. Biol.* **44,** 253–269.

Cresswell, P. (1985). *Proc. Natl. Acad. Sci. U.S.A.* **82,** 8188–8192.

Drapeau, G. R., Boily, Y., and Houmard, J. (1972). *J. Biol. Chem.* **247,** 6720–6726.

Fransson, L.-A., Carlstedt, I., Coster, L., and Malmstrom, A. (1984). *Proc. Natl. Acad. Sci. U.S.A.* **81,** 5657–5661.

Giacoletto, K. S., Sant, A. J., Bono, C., Gorka, J., O'Sullivan, D. M., Quaranta, V., and Schwartz, B. D. (1986). *J. Exp. Med.* **164,** 1422–1439.

Hart, G. W. (1976). *J. Biol. Chem.* **251,** 6513–6521.

Jones, P. P., Murphy, D. B., Hewgill, D., and McDevitt, H. O. (1979). *Immunochemistry* **16,** 51–60.

Karr, R. W., Kannapell, C. C., Stein, J. A., Gebel, H. M., Mann, D. L., Duquesnoy, R. J., Fuller, T. C., Rodey, G. E., and Schwartz, B. D. (1982). *J. Immunol.* **128,** 1809–1818.

Kimata, K., Okayama, M., Oohira, A., and Suzuki, S. (1974). *J. Biol. Chem.* **249,** 1646–1653.

Laemmli, U. K. (1970). *Nature (London)* **227,** 680–685.

Maizel, J. V., Jr. (1971). *Methods Virol.* **5,** 179–246.

Nakazawa, K., and Suzuki, S. (1975). *J. Biol. Chem.* **250,** 912–917.

Nakazawa, K., Suzuki, N., and Suzuki, S. (1975). *J. Biol. Chem.* **250,** 905–911.

O'Farrell, P. M. (1975). *J. Biol. Chem.* **250,** 4007–4021.

Pugsley, A. T., and Schnaitman, C. A. (1979). *Biochim. Biophys. Acta* **581,** 163–178.

Rodén, L. (1980). *In* "The Biochemistry of Glycoproteins and Proteoglycans" (W. J. Lennarz, ed.), pp. 267–371. Plenum, New York.

Rosamond, S., Brown, S., Gomez, C., Braciale, T. J., and Schwartz, B. D. (1987). *J. Immunol.* **139,** 1946–1951.

Rosenthal, A. S., and Shevach, E. M. (1973). *J. Exp. Med.* **138,** 1194–1212.

Sant, A. J., Schwartz, B. D., and Cullen, S. E. (1983). *J. Exp. Med.* **158,** 1979–1992.

Sant, A. J., Cullen, S. E., and Schwartz, B. D. (1984). *Proc. Natl. Acad. Sci. U.S.A.* **81,** 1534–1538.

Sant, A. J., Cullen, S. E., Giacoletto, K. S., and Schwartz, B. D. (1985). *J. Exp. Med.* **162,** 1916–1934.

Steinman, R. M., Silvers, J. M., and Cohn, Z. A. (1974). *J. Cell Biol.* **63,** 949–969.
Streicher, H. Z., Berkower, I. J., Busch, M., Gurd, F. R. N., and Berzofsky, J. A. (1984). *Proc. Natl. Acad. Sci. U.S.A.* **81,** 6831–6835.
Unanue, E. R. (1984). *Annu. Rev. Immunol.* **2,** 395—428.
Yamagata, T., Saito, H., Habuchi, O., and Suzuki, S. (1968). *J. Biol. Chem.* **243,** 1523–1535.

Chapter 9

Identification and Analysis of Glycoinositol Phospholipid Anchors in Membrane Proteins

TERRONE L. ROSENBERRY, JEAN-PIERRE TOUTANT,[1] ROBERT HAAS, AND WILLIAM L. ROBERTS

Department of Pharmacology
Case Western Reserve University School of Medicine
Cleveland, Ohio 44106

---

[1]Permanent address: Laboratoire de Physiologie Animale, INRA, Place Viala, 34060 Montpellier Cedex, France.

Copyright © 1989 by Academic Press, Inc.
All rights of reproduction in any form reserved.

I. Introduction

Integral membrane proteins maintain intimate contact with the hydrophobic phase of phospholipid bilayer membranes, require detergents for solubilization, and bind detergents in extracts. Such proteins generally interact with the bilayer through hydrophobic amino acid sequences in one or more transmembrane segments (Sabatini *et al.*, 1982), although specific instances of fatty acid acylation of residues in these segments also are known (Schmidt, 1983). Recently a new class of integral membrane proteins has emerged that are anchored, not by a transmembrane peptide segment, but exclusively by a glycoinositol phospholipid linked cova-

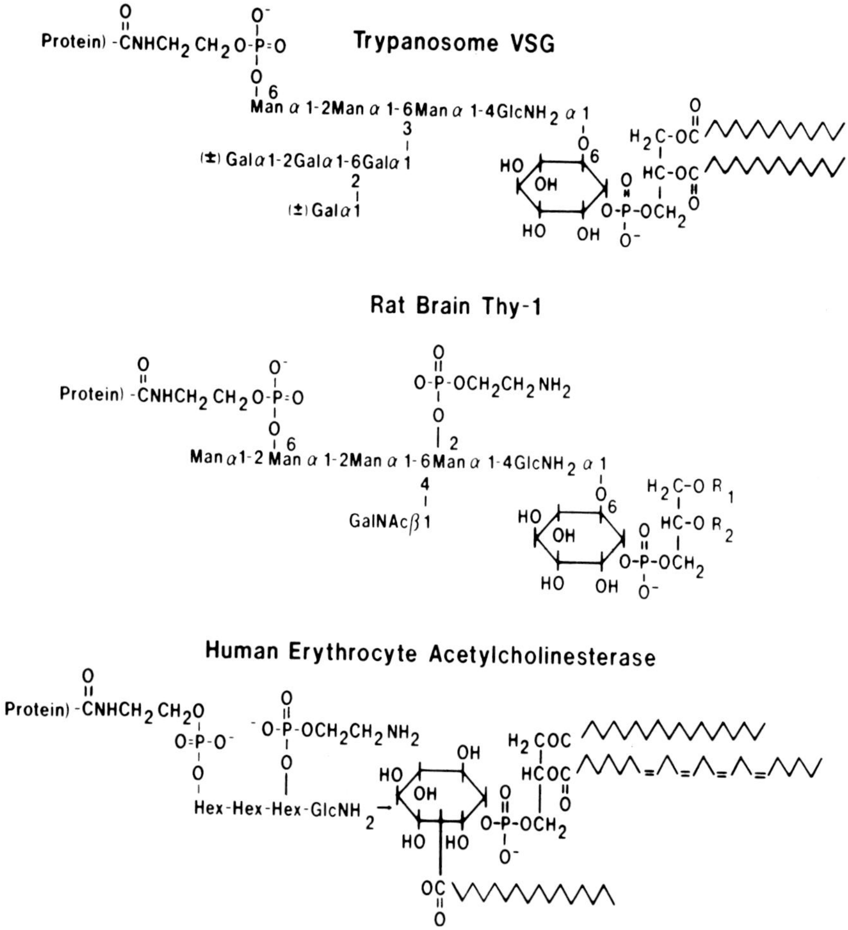

Trypanosome VSG

Protein)-C(O)NHCH$_2$CH$_2$O-P(O$^-$)=O-O-6Man α 1-2Man α 1-6Man α 1-4GlcNH$_2$ α 1-6 (inositol)

(±) Galα1-2Galα1-6Galα1 (→3 Man); (±) Galα1 (→2 Gal)

HO, OH, HO, OH; O-P(O$^-$)(=O)-OCH$_2$; HC-OC(=O); H$_2$C-OC(=O)

Rat Brain Thy-1

Protein)-C(O)NHCH$_2$CH$_2$O-P(O$^-$)=O-O-6

Man α1-2 Man α 1-2Man α 1-6Man α 1-4GlcNH$_2$ α 1-6 (inositol)

O-P(=O)-OCH$_2$CH$_2$NH$_2$ (→2); GalNAcβ1 (→4)

HO, OH, HO, OH; O-P(=O)(O$^-$)-OCH$_2$; HC-O R$_2$; H$_2$C-O R$_1$

Human Erythrocyte Acetylcholinesterase

Protein)-C(O)NHCH$_2$CH$_2$O-P(=O)(O$^-$)-O-Hex-Hex-Hex-GlcNH$_2$ → (inositol)

$^-$O-P(=O)-OCH$_2$CH$_2$NH$_2$

OH, HO, OH, HO, HO; O-P(=O)(O$^-$)-OCH$_2$; HCOC(=O); H$_2$COC; OC(=O)

lently to the protein C-terminus. In this chapter the term "anchor" will refer exclusively to such glycoinositol phospholipids. Detailed features of the three anchor structures that have been determined are outlined in Fig. 1.

Although direct documentation of glycoinositol phospholipid anchors has been obtained only in the last 4 years (Ferguson *et al.*, 1985b; see Low, 1987), some 30–40 proteins with such anchors have now been tentatively identified (Ferguson and Williams, 1988). Those cited in this paper include trypanosome variant surface glycoproteins (VSG), Thy-1, G_2 acetylcholinesterase (AChE), alkaline phosphatase, 5′-nucleotidase, decay-accelerating factor (DAF), a 120-kDa neural cell adhesion molecule (N-CAM 120), *Leishmania* p63 protease, *Dictyostelium discoideum* antigen 117, mammalian antigens Ly-6 and RT-6.2, and scrapie prion protein PrP27–30. All appear to reside on the extracellular face of the cell plasma membrane. In most cases the identification has been based on their susceptibility to release from the cell surface by purified bacterial phosphatidylinositol-specific phospholipase C (PIPLC). Biological functions that necessitate this anchoring mechanism have been the subject of considerable speculation, but few conclusions can yet be drawn. The discoveries of endogenous phospholipases C in trypanosomes (Cardoso de Almeida and Turner, 1983; Ferguson *et al.*, 1985a) and liver (Fox *et al.*, 1987) and phospholipase D in mammalian serum (Davitz *et al.*, 1987; Low and Prasad, 1988) that are specific for anchor inositol phospholipids have sug-

---

FIG. 1. Structures of glycoinositol phospholipid anchors. Anchors were released from the purified proteins by digestion with proteases and isolated by high-pressure liquid chromatography (HPLC). The structures of the anchors of variant surface glycoproteins (VSG) from *Trypanosoma brucei* and rat brain Thy-1 were obtained by a combination of nuclear magnetic resonance (NMR) spectroscopy, chemical modification, gas chromatography–mass spectrometry (GC–MS), and exoglycosidase digestions (Ferguson *et al.*, 1988; Homans *et al.*, 1988). The structure of the human erythrocyte (E^{hu}) acetylcholinesterase (AChE) anchor was deduced by lipid composition analyses and fast atom bombardment–mass spectrometry (FAB–MS) (Roberts *et al.*, 1988a, b). All three anchors contain a similar backbone of three linear hexose residues with a phosphodiester linkage at their nonreducing terminus to an ethanolamine in amide linkage to the C-terminal amino acid and a glucosamine with a free amino group at their reducing terminus linked to an inositol phospholipid. This phospholipid in VSG is simply dimyristoylphosphatidylinositol. The lipid groups R_1 and R_2 in Thy-1 are incompletely determined but include stearic acid. In E^{hu} AChE the phospholipid includes an alkylacylglycerol, 83% of which contains an 18 : 0 or 18 : 1 1-alkyl group and a 22 : 4, 22 : 5, or 22 : 6 2-acyl group, and an unusual direct palmitoylation of an inositol hydroxyl group. An additional phosphorylethanolamine is attached to the hexose adjacent to the glucosamine in the Thy-1 and the E^{hu} AChE anchor, and the Thy-1 anchor also contains *N*-acetylgalactosamine linked to this hexose. In contrast, in the VSG anchor a branching galactose side chain is found at this location. An additional mannose branching from the hexose closest to the C-terminal amino acid is found in 71% of the Thy-1 anchors.

gested a functional role for endogenous phospholipase cleavage. This suggestion is supported (1) by one example of an anchored latent protease in malarial parasites that is activated by phospholipase cleavage (Braun-Breton *et al.*, 1988), and (2) by evidence that insulin can activate endogenous phospholipases that cleave free glycoinositol phospholipids with anchorlike structures to generate insulin second messengers (see Low and Saltiel, 1988). In some cells, such as human erythrocytes, anchored proteins are at least partially resistant to phospholipase C cleavage. Resistance of E^{hu} AChE was shown to arise specifically from the palmitoylation of the inositol phospholipid noted in Fig. 1, because treatment with base removed this palmitoyl group and permitted release of alkyl- and alkylacylglycerol species by PIPLC with concomitant formation of inositol 1-phosphate (Roberts *et al.*, 1988a). The possibility that inositol palmitoylation may be a general mechanism for regulating the susceptibility of anchored proteins to endogenous phospholipase C is currently under investigation in our laboratory.

Additional information on proteins with glycoinositol phospholipid anchors may be found in excellent review articles by Low (1987), Cross (1987), Low and Saltiel (1988), and Ferguson and Williams (1988). This article will focus on criteria to demonstrate the presence of a glycoinositol phospholipid in a protein of interest.

II. Cleavage of Anchored Proteins by Exogenous Phospholipase C

Given the availability of an appropriate phospholipase, this criterion is one of the easiest to test. Slein and Logan (1963) first suggested that a bacterial phospholipase C could solubilize alkaline phosphatase from tissue slices, and later workers demonstrated that highly purified PIPLC from *Bacillus cereus* (Ikezawa *et al.*, 1976), *Staphylococcus aureus* (Low and Finean, 1977a), or *Bacillus thuringiensis* (Taguchi *et al.*, 1980) released alkaline phosphatase and other enzymes from cell surfaces. These bacterial PIPLC preparations have excellent phospholipase C specificity, are free of protease contamination, and have yet to cleave a protein that has failed to contain a glycoinositol phospholipid anchor on further analysis. However, it should be noted that some anchored proteins like E^{hu} AChE (Figs. 1 and 2) are resistant to PIPLC cleavage. Although many recent studies have been conducted with the *S. aureus* PIPLC provided by Dr. Martin Low, the *B. thuringiensis* enzyme is active over a wider range of conditions (Low *et al.*, 1988) and soon will be commercially available (American Radiolabeled Chemicals Inc., St. Louis, MO; Funakoshi Fine Chemicals, Tokyo; Calbiochem). The following sections outline procedures for demonstrating phospholipase cleavage of anchored proteins.

A. Release of Anchored Proteins from Intact Cells

Cleavage of anchored proteins with exogenous PIPLC has been achieved in a variety of culture media or isotonic buffers. In a typical experiment, cultured EL-4 murine thymoma cells (10^6 cells/ml) were incubated in culture medium (RPMI 1640–HEPES) with 20 μg/ml *S. aureus* PIPLC at 37°C for 1 hour (Low and Kincade, 1985; Low *et al.*, 1988). The loss of anchored Thy-1 antigen from the cells was monitored by flow cytometry. Other techniques to follow the release of anchored proteins include two-site immunoradiometric assay, applied to measure supernatant DAF (Davitz *et al.*, 1986; Medof *et al.*, 1986), immunoprecipitation following surface radioiodination, demonstrated with supernatant RT-6.2 antigen (Koch *et al.*, 1986), and appearance of enzyme activity in the incubation supernatant, useful with anchored proteins like alkaline phosphatase, AChE, or 5′-nucleotidase, which have intrinsic enzyme activity (Low and Finean, 1977b, 1978; Jemmerson and Low, 1987). Anchored proteins sufficiently abundant to be detected by biosynthetic labeling with [^{3}H]ethanolamine (Section IV,A) can be surveyed by polyacrylamide gel electrophoresis (PAGE) in sodium dodecyl sulfate (SDS) following release from the cell surface by PIPLC (C. Braun-Breton and T. L. Rosenberry, unpublished observations). Specific proteins radiolabeled by this procedure can be identified by immunoprecipitation with monospecific antisera prior to SDS–PAGE.

Anchored proteins also have been released selectively from membrane fractions by treatment with exogenous PIPLC. *S. aureus* PIPLC quantitatively solubilized dimeric G_2 AChE from torpedo electric organ synaptic plasma membranes and intact synaptosomes (Futerman *et al.*, 1985b). This PIPLC partially solubilized a 120-kDa form of neural cell adhesion molecule (N-CAM 120) from crude murine brain microsomes (He *et al.*, 1986) and alkaline phosphatase from a crude particulate fraction of human placenta (Malik and Low, 1986).

B. Loss of Anchored Protein Capacity for Nonionic Detergent Interaction

In all anchored proteins that have been studied, the inositol phospholipid is the only nonionic detergent-binding domain. Cleavage of the anchor by PIPLC removes the hydrophobic diacylglycerol or alkylacylglycerol and generates a hydrophilic protein that retains the glycan portion of the anchor. Techniques used to demonstrate this loss of detergent-binding capacity following incubation with PIPLC include charge-shift agarose gel electrophoresis (Cardoso de Almeida and Turner, 1983), sucrose gradient sedimentation (Futerman *et al.*, 1985a) and phenyl–Sepharose chromatography (Fatemi *et al.*, 1987; Davitz *et al.*, 1987), but the two procedures most widely employed have been phase partitioning with Triton X-114 and nondenaturing PAGE.

1. Phase Partitioning with Triton X-114

This technique is based on the observation that micelles of Triton X-114 condense to form a separate detergent phase above 20°C. Detergent-binding proteins are amphiphilic and partition preferentially into the detergent-condensed phase, while hydrophilic proteins associate with the detergent-depleted phase (Bordier, 1981). Triton X-114 (Serva) was precondensed to remove impurities by warming a 3% solution in 10 m*M* Tris-HCl, 150 m*M* NaCl (pH 7.5) to 37°C, centrifuging in a tabletop centrifuge, discarding the upper detergent-depleted phase, and diluting the 12% Triton X-114 lower phase to 3% with the same buffer (Bordier, 1988). The cycle was repeated three times. *Leishmania major* promastigotes were treated with $Na^{125}I$ and Iodogen, and the labeled anchored protease p63 was purified (Etges *et al.*, 1986). Purified ^{125}I-labeled p63 (1 μg, 600 cpm) was incubated with crude *B. cereus* PIPLC in 20 μl of the same buffer plus 0.1% Triton X-100 for 1 hour at 30°C, after which 300 μl of 2% Triton X-114 was added on ice. Phase separation was promoted at 37°C for 1 minute, followed by centrifugation at 10,000 *g* for 30 seconds at 25°C. The two phases were transferred to separate tubes, and the extent of PIPLC cleavage was determined by the proportion of radiolabel in the detergent-depleted phase (Etges *et al.*, 1986). This phase-partitioning method also has been used to demonstrate the PIPLC susceptibility of anchored proteins in crude cell extracts. A general survey of anchored proteins in yeast cells involved biosynthetic ^{35}S labeling of protein, extraction of labeled membrane proteins into the Triton X-114 detergent phase, transfer of labeled anchored proteins into the detergent-depleted phase following PIPLC treatment, and detection of the anchored proteins by trichloroacetic acid (TCA) precipitation and SDS–PAGE analysis (Conzelmann *et al.*, 1988). In studies of identified proteins, immunoprecipitation of surface-radioiodinated RT-6 (Koch *et al.*, 1986) or Thy-1 biosynthetically labeled with [^{35}S]methionine (Conzelmann *et al.*, 1987) and enzyme assay of alkaline phosphatase (Malik and Low, 1986) permitted direct assay of the distribution of the untreated and PIPLC-cleaved proteins in the two phases. In one case, that of trypanosome VSG, PIPLC cleavage of the

[^{3}H]myristate-labeled anchored protein was readily shown by phase partitioning but could not be demonstrated by direct release from intact trypanosomes, presumably because packing of the VSG is too tight to permit access to the anchor phospholipid (Low *et al.*, 1986).

2. Nondenaturing Polyacrylamide Gel Electrophoresis in Detergent

Cleavage of anchored proteins with PIPLC results in at most small shifts in mobility on SDS–PAGE. In contrast, loss of detergent binding on PIPLC cleavage can generate a several-fold change in the electrophoretic migration of these proteins in nondenaturing polyacrylamide gels that contain detergent. Electrophoresis of AChE extracted from heads of the insect *Pieris brassicae* was conducted in a water-cooled LKB Multiphor II horizontal apparatus with a 7.5% polyacrylamide gel (110 × 205 × 2 mm, 20 sample loading wells) at 10 V/cm for 3 hours at 20°C (Arpagaus and Toutant, 1985). Gel and electrode solutions contained 0.5% Triton X-100, 50 m*M* Tris-glycine (pH 8.9), and samples without Triton X-100 were mixed with this detergent (to 0.5%) before loading. Following electrophoresis, the gel was washed and stained for AChE activity. Use of this procedure to demonstrate PIPLC cleavage of bovine erythrocyte (E^{bo}) AChE and PIPLC resistance of E^{hu} AChE is illustrated in Fig. 2. Similar procedures have been applied to monitor cleavage of G_2 AChE from a variety of tissues (Bon *et al.*, 1988; Fournier *et al.*, 1988), and of alkaline phosphatase (Ogata *et al.*, 1988).

Assessment of PIPLC cleavage by the nondenaturing gel electrophoresis procedure is limited somewhat because it is most easily applied to isolated, radiochemically pure anchored proteins, or to anchored proteins like AChE and alkaline phosphatase with intrinsic enzyme activities that are specific and readily detected. Nondenaturing gel electrophoresis has been combined with nitrocellulose blotting and immunochemical staining to demonstrate PIPLC cleavage of an Ly-6 antigen in crude membrane extracts (Hammelburger *et al.*, 1987), but better methods of transfer to nitrocellulose from nondenaturing gels must be developed to make this a useful general procedure (see Bjerrum *et al.*, 1987).

C. Exposure of an Anchor Antigenic Determinant

Although release from intact cells or conversion from a detergent-binding to a hydrophilic form following treatment with PIPLC appear to be useful general criteria for the identification of proteins with glycoinositol phospholipid anchors, an assignment based only on these criteria may be

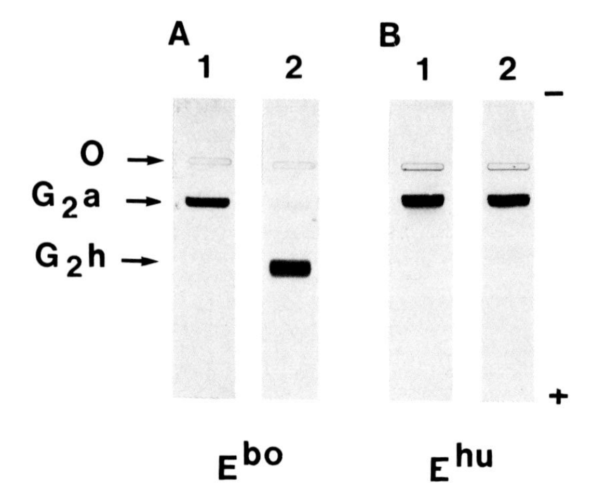

FIG. 2. PIPLC cleavage of E^{bo} AChE but not E^{hu} AChE revealed by nondenaturing gel electrophoresis (Toutant *et al.*, 1989). Purified samples (3 fmol in 15μl) of E^{bo} AChE (A) and E^{hu} AChE (B) were incubated with (lanes 2) or without (lanes 1) *B. thuringiensis* PIPLC (4 μg/ml) at 37°C for 1 hour and electrophoresed for 3 hours (Section II,B,2). The gel was then washed extensively in water and stained for AChE activity (Karnovsky and Roots, 1964) in a solution containing 5 m*M* sodium citrate, 3 m*M* $CuSO_4$, 0.5 m*M* $K_3[Fe(CN_6)]$, and 3.5 m*M* acetylthiocholine bromide in 100 m*M* sodium acetate buffer at pH 6.5. The origin is indicated by O; G_2a refers to the amphiphilic, detergent-binding AChE mobility, and G_2h to the cleaved, hydrophilic AChE mobility.

incorrect if the protein is involved in an extracellular aggregate on the cell surface. In this case PIPLC cleavage could involve another protein in the aggregate that mediates the membrane attachment of the protein of interest. This possibility may be ruled out by isolation of the protein and documentation of anchor components, but it also may be excluded by the appearance of a particular epitope called the "cross-reacting" determinant (CRD) following PIPLC treatment. Antisera to this determinant, prepared by immunization with trypanosome VSG produced by cleavage with an endogenous trypanosome phospholipase C, were shown to cross-react with the hydrophilic forms of several VSG classes but not with the intact detergent-binding VSG (Cardoso de Almeida and Turner, 1983). The epitope includes the anchor glucosamine and inositol phosphate produced by PIPLC cleavage (Holder, 1983; Zamse *et al.*, 1988). Anti-CRD antisera have been shown by SDS–PAGE and immunoblotting to cross-react with a variety of anchored proteins following PIPLC cleavage including *Paramecium* surface antigens (Capdeville *et al.*, 1986) and G_2 AChEs (Stieger *et al.*, 1986). It is noteworthy that the cross-reactivity of one of these AChEs was not detected by a radioimmunoassay (Ferguson *et al.*, 1985b), suggesting that structural variations may result in quantitative differences in anti-CRD antibody affinities to anchor structures (see Ferguson and Williams; Zamse *et al.*, 1988).

A fourth procedure for demonstrating phospholipase cleavage of anchored proteins involves the release of radiolabeled lipids from the protein on incubation with phospholipase and is described in Sections IV,A and V,A.

III. Other Phospholipases That Cleave Glycoinositol Phospholipid Anchors

The endogenous trypanosome phospholipase that generates the CRD from VSG was shown to be a phospholipase C by Ferguson *et al.* (1985a). This phospholipase was purified to homogeneity and found to cleave dimyristoylglycerol from intact VSG at 170 times the rate that it cleaved phosphatidylinositol (Bulow and Overath, 1986; Hereld *et al.*, 1986; Fox *et al.*, 1986). An enzyme with an even greater specificity for the VSG anchor has been purified from rat liver plasma membranes (Fox *et al.*, 1987). Like bacterial PIPLC, these anchor-specific phospholipases C do not require Ca ions and are fully active in buffers containing EGTA.

An anchor-degrading activity in mammalian serum cleaves between phosphate and inositol in a number of anchored proteins and thus corresponds to a phospholipase D (Davitz *et al.*, 1987; Low and Prasad, 1988). E^{hu} AChE is among the proteins cleaved (Roberts *et al.*, 1988a), and therefore the palmitoylation of inositol (Fig. 1) that renders this AChE resistant to PIPLC does not prevent reaction with this serum phospholipase D. The serum enzyme appears highly specific for glycoinositol phospholipids and shows no activity toward phosphatidylinositol or other nonglycosylated phospholipids.

Reaction of anchored proteins with these phospholipases C and D can be monitored by many of the techniques outlined in the previous section, and studies of cleavage by the endogenous phospholipases *in situ* could define physiological conditions of phospholipase activation. When applied exogenously, however, both trypanosome phospholipase C (C. Braun-Breton and T. L. Rosenberry, unpublished observations) and serum phospholipase D (Davitz *et al.*, 1987) appear to be much more efficient in detergent extracts than on intact cells. It should be noted that the hydrophilic protein arising from phospholipase D cleavage does not interact with anti-CRD antisera (Davitz *et al.*, 1987), as expected if the CRD epitope requires inositol phosphate on the residual anchor glycan.

IV. Radiolabeling of Anchor Components

Labeling procedures fall into two categories: biosynthetic incorporation of radiolabeled anchor precursors into cells in culture, and exogenous reaction of anchored proteins with radiolabeled reagents *in vitro*. Biosynthetic labeling calls for anchored proteins of sufficient abundance to permit detection of the incorporated label and requires documentation that the labeled precursor is incorporated into the anchor structure without metabolism to other precursors. Exogenous labeling relies on isolation of the anchored protein from protein contaminants and removal of excess radioactive reagents and also requires clear differentiation of labeled anchor components from labeled components outside the anchor.

A. Biosynthetic Incorporation of Anchor Precursors

The most abundant glycoinositol phospholipid anchored proteins are VSG, the predominant surface proteins that comprise 7–10% of the total cell protein in trypanosomes (Borst and Cross, 1982). The only lipid groups associated with VSG are the two anchor myristoyl groups indicated in Fig. 1, and [^{3}H]myristic acid is a good biosynthetic precursor. Trypanosomes corresponding to clone MITat 1.4 of *Trypanosoma brucei* strain 427 were purified from infected rat blood and cultured at 37°C in RPMI 1640 containing 25 m*M* HEPES (pH 7.4) at 5 × 10^7 cells/ml (Ferguson and Cross, 1984). [^{3}H]Myristic acid (1 mCi) was dissolved in 95% ethanol (20 μl) and mixed with 250 μl of defatted bovine serum albumin 20 mg/ml in 150 m*M* NaCl, 10 m*M* sodium phosphate, pH 7). Labeled stock was added to the cell culture (70 μCi/ml), aliquots were removed after 5–80 minutes of incubation, and cell pellets (2.5 × 10^6 cells) obtained by Microfuge centrifugation (10,000 *g* for 20 seconds) were dissolved in buffered SDS and analyzed by SDS–PAGE. The only labeled band outside the dye front corresponded to the position of VSG. Label incorporation into this band was linear over the incubation period and amounted to about 10 dpm/ng VSG protein. This label was completely released from purified [^{3}H]myristate-labeled VSG by purified *S. aureus* PIPLC as [^{3}H]dimyristoylglycerol (Ferguson *et al.*, 1985b; see Section V,A), a direct demonstration that the [^{3}H]myristic acid had been incorporated exclusively into the VSG anchor. Biosynthetic labeling of trypanosomes with [^{3}H]myristate also has revealed a free glycoinositol phospholipid that appears to be a precursor of the anchor in the VSG protein (Krakow *et al.*, 1986).

Several other potential precursors for glycoinositol phospholipid anchor labeling are suggested from the structures in Fig. 1, including anchor components like ethanolamine and inositol that are rarely linked covalently to proteins. [^{3}H]Ethanolamine, and to a lesser extent $^{32}PO_4$, have been biosynthetically incorporated into VSG (Rifkin and Fairlamb, 1985). Among the most abundant anchored proteins in mammalian cells is alkaline phosphatase, which approaches levels as high as 2% of total protein in the WISH human amniotic cell line (Howard *et al.*, 1987). Biosynthetic labeling of the alkaline phosphatase anchor in these cells as well as in the choriocarcinoma cells JEG-3 (Takami *et al.*, 1988) has been obtained with [^{3}H]ethanolamine, [^{3}H]inositol, and ^{3}H-labeled fatty acids. All three precursors were incorporated into alkaline phosphatase in WISH cells without metabolic conversion (Howard *et al.*, 1987), although interconversion among labeled fatty acids was observed. Results with JEG-3 cells are illustrated in Fig. 3, lanes 1, where label incorporation into the 66-kDa alkaline phosphatase SDS–PAGE band was observed with each precursor. Selectivity of anchor labeling with [^{3}H]stearic acid was demonstrated by complete loss of the labeled 66-kDa band upon incubation with PIPLC (Fig. 3D, lane 2). As expected from the anchor structures in Fig. 1, however, PIPLC did not release label from the 66-kDa band following incorporation of either [^{3}H]ethanolamine or [^{3}H]inositol (Fig. 3B and C, lanes 2). Selectivity of anchor labeling with these precursors was indicated by other procedures that release anchor fragments, papain digestion (Fig. 3B and C, lanes 3), and nitrous acid deamination (Section V,B and C). [^{3}H]Ethanolamine is efficiently taken up by most cultured cells and is incorporated into very few proteins even after overnight incubation, but unfortunately this precursor is not selective for anchored proteins exclusively. A predominant 45- to 50-kDa [^{3}H]ethanolamine-labeled band observed by SDS–PAGE in total extracts of murine lymphoma cells (Fatemi *et al.*, 1987) is present in many other cell types but corresponds to a hydrophilic, cytoplasmic protein that appears unrelated to anchors (Tisdale and Tartakoff, 1988). This protein has been identified as the protein synthesis elongation factor EF-1α (Rosenberry *et al.*, 1989).

Incorporation of biosynthetic precursors may provide information about the chemical composition of anchors, but evaluation of some anchor precursors is complicated by their simultaneous incorporation outside the anchor structure. This is potentially a problem with fatty acid incorporation, since myristoylation and palmitoylation can occur directly on amino acid side chains (see Sefton and Buss, 1987), but these particular modifications appear limited to intracellular and transmembrane proteins and so far have not been reported for any anchored protein. In contrast, sugars are prominent components of N-linked oligosaccharides, and

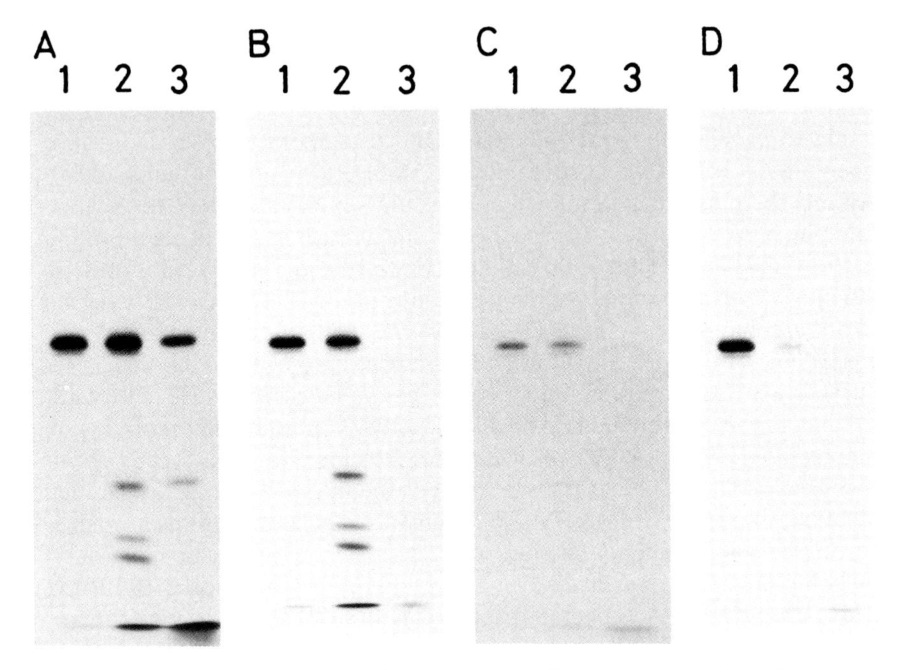

FIG. 3. Enzymatic release of anchor components from placental alkaline phosphatase (Takami *et al.*, 1988). JEG-3 choriocarcinoma cells, cultured in RPMI 1640 medium supplemented with antibiotics and 10% fetal calf serum were pretreated for 1 day with 1.5 m*M* sodium butyrate to induce placental alkaline phosphatase. Individual 60-mm dishes were labeled by overnight incubation with (A) L-[^{35}S]methionine (100 μCi, 1000–1300 Ci/mmol); (B) [1-^{3}H]ethanolamine (500 μCi, 12.0 Ci/mmol); (C) *myo*-[2-^{3}H]inositol (500 μCi, 17 Ci/mmol); or (D) [9,10-^{3}H]stearic acid (2 mCi, 23 Ci/mmol). Cells were washed twice with Dulbecco's phosphate-buffered saline, lysed in 0.5 ml of the same buffer containing 1% Triton X-100, 0.5% sodium deoxycholate, 0.1% SDS, and protease inhibitors, sonicated for 2 minutes, and centrifuged at 15,000 *g* for 30 minutes. Supernatants were cleared by immunoprecipitation with preimmune rabbit serum and immunoprecipitated with 3 μl of antiplacental alkaline phosphatase serum onto Pansorbin. Following a sequence of extensive washes, the immunoprecipitate was boiled for 4 minutes in 20–60 μl of buffered 1% SDS and 1% mercaptoethanol, and the Pansorbin was removed by centrifugation. Samples were digested with the following enzymes at 37°C following addition of Nonidet P-40 to minimize enzyme denaturation, precipitated with acetone/70 m*M* HCl, run on 9% Laemmli SDS–PAGE gels, and fluorographed. Lanes 1: undigested with enzyme; lanes 2: digested with *B. cereus* PIPLC (10 μg/ml) for 6 hours; lanes 3: digested with papain (10 μg/ml) for 1 hour.

only a fraction of the [^{3}H]mannose or [^{3}H]glucosamine incorporated into anchored proteins is expected at the anchor sites indicated in Fig. 1. One approach to demonstrating anchor incorporation of these sugars involves digestion with peptide *N*-glycosidase F, which selectively removes N-linked oligosaccharides. Treatment with this glycosidase converted alkaline phosphatase from a 66-kDa to a 59-kDa band on SDS–PAGE (Takami *et al.*, 1988), and biosynthetically incorporated [^{3}H]mannose and [^{3}H]glucosamine but not [^{3}H]galactose were retained in the 59-kDa band, a pattern consistent with the anchor composition expected from Fig. 1. However, this conclusion was tentative because the retained label was faint, not verified in the anchor, and not examined for metabolic conversion. A second approach to the localization of incorporated labeled sugars in the anchor involves the complete digestion of the isolated labeled anchored protein with pronase (see Section V,C) and the demonstration of a labeled fragment that binds nonionic detergents. Results with murine lymphoma Thy-1 suggested that small fractions of label incorporated from [^{3}H]mannose and [^{3}H]galactose were localized in the anchor, the latter possibly arising from metabolic conversion to glucosamine or *N*-acetylgalactosamine (Fatemi *et al.*, 1987), but evidence confirming that only labeled anchor fragments were hydrophobic was not pursued.

B. Exogenous Radiolabeling of Anchor Components

Since the inositol phospholipid appears to be the only nonionic detergent-binding domain in the anchored proteins that have been studied, agents that react preferentially with hydrophobic structures are reasonable candidates for use in selective anchor labeling. A reagent used successfully to label the anchors of several G_2 AChEs and E^{hu} DAF is 3-(trifluoromethyl)-3-(*m*-[^{125}I]iodophenyl)diazirine ([^{125}I]TID), which partitions into hydrophobic phases and on photoactivation produces a highly reactive carbene intermediate (Brunner and Semenza, 1981). At least 75% of the label incorporated into the AChEs is covalently attached to the inositol phospholipid (Steiger *et al.*, 1984; Roberts and Rosenberry, 1986; Roberts *et al.*, 1987; Haas *et al.*, 1988). A second procedure for radiolabeling anchor components is reductive radiomethylation, a reaction that adds labeled methyl groups to primary and secondary amines throughout the entire protein. Although most of the radiomethylation typically occurs outside the anchor on the N-terminus and the ε-amino groups of lysine residues, the free amino groups of anchor glucosamine and ethanolamine components (Fig. 1) also are radiomethylated and are identified by acid hydrolysis and analysis on an amino acid analyzer (Haas *et al.*, 1986).

1. [^{125}I]TID Labeling

This reagent is volatile, so all manipulations should be carried out in a fume hood and precautions taken to ensure that the reaction vessel is sealed during photolysis. To 0.5–2.0 ml of purified protein (40–2000 μg/ml) in 0.1% Triton X-100, 20 m*M* sodium phosphate (pH 7) in a 12 × 75-mm rubber-stoppered borosilicate glass test tube was added 10–50 μl of [^{125}I]TID (50–250 μCi, Amersham) in ethanol and the sample was agitated gently to ensure complete mixing. Photolysis was performed for 15–30 minutes at 350 nm in a Farrand Mark I spectrofluorometer from which the excitation slits were removed, and the sample was dialyzed extensively against the same buffer to remove noncovalently bound radioactivity. Acetylcholinesterase was repurified by affinity chromatography on acridinium resin (Roberts *et al.*, 1987); DAF was repurified by size exclusion chromatography on Sepharose CL-6B using a mobile phase containing 0.1% Triton X-100 (Medof *et al.*, 1986). Approximately 0.2–2.0% of the total radiolabel was incorporated into these proteins (3–30 dpm/ng protein). A 4-fold higher radiolabel incorporation was achieved when E^{hu} AChE was labeled in the absence of Triton X-100 (Roberts and Rosenberry, 1986).

2. Radiomethylation

Protein solutions (0.02–2 mg/ml) in 50–500 μl of 20–200 m*M* sodium phosphate (pH 7) were incubated with 10 m*M* radiolabeled HCHO (e.g., [^{14}C]HCHO, ICN, 40 mCi/mmol) and 50 m*M* $NaCNBH_3$ (Aldrich, recrystallized from acetonitrile as in Jentoft and Dearborn, 1979) for 15 minutes at 37°C. After extensive dialysis against phosphate buffer to remove unincorporated radioactivity, the labeled protein may be repurified as noted in the previous section. Alternatively, the labeled protein may be subjected to SDS–PAGE and the labeled band cut from the gel (Haas *et al.*, 1988). The dialyzed protein in solution or the gel slice is dried, hydrolyzed in 6 *N* HCl, and chromatographed on an amino acid analyzer, and the analyzer effluent is monitored by scintillation counting (Haas and Rosenberry, 1985). Radiolabeled peaks arising from methylated ethanolamine and glucosamine are identified by coelution with corresponding radiomethylated amine standards (Haas *et al.*, 1986). This procedure has been used to identify and quantitate ethanolamine and glucosamine with free amino groups in E^{hu} AChE (Haas *et al.*, 1986), DAF (Medof *et al.*, 1986), and Thy-1 (Fatemi *et al.*, 1987). It is noteworthy that radiomethylation does not label the anchor ethanolamine in amide linkage to the C-terminal amino acid (Fig. 1). Biosynthetic labeling with [^{3}H]ethanolamine, however, results in label incorporation into both ethanolamine sites. This was demonstrated by reductive methylation with unlabeled HCHO of Thy-1 that had been biosynthetically labeled with [^{3}H]ethanolamine (Fatemi *et al.*, 1987). One-half of the labeled ethanolamine was converted to dimethylethanolamine, while the remaining half was not modified, precisely the result expected for the reactivity of the two ethanolamine residues indicated in Fig. 1.

Both photolabeling with [^{125}I]TID and radiomethylation are gentle procedures that do not denature the enzymatic activity of AChE, and thus the labeled protein may be repurified by affinity chromatography to eliminate contaminants. Although extension of these procedures to radiolabeling of anchored proteins in crude extracts has not been investigated, this would appear feasible if sufficient label is used and the target protein is purified by immunoprecipitation and SDS–PAGE prior to further analysis.

V. Release of Fragments Characteristic of Anchors

Earlier sections have emphasized that cleavage by PIPLC or other anchor-specific phospholipases is one of the most useful criteria for the identification of an anchored protein. However, analysis of PIPLC-resistant anchors like that in E^{hu} AChE (Fig. 1) has relied on other cleavage procedures that produce characteristic anchor fragments. Furthermore, Fig. 1 emphasizes that anchor structures, while sharing a core of common components, exhibit variations that may have important functional consequences. These variations are revealed only by detailed analysis of anchor fragments. The following sections describe procedures for the generation of useful anchor fragments and outline techniques for their characterization. Analyses in general require either that components in the fragments be radiolabeled biosynthetically or exogenously and the preparation be radiochemically pure, or that the unlabeled protein be purified and free of carbohydrate and lipid contaminants.

A. Diradylglycerols Released by PIPLC

An anchored protein may be identified by its release from the cell surface or conversion from a detergent-binding to a hydrophilic form by PIPLC, as described in Section II. Further information about the components in the anchor can be obtained by analysis of the diradylglycerols (diacylglycerols or alkylacylglycerols) cleaved from the protein by this procedure. [^{3}H]Myristate-labeled VSG (Section IV,A) was extracted from trypanosomes in 1% sodium deoxycholate and purified on concanavalin A–Sepharose (Ferguson *et al.*, 1985b). Extraction of labeled VSG samples with toluene, diethyl ether, or chloroform/methanol (2 : 1) showed that only 3–4% of the radioactivity was due to contaminating lipids. Labeled VSG (25,000 dpm) was incubated in 25 m*M* HEPES–NaOH (pH 7.4), 0.1% sodium deoxycholate, and *S. aureus* PIPLC (2 μg/ml) in 0.4 ml total volume for 30 minutes at 37°C. The reaction was stopped by addition of 1.1 ml of toluene/acetic acid (10 : 1) followed by 0.6 ml of water. After vigorous vortex mixing and centrifugation to separate the phases, a sample of the upper phase was removed, dried, and mixed with a neutral lipid standard mixture for thin-layer chromatographic analysis (TLC) on activated silica gel G thin-layer plates (5 × 20 cm) with a solvent system of petroleum ether/diethyl ether/acetic acid (70 : 30 : 2). Lipid standards were located by exposure of the dried plates to iodine vapor, and radioactive lipids were detected by scraping sample lanes in strips 0.5 or 1 cm wide. The scrapings were taken for scintillation counting and demonstrated that 95% of the recovered radioactivity comigrated with the 1,2-dimyristoylglycerol standard. A similar TLC procedure was used to locate the ^{125}I-labeled lipid released by PIPLC from purified [^{125}I]TID-labeled E^{bo} AChE between 1,2- and 1,3-dioleoylglycerol standards (Roberts *et al.*, 1987). This study found that [^{125}I]TID derivatization had little effect on lipid TLC mobility. Further analysis of unlabeled lipid released from E^{bo} AChE with PIPLC by high-performance liquid chromatography (HPLC) and gas–liquid chromatography (GLC) revealed that it was predominantly an alkylacylglycerol, 1-stearyl-2-stearoylglycerol (Roberts *et al.*, 1988c).

Diradylglycerol components of E^{hu} AChE could not be released by PIPLC because of the inositol palmitoylation (Fig. 1) but were generated as diradylglycerol acetates by acetolysis in acetic acid/acetic anhydride (Ferguson *et al.*, 1985a; Roberts *et al.*, 1988a). HPLC and GLC analysis of the lipids released by acetolysis revealed a mixture in which some components appeared to arise by acetolysis-induced isomerization, but the major diradylglycerol acetate was shown to correspond to the alkylacylglycerol in Fig. 1 (Roberts *et al.*, 1988a).

Although it has not yet received widespread attention, the anchor-specific phospholipase D in mammalian serum (Section III) is attractive as a tool for the identification of anchored proteins and the characterization of anchor lipids. This enzyme is active on all anchored proteins that have been tested, including those with palmitoylated anchors like that in E^{hu} AChE (Fig. 1), which are resistant to PIPLC. Phosphatidic acids released by this enzyme may be extracted in organic solvents and detected by TLC (Davitz *et al.*, 1987; Low and Prasad, 1988).

B. Inositol Phospholipids Released by Nitrous Acid Deamination

Nitrous acid has long been known to cleave 2-amino-2-deoxy-D-glucosidic bonds through nitrosation of the sugar amino group followed by loss of N_2 and ring contraction of the D-glucosamine residue to 2,5-anhydromannose (see Williams, 1975). Since glucosamine is in glycosidic linkage to inositol in all anchors yet studied (Fig. 1), this cleavage results in release of the inositol phospholipid. However, the cleavage is not quantitative because some 2-deoxy-2-*C*-formyl-α-D-ribofuranoside is formed in which the glycosidic linkage is retained (see Williams, 1975). Typical extents of anchor cleavage range from 60 to 80% (Ferguson *et al.*, 1985b; Medof *et al.*, 1986; Low *et al.*, 1987). Furthermore, deamination and cleavage do not occur if the anchor glucosamine has been radio-methylated.

Detergent-depleted samples of [^{125}I]TID-labeled E^{hu} AChE (15–50 nmol) were dialyzed against water and reduced in volume in a Speedvac concentrator (Roberts *et al.*, 1988a). To the enzyme in 1.2 ml of 0.1 *M* sodium acetate (pH 3.5) was added 0.3 ml of 1 *M* sodium nitrite, and the pH of the solution was adjusted to 4.0 by the addition of 6 *N* HCl (6 μl). After incubation of the mixture at 50°C for 4 hours the phospholipid product was removed by extraction with 4 ml of chloroform/methanol (2 : 1) and two 3-ml portions of chloroform. The combined organic phases were dried and yielded 50–60% of the initial ^{125}I radioactivity. The TLC mobilities of the ^{125}I-labeled inositol phospholipids released from E^{bo} and E^{hu} AChE are compared in Fig. 4B. The mobility of the E^{bo} AChE phospholipid was very close to that of the [^{125}I]TID-labeled phosphatidylinositol standard, as expected for a plasmanylinositol susceptible to PIPLC (Section V,A). However, the E^{hu} AChE phospholipid had a much greater mobility, an observation that provided the first indication of the additional palmitoyl group on the inositol (Fig. 1). The structure of this palmitoylated inositol phospholipid was confirmed by fast atom bombardment–mass spectrometry (FAB–MS) (Roberts *et al.*, 1988b).

The cleavage of anchored proteins in immunoprecipitates with nitrous acid has been reported for alkaline phosphatase biosynthetically labeled with [^{35}S]methionine, [^{3}H]stearic acid, or [^{3}H]inositol (Takami *et al.*, 1988) and for *D. discoideum* antigen 117 labeled with tritiated fatty acids (Sadeghi *et al.*, 1988). Analysis by SDS–PAGE confirmed that nitrous acid treatment resulted in complete loss of the label from the fatty acid- or inositol-labeled proteins but in no effect on the methionine-labeled protein. Investigation by TLC of the labeled inositol phospholipids released by the treatment was not pursued.

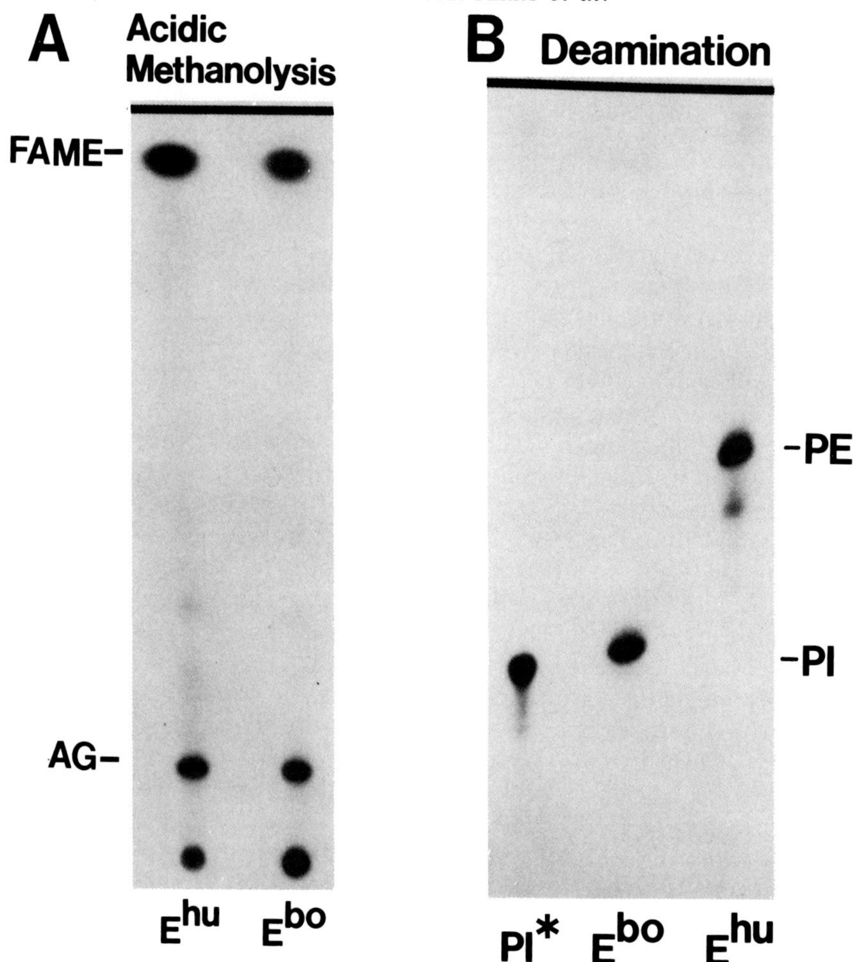

FIG. 4. TLC analysis of [^{125}I]TID-labeled fragments generated from E^{bo} and E^{hu} AChEs (Roberts *et al.*, 1987, 1988a). Samples of the [^{125}I]TID-labeled AChEs (50–100 pmol; Section IV,B,1) were subjected to acidic methanolysis (A) or nitrous acid deamination (B), and radiolabeled fragments were extracted and analyzed by silica TLC and autoradiography. Unlabeled lipid standards were visualized by exposure to iodine vapor. (A) Dried AChE samples were treated with methanolic HCl (Section VI,B) and extracted lipids were resuspended in chloroform (10 μl), spotted on a TLC plate, and developed in hexane/2-propanol (96 : 4). Standards were 1-hexadecylglycerol (AG) and methyl palmitate (FAME). Residual radioactivity at the origin may result from degradation of [^{125}I]TID. (B) AChE samples were treated with nitrous acid in a procedure similar to that in Section V,B. Extracted lipids were resuspended in chloroform/methanol (2 : 1, 10 μl) and applied to a silica TLC plate, which was developed in chloroform/methanol/water (65 : 25 : 4). The unlabeled standards were phosphatidylethanolamine (PE) and phosphatidylinositol (PI). PI* indicates [^{125}I]TID-labeled phosphatidylinositol spotted directly on the plate without nitrous acid treatment. In A and B, E^{bo} and E^{hu} designate the respective labeled AChE samples.

C. Complete Anchors Released by Proteolysis

Complete proteolysis of anchored proteins with pronase or proteinase K is a good general procedure for generating fragments containing intact anchors, although partial proteolysis may be tailored to individual proteins. For example, E^{hu} AChE can be digested selectively with papain to release a C-terminal dipeptide linked to the anchor while retaining an intact hydrophilic enzyme (Dutta-Choudhury and Rosenberry, 1984; Haas *et al.,* 1986). The hydrophobicity contributed by anchor lipids combined with the charged anchor phosphate and amine groups tend to reduce recoveries in subsequent purification of the anchor proteolytic fragment, and anchor lipids have been removed by digestion with PIPLC prior to proteolysis of VSG (Ferguson *et al.,* 1988) or alkaline phosphatase (Micanovic *et al.,* 1988; Ogata *et al.,* 1988) or immediately after proteolysis of Thy-1 (Homans *et al.,* 1988). Subsequent isolation of these delipidated proteolytic anchor fragments required multiple chromatographic steps to remove peptide contaminants, and preparations sufficient for the structural determinations of the VSG and Thy-1 anchors in Fig. 1 were obtained. Proteolysis without delipidation was applied to radiomethylated E^{hu} AChE (100–200 nmol) by incubation for 10 hours at 50°C in a 3-ml mixture of pronase (10 mg/ml), 0.1 *M* HEPES (pH 8.0), 15 m*M* $CaCl_2$, and 1% Triton X-100 (Roberts *et al.,* 1988b). Following the addition of SDS (to 1%) and a second aliquot of pronase (to 15 mg/ml total), digestion was continued for 7 hours at 50°C. The digestion mixture was applied to a Sephacryl S-200 column (1.5 × 80 cm) equilibrated in 20 m*M* sodium phosphate (pH 7) and 0.05% Triton X-100. The radiolabeled proteolytic anchor fragment associated with Triton X-100 micelles is the first fragment to elute from this column. Fractions from this peak can be further purified by chromatography on Sephadex LH-60 in ethanol/formic acid (Roberts and Rosenberry, 1986). While this preparation yielded an overall recovery of the anchor fragment of 30–40% and was adequate for the quantitation of several components, removal of residual Triton X-100 by HPLC following partial delipidation by base hydrolysis was required for analysis of the anchor structure by FAB–MS (Roberts *et al.,* 1988b).

Pronase proteolysis of biosynthetically radiolabeled Thy-1 was conducted following immunoprecipitation and SDS–PAGE (Fatemi *et al.,* 1987). Gel slices containing the Thy-1 band were digested, digest supernatants were applied to phenyl–Sepharose, and anchor proteolytic fragments were eluted from the washed resin with 2% SDS. The procedure was used to assess incorporation of biosynthetic precursors into the anchor (Section IV,A).

VI. Chemical Determination of Anchor Components

Previous sections have noted methods for the detection of individual anchor components. Glucosamine and ethanolamine with free amino groups can be quantitated in amounts as low as a few picomoles by the radiomethylation procedure in Section IV,B,2. Diradylglycerol and inositol phospholipid identification procedures were outlined in Section V. Two additional methods for the quantitation of anchor components are provided in the following sections.

A. Inositol

myo-Inositol is present in all anchors that have been examined. The first quantitation of anchor inositol was reported by Futerman *et al.* (1985c) for torpedo G_2 AChE, and this gas–liquid chromatography–mass spectrometry (GLC–MS) procedure has been incorporated with the following minor modifications for use in our laboratory. *myo*-Inositol can be quantitated by GLC with flame ionization detection (FID) or by GLC–MS with selected-ion monitoring (SIM). A purified sample of protein (>500 pmol for GLC with FID, >25 pmol for MS with SIM) was hydrolyzed at 115°C for 12 hours in 6 *N* HCl (100μl) containing 500 nmol Tris-HCl (to improve inositol recoveries). A known quantity of *scyllo*-inositol that approximated the amount of *myo*-inositol expected in the sample was included as an internal standard. After hydrolysis the sample was dried *in vacuo* prior to derivatization with *N,O*-bis(trimethylsilyl)trifluoroacetamide/trimethylchlorosi lane–pyridine (10 : 1 : 10) (20–50 μl) at 65°C for 1 hour. Aliquots of the derivatized sample (1μl) were analyzed on a Hewlett-Packard model 5890A capillary column gas chromatograph equipped for FID, with a split/splitless injector and a 30-m, 0.25-mm i.d., 0.25 μm d_f SPB-1 column (Supelco). Helium carrier gas at 15 psi was employed and the column was temperature-programmed from 100° to 280°C at 10°C/minute. The retention time of *scyllo*-inositol was 14.9 minutes while that of *myo*-inositol was 15.5 minutes, and the molar response factors of the two isomers were virtually identical. Mass spectrometry offers a more selective method for detecting *myo*-inositol; therefore less sample is required for accurate quantitation (Sherman *et al.*, 1970). Sample preparation and chromatographic conditions were identical to those described previously. The gas chromatograph was connected to a Hewlett-Packard 5970 mass selective detector by a capillary direct interface. Selected-ion monitoring of *m/z* 305 and 318 with a 100-millisecond dwell time for each ion was employed. The yield of the *m/z* 305 ion from *myo*-inositol was 120% of the same ion from *chiro*-inositol while that of the *m/z* 318 from *myo*- was 45% that of *chiro*-inositol. The ratios of the abundances of the *m/z* 305 and 318 ions from the two inositol isomers were linear with concentration over the applied range of 2–50 pmol.

myo-Inositol also can be quantitated in anchored proteins isolated by SDS–PAGE. An E^{hu} AChE protein band visualized by Coomassie blue staining was excised and hydrolyzed in 6 *N* HCl (200 μl) as before. After hydrolysis the sample was centrifuged and the supernatant was removed, dried *in vacuo,* silylated, and analyzed by GLC–MS as before. Typical recoveries of *myo*-inositol from the excised gel slices were 35% of the *myo*-inositol in the AChE sample applied to the gel.

B. Fatty Acids and Alkylglycerols

Acidic methanolysis not only releases ester-linked fatty acids by transesterification but also cleaves the glycerol–phosphate bond and thus permits determination of both fatty acid methyl esters and alkylglycerols. Anhydrous 1 *M* methanolic HCl can be prepared by the addition of acetyl chloride (0.38 ml; Aldrich gold label) to anhydrous methanol (5 ml; HPLC grade) at −70°C. Samples of unlabeled (3–10 nmol) or [^{125}I]TID-labeled E^{hu} AChE or E^{bo} AChE (Section IV,B,1) were dried *in vacuo* in 6 × 50-mm glass test tubes (Roberts *et al.*, 1988a). Heptadecanoic acid standard was included for subsequent GLC analysis. Anhydrous methanolic HCl (100 μl) was added, and the tubes were flame-sealed and incubated at 65°C for 16 hours. After cooling the sample, the tube was opened and 200 μl chloroform (HPLC grade) and 75 μl water were added. The sample was vortexed and the lower phase was recovered. Analysis of products from [^{125}I]TID-labeled AChEs by TLC revealed both fatty acid methyl esters and alkylglycerols (Fig. 4A). For quantitative GLC analysis, the aqueous phase was reextracted with chloroform (100 μl), and the combined lower phases were dried, acetylated by incubation with acetic anhydride/pyridine (1 : 1, 10 μl) for 30 minutes at 80°C, dried, and resuspended in 2,2,4-trimethylpentane (20 μl). An aliquot (1 μl) of the sample was analyzed on the 5890A gas chromatograph (Section VI,A) equipped with an SP-2380 fused-silica capillary column (15 m × 0.32 mm i.d., Supelco) with helium carrier gas at 5 psi. The oven temperature was programmed from 100° to 260°C at 10°C/minute. The retention times of fatty acid methyl esters were determined from standards (PUFA-2, Supelco). Alkylglycerol standards (16 : 0,18 : 0, and 18 : 1, Serdary Research Laboratories, Inc., London, ON Canada) were acetylated as described before.

C. Other Components

Anchor carbohydrate analyses can be performed after isolation of the proteolytic anchor fragment (Section V,C), as illustrated in a comprehensive study of the VSG anchor (Ferguson *et al.*, 1988). Combined GLC analyses of inositol, fatty acids, ethanolamine, phosphate, and glycerol in proteolytic anchor fragments from Thy-1 have been reported by Tse *et al.* (1985), and an example is presented in Fig. 5. Quantitation was achieved for all components except glycerol, for which considerable background amounts did not differ significantly from those in samples. Identification of ethanolamine, phosphate, and *myo*-inositol by GLC–MS was reported in SDS–PAGE gel slices containing the scrapie prion protein PrP27–30 (Stahl *et al.*, 1987).

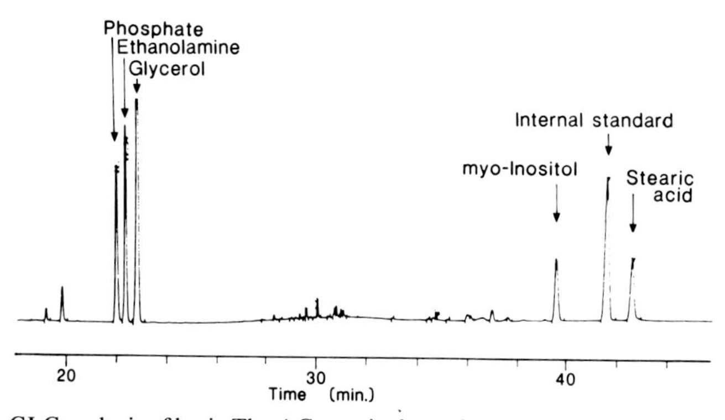

FIG. 5. GLC analysis of brain Thy-1 C-terminal tryptic peptide (Tse *et al.*, 1985). Purified rat brain Thy-1 (70 nmol) was reduced and alkylated with [^{14}C]iodoacetic acid and digested with trypsin (200 μg) in 100 m*M* NH_4HCO_3 and Brij 96 (5 μl) at 37°C for 24 hours. The tryptic fragment containing the radioalkylated cysteine C-terminus attached to the anchor was isolated by chromatography on Biogel P-10 (Section V,C) and purified by two-dimensional TLC on silica G60 plates (Tse *et al.*, 1985). The anchor peptide was recovered by scraping the radioactive spot visualized by autoradiography and extracting three times with 400 μl of 1-propanol/water (7 : 3), and the sample (25 nmol) was hydrolyzed in 5.7 *M* HCl at 100°C for 8 hours. Methyl nonadecanoate standard was added to the dried hydrolysate, and the mixture was treated with 2% methoxyamine in pyridine at 60°C for 30 minutes and then with *N,O*-bis(trimethylsilyl)trifluoroacetamide/trimethylchlorosilane (99 : 1) at 100°C for 1 hour. Aliquots of the sample (3–10% of total) were analyzed on a Hewlett-Packard gas chromatograph equipped with a nonpolar capillary column (0.32 mm id, 25 m). The temperature was programmed from 80°C (5 minutes initial) to 100°C at 5°C/minute; to 140°C at 2.5°C/minute; to 230°C at 15°C/minute; held at 230°C for 19 minutes; to 260°C at 15°C/minute; and held at 260°C for 18 minutes. The components indicated were identified by comparison of retention times with those of standards.

ACKNOWLEDGMENTS

We thank Dr. Y. Ikehara and Dr. A. F. Williams for permitting reproduction of published data. This investigation was supported by National Institutes of Health Grants N516577, DK 38181, and GM-07250 and by grants from the Muscular Dystrophy Association.

REFERENCES

Arpagaus, M., and Toutant, J.-P. (1985). *Neurochem. Int.* **7,** 793–804.

Bjerrum, O. J., Selmer, J. C., and Lihme, A. (1987). *Electrophoresis (Weinheim, Fed. Repub. Ger.)* **8,** 388–397.

Bon, S., Toutant, J.-P., Meflah, K., and Massoulié, J. (1988). *J. Neurochem.* **51,** 786–794.

Bordier, C. (1981). *J. Biol. Chem.* **256,** 1604–1607.

Bordier, C. (1988). *In* "Post-translational Modification of Proteins by Lipids: A Laboratory Manual" (U. Brodbeck and C. Bordier, eds.), pp. 29–33. Springer, Berlin.

Borst, P., and Cross, G. A. M. (1982). *Cell (Cambridge, Mass.)* **29,** 291–303.

Braun-Breton, C., Rosenberry, T. L., and Pereira da Silva, L. (1988). *Nature (London)* **332,** 457–459.

Brunner, J., and Semenza, G. (1981). *Biochemistry* **20,** 7174–7182.

Bulow, R., and Overath, P. (1986). *J. Biol. Chem.* **261,** 11918–11923.

Capdeville, Y., Baltz, T., Deregnaucourt, C., and Keller, A.-M. (1986). *Exp. Cell Res.* **167,** 75–86.

Cardoso de Almeida, M.-L., and Turner, M. J. (1983). *Nature (London)* **302,** 349–352.

Conzelmann, A., Spiazzi, A., and Bron, C. (1987). *Biochem. J.* **246,** 605–610.

Conzelmann, A., Riezman, H., Desponds, C., and Bron, C. (1988). *EMBO J.* **7,** 2233–2240.

Cross, G. A. M. (1987). *Cell (Cambridge, Mass.)* **48,** 179–181.

Davitz, M. A., Low, M. G., and Nussenzweig, V. (1986). *J. Exp. Med.* **163,** 1150–1161.

Davitz, M. A., Hereld, D., Shak, S., Krakow, J., Englund, P. T., and Nussenzweig, V. (1987). *Science* **238,** 81–84.

Dutta-Choudhury, T. A., and Rosenberry, T. L. (1984). *J. Biol. Chem.* **259,** 5653–5660.

Etges, R., Bouvier, J., and Bordier, C. (1986). *EMBO J.* **5,** 597–601.

Fatemi, S. H., Haas, R., Jentoft, N., Rosenberry, T. L., and Tartakoff, A. M. (1987). *J. Biol. Chem.* **262,** 4728–4732.

Ferguson, M. A. J., and Cross, G. A. M. (1984). *J. Biol. Chem.* **259,** 3011–3015.

Ferguson, M. A. J., and Williams, A. F. (1988). *Annu. Rev. Biochem.* **57,** 285–320.

Ferguson, M. A. J., Haldar, K., and Cross, G. A. M. (1985a). *J. Biol. Chem.* **260,** 4963–4968.

Ferguson, M. A. J., Low, M. G., and Cross, G. A. M. (1985b). *J. Biol. Chem.* **260,** 14547–14555.

Ferguson, M. A. J., Homans, S. W., Dwek, R. A., and Rademacher, T. W. (1988). *Science* **239,** 753–759.

Fournier D., Bergé, J.-B., Cardoso de Almeida, M.-L., and Bordier, C. (1988). *J. Neurochem.* **50,** 1158–1163.

Fox, J. A., Duszenko, M., Ferguson, M. A. J., Low, M. G., and Cross, G. A. M. (1986). *J. Biol. Chem.* **261,** 15767–15771.

Fox, J. A., Soliz, N. M., and Saltiel, A. R. (1987). *Proc. Natl. Acad. Sci. U.S.A.* **84,** 2663–2667.

Futerman, A. H., Fiorini, R.-M., Roth, E., Low, M. G., and Silman, I. (1985a). *Biochem. J.* **226,** 369–377.

Futerman, A. H., Low, M. G., Michaelson, D. M., and Silman, I. (1985b). *J. Neurochem.* **45,** 1487–1494.
Futerman, A. H., Low, M. G., Ackermann, D. E., Sherman, W. R., and Silman, I. (1985c). *Biochem. Biophys. Res. Commun.* **129,** 312–317.
Haas, R., and Rosenberry, T. L. (1985). *Anal. Biochem.* **148,** 154–162.
Haas, R., Brandt, P. T., Knight, J., and Rosenberry, T. L. (1986). *Biochemistry* **25,** 3098–3105.
Haas, R., Marshall, T. L., and Rosenberry, T. L. (1988). *Biochemistry* **27,** 6453–6457.
Hammelburger, J. W., Palfree, R. G. E., Sirlin, S., and Hämmerling, U. (1987). *Biochem. Biophys. Res. Commun.* **148,** 1304–1311.
He, H.-T., Barbet, J., Chaix, J.-C., and Goridis, C. (1986). *EMBO J.* **5,** 2489–2494.
Hereld, D., Krakow, J. L., Bangs, J. D., Hart, G. W., and Englund, P. T. (1986). *J. Biol. Chem.* **261,** 13813–13819.
Holder, A. A. (1983). *Mol. Biochem. Parasitol.* **7,** 331–338.
Homans, S. W., Ferguson, M. A. J., Dwek, R. A., Rademacher, T. W., Anand, R., and Williams, A. F. (1988). *Nature (London)* **333,** 269–272.
Howard, A. D., Berger, J., Gerber, L., Familletti, P., and Udenfriend, S. (1987). *Proc. Natl. Acad. Sci. U.S.A.* **84,** 6055–6059.
Ikezawa, H., Yamanegi, M., Taguchi, R., Miyashita, T., and Ohyabu, T. (1976). *Biochim. Biophy. Acta* **450,** 154–164.
Jemmerson, R., and Low, M. G. (1987). *Biochemistry* **26,** 5703–5709.
Jentoft, N., and Dearborn, D. G., (1979). *J. Biol. Chem.* **254,** 4359–4365.
Karnovsky, M. J., and Roots, L. (1964). *J. Histochem. Cytochem.* **12,** 219–222.
Koch, F., Thiele, H.-G., and Low, M. G. (1986). *J. Exp. Med.* **164,** 1338–1343.
Krakow, J. L., Hereld, D., Bangs, J. D., Hart, G. W., and Englund, P. T. (1986). *J. Biol. Chem.* **261,** 12147–12153.
Low, M. G. (1987). *Biochem. J.* **244,** 1–13.
Low, M. G., and Finean, J. B. (1977a). *Biochem. J.* **167,** 281–284.
Low, M. G., and Finean, J. B. (1977b). *FEBS Lett.* **82,** 143–146.
Low, M. G., and Finean, J. B. (1978). *Biochim. Biophys. Acta* **508,** 565–570.
Low, M. G., and Kincade, P. W. (1985). *Nature (London)* **318,** 62–64.
Low, M. G., and Prasad, A. R. S. (1988). *Proc. Natl. Acad. Sci. U.S.A.* **85,** 980–984.
Low, M. G., and Saltiel, A. R. (1988). *Science* **239,** 268–275.
Low, M. G., Ferguson, M. A. J., Futerman, A. H., and Silman, I. (1986). *Trends Biochem. Sci.* **11,** 212–215.
Low, M. G., Futerman, A. H., Ackermann, K. E., Sherman, W. R., and Silman, I. (1987). *Biochem. J.* **241,** 615–619.
Low, M. G., Stiernberg, J., Waneck, G. L., Flavell, R. A., and Kincade, P. W. (1988). *J. Immunol. Methods* **113,** 101–111.
Malik, A.-S., and Low, M. G. (1986). *Biochem. J.* **240,** 519–527.
Medof, M. E., Walter, E. I., Roberts, W. L., Haas, R., and Rosenberry, T. L. (1986). *Biochemistry* **25,** 6740–6747.
Micanovic, R., Bailey, C. A., Brink, L., Gerber, L., Pan, Y.-C. E., Hulmes, J. D., and Udenfriend, S. (1988). *Proc. Natl. Acad. Sci. U.S.A.* **85,** 1398–1402.
Ogata, S., Hayashi, Y., Takami, N., and Ikehara, Y. (1988). *J. Biol. Chem.* **263,** 10489–10494.
Rifkin, M. R., and Fairlamb, A. H. (1985). *Mol. Biochem. Parasitol.* **15,** 245–256.
Roberts, W. L., and Rosenberry, T. L. (1986). *Biochemistry* **25,** 3091–3097.
Roberts, W. L., Kim, B. H., and Rosenberry, T. L. (1987). *Proc. Natl. Acad. Sci. U.S.A.* **84,** 7817–7821.

Roberts, W. L., Myher, J. J., Kuksis, A., Low, M. G., and Rosenberry, T. L. (1988a). *J. Biol. Chem.* **263,** 18766–18775.
Roberts, W. L., Santikarn, S., Reinhold, V. N., and Rosenberry, T. L. (1988b). *J. Biol. Chem.* **263,** 18776–18784.
Roberts, W. L., Myher, J. J., Kuksis, A., and Rosenberry, T. L. (1988c). *Biochem. Biophys. Res. Commun.* **150,** 271–277.
Rosenberry, T. L., Krall, J. A., Dever, T. E., Haas, R., Louvard, D., and Merrick, W. C. (1989). *J. Biol. Chem.* **264,** 7096–7099.
Sabatini, D. D., Kreibich, G., Morimoto, T., and Adesnik, M. (1982). *J. Cell Biol.* **92,** 1–22.
Sadeghi, H., da Silva, A. M., and Klein, C. (1988). *Proc. Natl. Acad. Sci. U.S.A.* **85,** 5512–5515.
Schmidt, M. F. G. (1983). *Curr. Top. Microbiol. Immunol.* **102,** 101–129.
Sefton, B. M., and Buss, J. E. (1987). *J. Cell Biol.* **104,** 1449–1453.
Sherman, W. R., Eilers, N. C., and Goodwin, S. C. (1970). *Org. Mass Spectrom.* **3,** 829–840.
Slein, M. W., and Logan, G. F. (1963). *J. Bacteriol.* **85,** 369–381.
Stahl, N., Borchelt, D. R., Hsiao, K., and Prusiner, S. B. (1987). *Cell (Cambridge, Mass.)* **51,** 229–240.
Steiger, S., Brodbeck, U., Reber, B., and Brunner, J. (1984). *FEBS Lett.* **168,** 231–234.
Stieger, A., Cardoso de Almeida, M.-L., Blatter, M.-C., Brodbeck, U., and Bordier, C. (1986). *FEBS Lett.* **199,** 182–186.
Taguchi, R., Asahi, Y., and Ikezawa, H. (1980). *Biochim. Biophys. Acta* **619,** 48–57.
Takami, N., Ogata, S., Oda, K., Misumi, Y., and Ikehara, Y. (1988). *J. Biol. Chem.* **263,** 3016–3021.
Tisdale, E. J., and Tartakoff, A. M. (1988). *J. Biol. Chem.* **263,** 8244–8252.
Toutant, J.-P., Roberts, W. L., Murray, N. R., and Rosenberry, T. L. (1989). *Eur. J. Biochem.* **180,** 503–508.
Tse, A. G. D., Barclay, A. N., Watts, A., and Williams, A. F. (1985). *Science* **230,** 1003–1008.
Williams, J. M. (1975). *Adv. Carbohydr. Chem. Biochem.* **31,** 9–79.
Zamse, S. E., Ferguson, M. A. J., Collins, R., Dwek, R. A., and Rademacher, T. W. (1988). *Eur. J. Biochem.* **176,** 527—534.

Chapter 10

Low Temperature-Induced Transport Blocks as Tools to Manipulate Membrane Traffic

ESA KUISMANEN

Department of Biochemistry
University of Helsinki
Helsinki, Finland

JAAKKO SARASTE

Ludwig Institute for Cancer Research
Karolinska Institute
Stockholm, Sweden

I. Introduction

A number of studies carried out during recent years have used reduced temperature to map the intracellular pathways of protein transport. The effect of temperature on endocytosis and secretion has been known for some time (Jamieson and Palade, 1968; Lagunoff and Wan, 1974; Steinman *et al.*, 1974; Chu *et al.*, 1977; Rotundo and Fambrough, 1980), but

Copyright © 1989 by Academic Press, Inc.
All rights of reproduction in any form reserved.

particular intracellular steps involved were not identified. Subsequently the availability of well-characterized soluble and membrane-bound markers made it possible to discover thermosensitive steps in the endocytic pathway (Sandvig and Olsnes, 1979; Dunn *et al.*, 1980; Marsh *et al.*, 1983). Studies of virus glycoprotein transport have initiated the analysis of the function of distinct intracellular sites in the biosynthetic pathway at which protein movement is arrested at reduced temperatures. At 15°C proteins accumulate in a compartment between the endoplasmic reticulum (ER) and the Golgi complex (Saraste and Kuismanen, 1984; Balch *et al.*, 1986; Saraste *et al.*, 1986), whereas at 20°C accumulation occurs at the level of a late-Golgi compartment (Matlin and Simons, 1983; Saraste and Kuismanen, 1984; Griffiths *et al.*, 1985; for a review, see Griffiths and Simons, 1986). In this chapter we discuss the use of temperature experiments in studies of protein transport in the exocytic pathway. In particular, the focus is on the description of immunocytochemical techniques that we have used in protein detection in light and electron microscropy (LM, EM).

II. Low-Temperature Incubation

Temperature experiments with cultured cells grown on plastic or on glass coverslips require only simple equipment. Incubation is most conveniently carried out in water baths adjusted to the appropriate temperature. Relatively expensive commercial water baths are of course available, which can be adjusted to low temperatures. If this type of apparatus is not available, however, a simple way to construct a low-temperature incubation chamber is to use ~10-liter thick-walled Styrofoam boxes. If these containers are covered properly during the work, a stable low temperature (15°–20°C) can be maintained for longer periods. Bath temperature can be easily monitored with a thermometer punched through the Styrofoam cover, and adjusted, if necessary, by adding a few chucks of ice. When incubating cells on water baths we have found it convenient to use six-well (35-mm well diameter) dishes (e.g., Costar or Falcon), which float easily on the water bath.

In the absence of a 5% CO_2 atmosphere, additional buffering of the culture medium is required. Addition of a suitable organic buffer (e.g., 20 m*M* HEPES) to the medium is necessary and sufficient to maintain the physiological pH of the medium. The pH of the medium should be monitored during the experiments.

In temperature-shift experiments a rapid temperature change is obtained by replacing the medium with preconditioned medium and shifting the six-well plates to another water bath adjusted at the required temperature. If the cells are grown on coverslips as, for example, in immunofluorescence experiments, individual coverslips can be transferred to prewarmed medium using forceps. In experiments that involve harvesting of a number of samples at different time points, it is preferable to use individual culture dishes instead of six-well plates. The plates (Falcon) are marked on the bottom and floated on the inverted cover of the dish. An alternative is to use 10 × 10-cm parafilm squares, which can be used as supports for about four culture dishes. It should also be noted that these types of time course experiments require estimation of the times needed for the manipulations during shifting and harvesting of the samples.

III. 20°C and 15°C Transport Blocks

A. Monitoring by Glycosylation and Externalization

Matlin and Simons (1983) followed the processing of influenza virus hemagglutinin (HA) in MDCK cells and demonstrated the existence of the transport block operating at 20°C. In pulse–chase experiments HA was shown to acquire terminal sugar residues [resistance to endoglycosidase H; (Endo H); see Tarentino *et al.,* Chapter 5, this volume] at this temperature. However, under these conditions HA was not detected at the cell surface using a trypsin cleavage assay. Similar results were obtained with BHK-21 cells infected with wild-type or mutant (ts-1) Semliki Forest virus (SFV; Saraste and Kuismanen, 1984). Figure 1 shows the temperature dependency of SFV spike glycoprotein transport as measured using a ^{125}I-labeled protein A-binding assay (described in detail in Kääriäinen *et al.,* 1983). The cutoff temperature in the transport to the cell surface was found to be at ~20°C, which is similar to that observed by others for membrane and secretory proteins in different cell types.

Resistance to Endo H could also be used to analyze the processing of SFV spike glycoproteins at 15°C. Even after longer chase (≤4 hours) at this temperature, conversion to Endo H resistance failed to occur. This was of interest, since monitoring by immunofluorescence microscopy (Fig. 2B) as well as reversion experiments utilizing the surface expression assay (see Fig. 6) indicated that the exit of the proteins from the ER was not inhibited at 15°C.

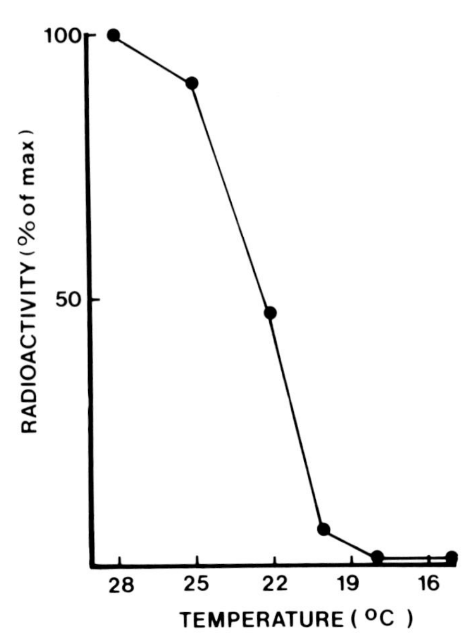

FIG. 1. Transport of SFV spike glycoproteins to the cell surface at reduced temperatures. BHK-21 cells were infected with a thermoreversible mutant (ts-1) of SFV at 39°C to accumulate newly synthesized virus glycoproteins in the ER. At 4 hours after infection the cells were shifted for 2 hours to the permissive temperature (28°C) or to different temperatures <28°C. The expression of the virus proteins at the cell surface was quantitated using a ^{125}I-labeled protein A-binding assay (see Section III,A). Reproduced from Saraste and Kuismanen (1984) with permission of Cell Press.

B. Immunofluorescence Microscopy

Immunofluorescence microscopy provides a powerful technique for the study of protein localization in eukaryotic cells (see Wang *et al.*, 1982; Reggio *et al.*, 1983). At present there exist markers for a number of different subcellular components, and by using double-labeling techniques one can study the localization of a molecule of interest in relation with the known markers. As a technique, immunofluorescence is relatively easy to carry out, although acquaintance with the use of the microscope itself is needed. Attention should be paid to the quality of the primary antibodies used. The quality of commercially available second-antibody conjugates is also variable.

We have applied immunofluorescence microscopy for monitoring the effect of temperature reduction on the intracellular transport of SFV spike glycoproteins in infected BHK-21 fibroblasts (see Fig. 2). By using

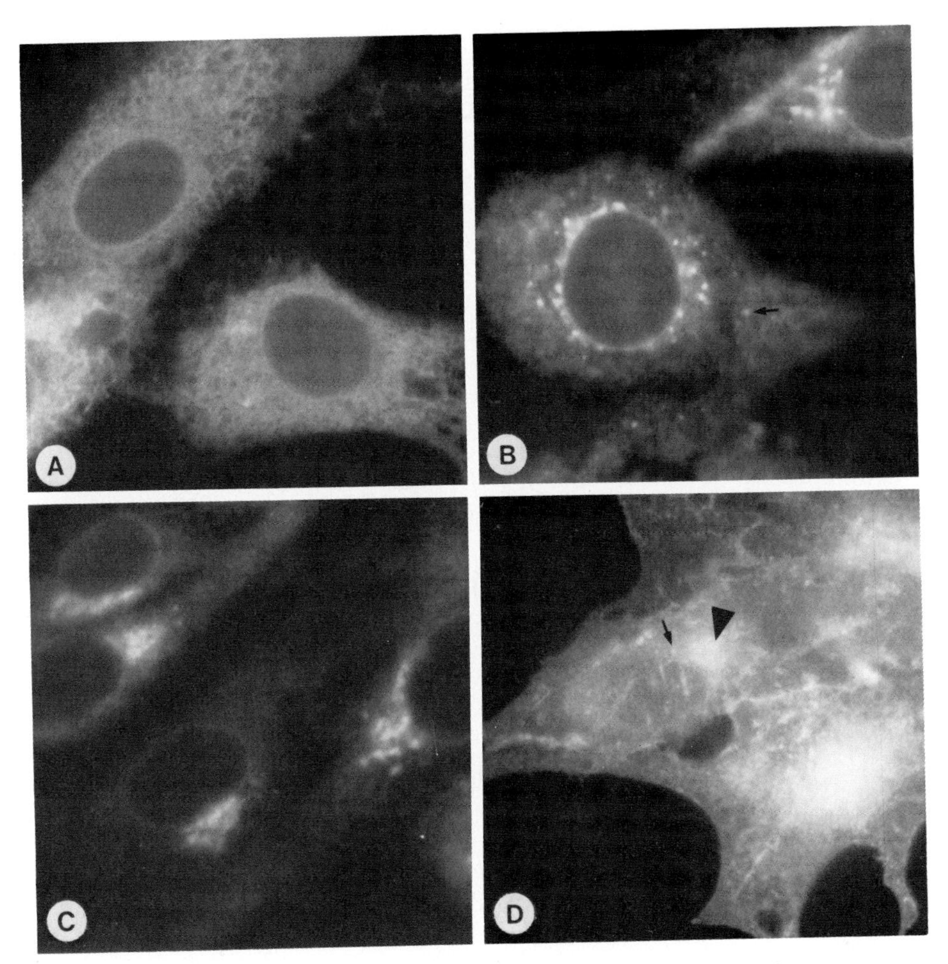

FIG. 2. Use of immunofluorescence microscopy in monitoring the effect of temperature on the transport of SFV spike glycoproteins in BHK-21 cells. Virus infection was carried out as described for Fig. 1. In all cases the cells were permeabilized with 0.05% Triton X-100 prior to staining with antibodies against the virus spike glycoproteins. (A) Accumulation of proteins in the ER (including the nuclear membrane) at the nonpermissive temperature (39°C). (B) After shift of cells to 15°C for 2 hours, labeling of pre-Golgi vesicles in the perinuclear region as well as in more peripheral regions (arrow) is evident. Also reticular ER staining is still prominent. The cell in the upper right shows labeled vesicles in the Golgi region. (C) Shift of cells to 20°C for 2 hours. The ER staining is largely depleted and the proteins accumulate in the Golgi complex. (D) Same as in (C), except that the cells were further incubated at 37°C for 60 minutes. Intensive staining of the plasma membrane is seen after reversion of the transport block. Intracellular staining of Golgi elements is also seen in the permeabilized cells (arrow, arrowhead). × 890.

a viral mutant with a thermoreversible defect in the spike glycoprotein(s), one can synchronize the movement of proteins between the rough ER and the plasma membrane (see Bergmann, Chapter 4, this volume). Immunofluorescence microscopy can, however, be applied also for the study of the biosynthetic pathway of endogenous cellular components. Intracellular pools of the protein can be depleted using cycloheximide treatment, and the transport of newly synthesized protein can be followed after removal of the drug (Keller *et al.*, 1986). Since protein synthesis continues—although at reduced levels—even when temperature is lowered to 20°C and 15°C (in BHK-21 cells at 30% and 15% of 37°C control level, respectively), the temperature-induced transport blocks allow the design of experiments (both biochemical and immunocytochemical) in which the synchronized transport of molecules between different intracellular sites can be studied.

1. The Procedure

1. Remove the medium and fix the cells for 30 minutes with 3% paraformaldehyde fixative.
2. Wash with phosphate-buffered saline (PBS) and quench with PBS containing 50 m*M* NH_4Cl.
3. Permeabilize for 30 minutes with 0.05% Triton X-100 (TX-100) in PBS.
4. Wash away the detergent with PBS containing 0.1% bovine serum albumin (BSA).
5. Incubate with primary antibody diluted in PBS–BSA containing 0.02% NaN_3.
6. Wash extensively with PBS–BSA.
7. Incubate with fluorochrome-coupled second antibody diluted in PBS–BSA.
8. Wash extensively with PBS–BSA followed by PBS.
9. Mount the coverslip in a small drop (5 μl) of glycerol-based mounting medium (supplemented with 0.1% phenylenediamine) on an objective glass.
10. Observe in a fluorescence microscope.

2. Remarks

This procedure describes the processing for immunofluorescence microscopy of adherent cells grown on coverslips. Round glass coverslips (12 mm diameter) are washed extensively with absolute ethanol (kept in ethanol until use), sterilized one by one by rapid flaming after removal of excess ethanol, and placed on 35-mm culture dishes. Cells are seeded at the desired density and grown (usually 1–2 days) to ~50% confluency. In the case of fibroblastic BHK-21 cells, for example, this allows optimal visualization of the subcellular organization in the individual well-spread cells.

A commonly used procedure in localizing membrane proteins includes fixation in 2–4% paraformaldehyde. Paraformaldehyde (2–4 g, Merck) is dissolved into 50 ml of distilled water to which two to three drops of 1 *M* NaOH have been added. Heating of the solution to 60°C (with stirring) is required. This should be carried out in the fume hood. The solution is cooled, and 50 ml of 0.2 *M* phosphate buffer (pH 7.2) are added. The fixing solution is ready for use after filtration through a 0.22-μm filter (Millipore, Corning Glass Works, Corning, NY).

Fixation times of 15–30 minutes are usually sufficient irrespective of the fixation temperature used. After fixation the cells are washed with PBS and used immediately or stored in PBS at 4°C. Prolonged storage is not recommended, since the fixation obtained with paraformaldehyde is reversible. For some antigens the addition of a low concentration of glutaraldehyde (0.05–0.1%) can be used without loss of antigenicity. In some cases glutaraldehyde-containing fixatives must be used to inhibit the extraction of the antigen of interest during permeabilization. Unreactive aldehyde groups must be efficiently quenched by washing with PBS supplemented with 50 m*M* NH_4Cl (or, for example, 50 m*M* lysine-HCl) for 15 minutes. Fixed, nonpermeabilized cells can be used directly for the visualization of surface antigens, although surface staining patterns are also easily distinguishable in permeabilized cells (see Fig. 2D).

For staining of intracellular membrane antigens, cells are commonly permeabilized with 0.05% TX-100 in PBS for 15–30 minutes (Laurila *et al.*, 1978). After permeabilization, a buffer (PBS) containing 0.1% BSA, or alternatively, 0.1% gelatin or ovalbumin, is used. When a previously uncharacterized antigen is studied, it is advisable to test different fixation and permeabilization methods. Also, fixation times can be varied to optimize the results. It should be noted that even some membrane proteins in fixed cells are extracted during permeabilization (Goldenthal *et al.*, 1985). Triton X-100 can be replaced by another milder detergent, saponin (see Brown and Farquhar, Chapter 25, Vol. 31), which acts by removing cholesterol from intracellular membranes (see, e.g., Wassler *et al.*, 1987).

If affinity-purified antibodies are not available the IgG fraction of the serum is preferred for obtaining optimal results. IgG—as well as most second-antibody conjugates [fluorescein isothiocyanate (FITC) and tetramethyl rhodamine isothiocyanate (TRITC)—can be conveniently stored in PBS containing 50% glycerol at −20°C. For incubation with fixed, permeabilized cells, antibodies are diluted in PBS–BSA containing 0.02% sodium azide. This antibody solution can be stored at 4°C and used continuously for longer periods. If the antibody is stored in a microfuge tube, preclearing by centrifugation (5 minutes at ~15,000 *g*) is easy to carry out prior to every use.

It is not necessary to remove the coverslips from the dishes for the staining procedure to be carried out. Buffer is removed by suction, which is also used to dry the plate surface surrounding the coverslip. One should be careful to avoid the drying of the coverslip at this or any subsequent stage of the treatment. A 20- 30-μl quantity of antibody solution is sufficient to cover the whole surface of a 12-mm coverslip. If the supply of the antibody is sparse, one may consider an alternative method. Here a small drop (5–10 μl) of diluted antibody is applied on a piece of parafilm, and the coverslip is placed inverted onto the drop using forceps. If kept in a moist chamber the cells do not dry out. After incubation the coverslip is floated by carefully pipeting buffer under it, it is then removed using forceps.

Light-induced fading of the fluorescence signal can be inhibited by using special mounting media. We prepare the phenylenediamine-containing mounting medium (Platt and Michael, 1983) as follows: 10 mg of *p*-phenylendiamine (Sigma, cat. no. P-6001) are dissolved into 1 ml of distilled water. This can be speeded up by sonication. Then, 1 ml of PBS and 8 ml of glycerol are added, and the solution is mixed thoroughly. The resulting solution should be light brown. It is stored at −20°C (in the dark) and can be used within 1–2 months.

C. Immunoelectron Microscopy

In the following we describe a preembedding staining procedure for EM based on the use of saponin as a permeabilizing agent (Ohtsuki *et al.*, 1978; Tougard *et al.*, 1980; Hedman, 1980; for recent reviews, see Reggio *et al.*, 1983; Kerjaschki *et al.*, 1986; see also Brown and Farquhar, Chapter 25, Vol. 31). The general outline of the method is very similar to the immunofluorescence technique described previously. Different fixation and permeabilization conditions are used to ensure the preservation of the cell morphology. Also, incubation times and temperatures are different. Because of the similarity of the two techniques the conditions of the EM procedure can be tested first at the LM level. Since the diffusion of the peroxidase reaction product is an often-encountered problem, we have modified the general procedure by including a gelatin-embedding step and carrying out the reaction on ice (Saraste and Hedman, 1983; Kuismanen *et al.*, 1985). During the work, attention should be paid to the toxicity of some of the reagents.

1. The Procedure

1. Remove the culture medium and add the fixative (paraformaldehyde–glutaraldehyde in 0.1 *M* phosphate buffer supplemented with 0.05% saponin, pH 7.2). The fixative is added at the desired temperature, and incubattion is continued at 4°C. Fixation time is 30–60 minutes.
2. All subsequent steps (until step 7) are carried out at 4°C. Wash the cells three times with PBS containing 0.2% BSA and 0.05% saponin (DBS) to which 50 m*M* lysine-HCl has been added. Wash for a total of 60 minutes with several changes.
3. Wash once with DBS. Add antibody diluted in DBS, and incubate for 2 hours. Wash for 2 hours in DBS with several changes.
4. Add protein A–peroxidase conjugate diluted in DBS, and incubate for 2 hours.
5. Wash as in step 3 with DBS, followed by several rapid washes with PBS.
6. Postfix the cells for 30 minutes with 2.5% glutaraldehyde in 0.1 *M* sodium cacodylate (or phosphate) buffer, pH 7.2.
7. Wash with PBS containing 50 m*M* lysine-HCl for 30 minutes with three changes.
8. Wash with PBS at room temperature, three times, 5 minutes each.
9. Incubate for 30 minutes in PBS containing 2% gelatin at room temperature.
10. Transfer the plates one by one on a metal support placed on ice, and after ~30 seconds remove the gelatin solution by suction from one side of the plate. Wash the gelatin-embedded cells once with ice-cold PBS.
11. Add ice-cold peroxidase substrate solution [0.5 mg/ml diaminobenzidine (DAB), 0.02% H_2O_2 in PBS].
12. Stop reaction by washing three times with ice-cold PBS.
13. Fix the cells with 1% OsO_4 in 0.1 *M* phosphate buffer, pH 7.2, for 90 minutes on ice.
14. Wash three times with ice-cold PBS.
15. Dehydrate with ethanol and embed in Epon 812.

2. Remarks

To avoid excess use of (valuable) reagents we have routinely cultivated the cells for EM on 35-mm dishes. As mentioned earlier, immunofluorescence pilot experiments can be carried out to find the right fixation–permeabilization conditions for optimal detection. The glutaraldehyde

(0.005–0.1%) as well as saponin concentrations (0.005–0.2%) can be varied. Saponin is present in all solutions during the staining procedure. Some antigens will become undetectable in the presence of even low concentrations of glutaraldehyde. As the second antibody instead of protein A–peroxidase, Fab–peroxidase can be used. All these conjugates are commercially available. Different dilutions of the primary and secondary antibodies must be tested to obtain optimal reaction. The DAB substrate solution (step 11) should be filtered before use (e.g., use 0.22-μm Millipore filter) and the pH should be adjusted to 7.4. One should observe the peroxidase reaction carefully, preferably using an inverted microscope. Instead of using plain OsO_4, reduced OsO_4 may be used (step 13). For details of the effect of this fixative on the membrane contrast obtained, see, for example, Willingham and Rutherford (1984). The pH of the OsO_4 solution should be adjusted.

The gelatin–PBS solution is prepared by dissolving 2 of gelatin (reagent grade is sufficient) in 100 ml of PBS. The solution is centrifuged in a tabletop centrifuge at 3000 rpm for 10 minutes to remove undissolved aggregates. Tests can be carried out to find the right timing in carrying out the gelatin-embedding step (step 10). The purpose is to embed the cells in as thin a semisolid layer of gelatin as possible. The result can best be evaluated when the ice-cold PBS is pipeted slowly on the plate from which excess gelatin solution has been removed.

In processing the cells for EM (step 15; follow routine procedures), the propyleneoxide step during dehydration is omitted, since this reagent will dissolve the plastic. After removal of the last ethanol solution, the Epon should be added to the plates as quickly as possible to avoid drying of the cells. About 1–2 ml of the embedding medium per a 35-mm plate is sufficient to cover the surface completely. To obtain horizontal sections of the cells (see Fig. 3), five or six gelatin capsules filled to the rim with Epon are inverted on the plates. After polymerization the easiest way to remove the embedded cells from the dish is to submerge the plate partly in liquid nitrogen. The desired areas of sectioning can be selected and marked by using a needle and a stereomicroscope equipped with a high-magnification objective. If only a minority of the cells are reactive (e.g., after transfection or microinjection), those can now be detected. The marked capsules are removed using a jeweler's saw and trimmed for horizontal sectioning.

D. Reversion of the Transport Blocks

An important and useful property of the low-temperature-induced transport blocks is that they are readily reversible. Figure 4 shows that the SFV spike glycoproteins, accumulated at intracellular sites during in-

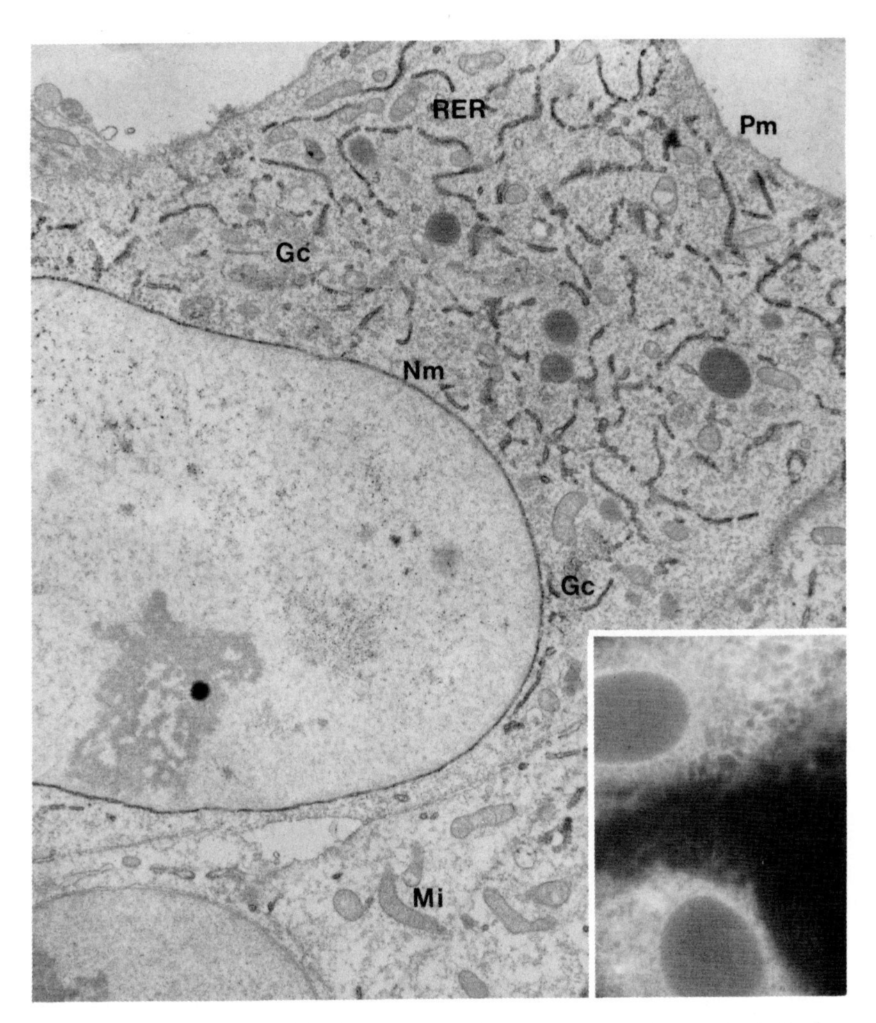

FIG. 3. Localization of SFV spike proteins by the immunoperoxidase method. The low-magnification micrograph shows mutant ts-1-infected cells grown for 4 hours at 39°C. Cells were fixed *in situ* and processed for EM without detaching them from the dish. In thin sectioning, the embedded cell layer was cut horizontally. This allows direct comparison with immunofluorescence results (inset). Peroxidase staining is restricted to the cisternae of the rough endoplasmic reticulum (RER) and the nuclear membrane (Nm), whereas elements of the Golgi complex (Gc) and the plasma membrane (PM) are not reactive. Mi, Mitochondria. × 7300 (inset × 920).

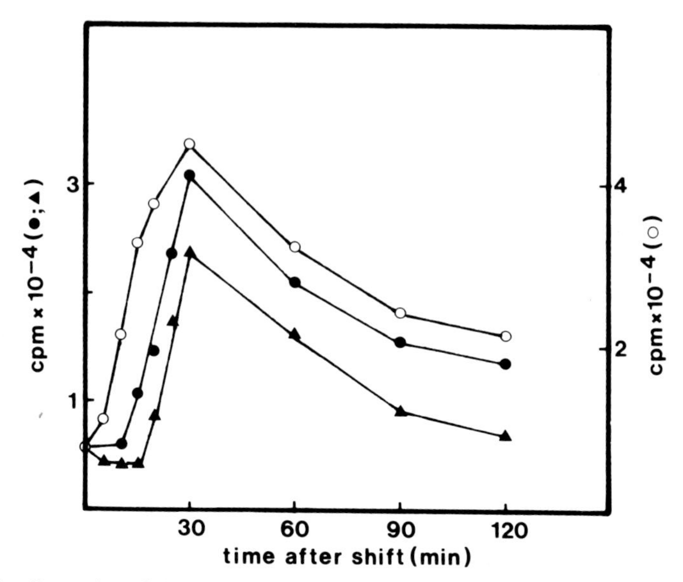

FIG. 4. Reversion of the 15°C and 20°C transport blocks. Virus infection was carried out as described for Fig. 1. At 4 hours the cells were shifted to 15°C and 20°C for 2 hours and then returned to the nonpermissive temperature (39°C) at which exit from the ER is inhibited. Duplicate plates were harvested at different times after reversion, and the amounts of spike glycoproteins transported to the cell surface were quantitated using a radioimmunoassay (see text for details). Note the differences observed in the kinetics of transport of the proteins to the cell surface after reversion from 20°C (○) and 15°C (●) as compared to the control curve (▲), which gives an estimate for transport from RER to the cell surface. In this experiment cells were shifted to the permissive temperature (28°C) for 2.5 minutes and then returned to 39°C. The decrease in the amount of surface-associated glycoproteins seen after 30 minutes is due to internalization. Adapted from Saraste and Kuismanen (1984) with permission of Cell Press.

cubation at 15°C or 20°C, are rapidly transported to the plasma membrane following the return of the cells to physiological temperature (see also Fig. 2D). We have previously made use of this reversibility and the immunoelectron-microscopic procedure described previously to study the movement of proteins between the 15°C block site (pre-Golgi elements) and trans-Golgi in order to obtain information of the transport mechanism that operates between these two compartments (Saraste and Kuismanen, 1984; see Fig. 5). Similarly, the reversible accumulation of proteins at the 20°C block site has been used to identify those vesicles that operate between the trans-Golgi compartment and the plasma membrane (Griffiths *et al.*, 1985).

Figures 6 and 7 summarize results from experiments in which the reversibility of the 20°C block was used to investigate the pathway of the

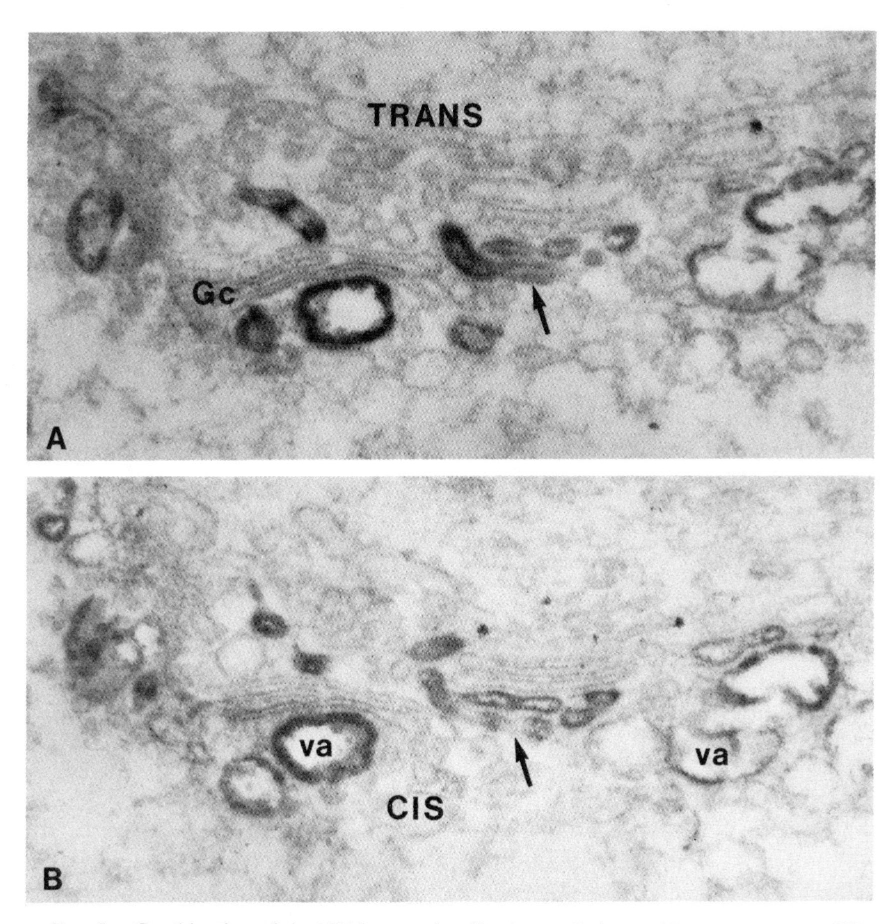

Fig. 5. Combination of the EM immunolocalization technique with a temperature shift to study the entry of SFV spike glycoproteins into the Golgi stack. (A) and (B) Consecutive serial sections. Cells infected with SFV ts-1 were shifted from 39°C to 15°C for 2 hours (see Fig. 2B) and then transferred to 20°C for 10 minutes. The viral proteins are localized to vacuoles (va) and tubular or lamellar elements (arrows) seen predominantly on the entry (cis) face of the cisternal stacks (Gc). Variable amounts of label are seen in the stack. The trans face of the Golgi is occupied by numerous unlabeled vesicles. × 60,000.

maturation of SFV. Two of the viral glycoproteins (E2 and E3) are synthesized as a precursor (p62), which is proteolytically processed in a late stage of transport (Green *et al.*, 1981). The mechanism involved is similar, if not identical, to the cleavage of prohormones (for a review see Simons and Warren, 1982). Cells infected with the wild-type strain of SFV were pulse-labeled for 10 minutes with [^{35}S]methionine at 37°C, and chased for

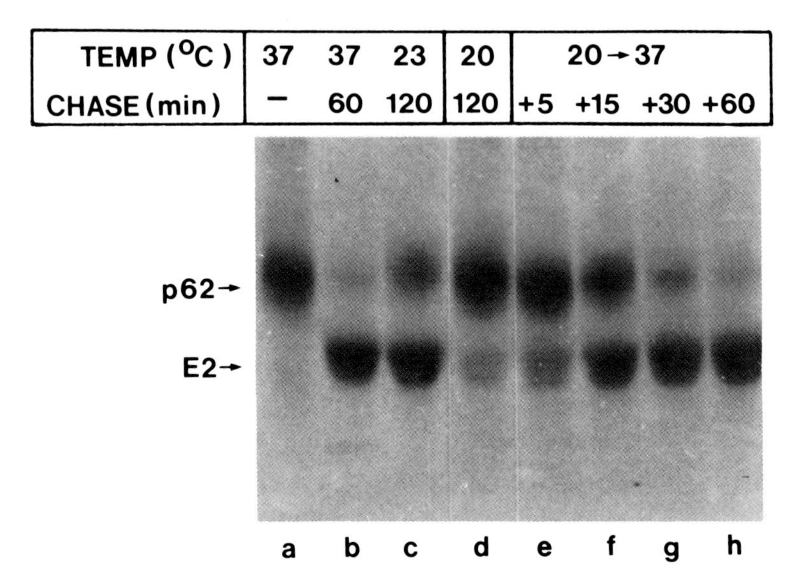

FIG. 6. Effect of temperature on the cleavage of the SFV spike protein precursor p62. See the text for experimental details. p62 and E2 (one of the cleavage products) were immunoprecipitated with anti-E2 antibodies after a 10-minute pulse (lane a) or after chase at different temperatures. The immunoprecipitates were analyzed by SDS–PAGE using a 10% gel. Shown is an exposure of a dried gel. Conversion of p62 to E2 occurs efficiently during chase at 37°C (lane b) and 23°C (lane c) but is almost completely inhibited at 20°C (lane d). Reversion from 20°C to 37°C results in rapid processing of the precursor (lanes e–h).

2 hours at 20°C to accumulate the proteins in the late-Golgi compartment (Matlin and Simons, 1983). Immunoprecipitation with specific antibodies and the analysis with Endo H showed that the virus spike proteins acquired resistance to the enzyme. Interestingly, the cleavage of the p62 precursor to E2 and E3 was almost completely inhibited at 20°C. At a slightly higher temperature (23°C), at which transport to the cell surface occurs (Fig. 1), processing was almost complete (Fig. 6, lanes c and d).

Following reversion of the cells from 20°C to 37°C, the cleavage of p62 took place with a half-time of ~15 minutes (Fig. 6, lanes e–h; Fig. 7A). In contrast, an ~30-minute lag phase was observed in the release of radiolabeled virus to the medium. The release of labeled virus occurred at maximal rate at ~60 minutes after shift from 20°C to 37°C (Fig. 7C). The synchronous release of the virus spike proteins from the 20°C block site allowed the dissection of late events in the maturation of SFV. These experiments provide support to the idea that the cleavage of p62 is an

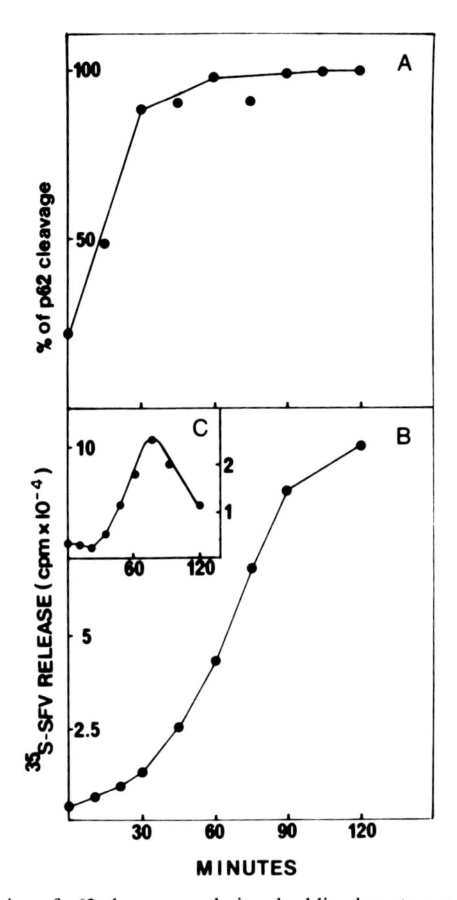

FIG. 7. Dissection of p62 cleavage and virus budding by a temperature-shift experiment. The experiment was carried out as described for Fig. 6 (lanes d–h). (A) Kinetics of p62 cleavage after reversion from 20°C to 37°C. p62 and E2 were immunoprecipitated as in Fig. 6. After exposure of the gel the bands were excised and the radioactivities were quantitated by scintillation counting. (B) and (C) Release of radiolabeled SFV from cells shifted from 20°C to 37°C. At different times after the temperature shift, the medium was harvested and cleared from cell debris by centrifugation at 10,000 *g* for 15 minutes. Radiolabeled virus was pelleted by centrifugation at 100,000 *g* for 60 minutes through a 30% sucrose cushion. Analysis of the pellets by SDS–PAGE was used to verify that they contained only radiolabeled viral proteins. (C) The data are plotted to show the rate of virus release after the temperature shift.

intracellular event that occurs in parallel with the transport from distal Golgi elements to the plasma membrane and that precedes virus budding.

IV. Mechanism of the Temperature Arrest

Although experiments by several investigators using different cell types have documented the existence and versatility of the low-temperature transport blocks, the mechanisms involved are still unknown. The inhibition of transport at reduced temperatures is not a result of energy depletion (Balch *et al.,* 1986; Tartakoff, 1986), and protein synthesis takes place even at 15°C (Craig, 1979). Also the ATP- and temperature-sensitive steps in transport appear to be different (Jamieson and Palade, 1968; Clarke and Weigel, 1985; Balch *et al.,* 1986; Doms *et al.,* 1987). If temperature reduction exerts a specific effect on vesicular transport, the mechanism could, for instance, involve any of the following processes: vesicle budding, vesicle movement, or fusion of the vesicle with the target membrane. However, the available information is insufficient to support any of these alternatives. It is known, in fact, that vesicle formation at the plasma membrane (see, e.g., Marsh *et al.,* 1983) and in the transitional regions of the ER (Saraste and Kuismanen, 1984; Saraste *et al.,* 1986; Tartakoff, 1986) continues at temperatures >10°C. Also, incubation of cells even at 15°C for longer periods of time does not appear to greatly affect the integrity of the microtubular system (E. Kuismanen and J. Saraste, unpublished results).

Two interesting aspects of the temperature blocks may help us to understand the underlying mechanism involved in temperature-induced transport arrest. During longer incubation at 20°C an increase in the net membrane surface of the trans-Golgi compartment is observed (see the discussion in de Curtis *et al.,* 1988). This happens in parallel with a progressive leakiness of the block (Saraste and Kuismanen, 1984; Griffiths *et al.,* 1985; Fries and Lindstrom, 1986). Also, in the case of the 15°C temperature block, an apparent "piling up" of pre-Golgi membranes is observed (Saraste and Kuismanen, 1984; Saraste *et al.,* 1986; J. Saraste and E. Kuismanen, unpublished results). Since the most obvious consequence of temperature reduction is the slowing down of processes, one can expect that a cumulative effect is observed at special sites where multiple processes converge. In the case of membrane traffic we know, for instance, that the trans-Golgi compartment is a location where multiple vesicular transport pathways come together (Griffiths *et al.,* 1985; Taatjes and Roth, 1986; Orci *et al.,* 1987; Van Deurs *et al.,* 1987). Temperature reduction could preferentially affect compartments that represent "crossroads" sites in the network of membrane traffic pathways.

V. Perspectives

It has become apparent that the reversible temperature-induced transport blocks provide valuable tools for the morphological and biochemical analysis of membrane traffic. Distinct threshold temperatures exist for different steps in transport pathways, but temperature reduction can also be used to slow down events between different sites and thereby to increase resolution. Low-temperature incubation can be used in combination with cell fractionation (see, for example, Fries and Lindstrom, 1986; Saraste *et al.,* 1986; Tartakoff, 1986), to provide kinetic markers for organelle isolation (de Curtis *et al.,* 1988), or to dissect different steps in protein modification (see, e.g., Bauerle and Huttner, 1987; Pelham, 1988). It seems that the potential of this type of experimental approach is far from exhausted.

Acknowledgments

J. S. is grateful to Hannele Ruohola for helpful discussions. This work was partly supported by European Molecular Biology Organization Long Term Fellowships ALTF 317–1984 (to E. K.) and ALTF 270-1983 (to J. S.).

References

Bauerle, P. A., and Huttner, W. B. (1987). *J. Cell Biol.* **105,** 2655–2664.
Balch, W. E., Elliot, M. M., and Keller, D. S. (1986). *J. Biol. Chem.* **261,** 14681–14689.
Chu, L. L. H., MacGregor, R. R., and Cohn, D. V. (1977). **J. Cell Biol. 72,** 1–10.
Clarke, B. L., and Weigel, P. H. (1985). *J. Biol. Chem.* **260,** 128–133.
Craig, N. (1979). *J. Cell. Physiol.* **100,** 323–334.
de Curtis, I., Howell, K. E., and Simons, K. (1988). *Exp. Cell Res.* **175,** 248–265.
Doms, R. W., Keller, D. S., Helenius, A., and Balch, W. E. (1987). *J. Cell Biol.* **105,** 1957–1969.
Dunn, W. A., Hubbard, A. L., and Aronson, N., Jr. (1980). *J. Biol. Chem.* **12,** 5971–5978.
Fries, E., and Lindstrom, I. (1986). *Biochem. J.* **237,** 33–39.
Goldenthal, K. L., Hedman, K., Chen, J. W., August, J. T., and Willingham, M. C. (1985). *J. Histochem. Cytochem.* **33,** 813–820.
Green, J., Griffiths, G., Louvard, D., Quinn, P., and Warren, G. (1981). *J. Mol. Biol.* **152,** 663–698.
Griffiths, G., and Simons, K. (1986). *Science* **234,** 438–443.
Griffiths, G., Pfeiffer, S., Simons, K., and Matlin, K. (1985). *J. Cell Biol.* **101,** 949–964.
Hedman, K. (1980). *J. Histochem. Cytochem.* **28,** 1233–1241.
Jamieson, J. D., and Palade, G. E. (1968). *J. Cell Biol.* **39,** 589–603.

Kääriäinen, L., Virtanen, I., Saraste, J., and Keränen, S. (1983). *In* "Methods in Enzymology" (S. Fleischer and B. Fleischer, eds.). Vol. 96, pp. 453–465. Academic Press, New York.
Keller, G.-E., Glass, C., Louvard, D., Steinberg, D., and Singer, S. J. (1986). *J. Histochem. Cytochem.* **34,** 1223–1230.
Kerjaschki, D., Sawada, H., and Farquhar, M. G. (1986). *Kidney Int.* **30,** 229–245.
Kuismanen, E., Saraste, J., and Pettersson, R. (1985). *J. Virol.* **55,** 813–822.
Lagunoff, D., and Wan, H. (1974). *J. Cell Biol.* **61,** 809–811.
Laurila, P., Virtanen, I., Wartiovaara, J., and Stenman, S. (1978). *J. Histochem. Cytochem.* **26,** 251–258.
Marsh, M., Bolzau, E., and Helenius, A. (1983). *Cell (Cambridge, Mass.)* **32,** 931–940.
Matlin, K., and Simons, K. (1983). *Cell (Cambridge, Mass.)* **34,** 233–243.
Ohtsuki, I., Manzi, R. M., Palade, G. E., and Jamieson, J. D. (1978). *Biol. Cell.* **31,** 119–126.
Orci, L., Ravazzola, M., Amherdt, M., Perrelet, A., Powell, S. K., Quinn, D. L., and Moore, H. H. (1987). *Cell (Cambridge, Mass.)* **51,** 1039–1051.
Pelham, H. R. B. (1988). *EMBO J.* **7,** 913–918.
Platt, J. L., and Michael, A. F. (1983). *J. Histochem. Cytochem.* **31,** 840–842.
Reggio, H., Webster, P., and Louvard, D. (1983). *In* "Methods in Enzymology" (S. Fleischer and B. Fleischer, eds.), Vol. 98, pp. 379–395. Academic Press, New York.
Rotundo, R. L., and Fambrough, D. M. (1980). *Cell (Cambridge, Mass.)* **22,** 595–602.
Sandvig, K., and Olsnes, S. (1979). *Exp. Cell Res.* **121,** 15–25.
Saraste, J., and Hedman, K. (1983). *EMBO J.* **2,** 2001–2006.
Saraste, J., and Kuismanen, E. (1984). *Cell (Cambridge, Mass.)* **38,** 535–549.
Saraste, J., Palade, G. E., and Farquhar, M. G. (1986). *Proc. Natl. Acad. Sci. U.S.A.* **83,** 6425–6429.
Simons, K., and Warren, G. (1982). *Adv. Protein Chem.* **36,** 79–132.
Steinman, R. M., Silver, J. M., and Cohn, Z. A. (1974). *J. Cell Biol.* **63,** 949–969.
Taatjes, D. J., and Roth, J. (1986). *Eur. J. Cell Biol.* **42,** 344–350.
Tartakoff, A. M. (1986). *EMBO J.* **5,** 1477–1482.
Tougard, C., Picart, R., and Tixier-Vidal, A. (1980). *Am. J. Anat.* **158,** 471–490.
Van Deurs, B., Petersen, O. W., Olsnes, S., and Sandvig, K. (1987). *Exp. Cell Res.* **171,** 137–152.
Wang, K., Feramisco, J. R., and Ash, J. F. (1982). *In* "Methods in Enzymology" (D. W. Frederiksen and L. W. Cunningham, eds.), Vol. 85, pp. 515–562. Academic Press, New York.
Wassler, M., Jonasson, I., Persson, R., and Fries, E. (1987). *Biochem. J.* **247,** 407–415.
Willingham, M. C., and Rutherford, A. V. (1984). *J. Histochem. Cytochem.* **32,** 455–460.

Part II. Monitoring and Regulating the Progress of Transport—Endocytic and Transcytotic Path

As in the preceding section, a number of cell preparations have been of major importance for studies of endocytosis, and certain of these have been well described in previous methodology volumes (Dunn *et al.*, 1983; Goldstein *et al.*, 1983; Herzog, 1983; Marsh *et al.*, 1983). One of the basic issues that these systems are ideal for addressing is the nature of the covalent determinants of a surface protein that cause it to undergo endocytosis, recycling, or degradation. The extraordinary extent and rapidity of recycling of cell surface proteins other than those that bind metabolically important ligands raise unanswered questions with regard to the physiological importance of cyclic transport.

The chapters included here provide a broad methodological overview of procedures for ligand radiolabeling and quantitative analysis of ligand–receptor interactions (Owensby *et al.*, Chapter 12), a description of methods for photoaffinity labeling and ligand–receptor crosslinking and derivatization (Ji and Ji, Chapter 11), and methods for inferring the course of transport by analysis of glycoprotein glycan remodeling (Snider, Chapter 14). Chapter 13 on IgA transport by Breitfeld *et al.* should be read in conjunction with Chapter 2 on cell polarity by Rodriguez-Boulan *et al.* in Part I. Volumes 29 and 30 as well as Chapter 16 by Wilson and Murphy (Vol. 31) should be consulted for discussion of the use of endocytic tracers for pH determination. Chapter 19 by Leserman *et al.* in Part III describes novel procedures for exploiting endocytosis of surface-bound liposomes.

Electron-microscopic procedures for studying endocytosis have recently been reviewed (Herzog and Farquhar, 1983). A variety of methods for studying endocytosis and phagocytosis have also been presented in Volumes 96 and 132 of *Methods in Enzymology*.

References

Dunn, W., Wall, D., and Hubbard, A. (1983). Use of isolated, perfused liver in studies of receptor-mediated endocytosis. *In* "Methods in Enzymology" (S. Fleischer and B. Fleischer, eds.), Vol. 96, pp. 225–240. Academic Press, New York.

Goldstein, J., Basu, S., and Brown, M. (1983). Receptor-mediated endocytosis of low density lipoprotein in cultured cells. *In* "Methods in Enzymology" (S. Fleischer and B. Fleischer, eds.), Vol. 96, pp. 241–259. Academic Press, New York.

Herzog, V. (1983). Preparation of inside-out thyroid follicles for studies of transcellular transport. *In* "Methods in Enzymology" (S. Fleischer and B. Fleischer, eds.), Vol. 96, pp. 447–457. Academic Press, New York.

Herzog, V., and Farquhar, M. (1983). Use of electron-opaque tracers for studies on endocytosis and membrane recycling. *In* "Methods in Enzymology" (S. Fleischer and B. Fleischer, eds.), Vol. 96, pp. 203–224. Academic Press, New York.

Marsh, M., Helenius, A., Matlin, K., and Simons, K. (1983). Binding, endocytosis and degradation of enveloped animal viruses. *In* "Methods in Enzymology" (S. Fleischer and B. Fleischer, eds.), Vol. 96, pp. 260, 265. Academic Press, New York.

Chapter 11

Affinity Labeling of Binding Proteins for the Study of Endocytic Pathways

TAE H. JI, RYUICHIRO NISHIMURA, AND INHAE JI

Department of Molecular Biology
University of Wyoming
Laramie, Wyoming 82071

Copyright © 1989 by Academic Press, Inc.
All rights of reproduction in any form reserved.

I. Introduction

The goal of affinity labeling is formation of covalent bonds between a ligand of interest and its binding site, so that the binding site can be characterized using such easy methods as electrophoresis, without an expensive and difficult purification. There are two methods to achieve this goal: macromolecular-photoaffinity labeling and affinity crosslinking (Ji, 1983).

In macromolecular-photoaffinity labeling, one begins with a macromolecular ligand, which ordinarily binds reversibly with a biologically relevant site. The basic approach is as follows (Fig. 1): Macromolecular ligand such as peptide hormone is radioactively labeled and derivatized with photoactivable reagents in the dark, and the hormone derivative is allowed to bind to its receptor. Subsequently, the hormone–receptor complex can be covalently linked by photolysis. This will enable one to identify and to examine the resultant radioactive crosslinked complex.

Macromolecular-photoaffinity labeling utilizes heterobifunctional reagents that contain two dissimilar reactive functional groups, frequently one photosensitive (e.g., azide) and one conventional [e.g., *N*-hydroxysuccinimide (NHS) ester and imidate]. By taking advantage of the differential reactivities of these two functional groups, two different molecules can be covalently linked to produce specifically a particular complex both selectively and sequentially.

In affinity crosslinking, radioactively labeled ligands are allowed to bind to the receptor, and the resultant ligand–receptor complex, along with all other protein molecules present in the system, is subjected to

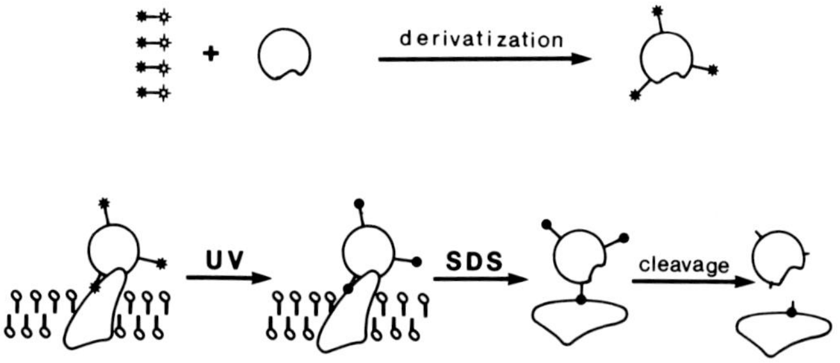

FIG. 1. Macromolecular photoaffinity labeling. A macromolecular ligand such as peptide hormone is derivatized with photoactivable reagents, and the hormone derivative is allowed to bind to its receptor. Subsequently, the hormone–receptor complex can be covalently linked by photolysis. After solubilization, the crosslinked ligand–receptor complex can be analyzed or cleaved and analyzed.

covalent crosslinks by introduction of homobifunctional reagents that are often called crosslinkers.

Macromolecular-photoaffinity labeling, which was initially developed to identify concanavalin A (Con A) receptors (Ji, 1976, 1977), has been extended to the crosslinking of various polypeptide ligands to their specific binding sites; examples include studies of epidermal growth factor (Das *et al.*, 1977), insulin (Yip *et al.*, 1978; Jacobs *et al.*, 1979; Hofmann *et al.*, 1981), fibronectin (Perkins *et al.*, 1979), bungarotoxin (Witzemann *et al.*, 1979; Nathanson and Hall, 1980), lutropin (Ji and Ji, 1980, 1981; Ji *et al.*, 1981), thrombin (Carney *et al.*, 1979), antigloboside antibody (Lingwood *et al.*, 1980), calmodulin (Andreasen *et al.*, 1981), glucagon (Johnson *et al.*, 1981), multiplication-stimulating activity (Massague *et al.*, 1981a), nerve growth factor (Massague *et al.*, 1981b), somatomedin (Bhaumick *et al.*, 1981), parathyroid hormone (Coltrera *et al.*, 1981), follitropib (Shin and Ji, 1985), interleukin-3 (Sorenson *et al.*, 1987), melanotropin (Scimonelli and Eberle, 1987), and vasoactive intestinal peptide (Robichon *et al.*, 1987).

The advantages of using photoactivable reagents are that the crosslinking reaction can be simply controlled by photolysis, and that the reactive intermediates formed during photolysis react nonspecifically with a wide variety of chemical bonds, including the ubiquitous C—H and O—H bonds. These reagents are especially convenient when rapid crosslinking is desirable, either to minimize nonspecific crosslinking upon accidental collisions between molecules or to investigate dynamic aspects of transitory polypeptide binding to binding sites (Kiehm and Ji, 1977).

The maximum distance between two reactive groups of a reagent obviously has a significant effect on the success of crosslink formation. For example, a bisimidate shorter than 5 Å usually yields few or no crosslinks, whereas extensive crosslinking can frequently be achieved when the length is 11–22 Å (Ji, 1974; Aizawa *et al.*, 1977). Beyond this range, greater length may not necessarily be advantageous (Aizawa *et al.*, 1977). Experiments in which the length of a homologous series of crosslinking reagents is varied can in principle be used to determine the distance between neighboring components, but the intrinsic flexibility (dynamic structure) of most reagents as well as their macromolecular targets make this difficult to accomplish.

Radioactive bifunctional reagents facilitate the identification of crosslinked products and reacted residues as well as help in the elucidation of precursor–product relationships. A limited number of radioactively labeled reagents (Hartman and Wold, 1966; Trommer *et al.*, 1977) have been synthesized. Many heterobifunctional reagents containing ^{125}I can be prepared by simple radioiodination of the parent reagents immediately prior to use (Ji and Ji, 1985).

II. Photoaffinity Labeling

A. Advantages and Disadvantages of Photoaffinity Labeling and Affinity Crosslinking

The primary advantage of photoaffinity labeling over affinity crosslinking is that formation of a covalently crosslinked complex is limited to the ligand and the receptor in photoaffinity labeling, whereas in affinity crosslinking formation of crosslinked complexes is random, causing undesirable crosslinks between any protein molecules present in samples. As a result, photoaffinity labeling will allow one to identify the receptor directly and to follow biological pathways of crosslinked ligand–receptor complexes (Ji, 1983) if cells are alive and the system is still active. In contrast, it is not easy to characterize the receptor alone in affinity-crosslinking studies, particularly in a system of multicomponents such as cell membranes and the cytosol. In addition, it is extremely difficult, if not impossible, to study the biological pathways of affinity-labeled ligand–receptor complexes.

In affinity crosslinking, every molecule in samples is subjected to crosslinking, and disruptions to the entire biological system are suspected. Therefore, studies on biological pathways of crosslinked complexes may not be credible. One major advantage of affinity crosslinking over photoaffinity labeling is a higher yield of crosslinked ligand–receptor complexes. Biological pathways of ligand–receptor complexes, which are not crosslinked, can be examined by crosslinking them at different steps.

B. Radioactively Labeled Photosensitive Reagents

Radioactively labeled reagents are normally prepared in two ways. The first and most convenient way is to label cold reagents radioactively just before coupling to ligands. An example is radioiodinatable 4-azidosalicylyl derivatives (I. Ji and Ji, 1981; T. H. Ji and Ji, 1985). The other method is to label radioactively an intermediate, not the final product, of a reagent during the synthesis. Since the processing of radioactive intermediates is always difficult, this method is less desirable. On the other hand, this may be the only way to label these reagents radioactively, because it is difficult to label the final products radioactively. This type consists of ^{3}H-, ^{35}S-, or ^{125}I-labeled reagents, some of which are commercially available. A major problem with these radioactive photoactivable reagents can be the low specificity of radioactivities. The efficiency of radioiodination has been improved from 3% to >63% (Ji and Ji, 1985), and the procedure will be described.

C. Radioiodination of Photoactivable Reagents, Derivatives of the *N*-Hydroxysuccinimide (Succinimidyl) Ester of 4-Azidosalicylic Acid (NHS–ASA)

1. Preparation

Acetonitrile; ethylacetate; dimethyl sulfoxide (DMSO); 0.5 *M* dibasic sodium phosphate (pH 7.0), adjusted with phosphoric acid; 10% sodium chloride; 2.5 nmol chloramide-T in 10 μl of acetonitrile and dimethylformamide (DMF) (9 : 1, v/v); dry nitrogen gas; reagent.

2. Procedure

1. Dissolve 167 nmol of reagent in 150 μl of acetonitrile.
2. Add 5 μl of sodium phosphate to 15 μl of the reagent solution and mix.
3. Quickly add 5 mCi of carrier-free [^{125}I]sodium iodide in 10 μl of 0.1 *M* sodium hydroxide.
4. Quickly add 2.5 nmol chloramine-T in 10 μl of a mixture of acetonitrile and DMF (9 : 1, v/v).
5. Mix and incubate for 2 minutes at 25°C.
6. Add 100 μl of ethylacetate and then 100 μl of 10% sodium chloride, and mix.
7. Collect the ethylacetate layer.
8. Add 100 μl of 10% sodium chloride to the ethylacetate layer and mix.
9. Collect the ethylacetate layer.
10. Dry the ethylacetate layer under reduced pressure of nitrogen gas.
11. Redissolve the dried reagent in 2 μl of DMSO.
12. This reagent solution can be used for derivatization of peptides possessing free amino groups.

3. Solutions to Problems

The minimum concentration of NHS derivatives in the final reaction mixture containing ligands and the reagents is 50 μ*M*, and the reagent concentration has to be >10-fold of the concentration of free amino groups in ligands. These two conditions have to be met for successful photoaffinity labeling. When the reagent concentration is >1 m*M*, derivatized ligands often lose the binding activity. When it is difficult to estimate the number of free amino groups of ligands, use the weight of ligands as suggested in the instruction, which was prepared on the basis of the average number of amino groups of proteins.

The problem most frequently encountered in handling heterobifunctional as well as homobifunctional reagents is the instability of reagents in aqueous solution. Particularly, those reagents possessing imidate and succinimide (succinimidyl) groups can undergo rapid hydrolysis within a minute; therefore, the exposure to aqueous solutions should be limited to a minimum.

Arylazides, the most frequently used photosensitive group, are unstable in solutions containing reducing agents such as 2-mercaptoethanol (2-ME) and dithiothreitol (DTT).

Radioiodination requires a well-ventilated hood and protective shields. Contamination by ^{125}I during transfer can be prevented or reduced to a minimum by employing cotton-plugged Pasteur pipets with capillary tips. The ^{125}I solution can be easily picked up by capillary action and transferred without contamination.

D. Derivatization of Peptide Ligands

1. Preparation

Sodium phosphate, pH 7.4; gel permeation column for purification of derivatized ligands.

2. Procedure

1. Dissolve 7 μg of peptide ligands possessing free amino groups in 13 μl of 0.1 *M* sodium phosphate.
2. Add 2 μl of the radioiodinated reagent solution described in Section II,C, and mix.
3. Incubate for 15 minutes at 25°C.
4. Purify derivatized peptide ligands by column chromatography or other methods.

3. Solutions to Problems

It is not necessary to work in the dark, because most of photosensitive reagents of arylazide derivatives undergo very slow photodecomposition in aqueous buffers under white fluorescent light, with a decay rate of <0.7% per hour. Direct and indirect sunlight, however, cause a significant amount of decomposition (Ji, 1977).

E. Derivatization with Nonradioiodinatable Photosensitive Reagent and Radioiodination of the Peptide Derivative

When radioiodinatable photosensitive reagents are not successful, ligands can be derivatized with nonradioiodinatable photosensitive reagent such as the family of NHS–4-azidobenzoic acid; the ligands are radioiodinated and then used for photoaffinity labeling. The length of reagents has a significant effect on the potency of the derivatized ligands and the success of crosslink formation. Among the family of NHS–4-azidobenzoic acid (NHS–AB), NHS–4-azidobenzoyl glycine (NHS–ABG), and NHS–4-azidobenzoylglycyl glycine (NHS–ABGG), NHS–ABG was best and NHS–AB was worst in preserving the biological activity of the derivatized ligands (Ji *et al.*, 1981).

1. Preparation

Phosphate-buffered saline (PBS): 0.9% sodium chloride and 10 m*M* sodium phosphate, pH 7.4; 0.1 *M* sodium phosphate, pH 7.5; chloramine-T, 0.3 mg/ml of PBS; sodium metabisulfite, 0.66 mg/ml of PBS; 50 m*M* NHS–ABG in DMSO; column to isolate peptide derivatives.

2. Procedure

1. Dissolve 10 μg of peptide ligand in 40 μl of 0.1 *M* sodium phosphate.
2. Dilute 50 m*M* NHS–ABG in DMSO with 0.1 *M* sodium phosphate to make a desired concentration, normally 1 m*M*, of the reagent. Because of the instability of the reagent in aqueous solution, the reagent has to be used immediately.
3. Promptly add 10 μl of the diluted reagent to the peptide ligand and mix.
4. Incubate for 15 minutes at 25°C with occasional shaking.
5. Add 1 mCi [^{125}I]sodium iodide in 10 μl 0.1 *M* sodium hydroxide.
6. Add 7 μl chloramine-T solution and mix.

7. Incubate for 20–60 seconds at 25°C.
8. Add 7 μl sodium metabisulfite and mix.
9. Isolate the derivatized and radioiodinated ligand by column chromatography or by other methods.

F. Photolysis Conditions

A common method used to photolyze azides is irradiation with a short-wavelength ultraviolet (UV) lamp such as the Mineralight USV-11. The half-time of photolysis with this source varies with the reagent employed but is on the order of 10–20 seconds in aqueous solution (Ji, 1977). Because of quenching by other absorbing groups, the rate of photolysis is substantially reduced. When photoactivable reagents are mixed with proteins or cell membranes, complete photolysis requires 3–5 minutes of irradiation from a UV lamp located 15 cm from a sample (Ji, 1983). An alternative method with several advantages is flash photolysis for an extremely short period, normally on the order of milliseconds. Regular flash source units are expensive and require a high-voltage (several thousand volts) power supply. It has been found that inexpensive electronic flash units made for cameras discharge an intense flash of milliseconds' duration in the UV and visible range (Ji, 1983; Kiehm and Ji, 1977). The intensity of xenon flash is somewhat variable, but normally the flashes are capable of photolysing ~20–40% of 10^{-4}–10^{-5} *M* arylazides in aqueous buffer (Ji, 1979). In a typical experiment, 1–10 flashes suffice to photolyse reagent molecules associated with the erythrocyte membrane (Middaugh and Ji, 1980). The xenon lamps in electronic flash units for cameras produced in the past 10 years are coated with UV-absorbing film and may not efficiently photolyse reagents. However, xenon lamps that do not absorb UV are commercially available.

The absorption maxima of arylazides are in the range 265–275 nm, with molar-extinction coefficients of ~2×10^4mol/cm (Ji, 1983; Vanin and Ji, 1981; Bayley and Knowles, 1983). This absorption band is probably the one responsible for the activation of arylazide by photolysis. Other absorption bands at longer wavelengths (300–460 nm) appear when the phenyl group is substituted with nitro or hydroxyl groups (Ji and Ji, 1985; Huang and Richards, 1977). An attractive feature of these modified azides is that they can be photolysed with long-wavelength UV or visible light with a resultant minimization of radiation damage to the target (Mas *et al.*, 1980; Staros and Richards, 1974). In these cases, however, irradiation times must be on the order of minutes (Mas *et al.*, 1980; Staros and Richards, 1974) or even hours (Fleet *et al.*, 1972). The need for these longer irradiation periods is due to the relatively low molar-extinction coefficients (4000–5000 mol/cm) of the substituted arylazides (Ji, 1979; Staros and Richards, 1974). Such long irradiation times are undesirable because of the possibility of photodamage to sensitive biological structures. In addition, long irradiations can heat a sample. For example, irradiation with an argon laser (488 nm) for only 10 minutes with 1 W of power raises the sample temperature. Despite the significant increase in temperature produced by this procedure, little photolysis resulted (Matheson *et al.*, 1977). Although a comparative systematic study has not yet been attempted to determine whether sustained irradiation at a long wavelength or short pulses at reduced wavelengths is more effective in minimizing damage to biological materials, some data are available. Xenon flashes from electronic camera flash units do not cause any change in the absorption spectrum of Con A or in the activities of several erythrocyte membrane enzymes (Kiehm and Ji, 1977). When RNase A is subjected to a flash from a 7L6 xenon flash tube at 2 kV (which is expected to be considerably more intense than a flash from an electronic camera flash unit), the antigenic activity remains intact with enzyme activity 90% of the original. The circular dichroism spectrum >235 nm does, however, show slight changes (Matheson *et al.*, 1977). Flash photolysis generally produces a better defined crosslinking pattern on polyacrylamide gels than does irradiation with a UV lamp. A special device has been developed to prevent heating of samples during irradiation (Staros and Richards, 1974).

G. Photolysis

Samples can be irradiated from the top or bottom. In either case, a UV-transparent container is desirable. However, normal glass containers can be used when samples are irradiated from the top. Since UV can damage nucleic acids and proteins, and could kill cells, a shorter exposure is desirable. A larger surface area and shallower depth of sample solutions are helpful. Normally sample solution >1 cm deeper can be detrimental. Samples in spot plates are efficiently photolysed. Higher concentrations of sample materials or cells will reduce the efficiency of photolysis. Keep samples on ice if heating during UV irradiation becomes a problem.

Two types of UV sources are in common use: short-wave UV lamps and xenon flash lamps. With a short-wave UV lamp, which has a maximum at 254 nm, irradiation from a distance <10 cm for 0.5–5 minutes is sufficient. Flash photolysis with a xenon flash lamp, <3 cm away from a sample, requires repeated flashes for the extensive photoactivation of reagents (Kiehm and Ji, 1977).

Advantages of using xenon flash lamps are short duration of flash (<0.2 milliseconds; Kiehm and Ji, 1977) and less heating.

H. Cleavable Bonds

The use of cleavable reagents permits the reversal of crosslinking. Cleavage of crosslinks can be employed as a control to demonstrate true crosslink formation, to establish precursor–product relationships, to distinguish the effects of crosslinking from those of simple chemical modifications of reactive groups, and to establish the position of individual crosslinks. A number of cleavable groups have been incorporated into crosslinking reagents for this purpose, including ester (-C[=O]—O-) bridge, disulfide (-S-S-), glycol (-CH[OH]-CH[OH]-), and azo (-N=N-).

1. Disulfide Bonds

Cleavage of disulfide bonds can be achieved with remarkable efficiency under mild conditions, such as reaction with 2-ME, DTT, or dithioerythritol and concentrations of 10–100 m*M* reducing agent between pH 7 and 9 at 25°–37°C for 5–30 minutes. Common buffers such as Tris and phosphate, and detergents such as sodium dodecyl sulfate (SDS) and Triton X-100, do not interfere with this type of cleavage. By addition of a reducing agent to the electrophoretic buffer, complexes crosslinked with reagents containing a disulfide bond can also conveniently be cleaved during electrophoresis. Occasionally acrylamide gels containing the reducing agent have also been employed, but the presence of a sulfhydryl reagent interferes with gel polymerization.

There are several disadvantages in the use of a reagent that contains a disulfide bond. They are susceptible to disulfide exchange, with the resulting possibilities of denaturation and crosslinking of noninteracting molecules, their use precludes the application of reducing agents during the solubilization of membranes and isolation of crosslinked complexes, and they cannot be used in a system that is sensitive to oxidation and would normally be kept under reducing conditions. Disulfide exchange and complete reduction usually require substantial amounts of free sulfhydryl groups (Liu, 1977). Kinetic studies have shown that disulfide exchange rates are proportional to the concentration of the deprotonated form (RS-) rather than the protonated form (RSH) (Fava *et al.*, 1957; Eldjarn and Pihl, 1957). Therefore, disulfide exchange reactions can be decreased by lowering the pH of the reactions below the pK of sulfhydryl groups (~8.5). To avoid the problems involved with disulfide reagents, other cleavable crosslinking reagents have been introduced.

2. Ester Bridges

Esters are cleaved either in alkaline solution or by reaction in 1 *M* hydroxylamine. They can be cleaved in 100 m*M* sodium phosphate adjusted to pH 11.6 with Tris, 6 *M* urea, 0.1% SDS, and 2 m*M* DTT for 2 hours at 37°C. The presence of DTT is not necessary for the cleavage, which occurs almost quantitatively.

Alternatively, esters can be cleaved by reaction with 1 *M* hydroxylamine, 50 m*M* Tris (pH 7.5–8.5), 25 m*M* $CaCl_2$, and 1 m*M* benzamidine for 3–6 hours at 25°–37°C (Abdella *et al.*, 1979). It is theoretically possible to hydrolyze esters under both acidic and alkaline conditions. The rate of ester hydrolysis is controlled by the rate of formation of an intermediate with a tetrahedral carbon atom formed by a carbonyl group and water to hydroxide ion. If the carbonyl group is surrounded by bulky groups, hydrolysis may be relatively slow.

3. Glycol Bridges

A glycol bridge can be cleaved simply by soaking gels in 15 m*M* sodium periodate at pH 7.5 for 4–5 hours at 25°C. Sodium dodecyl sulfate and buffers such as triethanolamine and sodium phosphate do not interfere with the cleavage, but Tris reacts with sodium periodate and cannot be used. Disadvantages encountered with the use of glycol reagents are reduced rates of cleavage relative to that obtained with disulfides, difficulty in obtaining complete cleavage, and lack of cleavage specificity. Carbohydrate portions of glycoproteins containing carboxyl, amine, or ketone groups next to hydroxyls can also be split (Barker, 1971). The reactivity of aldehydes produced by periodate oxidation and the possibility of Schiff base formation with protein amino groups present a further complication.

4. Azo Bridges

Azo bridges can be cleaved by reduction with 0.1 *M* sodium dithionite in 0.15 *M* NaCl, buffered in pH 8.0 with 0.1 *M* $NaHCO_3$ for 25 minutes (Jaffe *et al.*, 1980). Unlike the cleaving conditions for glycol bridges, carbohydrates will not be destroyed during this reduction.

I. Cleavage of Crosslinks

Cleavage of two popular cleavable bonds, esters and disulfides, will be described. Depending on needs, cleaved products are analyzed in three different ways. The simplest method is to crosslink samples, solubilize, cleave, and electrophorese them (Fig. 2). This indiscriminate cleavage, prior to separation of crosslinked complexes, can be used to demonstrate formation of crosslinks, to identify crosslinked complexes, and, sometimes, to determine precursor–product relationships in crosslinking.

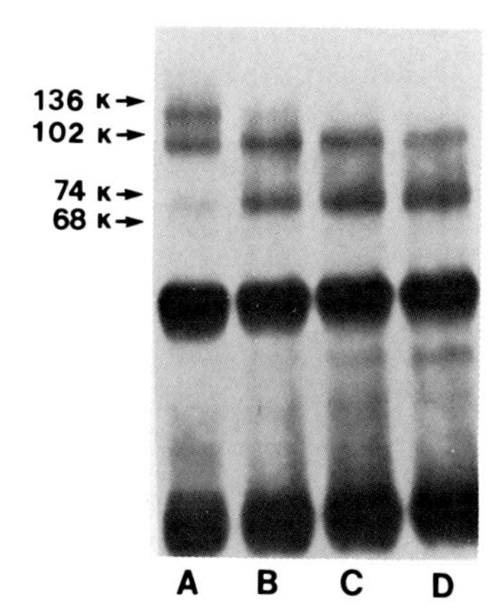

FIG. 2. One-dimensional analysis of cleaved components of crosslinked complexes. Crosslinked complexes (lane A) are treated for increasing extents of cleavage (from B to D) and electrophoresed. Note that smaller components are released.

In the second approach, samples are crosslinked and electrophoresed. After the identification of crosslinked complexes on a gel, bands of crosslinked complexes are cut, incubated in a solution containing cleaving agents (cleaving solution), placed on top of fresh gel, and electrophoresed. In this one-dimensional analysis of cleaved components, each of crosslinked complex bands can be examined separately for the composition. The excision of crosslinked complex bands, however, may not be free of contamination and cleaved components could be contaminated.

In the third approach, crosslinked samples are electrophoresed. From the first-dimensional gel, an entire lane is excised, incubated in the cleaving solution, placed on top of fresh slab gel, and electrophoresed. In this two-dimensional analysis of cleaved components, all of the crosslinked complexes can be analyzed simultaneously (Fig. 3). Since this type of cleavage produces the most useful information, it will be described here.

1. Cleavage of Esters in Complexes Crosslinked by Reagents Containing Esters

a. *Procedure*

1. Excise the first-dimension gel lane carefully so that the cut surfaces are straight and smooth. Otherwise air bubbles will be trapped between the first-dimension gel and top of the second-dimension gel. These air bubbles produce an uneven interface, which in turn will distort the resolution of two-dimensional analysis.

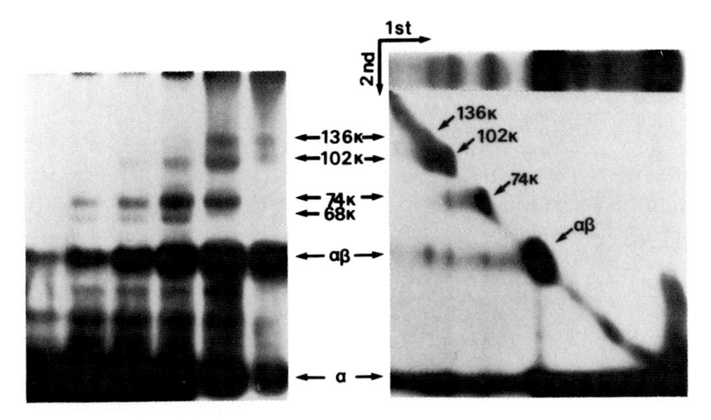

FIG. 3. Two-dimensional analysis of cleaved components of crosslinked complexes that were produced by reagents containing ester bonds. (Left) Samples were treated for increasing extents of crosslinking, from left to right. (Right) One lane of the crosslinked samples was excised from the first-dimension gel, treated in a cleaving solution, mounted on top of fresh slab gel, and electrophoresed. Note that smaller components are released from crosslinked complexes.

2. Place the gel strip on a polyethylene film such as Cling Wrap and Saran Wrap.
3. Prepare the alkaline cleaving solution containing 100 m*M* DTT, 0.1% SDS, 6 *M* urea, and 100 m*M* tribasic sodium phosphate, pH 11.5 adjusted with Tris.
4. Spread the cleaving solution below and above the gel strip. An excess amount of solution will extract the proteins out of the gel, while too little may not be able to cleave crosslinks.
5. Wrap the gel strip.
6. Incubate at 37°C for 2 hours.
7. Keep the gel strip wet by adding the alkaline solution every 30 minutes.
8. Soak the gel strip in deionized water for 5 minutes in order to remove DTT, which will prevent polymerization of acrylamide.
9. For second two-dimensional analysis, a new slab gel will be cast around the treated gel strip.
10. Place the gel strip on a slab gel plate, 1 cm below, in parallel to, the notched side, as shown in Fig. 4.
11. Assemble gel plates.
12. Cast new gel around and 3–5 mm above the first-dimension gel strip, while holding gel plate assembly slanted so that air bubbles will move upward and not be trapped below the gel strip.

13. At this time a sample well can be molded between the first-dimension gel strip and a spacer, if standard proteins or other control samples need to be run along with the cleaved samples.
14. Layer a tracking dye solution and electrophorese.

b. Problems and Solutions. The gel may not sufficiently polymerize, particularly around the first-dimension gel. This is a result of the inhibition of polymerization by residual DTT in the treated gel strip. If this problem occurs, add additional ammonium persulfate solution during the last stage of gel casting when gel solution reaches the first-dimension gel strip. Air bubbles trapped at the interface between the first-dimension gel strip and the new slab gel can distort the resolultion of two-dimensional analysis. There are a number of ways to prevent this. Cut the first-dimension gel strip even, straight, and smooth. Leave a 5- to 10-mm space between the side spacers and the gel strip (Fig. 4), in order for air to move out during gel casting. Laying the gel strip at a 10° angle and holding the gel plate assembly slanted during gel casting are helpful for not trapping air bubbles.

2. Cleavage of Disulfides in Complexes Crosslinked by Reagents Possessing a Disulfide Bond

Cleavage of disulfide bonds is similar to that of esters, and accomplished by either the alkaline cleaving solution or the reducing solution,

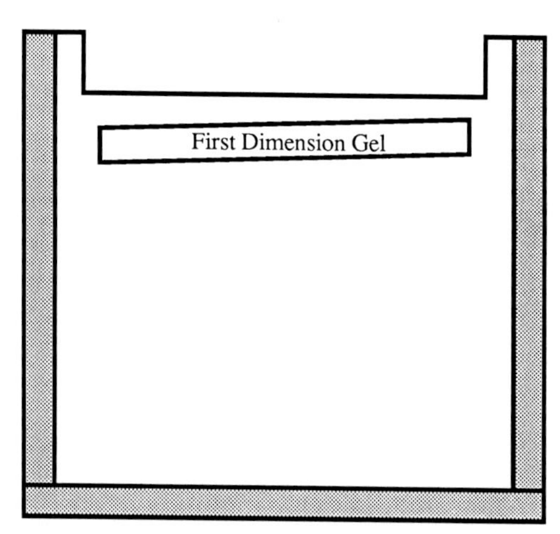

FIG. 4. Assembly of two-dimensional cleaving gel. Treat a gel lane in a cleaving solution and mount on a slab gel plate. Assemble gel plates and cast a new gel around and 3–5 mm above the first-dimension gel strip, while holding gel plate assembly slanted so that air bubbles will move upward and not be trapped below the gel strip.

containing 100 m*M* DTT, 6 *M* urea, and 0.1% SDS, pH 7.5 adjusted with HCl.

J. Analysis of Crosslinked Complexes

If noncleavable reagents are employed in a crosslinking study, components of crosslinked complexes cannot directly be identified. Instead, individual components are implied to be involved in crosslinked structures when (a) the loss of a particular component is associated with the concomitant appearance of the crosslinked complex and (b) the molecular weight of the complex is equivalent to the sum of the molecular weights or multiples of the molecular weights of the lost components.

New bands that appear after treatment with a crosslinking reagent do not necessarily represent crosslinked products, since they could also result from intrachain crosslinking, monofunctional chemical modifications, or nondissociable aggregation. To address the problem of intrachain crosslinking, cleavable reagents are particularly helpful. New bands produced by intrachain crosslinks usually demonstrate faster migration on SDS–polyacrylamide gel electrophoresis (PAGE) as a consequence of a more compact protein structure. They should regain their slower electrophoretic mobility upon cleavage. The question of chemical modification can be probed by reacting the sample with a monofunctional analog of the bifunctional reagent. Membrane components modified in this manner can be examined for altered mobility on SDS–PAGE. The potential problem of aggregation can be explored by variation of the solubilization reagent and extraction conditions.

Bands that diminish in intensity or disappear upon crosslinking are probable candidates for crosslinked molecules. Such bands, however, could also be the result of extraction, proteolysis, or reduced staining facilitated by the crosslinking conditions. Careful attention to extracted components, proteolytic processes, and total intensity of staining can substantially lessen ambiguity in these types of artifacts. Because not all membrane components are readily visualized by a single staining procedure (Lenard, 1970; Fairbanks *et al.*, 1971), the possibility of unstained components should also be considered for their potential involvement in crosslinked products. This is most simply accomplished by employing multiple staining methods.

Cleavage of crosslinks to identify protomers of crosslinked products is now a widely employed procedure. Once the identification of protomers is made, the composition and/or stoichiometry of crosslinked products is usually evaluated by (a) molecular weights estimated on the basis of relative electrophoretic mobility, (b) the alignment of cleaved components on two-dimensional gels (Wang and Richards, 1974), and (c) the relative intensities of the bands. Although these tests may be sufficient in a simple system, they are indirect and certainly not foolproof. Therefore, supporting evidence such as immunostaining is highly desirable. Two approaches that have been employed with a moderate degree of success involve specific staining with either lectins (Robinson *et al.,* 1975; Tanner and Anstee, 1976) or antibodies (Olden and Yamada, 1977; Gordon *et al.,* 1978).

III. Affinity Crosslinking

In affinity crosslinking, radioactively labeled ligands are allowed to bind to the receptor and the resultant ligand–receptor complex, along with all other protein molecules present in the system, is subjected to covalent crosslinks by the introduction of homobifunctional reagents called crosslinkers. Advantages of affinity crosslinking over photoaffinity labeling are that this is simpler and the yield of crosslinked complexes higher. Biological pathways of ligand–receptor complexes could be examined by crosslinking them at different stages.

Disadvantages are as follows. Formation of crosslinked complexes is random, causing undesirable crosslinks between all protein molecules present in samples. It is not easy to characterize the receptor exclusively, particularly in a complex system such as cell membranes and the cytosol. In addition, it is extremely difficult, if not impossible, to study biological pathways of affinity-labeled ligand–receptor complexes. Since all molecules in samples are subjected to crosslinking, which could disrupt the entire biological system, studies on biological pathways of crosslinked complexes may not be credible.

1. Preparation

Ligands are labeled with markers such as radioisotopes. If ligands are not labeled with a marker, ligand–receptor complexes may be identified after crosslinking with antibodies or other affinity labels.

2. Procedure

1. Dissolve reagents.
2. Add crosslinking reagents to cells, membranes, or molecules in solution at a desired time. Reagent concentrations sufficient for crosslinking are normally in the range of 100 μM to 3 mM.
3. Incubate samples for 1–60 minutes at room temperature.

4. If it is necessary to stop crosslinking reaction, add 50-fold molar excess of agents to quench crosslinking reagents, such as glycine for imidates and NHS.
5. If it is desirable to remove excess reagents, wash the samples.
6. Solubilize samples in detergents.
7. Electrophorese samples.
8. Crosslinked complexes can be analyzed and cleaved as described in the previous sections.

3. Problems and Solutions

In crosslinking studies, the primary objective is to identify immediate neighbors of a molecule. Since crosslinking reactions are always random, a molecule will be crosslinked first to its immediate neighbors and then to distant neighbors. When the reaction goes too extensively, a molecule could be crosslinked, directly or indirectly, to many other molecules in addition to its immediate neighbors. The analysis of such extensively crosslinked complexes is difficult. The extent of crosslinks can be controlled by varying the reagent concentration and reaction time.

Some reagents are more soluble in water than others. If the reagents are insoluble in water and need to be dissolved in organic solvents, the final concentration of the organic solvents in sample solutions should be sufficiently diluted in order not to disrupt the samples. For example, 5% DMSO may help to permeabilize cell membranes, while ethanol could precipitate water-soluble molecules. On the other hand, reagents such as imidates and NHS undergo rapid hydrolysis in aqueous buffers. In such cases, reagents should be used immediately after solubilization in an aqueous buffer.

IV. Reagents

A. Amino Reagents

1. *N*-Hydroxysuccinimide Esters

N-Hydroxysuccinimide (NHS) esters were first developed by Anderson *et al.* (1964). They have become one of the most popular reagents because of their mild reaction conditions as well as their commercial availability and ease of synthesis. They react preferentially with primary (including aromatic) amino groups (Cuatrecasas and Parikh, 1972). The imidazole group of histidine is known to compete with primary amines for reaction, but the reaction product is unstable and readily hydrolyzed. As a consequence, imidazole groups accelerate the hydrolysis of NHS esters (Cuatrecasas and Parikh, 1972). Reaction of NHS esters occurs at pH 6–9 (most efficient at pH 7–8) in most buffers that do not contain free amino groups. The reaction is rapid, with the most accessible protein amino groups being attacked within 10–20 minutes (Cuatrecasas and Parikh, 1972; Lomant and Fairbanks, 1976; Vanin and Ji, 1981). A 10-fold molar excess of an NHS ester in the concentration range of 50–500 μM (Vanin and Ji, 1981) is usually sufficient to acylate amino groups quantitatively (Lomant and Fairbanks, 1976); they also react well over a wide range of temperature (4°–25°C) (Cuatrecasas and Parikh, 1972; Lomant and Fairbanks, 1976; Vanin and Ji, 1981). Unfortunately, these compounds are not readily soluble in aqueous buffers at concentrations >1 mM and therefore must be dissolved initially in an organic solvent such as acetone or DMSO.

The principal product of the reaction with an amine is an amide; thus, the positive charge of the original amino group is lost. The reaction involves the nucleophilic attack of an amine on the acid carboxyl of an NHS ester to form an amide, releasing the NHS. Since unprotonated amines are required, one might expect the reaction under alkaline conditions to be considerably favored. However, NHS esters hydrolyze faster under these conditions, rapidly reducing the effective concentration of the reagent; half-times of hydrolysis at pH 7.5 and 8.6 are 4–5 hours (Lomant and Fairbanks, 1976) and <10 minutes, respectively. Conflicting values for the half-lives have also been reported (Abdella *et al.*, 1979).

N-Hydroxysuccinimide esters are stable for several months at 4°–25°C under anhydrous conditions. These compounds are generally synthesized by the procedure of Anderson *et al.* (1964). This involves an anhydrous coupling of NHS to a carboxyl group employing a dicyclocarbodiimide. Solvents commonly used for this reaction are dioxane, acetonitrile, DMF, or DMSO.

2. Imidates (Imidoesters)

Imidates are readily soluble in aqueous buffers and have greater amino specificity than the NHS esters. They also display reactivity in relatively neutral solutions of pH 7–10, but unlike NHS esters, an alkaline pH favors the reaction because of the more rapid hydrolysis at neutral pH (Browne and Kent, 1975; Hunter and Ludwig, 1962) (i.e., reaction with ligand at neutral pH is in major competition with hydrolysis). To minimize this degradation problem, incremental additions of reagents are recommended (Ji, 1983).

The complete reaction of available amino groups in the erythrocyte membrane requires a 100-fold molar excess of imidate (Ji, 1983). It takes 10–20 minutes to react with half of the available amino groups of Con A (Ji, 1977) and <10 minutes for phosphofructokinase (Lad and Hammes, 1974). The reaction is temperature-dependent; the rate decreases several-fold at 25°C compared to the rate at 39°C, and it decreases further at temperatures near 0°C, at which the reaction requires considerably longer times (Ji, 1983). The reaction mechanism of imidates is discussed in detail by Peters and Richards (1977). Primary amines attack imidates nucleophilically to produce an intermediate that breaks down to amidine at high pH or to a new imidate at low pH. The new imidate can react with another primary amine, thus crosslinking two amino groups, a case of putatively monofunctional imidate reacting bifunctionally. The principal product of reaction with primary amines is an amidine that is a stronger base than the original amine. The positive charge of the original amino group is therefore retained, a potential advantage over NHS esters. Amidines are normally resistant to mild acid hydrolysis (Wold, 1969) but are not completely stable to the conditions used for amino acid hydrolysis (Peters and Richards, 1977). They are also reported to be cleaved by hydrazine (Means and Feeney, 1971) or ammonia (Hunter and Ludwig, 1962). This cleavage readily occurs with amino acids and small peptides but occurs with only low efficiency with proteins (Ji, 1983; Peters and Richards, 1977; Wold, 1969).

Among a variety of synthetic procedures for imidates, the Pinner method (Peters and Richards, 1977) is the most frequently employed. In this synthesis, a nitrile is allowed to react with a primary alcohol in the presence of HCl under anhydrous conditions. This is usually done by mixing the appropriate nitrile and alcohol in a dry solvent and then bubbling in dry HCl gas. Other synthetic procedures for the formation of imidates are discussed by McElvain and Schroeder (1949).

3. ACYLAZIDES

These compounds have also been used as amino-specific reagents (Lutter *et al.*, 1974) in reactions in which nucleophilic amines attack reagent acid carboxyl groups. Therefore, this reaction is expected to occur more favorably in alkaline than in neutral conditions. Ribosomal subunits have been crosslinked by incubation with 50 m*M* bisacylazides in 50 m*M* triethanolamine (pH 8.5) at 22°C for 30 minutes (Lutter *et al.*, 1974).

4. Arylhalides

1,5-Difluro-2,4-dinitrobenzene has been used for crosslinking aminophospholipids in membranes (Marinetti and Love, 1976). Crosslinking of lipids in micelles occurs extensively with this reagent at 50–200 μ*M* in 120 m*M* sodium carbonate (pH 8.5) and 40 m*M* sodium chloride after incubation for 2 hours at 23°C. Crosslinking in the erythrocyte membrane requires higher concentrations of the reagent (55–800 μ*M*). In the same system, 4,4-difluoro-3,3-dinitrodiphenyl sulfone at 386 μ*M* in 5 m*M* sodium phosphate (pH 8.0) and 0.15 sodium chloride was more efficient than 1,5-difluoro-2,4-dinitrobenzene.

5. Isocyanates

These reagents generally react with primary amines to form stable bonds. Their reactions with sulfhydryl, imidazole, and tyrosyl groups give relatively unstable products (Hartman and Wold, 1966). In aqueous solutions, hydrolysis is appreciable, since the half-life of aliphatic isocyanates at pH 7.6 is <2 minutes (Wold, 1969). Crosslinking of anti-β-galactosidase A antibody to the enzyme has been observed after incubation in 1 m*M* hexamethylene diisocyanate and 20 m*M* phosphate buffer (pH 6.5) for 15 minutes.

B. Sulfhydryl Reagents

Maleimides react preferentially with sulfhydryl groups (Means and Feeney, 1971; Hunter and Ludwig, 1962; McElvain and Schroeder, 1949; Lutter *et al.,* 1974; Marinetti and Love, 1976; Marfey and Tsai, 1975; Snyder *et al.*, 1974; Vallee and Riordan, 1969). They also react at a much slower rate with amino groups and the imidazole group of histidine (Brewer and Riehm, 1967). For example, at pH 7 the reaction rate of simple thiols is 1000-fold greater than that of the corresponding amine (Brewer and Riehm, 1967). Furthermore, the reaction of maleimides with the imidazole group of histidine requires very vigorous conditions. Therefore, under normal crosslinking conditions, the maleimide group can be considered a sulfhydryl-specific reagent. Reaction of maleimides with glutathione is complete within a minute in sodium phosphate buffer (pH 7.0). Somewhat longer reaction times are necessary with proteins.

Appropriately positioned sulfhydryl groups can be crosslinked directly to each other by oxidation. The *bis-1,10-phenanthroline complex of cuprin ion* has been widely employed for this purpose. Membrane proteins as well as model compounds such as cysteine, 2-ME thioglycolic acid, and reduced lipoic acid all form disulfides very rapidly in 25–50 μ*M* phenanthroline and 5–10 μM $CuSO_4$ at 0°–4°C (Ji, 1983; Peters and Richards, 1977; Huang and Richards, 1977). In membrane systems, 5 minutes is sufficient for complete oxidation at the indicated concentrations. The crosslinking reaction can be inhibited by metal ion (Zn^{2+}, Ni^{2+}, Co^{2+}) at 100 μM as well as by copper-chelating agents (EDTA, histidine).

Thiophthalimides have also been introduced as crosslinking reagents (Ji, 1983; Vanin and Ji, 1981). These compounds can react with sulfhydryl groups to form disulfides (Brewer and Riehm, 1967). Thiophthalimides are synthesized using a two-step procedure in which a disulfide or thiol is treated with chlorine gas to form a sulfenyl chloride, which is allowed to react with phthalimide to produce the thiophthalimide derivative (Behforouz and Kenwood, 1969).

Disulfide dioxide derivatives are another addition to the class of sulfhydryl-specific crosslinking reagents (Huang and Richards, 1977). They are formed by treating a disulfide with two equivalents of peracetic acid. Monooxide disulfide derivatives have previously been shown to react specifically with sulfhydryl groups. Employing dioxide derivatives, Huang and Richards (1977) showed that *N*-ethylmaleimide completely inhibits any reaction, suggesting sulfhydryl specificity.

The introduction of extrinsic sulfhydryl groups into molecules and subsequent oxidation is another approach to crosslinking by disulfide formation. Oxidation products may be of a reagent–reagent, reagent–intrinsic sulfhydryl of secondary intrinsic sylfhydryl–intrinsic sulfhydryl nature. Extrinsic sulfhydryl groups have been successfully introduced into ribosomes by the use of substituted imidates or *N*-(3-fluoro-4,6-dinitrophenyl) cysteine (which requires subsequent reduction) (Mas *et al.*, 1980).

C. Guanidino Reagents

Phenylglycoxal reacts primarily with the guanidino group of arginine residues of proteins (Ngo *et al.*, 1981) and guanine nucleotides (Vanin *et al.*, 1981). Histidine and cysteine also react, albeit to much lesser extent (Vanin *et al.*, 1981). The reaction products are stable under acidic conditions but decompose slowly at neutral and alkaline pH, regenerating the original guanidino group (Takahashi, 1968). Arginine reacts extensively in 6 mM glycoxal, 0.1 M sodium phosphate (pH 7.0–7.5) after 30–60 minutes of reaction at 25°C (Ngo *et al.*, 1981).

D. Indole Reagents

Sulfenyl halides have been shown to react with tryptophan and cysteine, producing a thioester derivative and a disulfide, respectively

(Means and Feeney, 1971). To a minor extent, methionine may undergo oxidation in the presence of sulfenyl chloride (Demoliou and Epand, 1980). In 50–100% acetic acid, 90% of tryptophan molecules react in 20 minutes (Demoliou and Epand, 1980).

E. Carboxyl Reagents

Carbodiimides react with carboxyl groups to produce *O*-acylisoureas, which in turn react with nucleophiles such as amino groups. In the presence of diamines (ethyl diamine or cleavable cystamine), adjacent carboxyl groups can be crosslinked via the diamine. In addition to these types of crosslinks, carbodiimides are also capable of condensing carboxyl and intrinsic amino groups. It is also possible to couple a photoactivable heterobifunctional reagent carrying an amino group at one end to a carboxyl group of a macromolecule. There are a number of water-soluble as well as insoluble carbodiimides. Incubation of lutropin in 25 m*M* 1-ethyl-3-(3-dimethylaminopropyl) carbodiimide (pH 4.75 adjusted with HCl) crosslinks the subunits (α to β) of the hormone within 1 hour at room temperature (Parsons and Pierce, 1979).

F. Nonspecific Reagents

Formaldehyde, the simplest of crosslinking reagents, has the broadest reaction specificity. In addition to amines and amides, it reacts with the side chains of cysteine, tyrosine, histidine, tryptophan, and arginine. Although formaldehyde contains a single functional group, it can react bifunctionally and therefore crosslink. At concentrations of 5–15 m*M* formaldehyde (pH 6–8), a number of red blood cell membrane proteins are crosslinked after incubation for 30 minutes at room temperature (French and Edsall, 1945). This concentration is considerably lower than that commonly used to fix tissues and protein crystals. Bifunctional reaction involves the attack of a nucleophile onto the aldehyde to form a quaternary ammonium salt, after which the loss of water produces an immonium cation. This cation is then attacked by another nucleophile producing a methylene-bridged crosslink. Formaldehyde is commercially available in 37–40% aqueous solutions (formalin) and exists as a series of low molecular weight polymers $H(OCH_2)_nOPH$ (Means and Feeney, 1971). When diluted in aqueous solution, however, it reverts to a hydrated monomeric form (Means and Feeney, 1971; French and Edsall, 1945).

Because of its lack of specificity, formaldehyde will sometimes crosslink two different reactive groups. For example, a methylene-bridged lysine–tyrosine complex has been isolated from acid hydrolysates of formaldehyde-treated tetanus and diphtheria toxins (Blass *et al.*, 1965). Glutaraldehyde, on the other hand, is more specific than formaldehyde, although it also displays reactivity with several amino acid side chains, including those of lysine, cysteine histidine, and tyrosine. Incubation of the target macromolecule at concentrations of 1–3 m*M* (pH 6–8) for 20 minutes at room temperature is usually sufficient for crosslinking.

Unlike the other crosslinking reagents discussed, there exist a large number of different polymeric forms of glutaraldehyde in dilute solution. As a result, the distance between two crosslinked groups cannot be estimated. Commercial glutaraldehyde solution contains an equilibrium mixture of the monomeric and polymeric forms of the cyclic hemiacetal. Commercial solutions are usually found to have a pH near 3. As the pH increases (as under typical crosslinking conditions), however, the cyclic polymers undergo a dehydration to form α,β-unsaturated aldehyde polymers. As with formaldehyde, crosslinking also increases at higher pH. When protein amino groups react with aldehydes of the polymer, Schiff bases are formed that are stable only when conjugated to another double bond. The resonant interaction of the Schiff base with this double bond provides stability that prevents hydrolysis of the Schiff linkage. When amines are present in excess, they can attack the ethylenic double bond and form a stable Michael addition product (Peters and Richards, 1977).

G. Photoactivable Groups

Photoactivable moieties differ dramatically from ordinary chemical groups in their use. Photoactivable groups are completely inert in the dark. They are converted to reactive species, however, upon absorption of a photon of appropriate energy. Currently employed photoactivable groups can be classified into the categories of precursors of either nitrenes or carbenes. Nitrenes are generated upon heating or photolysis of azides. Electron-deficient nitrenes are extremely reactive with broad reaction specificities. They are therefore considered to be nonspecific reagents. They can potentially react with a variety of chemical bonds including N—H, O—H, C—H, and C═C. Aryl nitrenes appear to react better with N—H and O—H than C—H bonds (Bayley and Knowles, 1980, 1983). Several potential problems exist in using azides. They can be reduced to amino groups (Cartwright *et al.*, 1976). In this process, dithiols such as DTT are more effective than monothiols like 2-ME (Vanin and Ji, 1981; Staros *et al.*, 1978). The half-life of arylazides is 5–15 minutes in 10 m*M* DTT (pH 8.0) but >24 hours in 50 m*M* 2-ME (pH 8.0) (Staros *et al.*, 1978). Nitrenes once formed can undergo ring expansion. The azido group ortho to a ring nitrogen can rearrange to form a tetrazole that is much less photosensitive than the azide (Bayley and Knowles, 1983).

Three types of azides are currently available, the aryl, alkyl, and acyl derivatives. Only the arylazides, however, have been used as photoactivable groups in crosslinking reagents. Alkylazides are not suitable for crosslinking because their photolysis products, alkylnitrenes, are very susceptible to rearrangement to form inactive imines. Acylazides as well as sulfonylazides and phosphorylazides produce very reactive nitrenes upon photolysis, but they are not useful as photoactivable reagents because of their sensitivity to nucleophilic attack even in the absence of photolysis (Bayley and Knowles, 1983). In fact, acylazides have been used as nonphotoactivable crosslinking reagents (Lutter *et al.*, 1974) (see Section IV,A). An additional problem in the use of acylazides is that the resultant acylnitrene can undergo intramolecular rearrangement to form a cyanate.

The reactivity of arylnitrenes is increased by the presence of electron-withdrawing substituents such as nitro (Bayley and Knowles, 1983) or hydroxyl (Ji and Ji, 1985) groups in the ring. Such increased reactivity, however, is not necessarily desirable, since this will result in increased reaction with water. Electron-withdrawing substituents also push the absorption maximum of arylazides to longer wavelengths. Unsubstituted arylazides have an absorption maximum in the range of 260–280 nm, while hydroxyl (Ji and Ji, 1985) or nitroarylazides absorb significant light beyond 305 nm (Peters and Richards, 1977; Bayley and Knowles, 1983), despite the presence of major absorption peaks below 275 nm.

Arylnitrenes have a half-life on the order of 10^{-2}–10^{-4} seconds (Resier *et al.*, 1968; DeGraff *et al.*, 1974). Crosslinking reactions are expected to be terminated within this very short time period. Extremely rapid crosslinking can therefore be accomplished by employing photolytic periods of <1 millisecond (Baron *et al.*, 1973). The possible existence of undefined long-lived nitrene photoproducts has been raised (Huang and Richards, 1977; Mas *et al.*, 1980), but the presence of such intermediates has yet to be convincingly demonstrated.

The second major class of photoactivable crosslinking reagents consists of the diazo compounds, which form an electron-deficient carbene upon photolysis. These carbenes undergo a variety of reactions including insertion into C—H bonds, addition to double bonds (including aromatic systems), hydrogen abstraction, and coordination to nucleophilic centers to give carbon ions (Baron *et al.*, 1973). A limitation to their use is that they are highly reactive with water. A second limitation in the use of carbenes is that α-ketocarbenes can rearrange to produce ketenes. Furthermore, the parent diazocarbonyl compounds are unstable at low pH

(Bayley and Knowles, 1983). α,β-Unsaturated ketones, unlike the diazocarbonyl compounds, do not form an electron-deficient species when activated. Upon photolysis, these derivatives generate a diradical. This triplet state functions as an efficient hydrogen abstractor and reacts preferentially with C—H bonds rather than with the O—H bonds of water (Bayley and Knowles, 1983). These types of heterobifunctional reagents have not been used for macromolecular photoaffinity labeling, but in view of their properties they should be considered.

V. Random-Collisional Crosslinks

In solutions and fluid membranes, mobile molecules are thought to collide at high frequencies. Concern has been expressed as to whether these colliding molecules are crosslinked during chemical crosslinking, thereby producing nonspecifically crosslinked complexes and creating problems in interpreting such results. Our studies have demonstrated that the probability of random-collisional crosslinks under normal conditions is insignificant (Ji and Middaugh, 1980). The reasons are as follows. The duration of random-collisional contacts between two molecules is expected to be too short for the crosslinking reaction to be completed. Furthermore, crosslinks require that two molecules assume specific orientations, that reactive groups in each molecule are exposed to each other, and that crosslinking reagents are available at the vicinity. All of these events must occur simultaneously for random-collisional crosslinks of two molecules to occur.

Acknowledgments

This work was supported by NIH Grant HD-18702. T. H. Ji is the recipient of an American Cancer Society Faculty Research Award AFR-262.

References

Abdella, P. M., Smith, P. K., and Royer, G. P. (1979). *Biochem. Biophys. Res. Commun.* **87,** 734–742.

Aizawa, G. P., Kurimoto, F., and Yokono, O. (1977). *Biochem. Biophys. Res. Commun.* **75,** 870–878.

Anderson, O., Zimmerman, J. E., and Callahan, F. M. (1964). *J. Am. Chem. Soc.* **86,** 1839–1842.

Andreasen, T. J., Keller, C. H., LaPorte, D. C., Edelman, A. M., and Storm, D. R. (1981). *Proc. Natl. Acad. Sci. U.S.A.* **78,** 2782–2785.

Barker, R. (1971). "Organic Chemistry of Biological Compounds." Prentice-Hall, Englewood Cliffs, New Jersey.

Baron, W. J., DeCamp, M. R. Hendrick, M. E., Jones, M., Levin, R. H., and Sohn, M. B. (1973). *In* "Carbenes" (M. Jones, and R. A. Moss, eds.), Vol. 1, pp. 1–151. Wiley, New York.

Bayley, H., and Knowles, J. R. (1980). *Biochemistry* **17,** 2414–2419.

Bayley, H., and Knowles, J. R. (1983). *In* "Methods in Enzymology" (S. Fleischer and B. Fleischer, eds.), Vol. 96, pp. 69–114. Academic Press, New York.

Behforouz, M., and Kenwood, J. E. (1969). *J. Org. Chem.* **34,** 51–55.

Bhaumick, B., Bala, R. M., and Hollenberg, M. D. (1981). *Proc. Natl. Acad. Sci. U.S.A.* **78,** 4279–4283.

Blass, J., Bizzini, B., and Raynaud, M. (1965). *C. R. Hebd. Seances Acad. Sci.* **261,** 1448–1454.

Brewer, C. F., and Riehm, J. P. (1967). *Anal. Biochem.* **18,** 248–255.

Carney, D. H., Glenn, K. C., Cunningham, D. D., Das, M., Fox, C. F., and Fenton, J. W. (1979). *J. Biol. Chem.* **254,** 6244–6247.

Cartwright, I. L., Hutchinson, D. W., and Armstrong, V. W. (1976). *Nucleic Acids Res.* **3,** 2331–2339.

Coltrera, M. D., Potts, J. T., and Rosenblatt, M. (1981). *J. Biol. Chem.* **256,** 10551–10554.

Cuatrecasas, P., and Parikh, I. (1972). *Biochemistry* **11,** 2291–2299.

Das, M., Miyakawa, T., Fox, C. F., Pruss, R. M., Aharonov, A., and Hershman, H. R. (1977). *Proc. Natl. Acad. Sci. U.S.A.* **74,** 2790–2794.

DeGraff, B. A., Gillespie, D. W., and Sundberg, R. J. (1974). *J. Am. Chem. Soc.* **96,** 7491–7496.

Demoliou, C. D., and Epand, R. M. (1980). *Biochemistry* **19,** 4539–4546.

Eldjarn, L., and Pihl, A. (1957). *J. Am. Chem. Soc.* **79,** 4589–4593.

Fairbanks, G., Steck, T. L., and Wallach, D. F. H. (1971). *Biochemistry* **10,** 2606–2617.

Fava, A., Iliceto, A., and Camera, E. (1957). *J. Am. Chem. Soc.* **79,** 833–838.

Fleet, G. W. J., Knowles, J. R., and Porter, R. R. (1972). *Biochem. J.* **128,** 499–508.

French, D., and Edsall, J. T. (1945). *Adv. Protein Chem.* **2,** 277–335.

Gordon, W. E., Bushnell, A., and Burridge, K. (1978). *Cell (Cambridge, Mass.)* **13,** 249–261.

Hartman, F. C., and Wold, F. (1966). *J. Am. Chem. Soc.* **88,** 3890–3891.

Hofmann, C., Ji, T. H., Miller, B., and Steiner, D. F. (1981). *J. Supramol. Struct. Cell. Biochem.* **15,** 1–13.

Huang, C. K., and Richards, F. M. (1977). *J. Biol. Chem.* **252,** 5514–5521.

Hunter, M. J., and Ludwig, M. L. (1962). *J. Am. Chem. Soc.* **84,** 3491–3504.

Jacobs, S., Hazum, E., Schecter, I., and Cuatrecasas, P. (1979). *Proc. Natl. Acad. Sci. U.S.A.* **76,** 4918–4921.

Jaffe, C. L., Lis, H., and Sharon, N. (1980). *Biochemistry* **19,** 4423–4429.

Ji, I., and Ji, T. H. (1980). *Proc. Natl. Acad. Sci. U.S.A.* **77,** 7167–7170).

Ji, I., and Ji, T. H. (1981). *Proc. Natl. Acad. Sci. U.S.A.* **78,** 5465–5469.

Ji, I., Yoo, B. Y., Kaltenbach, C., and Ji, T. H. (1981). *J. Biol. Chem.* **256,** 10853–10858.

Ji, T. H. (1974). *J. Biol. Chem.* **249,** 7841–7847.

Ji, T. H. (1976). *In* "Membranes and Neoplasia: New Approaches and Strategies" (V. T. Marchesi, ed.), pp. 171–178. Liss, New York.

Ji, T. H. (1977). *J. Biol. Chem.* **252,** 1566–1570.

Ji, T. H. (1979). *Biochim. Biophys. Acta* **559,** 39–69.

Ji, T. H. (1983). *In* "Methods in Enzymology" (C. H. W. Hirs and S. N. Timasheff, eds.), Vol. 91, pp. 580–609. Academic Press, New York.

Ji, T. H., and Ji, I. (1985). *Anal. Biochem.* **151,** 348–349.
Ji, T. H., and Middaugh, C. R. (1980). *Biochim. Biophys. Acta* **603,** 371–374.
Johnson, G. L., MacAndres, V. I., and Pilch, P. F. (1981). *Proc. Natl. Acad. Sci. U.S.A.* **78,** 875–878.
Kiehm, D. J., and Ji, T. H. (1977). *J. Biol. Chem.* **252,** 8524–8531.
Lad, P. M., and Hammes, G. G. (1974). *Biochemistry* **13,** 4530–4537.
Lenard, J. (1970). *Biochemistry* **9,** 1129–1132.
Lingwood, C. A., Hakomori, S., and Ji, T. H. (1980). *FEBS Lett.* **112,** 265–268.
Liu, T. Y. (1977). *In* "The Proteins" (H. Neurath and R. L. Hill, eds.), 3rd ed., Vols. 3, pp. 239–402. Academic Press, New York.
Lomant, A. J., and Fairbanks, G. (1976). *J. Mol. Biol.* **104,** 243–261.
Lutter, L. C., Ortanderl, F., and Fasold, H. (1974). *FEBS Lett.* **48,** 288–292.
Marfey, S. P., and Tsai, K. H. (1975). *Biochem. Biophys. Res. Commun.* **65,** 31–38.
Marinetti, G. V., and Love, R. (1976). *Chem. Phys. Lipids* **16,** 239–254.
Mas, M. T., Wang, J. K., and Hargrave, P. A. (1980). *Biochemistry* **19,** 684–692.
Massague, J., Guillette, B. J., and Czech, M. P. (1981a). *J. Biol. Chem.* **256,** 2122–2125.
Massague, J., Guillette, B. J., Czech, M. P., Morgan, C. J., and Bradshaw, R. A. (1981b). *J. Biol. Chem.* **256,** 9419–9424.
Matheson, R. R., van Wart, H. E., Burgess, A. W., Weinstein, L. I., and Scheraga, H. A. (1977). *Biochemistry* **16,** 396–403.
McElvain, S. M., and Schroeder, J. P. (1949). *J. Am. Chem. Soc.* **71,** 40–46.
Means, G. E., and Feeney, R. E. (1971). "Chemical Modification of Proteins." Holden-Day, San Francisco, California.
Middaugh, C. R., and Ji, T. H. (1980). *Eur. J. Biochem.* **110,** 587–592.
Nathanson, N. M., and Hall, Z. W. (1980). *J. Biol. Chem.* **255,** 1698–1703.
Ngo, T. T., Yam, C. F., Lenhoff, H. M., and Ivy, J. (1981). *J. Biol. Chem.* **256,** 11313–11318.
Olden, K., and Yamada, K. M. (1977). *Anal. Biochem.* **78,** 483–490.
Perkins, M. E., Ji, T. H., and Hynes, R. O. (1979). *Cell (Cambridge, Mass.)* **16,** 941–952.
Peters, K., and Richards, F. M. (1977). *Annu. Rev. Biochem.* **46,** 523–551.
Resier, A., Willets, F. W., Terry, G. C., Williams, V., and Morley, R. (1968). *Trans. Faraday Soc.* **64,** 3265–3275.
Robichon, A., Kuks, P. F. M., and Besson, J. (1987). *J. Biol. Chem.* **262,** 11539–11545.
Robinson, P. J., Bull, F. G., Anderton, B. H., and Roitt, I. M. (1975). *FEBS Lett.* **58,** 330–333.
Scimonelli, T., and Eberle, A. N. (1987). *FEBS LETT.* **226,** 134–138.
Shin, J., and Ji, T. H. (1985). *J. Biol. Chem.* **260,** 14020–14025.
Snyder, P. D., Wold, F., Bernlohr, R. W., Dullum, C., Desnick, R. J., Krivit, W., and Condie, R. M. (1974). *Biochim. Biophys. Acta* **350,** 432–436.
Sorenson, P., Farber, N. M., and Krystal, G. (1987). *J. Biol. Chem.* **261,** 9094–9097.
Staros, J. V., and Richards, F. M. (1974). *Biochemistry* **13,** 2720–2726.
Staros, J. V., Bayley, H., Standring, D. N., and Knowles, J. R. (1978). *Biochem. Biophys. Res. Commun.* **80,** 568–572.
Steck, T. L. (1972). *J. Mol. Biol.* **66,** 295–305.
Takahashi, K. (1968). *J. Biol. Chem.* **243,** 6171–6179.
Tanner, M. J. A., and Anstee, D. J. (1976). *Biochem. J.* **153,** 265–270.
Trommer, W. E., Friebel, K., Kiltz, H.-H., and Kolkenbrook, H.-J. (1977). *In* "Protein Crosslinking" (F. Mendel, ed.), Part A, pp. 187–195. Plenum, New York.
Vallee, B. L., and Riordan, J. F. (1969). *Annu. Rev. Biochem.* **38,** 733–794.
Vanin, E. F., and Ji, T. H. (1981). *Biochemistry* **20,** 6754–6760.

Vanin, E. F., Burkhard, S. J., and Kaiser, I. I. (1981). *FEBS Lett.* **124,** 89–92.
Wang, K., and Richards, F. M. (1974). *J. Biol. Chem.* **244,** 8005–8018.
Witzemann, V. W., Muchmore, D., and Raftery, M. A. (1979). *Biochemistry* **24,** 5511–5518.
Wold, F. (1969). *In* "Methods in Enzymology" (R. B. Clayton, ed.), Vol. 15, pp. 623–651. Academic Press, New York.
Yip, C. C., Yeung, C. W. T., and Moule, M. L. (1978). *J. Biol. Chem.* **253,** 1743–1745.

Chapter 12

Quantitative Evaluation of Receptor-Mediated Endocytosis

DWAIN A. OWENSBY

The Cardiovascular Division
Department of Medicine
Washington University School of Medicine
St. Louis, Missouri 63110

PHILLIP A. MORTON AND ALAN L. SCHWARTZ

Hematology–Oncology Division
Departments of Pediatrics and Pharmacology
Washington University School of Medicine
St. Louis, Missouri 63110

Copyright © 1989 by Academic Press, Inc.
All rights of reproduction in any form reserved.

I. Introduction

The study of several model systems of receptor-mediated endocytosis has established its general characteristics, delineated the pathways involved, and begun to elucidate the underlying mechanisms responsible for its regulation.

The assays, experimental procedures, and analyses described in this chapter are designed to allow a quantitative evaluation of the process of receptor-mediated endocytosis. The emphasis is directed toward both a theoretical and a practical approach to these issues.

II. Choice of Ligand–Receptor Probe

Receptors generally represent only a very small fraction of the cell's constituent molecules. As a result, the analysis of ligand–receptor binding and endocytosis almost invariably requires sensitive assays for ligand–receptor interaction, generally achieved by incorporation of radioactive atoms into ligands. Alternatively, antireceptor antibodies may be utilized as receptor probes, and also may be radiolabeled to enhance their detection. However, radiolabeling potentially can compromise important ligand parameters including bioactivity, specificity, and intracellular trafficking. In addition, antireceptor antibodies may perturb internalization, trafficking, or recycling of receptors. As a first approach to the identification of receptors and evaluation of receptor-mediated endocytosis, most investigators utilize radiolabeled ligands. In the next section we discuss several methods for introducing radioatoms into protein ligands for the study of receptor-mediated endocytosis, together with controls to assure retention of selected physicochemical and biological properties following ligand modification.

III. Methods for Radiolabeling of Ligand

Radioactive atoms can be introduced into ligands with metabolic, direct, or indirect labeling procedures (Table I). Radioiodine is the most commonly used isotope for labeling protein ligands because of the high specific activity obtainable and the relative ease of detection of γ radiation. Radioiodination methods can be divided into two groups: direct and conjugation methods. In direct methods, radioiodine is incorporated directly into tyrosine (or rarely, histidine) residues. In contrast, indirect

TABLE I

METHODS FOR RADIOLABELING OF LIGAND OR ANTIRECEPTOR ANTIBODIES

Radioatom	Method	Site of incorporation	References
^{35}S	Metabolic	Met, Cys	—
^{125}I	Direct		
	Chloramine-T	Tyr, His	Greenwood *et al.* (1963)
	Iodogen	Tyr	Fraker and Speck (1978)
	Lactoperoxidase	Tyr	Marchalonis (1969); David and Reisfeld (1974)
	Conjugation		
	Bolton–Hunter	NH_2 Groups	Bolton and Hunter (1973)
^{3}H	Galactose oxidase	Gal, GalNAc	Morell and Ashwell (1972)
	Periodic acid	N-acetyl neuraminic acid	Van Lenten and Ashwell (1972)

methods result in conjugation of a radioiodine-containing group to a specific NH_2 side chain, generally lysine, or to the N-terminal amino group. Although direct labeling methods generally provide higher specific activity and are easier to perform than conjugation methods, they often result in some degree of chemical damage to the protein. Potential problems associated with radioiodination of proteins include radiation damage, loss of radioiodide, protein aggregation, and loss of immunoreactivity.

Direct iodination of a protein ligand in the presence of the oxidant chloramine-T (Greenwood *et al.*, 1963) technically is simple and rapid. The method is illustrated by the technique used to label tissue-type plasminogen activator (t-PA) (Owensby *et al.*, 1988). Carrier-free $Na^{125}I$, 1 mCi in 10 μl of 0.1 *M* NaOH, was added to t-PA, 100 μg in 50 μl of 0.1 *M* sodium phosphate buffer (pH 7.5) containing 0.01% Tween 80 to minimize ligand adsorption to labware. To initiate the reaction, chloramine-T 10 μl (10 mg/ml in the same buffer) was added and the reaction mixture was swirled gently. The reaction was terminated after 30 seconds at room temperature by adding 25 μl of quencher (0.1 *M* Tris, pH 7.8–8.0, containing sodium metabisulfite, 20 mg/ml). After an additional 2 minutes at room temperature, the reaction volume was transferred to a prewashed (in 0.1 *M* sodium phosphate–0.01% Tween 80, pH 7.5) disposable column of Sephadex G-10 (bed volume 10 ml). ^{125}I-Labeled-t-PA (^{125}I-t-PA) eluted in the void volume.

In an attempt to lessen potential oxidation damage to proteins during radiolabeling, Fraker and Speck (1978) introduced an alternative direct-iodination procedure utilizing the milder oxidant 1,3,4,6-tetrachloro-3α,6α-diphenylglycoluril (Iodogen). The reagent is applied to the surface of the reaction vessel prior to addition of the ligand and Na^{125}I. This method requires longer reaction times but may result in less oxidative damage to the ligand. Radioiodine also may be incorporated directly into ligands by use of lactoperoxidase (Marchalonis, 1969), although this method results in low specific activity and also requires exposure of ligand to oxidizing conditions.

The conjugation method of radioiodination was developed specifically to avoid exposure to oxidants and thereby to overcome problems of oxidative damage to proteins. Furthermore, it allows radiolabeling of ligands that lack tyrosine residues. The most commonly used method employs *N*-succinimidyl-3-(4-hydroxy 5-[^{125}I]iodophenyl)propionate (Bolton and Hunter, 1973). This method is useful in generating radioligands when retention of enzymatic or immunological reactivity is particularly important. Although the method technically is more complex and generally results in lower specific activity than direct methods, these disadvantages may be overcome partially by using the commercially available mono- or di-iodo derivative of the reagent.

Glycoprotein ligands with terminal galactose residues may be ^{3}H-labeled following exposure to galactose oxidase and tritiated borohydride (Morell and Ashwell, 1972). In addition, glycoproteins containing sialic acid residues may be ^{3}H-labeled after exposure to periodic acid (Van Lenten and Ashwell, 1972).

In some cases, ligands incorporate metal atoms that may be replaced by radioactive isotopes. Examples include the use of ^{59}Fe for transferrin, ^{57}Cu for ceruloplasmin, and ^{60}Co for cobalamin. However, introduction of a radioactive isotope into the ligand may impart properties that result in behavior different from that observed for the authentic physiological ligand. Furthermore, during receptor-mediated endocytosis, the fate of the radioactive moiety does not necessarily reflect the fate of the polypeptide components of the ligand.

A. Fidelity of Radioligand

After a ligand is radiolabeled, retention of several physicochemical properties should be ascertained. For example, a similar elution profile for authentic and radiolabeled ligand on gel filtration chromatography, and a similar electrophoretic mobility on sodium dodecyl sulfate–polyacrylamide gel electrophoresis (SDS–PAGE) under reducing and nonreducing conditions would suggest that significant aggregation or other radiation damage did not occur. Substantial protein degradation to low molecular weight fragments can be detected by solubility in 4% phosphotungstic acid–20% trichloroacetic acid. If specific antiligand antibody is available, retention of immunochemical determinants can be established by solid-phase immunoassay, enzyme-linked immunosorbent assay (ELISA), SDA–PAGE and immunoblotting, or immunoprecipitation to confirm the similarity of labeled and parent ligands. If the ligand possesses enzymatic or other functional activity, retention of these functions should be demonstrated to show that significant oxidation or radiation damage did not occur during radiolabeling.

B. Determination of Specificity of Radioligand for Receptor

The specificity of a ligand for a saturable receptor population most often is characterized by competition studies. Binding of radioligand at a selected concentration, generally $\leq K_d$, is evaluated in the presence or absence of excess unlabeled potential competitor(s). In the example shown in Table II, binding of ^{125}I-t-PA to the surface of Hep G2 cells was examined in the presence or absence of variants of t-PA, as well as ligands for other receptors known to be expressed by these cells. In addition, the effect of various agents known to inhibit nonspecific binding of proteins to cells was examined. In the example shown, only t-PA was effective in competing for ligand-binding sites. The capability of specific ligand variants to compete for receptor-binding sites can elucidate determinants of ligand binding (e.g., the role of posttranslational protein glycosylation, acylation, or phosphorylation on ligand–receptor interaction). Such studies may result in the identification of variants of higher binding affinity than the native ligand and may define structure–function determinants.

IV. Receptor Populations

The entire cell complement of receptors for a specific ligand is composed of several distinct but overlapping subpopulations (Fig. 1). These populations differ with respect to their location (cell surface versus cell

TABLE II

SPECIFICITY OF ^{125}I-t-PA BINDING TO HEP G2 CELLS[a]

Competitor[b]	^{125}I-t-PA bound (fmol/10^6 cells)
None	105 ± 5
t-PA$_{CHO}$	36 ± 2
t-PA$_{HCF}$	20 ± 1
ASOR	104 ± 6
Ferrotransferrin	104 ± 4
Insulin	112 ± 5
Epidermal growth factor	105 ± 3
Mannosylated BSA	132 ± 4
Hemoglobin-haptoglobin	96 ± 2
BSA	119 ± 2
Human serum albumin	114 ± 4
Ovalbumin	123 ± 5
Cytochrome *c*[c]	117 ± 1
Cytochrome *c*[c,d]	127 ± 4

[a]Cells were incubated for 2 hours at 4°C with 1 ml of ^{125}I-t-PA, 3 n*M*, in binding media in the absence or presence of potential competitors. Competitors were present at 300 n*M*. Results are expressed as mean ± SD of triplicate determinations.

[b]t-PA$_{CHO}$ is of Chinese hamster ovary cell origin. t-PA$_{HCF}$ is of human colon fibroblast origin. BSA, Bovine serum albumin.

[c]Binding was assessed in minimum essential medium containing 20 m*M* HEPES, 0.01% Tween 80, and 0.1 mg/ml BSA.

[d]Cytochrome *c* was present at 30 μ*M*.

interior), their functional capacity for binding ligand (occupied with endogenous ligand or other substrate, or unoccupied and available for ligand binding, or unoccupied and quiescent), and their capability for recycling (either with or without ligand). General approaches for assessing ligand–receptor binding as well as techniques for evaluating specific receptor populations separately are discussed. Examples are taken from our laboratory's experience in characterizing receptor-mediated endocytosis of asialoorosomucoid (ASOR) and t-PA by Hep G2 cells.

A. Equilibrium Methods for Quantitating Ligand Binding

Quantitative evaluation of unoccupied functional receptors involves determining the amount of ligand–receptor complex formed as a function both of time and of concentration of ligand added. In simplest terms, li-

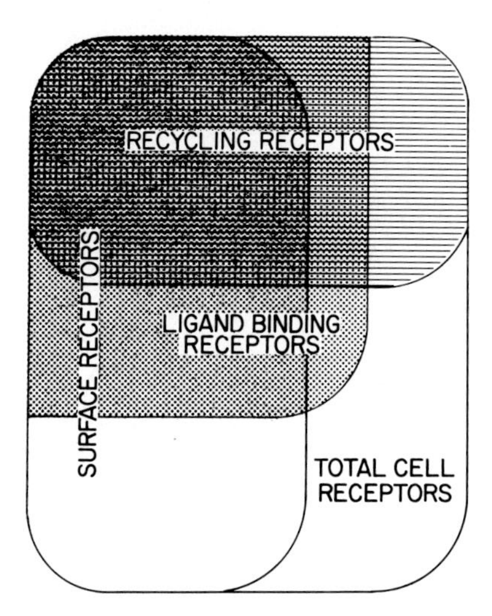

FIG. 1. Venn diagram of receptor site and state. A cell's complement of receptors includes intracellular as well as surface molecules. Because of latency or prior occupancy, a subpopulation of receptors cannot bind exogenously added ligand. The recycling receptor pool represents only a fraction of the total number of receptors, and includes those that recycle independently of ligand.

gand–receptor interaction can be represented as a reversible bimolecular reaction:

$$[L] + [R] \underset{k_{off}}{\overset{k_{on}}{\rightleftharpoons}} [LR] \tag{1}$$

where [L] is the concentration of free ligand, [R] the concentration of unoccupied receptor, [LR] the concentration of ligand–receptor complex, k_{on} the rate constant for association of ligand with receptor, and k_{off} the rate constant for dissociation of ligand–receptor complex. At equilibrium, the equilibrium constant K is given by

$$K = \frac{[LR]}{[L][R]} = \frac{k_{on}}{k_{off}} \tag{2}$$

and the commonly used measure of ligand affinity for receptor, the equilibrium dissociation constant K_d, is given by

$$K_d = \frac{1}{K} = \frac{[L][R]}{[LR]} = \frac{k_{off}}{k_{on}} \tag{3}$$

Furthermore, the total or maximal number of functional receptors $[\mathrm{R}]_{\mathrm{max}}$ in the population under consideration is the sum of those receptors that are occupied with added ligand [LR] and those that are unoccupied [R]:

$$[\mathrm{R}]_{\mathrm{max}} = [\mathrm{LR}] + [\mathrm{R}] \tag{4}$$

or

$$[\mathrm{R}] = [\mathrm{R}]_{\mathrm{max}} - [\mathrm{LR}] \tag{5}$$

By substituting this expression for [R] into Eq. (3),

$$K_{\mathrm{d}} = \frac{[\mathrm{L}]\,([\mathrm{R}]_{\mathrm{max}} - [\mathrm{LR}])}{[\mathrm{LR}]} \tag{6}$$

and solving for [LR]

$$[\mathrm{LR}] = \frac{[\mathrm{R}]_{\mathrm{max}}\,[\mathrm{L}]}{K_{\mathrm{d}} + [\mathrm{L}]} \tag{7}$$

it can be seen that [LR] is a rectangular hyperbolic function of [L]. By simple rearrangement, this expression can be rewritten in a linear (Scatchard, 1949) form

$$\frac{[\mathrm{LR}]}{[\mathrm{L}]} = -\frac{1}{K_{\mathrm{d}}}[\mathrm{LR}] + \frac{[\mathrm{R}]_{\mathrm{max}}}{K_{\mathrm{d}}} \tag{8}$$

in which the ratio of receptor-bound to free ligand [LR]/[L] is a linear function of receptor-bound ligand [LR]. The slope and intercept on the abscissa allow direct determination of ligand-binding affinity K_{d} and the maximal number of functional receptors $[\mathrm{R}]_{\mathrm{max}}$, respectively.

Inherent to this approach for determining ligand-binding parameters are several assumptions. First, equilibrium must be achieved. Second, both ligand and receptor molecules must be homogeneous populations and be stable during the binding interval. Third, ligand bound specifically to the receptor of interest must be detectable and differentiable from ligand bound nonspecifically to cells or labware. Fourth, the binding affinity of ligand for its receptor must be sufficiently high that little or no dissociation of ligand–receptor complex occurs during maneuvers to separate unbound and bound ligand after equilibrium is reached. Several experimental approaches have been devised to ensure that these criteria are satisfied. Furthermore, alternative methods for analyzing simple bimolecular ligand–receptor binding data as well as methods for analyzing more complex binding phenomena may be employed (for reviews see Limbird, 1986; Levitzky, 1985).

B. Functional Unoccupied Surface Receptor

The Scatchard approach for determining ligand–receptor binding parameters has been used in characterizing the asialoglycoprotein receptor (ASGP-R) on Hep G2 cells (Schwartz *et al.*, 1981). To determine number and affinity of unoccupied functional receptors on the cell surface, equilibrium binding of radioiodinated ASOR, ^{125}I-ASOR (Schwartz *et al.*, 1980), was evaluated.

Cells were grown as monolayers to near confluence in multiwell plates ($\sim 10^6$ cells/well). Choice of well size was determined by assuming that each cell contained 10^5 functional surface receptors, such that with a ligand specific activity of 5×10^4 cpm/pmol and 50 pmol of ligand added per well, sufficient numbers of cells would be present at confluency to allow detection of specific receptor-mediated ligand binding. Immediately prior to an experiment, cultures were transferred to a cold room at 4°C, media were aspirated, and the monolayers were washed by brief emersion in four sequential 1-liter volumes of cold phosphate-buffered saline (PBS) containing Ca^{2+} 1.7 m*M*. Individual wells then were aspirated to dryness.

Ligand binding was evaluated at 4°C to minimize receptor synthesis, degradation, or internalization during the binding interval. To initiate binding, 1 ml of Earle's minimal essential medium (serum-free) containing the desired amount of ^{125}I-ASOR was added to each well. Binding media also contained HEPES (20 m*M*, pH 7.3) and cytochrome *c* (0.1 mg/ml) to minimize nonspecific binding of ligand to cells and plasticware. Total ligand binding was determined in the presence of ^{125}I-ASOR. Nonspecific binding was determined in the presence of excess unlabeled ASOR (5 μ*M*). Specific binding was defined as the difference between total and nonspecific binding. After the binding interval, media containing unbound ligand were aspirated. The adherent cells with their complement of bound ligand were washed by brief immersion in four sequential 1-liter volumes of cold PBS. Cells then were lysed in NaOH (1 *M*, 1 ml/well) and returned to room temperature. After 10 minutes the tissue culture plates were agitated vigorously and the contents were transferred to a tube for quantitation of cell-associated radioactivity by γ counting. Protein content of an aliquot was determined with the Bradford assay (Bradford, 1976). In separate wells, cell number (determined by counting a trypsinized cell suspension in a hemacytometer) was correlated with cell lysate protein content. Radioactivity per well was normalized to 10^6 cells.

Data derived from a representative 90-minute binding isotherm at 4°C are depicted in Fig. 2. Total binding increases in a curvilinear fashion with increasing concentrations of added ligand. Nonspecific binding increases linearly over the same concentration range. The derived specific binding

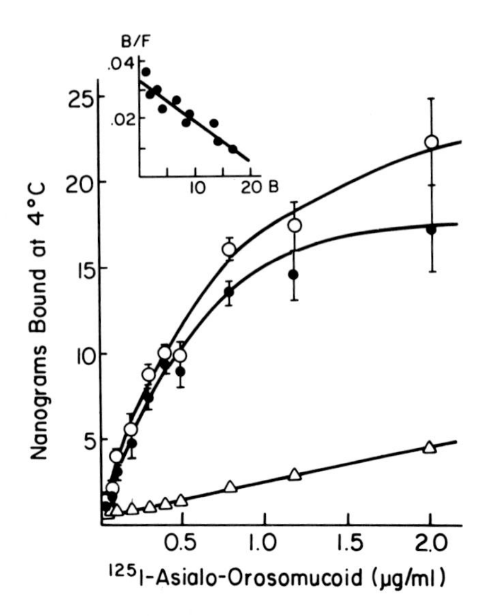

FIG. 2. Saturation binding of ^{125}I-ASOR to Hep G2 cells. Hep G2 cells were incubated for 90 minutes at 4°C with selected concentrations of ^{125}I-ASOR in the absence or presence of excess unlabeled ASOR to allow surface binding as described in the text. Total binding to 10^6 cells (○); nonspecific (△); specific (●). The Scatchard transformation of the specific binding data is shown in the inset. Adapted from Schwartz *et al.* (1981).

has the shape of a rectangular hyperbola as predicted by Eq. (7). The Scatchard linear transformation [see Eq. (8)] of the specific binding data is shown in the inset. The slope and x-intercept allow estimates of $K_d = 7$ nM and $[R]_{max} = 150{,}000$ surface sites per cell to be determined. Separate experiments characterizing the time course of specific binding demonstrated that equilibrium was achieved. In addition, virtually all of the bound ligand could be removed from the cell by incubating monolayers further either with EDTA or with trypsin, indicating that bound ligand was confined to the cell surface. Thus, binding parameters for unoccupied functional receptors on the cell surface can be determined.

C. Total Functional Surface Receptors

A similar approach allows quantitation of the total number of functional receptors on the cell surface, that is, those that are unoccupied as well as those occupied by ligand acquired either from the cell itself or from the serum-containing growth medium. In an initial preparative step bound ligand is removed from the occupied receptors before a binding isotherm is determined. For several ligand–receptor systems, the presence of a divalent cation (Ca^{2+} or Mg^{2+}) is an absolute requirement for ligand binding. Accordingly, chelation of Ca^{2+} or Mg^{2+} with excess EDTA promotes rapid dissociation of ligand from its receptor in such systems. This method has been used in comparing specific binding of ^{125}I-ASOR to the Hep G2 cell ASGP receptor (known to require Ca^{2+}), both with and without pretreatment of cells for 2 minutes with EDTA (5 m*M*) in calcium-free PBS at 4°C. Results obtained suggested that in growing cells 88% of surface receptors were unoccupied and 12% were occupied (Schwartz *et al.*, 1982). Similar strategies have included use of polyanionic molecules or low pH to dissociate bound ligand from receptors before determining binding parameters (Goldstein *et al.*, 1983).

D. Unoccupied Functional Intracellular and Surface Receptors

Intracellular receptors are inaccessible to exogenously added ligand at 4°C. Nevertheless, the cell's total complement of unoccupied functional receptors can be estimated by evaluating ligand binding to cells whose membranes have been rendered permeable to macromolecules. This approach has been used for estimating the total complement of unoccupied functional ASGP-R in Hep G2 cells. Washed monolayers were preincubated for 30 minutes at 4°C with the nonionic detergent saponin (0.1% w/v) in binding medium (Simmons and Schwartz, 1984; Fallon and Schwartz, 1986). Thereafter, detergent was removed, the cells were washed in PBS, and a binding isotherm was determined as described earlier. Saponin treatment was shown to render cell membranes permeable to macromolecules as large as 200 kDa (Shepherd *et al.*, 1984), but not to cause solubilization and loss of integral membrane proteins such as the ASGP-R. Furthermore, this treatment was shown not to cause loss of prebound ligand from receptor nor to alter the affinity of ligand for receptor. Following saponin pretreatment, both total and nonspecific binding of ^{125}I-ASOR to Hep G2 cells increase in comparison with control cells. Parameters derived from binding data following saponin treatment indicate that ~60% of unoccupied functional receptors are intracellular.

E. Effect of Perturbations on Receptor Populations

The equilibrium-binding method just outlined also may be used to assess the effect of perturbations intended to alter the number or affinity of functional receptors. For example, the effect of selected lysosomotropic amines on the ASGP-R at the Hep G2 cell surface was evaluated (Schwartz *et al.*, 1984). Growing Hep G2 cells were incubated for 30 minutes at 37°C in serum-free medium containing primaquine (100 μM). Cells then were washed in PBS at 4°C and a standard ^{125}I-ASOR binding isotherm was determined in the continued presence of primaquine. As shown in Fig. 3, the magnitude of specific binding was lower for primaquine-treated cells than for control cells. Scatchard analysis of the binding data is depicted in the inset. The identical slopes of the two lines indicate that binding affinity of receptor for ligand is unchanged following exposure to primaquine. In contrast, the maximal number of unoccupied surface binding sites is reduced by two-thirds, as reflected by the respective intercepts on the abscissa.

F. Nonfunctional Receptors

Nonfunctional receptors (e.g., immature or otherwise quiescent forms) cannot be evaluated using ligand-binding techniques. Quantitation of this pool of receptors relies largely on immunochemical techniques (i.e., using antireceptor antibody as ligand) with approaches similar to those described earlier and in other chapters in this volume.

G. Nonequilibrium Methods for Evaluating Ligand–Receptor Interaction

Nonequilibrium methods also can be used to determine both binding affinity and individual rate constants for ligand–receptor association or dissociation. Recalling Eq. (1), experimental conditions can be chosen such that added ligand [L] is in vast excess over unoccupied receptors [R]. Accordingly, [L] remains essentially constant and the reaction may be considered "pseudo-first-order" with respect to concentration of unoccupied receptors [R]. The time course of appearance of bound ligand [LR] is represented by

$$\frac{d[\mathrm{LR}]}{dt} = k_{\mathrm{on}}[\mathrm{L}][\mathrm{R}] - k_{\mathrm{off}}[\mathrm{LR}] \tag{9}$$

Substituting the expression for [R] at time t, $[\mathrm{R}]_t$, from Eq. (5) and integrating from time 0 to t, one obtains

$$\ln\left(1 - \frac{[\mathrm{LR}]_t}{[\mathrm{R}]_{\mathrm{max}}}\right) = -(k_{\mathrm{on}}[\mathrm{L}] + k_{\mathrm{off}})\,t \tag{10}$$

$$\ln\left(1 - \frac{[\mathrm{LR}]_t}{[\mathrm{R}]_{\mathrm{max}}}\right) = -k_{\mathrm{obs}}\,t \tag{11}$$

where the observed overall rate constant k_{obs} is given by

$$k_{obs} = k_{on}[L] + k_{off} \tag{12}$$

Experimentally, the amount of bound ligand $[LR]_t$ is determined as a function of time t. Maximal number of receptors $[R]_{max}$ can be estimated separately by allowing equilibrium saturation binding of receptors. Equation (11) predicts $\ln(1 - [LR]_t/[R]_{max})$ to be a linear function of time, with the slope of the line being $-k_{obs}$. Furthermore, as seen in Eq. (12), k_{obs} is expected to be a linear function of the concentration of added ligand [L].

This method has been used to delineate the rate constants for ^{125}I-ASOR binding and dissociation at 4°C from unoccupied ASGP-R on the surface of Hep G2 cells (Fig. 4) (Schwartz *et al.*, 1981). As shown in Fig. 4A, specific binding of ^{125}I-ASOR was determined as a function of time for selected concentrations of added ligand. In Fig. 4B, the data from Fig. 4A were plotted according to Eq. (11). The slopes of the lines obtained allowed estimation of a k_{obs} value for each ligand concentration. In Fig. 4C, the k_{obs} values obtained from Fig. 4B were plotted as a function of

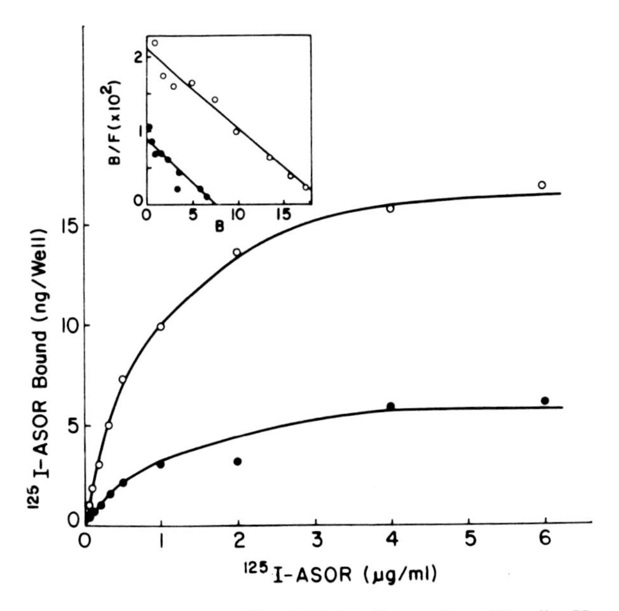

FIG. 3. Effect of primaquine on ^{125}I-ASOR binding to Hep G2 cells. Hep G2 cells were preincubated for 30 minutes at 37°C in the absence (○) or presence (●) of primaquine 100 μM before determining capacity for specific surface binding of ^{125}I-ASOR as described in the text. The Scatchard transformation of each set of specific binding data is shown in the inset.

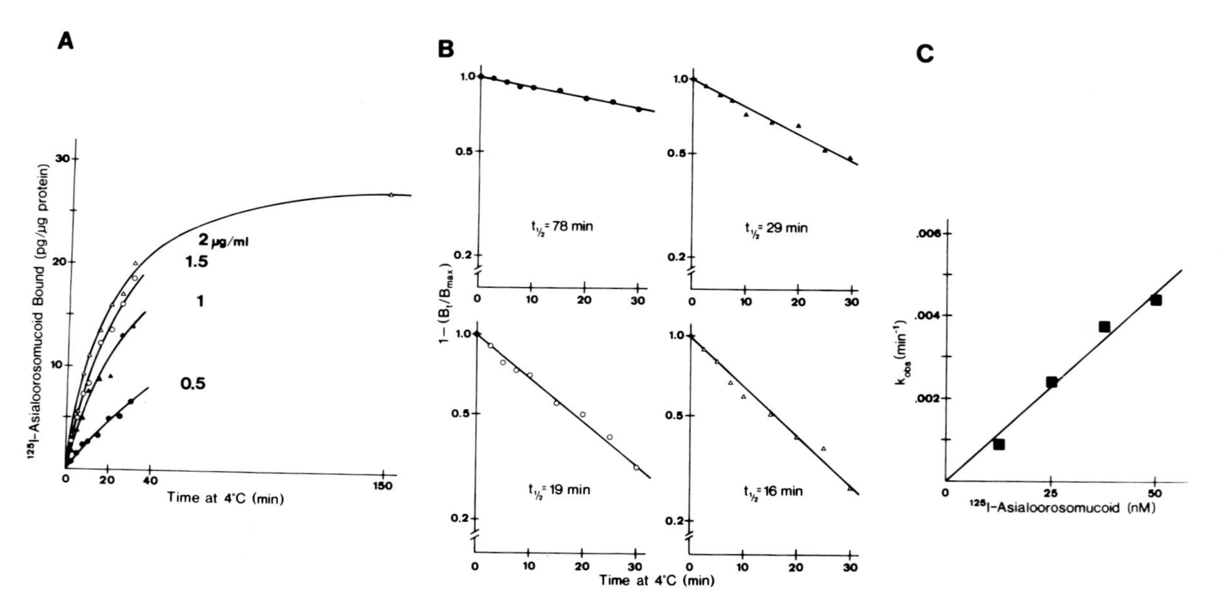

FIG. 4. Rate of ^{125}I-ASOR binding to Hep G2 cells. (A) Specific binding of ^{125}I-ASOR to Hep G2 cells was determined at selected intervals at ligand concentrations of 0.5 (●), 1.0 (▲), 1.5 (○), or 2.0 μg/ml (△). (B) Specific binding data from (A) were tranformed to be a linear function of time to determine kinetic parameters as described in the text. The slope of each line reflects the characteristic observed rate constant for ligand binding at that ligand concentration. (C) The observed rate constants from (B) are a linear function of ligand concentration. Slope and *y*-intercept reflect true k_{on} and k_{off} values, respectively. Adapted from Schwartz *et al.* (1981).

ligand concentration according to Eq. (12). The rate constants k_{on} (10^6 minutes/mol and k_{off} (10^{-3} per minute) were determined from the slope and intercept on the ordinate, respectively. The measure of binding affinity $K_d[= k_{off}/k_{on}$, Eq. (3)] obtained from these data was in close agreement with the value determined under equilibrium conditions (Fig. 2). Thus, with concordance of results derived from two independent approaches for determining K_d, confidence in the validity of the data is enhanced.

V. Ligand Internalization—Single Cohort Kinetics

At 4°C ligand binding to surface receptors on intact cells can be characterized independently of concomitant ligand uptake and degradation. However, at 37°C ligand internalization occurs rapidly following surface binding. Ligand internalization can be evaluated with methods that allow the distribution of a single cohort of prebound cell surface ligand molecules to be determined as a function of time at 37°C. The technique involves establishing a surface cohort of receptor-bound ligand molecules at 4°C. Unbound labeled ligand subsequently is removed and replaced by an excess of unlabeled ligand. Cells then are warmed to 37°C to allow ligand internalization. At selected intervals media are removed and cells are chilled rapidly to 4°C to arrest ligand movement. Cell-associated labeled ligand is partitioned into cell surface and intracellular fractions with protease treatment (or alternative "dissociating" conditions) to remove residual surface-bound ligand. Media are assessed for presence of intact or degraded ligand.

To illustrate, this technique has been used to characterize receptor-mediated endocytosis of ^{125}I-t-PA by Hep G2 cells (Owensby *et al.*, 1988). Cells were incubated for 2 hours at 4°C with ^{125}I-t-PA(3nM) in the absence or presence of unlabeled t-PA (1 μM) to allow specific binding as described before. Unbound ligand was removed by aspiration, and cells with surface-bound ligand were washed in cold PBS. At time 0, 1 ml of prewarmed (37°C) serum-free medium containing unlabeled t-PA (1 μM) was added to each well and the incubation was continued at 37°C. At chosen intervals, media were aspirated and retained. Then, cells quickly were immersed in four sequential 1-liter volumes of PBS at 4°C to cause rapid cooling and to rinse the cell surface. Subsequently, cells were exposed to pronase (0.25% w/v) in serum-free medium at 4°C for 30 minutes to remove residual surface-bound ligand and to detach cells from the wells. Cell suspensions were centrifuged at 13,000 g for 5 minutes. Supernatants containing pronase-sensitive ligand from the cell surface were removed from cell pellets containing pronase-resistant intracellular ligand. Radioactivity of each cell fraction, as well as of the retained media, was quantitated. Aliquots of media also were precipitated by 4% phosphotungstic acid–20% trichloroacetic acid to determine amount of acid-soluble released degraded ligand fragments. The time course of ligand distribution among these defined compartments is shown in Fig. 5. Each symbol reflects the amount and location of specifically bound ligand, that is, the difference between signals detected when initial ligand binding occurred in the absence or presence of excess unlabeled t-PA. Specifically bound ^{125}I-t-PA disappeared rapidly from the cell surface with 50% loss occurring within 5 minutes. Concomitantly, ligand was internalized, reaching a peak level at 12–15 minutes and then declining. Ligand also dissociated from the cell surface and appeared in medium in acid-precipitable form reflecting intact ligand. Degraded low molecular weight ligand fragments were released into the medium only after a delay of 15–20 minutes. Their

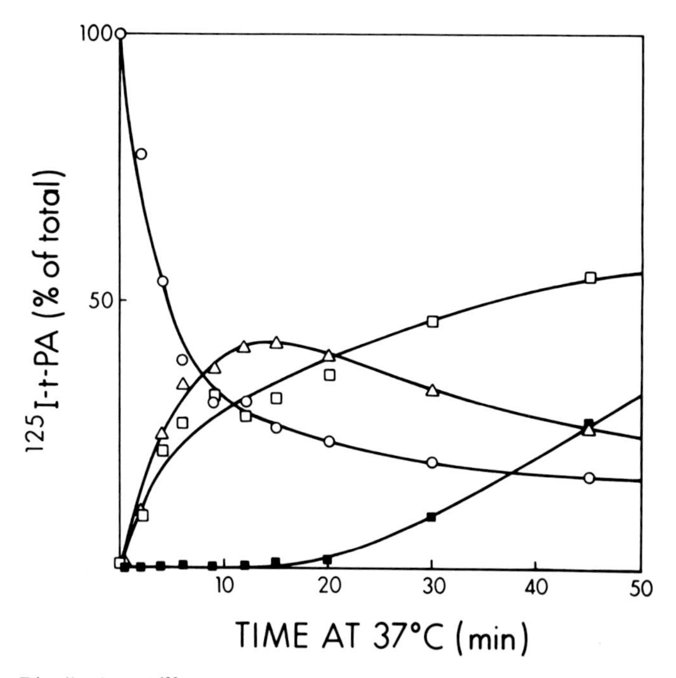

FIG. 5. Distribution of ^{125}I-PA during a single cycle of endocytosis in Hep G2 cells. Hep G2 cells were incubated with ^{125}I-t-PA at 4°C to allow surface binding of ligand. Unbound ligand was removed and the incubation was continued at 37°C for selected intervals to allow ligand uptake. After each interval, media samples were removed and analyzed for degraded ligand (■), and cells were analyzed for intracellular (△) and surface-bound (○) ligand as described in the text. Medium (□). Adapted from Owensby *et al.* (1988).

appearance coincided with the decline in intracellular ligand. This distribution pattern is typical of many ligands whose internalization is receptor-mediated.

VI. Steady-State Rate of Ligand Uptake and Degradation

The capacity of functional unoccupied receptors for ligand uptake and degradation can be evaluated quantitatively at 37°C. Both the interval required to achieve steady state and the maximal rate of ligand catabolism must be determined.

The method used to characterize steady-state uptake and degradation of ^{125}I-t-PA by Hep G2 cells (Owensby *et al.*, 1988) illustrates the experimental technique. Tissue culture wells containing washed cell monolayers were incubated at 37°C with ^{125}I-t-PA (3 n*M*), in the absence or presence of unlabeled t-PA (1 μ*M*). At selected intervals, aliquots of medium were removed and assayed for the presence of released low molecular weight degraded (acid-soluble, see earlier) ligand fragments. At each interval, cells were washed in PBS and lysed to determine cell-associated radioactivity. Results are shown in Fig. 6. Initial ligand uptake is approximately linear for 20–30 minutes. Thereafter, cell-associated ligand plateaus at a steady-state level that is maintained for several hours. During

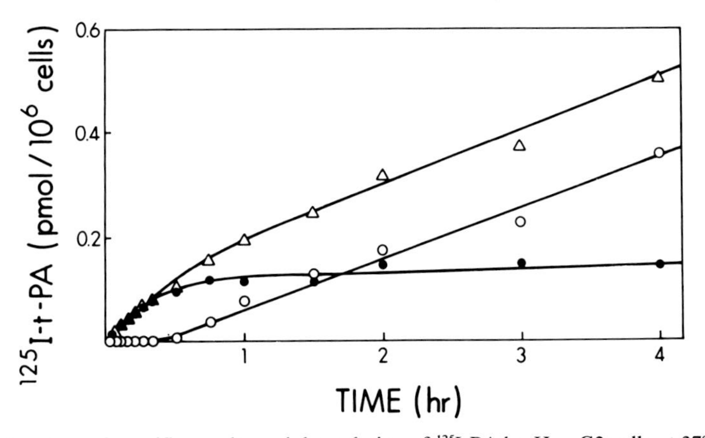

FIG. 6. Rate of specific uptake and degradation of ^{125}I-PA by Hep G2 cells at 37°C. Hep G2 cells were incubated with ^{125}I-t-PA 3 n*M* in the absence or presence of excess unlabeled t-PA for selected intervals at 37°C. After each interval, media samples were removed and analyzed for degraded ligand (○), and cells were analyzed for specific uptake of ligand (●) as described in the text. Total (△). Adapted from Owensby *et al.* (1988).

the initial 20–30 minutes of incubation, the media are devoid of released low molecular weight degraded ligand fragments. Only after this delay does degraded ligand appear in the media. It then accumulates linearly in parallel with overall catabolism (uptake and degradation). Steady state is reached at 1 hour. The slope of lines depicting steady-state ligand uptake and degradation (total), or degradation alone, provides an estimate of the rate of ligand catabolism for a particular ligand concentration. In separate steady-state incubations, uptake and degradation were observed to approach saturation with increasing concentrations of added ligand. Maximal rate of uptake and degradation, 1.2 pmol/10^6 cells per hour, was obtained by dividing the saturation value by the incubation interval.

VII. Receptor Recycling

The existence of a population of recycling receptors can be inferred from the relationship between maximal rate of ligand uptake and degradation and total cellular binding capacity in cycloheximide-treated cells in which no protein synthesis occurs. Saturation binding of ^{125}I-ASOR at 4°C by Hep G2 cells, and maximal uptake and degradation of ligand at 37°C, were unaffected in the presence of cycloheximide 100 μg/ml for at least 2 hours (Schwartz *et al.*, 1982). If the Hep G2 cell's total complement of functional ASGP-R (225,000 per cell) participates in ligand internalization (15,000 molecules/minute/cell at 50 n*M* ^{125}I-ASOR), receptor capacity would be exceeded after only 15 minutes. Continuation of ligand uptake and degradation well beyond this interval, despite the absence of new receptor synthesis, implies receptor recycling.

Kinetic parameters for individual steps (binding of ligand to receptor, internalization of ligand and receptor, dissociation of ligand from receptor, and return of receptor to the cell surface) during receptor-mediated endocytosis of ^{125}I-ASOR by Hep G2 cells at 37°C can be determined (Schwartz *et al.*, 1982). Rate of ligand binding to surface receptors can be measured in the presence of metabolic inhibitors that abolish ligand uptake. For Hep G2 cells this was achieved with NaN_3 (10 m*M*). Analysis of binding data collected at 37°C was analogous to the method described earlier for evaluating binding in isotherms at 4°C. Rate of internalization of a saturating cohort of ^{125}I-ASOR molecules (prebound at 4°C) was determined by allowing ligand internalization for selected intervals at 37°C followed by rapid cooling to 4°C, removal of residual surface ligand with EDTA, and quantitation of cell-associated ligand. Rate constants and mean time the receptor spends in each of these states (see Section IX), together with the overall cycle time, allows calculation of rate parameters for dissociation of ligand and return of unoccupied receptor to the cell surface.

VIII. Ligand Recycling

Following endocytosis a fraction of an internalized cohort of ligand molecules is spared degradation and returns undegraded to the cell surface or to the medium. Thus, ligand as well as receptor is capable of recycling.

Methods for determining kinetic parameters relevant to this pathway of ligand movement are illustrated for ^{125}I-ASOR internalized by Hep G2 cells (Simmons and Schwartz, 1984). Cells were allowed to bind ^{125}I-ASOR to saturation at 4°C as described before. Unbound ligand was removed, and cells were warmed to [illegible]°C for 5 minutes to allow ligand internalization. Cells then were chilled quickly to 4°C to arrest further uptake or degradation. Residual surface-bound ligand was removed by washing with EDTA (10 m*M* in PBS, pH 5), for 5 minutes at 4°C. The incubation was resumed by adding unlabeled ASOR (50 n*M*) at 37°C. After selected intervals ranging from 15 minutes to [illegible] hours, aliquots of medium were removed and analyzed. Media-associated radioactivity comprised not only acid-soluble degraded fragments of ^{125}I-ASOR, but also acid-precipitable ligand (>90% intact as assessed by SDS–PAGE and autoradiography). As shown in Fig. 7, the time course for reappearance of acid-precipitable radioactivity in medium approaches a constant value, ~30% of total (total = sum of cell-associated and media-associated radioactivity). Furthermore, the logarithm of one minus fraction of maximal exocytosis, $\ln(1 - C_t/C_{max})$, is a linear function of time as expected for a pseudo-first-order process. The rate constant can be estimated from the slope.

IX. Simple Kinetic Models

During the past several years a number of mathematical models have been designed to unify and simplify the overall processes of receptor-mediated endocytosis and receptor recycling (Wiley and Cunningham, 1981; Bridges *et al.*, 1982; Schwartz *et al.*, 1982). In general, each model is based on the established rate constants for ligand–receptor binding, internalization, dissociation (and delivery to the degradative compart-

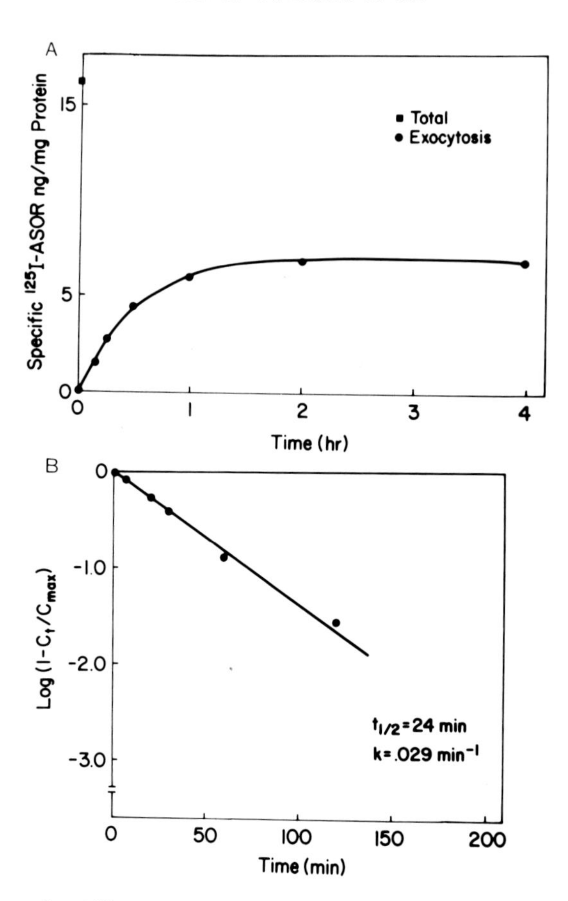

FIG. 7. Exocytosis of ^{125}I-ASOR previously internalized by Hep G2 cells. (A) Hep G2 cells were allowed to internalize a prebound surface cohort of ^{125}I-ASOR molecules for 5 minutes at 37°C. Next cells were chilled to 4°C to arrest ligand movement and to allow removal of residual ligand from the cell surface. In the presence of excess unlabeled ASOR, cells were rewarmed to 37°C for selected intervals. After each interval, media samples were removed and analyzed for intact ligand as described in the text. (B) Exocytosis data from (A) were transformed to be a linear function of time to determine kinetic parameters as described in the text. Adapted from Simmons and Schwartz (1984).

ment), and receptor return to the cell surface. Furthermore, most of the models have been defined on the basis of ligand behavior with the assumption that the ligand is receptor-associated for many of the steps in the pathway (a reasonable assumption, as discussed earlier). In addition, all three models are internally rather consistent, despite significant differences in the systems initially tested.

Wiley and Cunningham (1981) introduced a *steady-state mathematical model* for the binding, uptake, and degradation of polypeptides (epidermal growth factor) by fibroblasts. Central to this model is the concept of the "endocytic rate constant (K_e)," which is the probability of an occupied receptor being internalized in 1 minute at 37°C. This constant, independent of ligand–receptor affinity and receptor number, is a reflection of internalization of the ligand–receptor complex and thus is independent of the extracellular ligand concentration. This rate constant combined with the steady-state equation allows calculation of many of the other rate constants, including the rate of insertion of new receptors at the cell surface. However, application of this model was limited to steady-state analysis. Wiley and Cunningham (1982) later modified their original model to include nonsteady conditions, such that K_e could be determined from initial rates and approach to the steady state. A detailed analysis of this model has been presented by Limbird (1986).

Bridges, Harford, Ashwell, and Klausner have mathematically modeled the *fate of a single cohort of ligand* molecules (i.e., ASOR) in freshly isolated rat hepatocytes (Bridges *et al.*, 1982; Harford *et al.*, 1983) using the ConSamm computer program (Berman *et al.*, 1962). The steps in ligand uptake and processing were connected by linear first-order differential equations and closely simulated the experimental data (Bridges *et al.*, 1982). However, analysis of the same ligand–receptor system in monolayer cells yielded experimental data that fit more closely an expanded linear model (Harford *et al.*, 1983). The current model includes both multiple intracellular processing pathways as well as steps in which there appears to be a delay in ligand flow (Wolkoff *et al.*, 1984).

Schwartz *et al.* (1982) presented a simple kinetic model to determine the cycling time of cell surface receptors during endocytosis. Initially this was applied to the ASGP-R in hepatoma cells (Schwartz *et al.*, 1982), and thereafter extended to the transferrin receptor (Ciechanover *et al.*, 1983). A scheme for this model is presented in Fig. 8 which denotes the kinetic parameters evaluated.

k_1 is the rate constant for binding a ligand, L, to an unoccupied surface receptor, $(R)_s$.

k_2 is the first-order rate constant for internalization of the surface receptor–ligand complex $(LR)_s$.

k_3 is the rate constant of dissociation of the ligand from the internalized receptor, $(LR)_i$.

k_4 is the first-order rate constant for transfer of unoccupied intracelluar receptor, $(R)_i$, to the surface, $(R)_s$. In practice, the rate of dissociation of ligand from intracellular ligand–receptor complexes, k_3, cannot be separated from the return of the receptor to the surface, k_4, and thus a single rate constant, k_x, is employed to encompass both reactions. By definition, $k_x^{-1} = k_3^{-1} + k_4^{-1}$.

In the formulation of this model several simplifying assumptions were made. First, it is assumed that binding of ligand to surface receptors is essentially irreversible and that dissociation of the surface $(LR)_s$ complex is insignificant (see earlier). Second, it is assumed that all functional receptors behave identically and that there is no alternation in functional receptor number during short periods of incubation. In particular, there is little synthesis or degradation of receptor molecules during this period. Thus, the total number of receptors is constant and is composed of unoccupied surface receptors, $(R)_s$, occupied surface receptors, $(LR)_s$, and internal receptors, occupied and free, $(LR)_i$. Third, all ligand-binding receptors are capable of recycling and are equivalent. Therefore,

$$(R)_{total} = (R)_s + (LR)_s + (LR)_i = 1$$

Finally, it is assumed that the system has reached a steady state and, thus, at steady state:

$$\frac{d(R)_s}{dt} = 0 = (k_x)(LR)_i - (k_1)(L)(R)_s$$

$$\frac{d(LR)_s}{dt} = 0 = (k_1)(L)(R)_s - (k_2)(LR)_s$$

Since the concentration of ligand does not change significantly over the time of incubation, (k_1) (L) can be written as one rate constant, (k_1L). Therefore,

$$(k_1L)(R)_s = (k_x)(LR)_i = (k_2)(LR)_s$$

The half-times, $t_{1/2}$, for each of the reactions just given are equal to ln $2/k$. However, the mean time a receptor spends in each state is equal to $t_{1/2}/\ln 2$ or k^{-1}. Thus, the mean time a receptor molecule spends as a surface receptor, $(R)_s$, is equal to $(k_1L)^{-1}$, and the mean time required for a receptor to traverse the entire cycle, T_c, is defined by

$$T_c = (k_1)^{-1} + (k_2)^{-1} + (k_x)^{-1}$$

In the case of ASGP endocytosis, determination of R_s and $(LR)_s$ at steady-state uptake of ligand revealed $R_s = 0.58$, $(LR)_s = 0.14$, and therefore, $(LR)_i = 0.28$. By direct determination $k_1L = 0.115$ per minute and $k_2 = 0.46$ per minute. At steady state this allows a solution for k_x, the rate coefficient for return of receptor to the cell surface, calculated to be

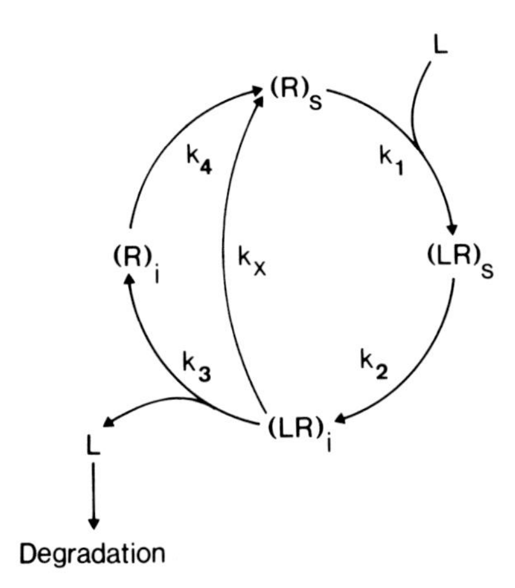

FIG. 8. A kinetic model for receptor-mediated endocytosis. L, Ligand; $(R)_s$, unoccupied surface receptors; $(LR)_s$, occupied surface receptors; $(LR)_i$, occupied internal receptors; $(R)_i$, unoccupied internal receptors; k_1, rate constant for binding; k_2, rate constant for internalization; k_3, rate constant for dissociation of ligand and receptor within the cell; k_4, rate constant for reappearance of receptor to cell surface; k_x, overall rate constant for dissociation of receptor–ligand complex and return of internal receptor to cell surface ($1/k_x = 1/k_3 + 1/k_4$). Adapted from Schwartz *et al.* (1982).

~0.24 per minute. The mean time of the entire receptor cycle is defined as $T_c = 1/k_1 + 1/k_2 + 1/k_x$ and is calculated to be ~15 minutes, at the ligand concentration of 2 μg/ml. Independetly, the overall receptor cycling time can be calculated from the number of total functional receptors divided by the rate of ligand uptake. This yielded a similar value in the example discussed earlier (see Section VII and Schwartz *et al.*, 1982). At *maximal* rates of ligand uptake, the total cycle time is reduced to 7–8 minutes in both cases.

p. 327-328

REFERENCES

Berman, M., Shahn, E., and Weiss, M. F. (1962). *Biophys. J.* **2;** 275–290.

Bolton, A. E., and Hunter, W. M. (1973). *Biochem. J.* **133;** 529–538.

Bradford, M. M. (1976). *Anal. Biochem.* **72;** 248–254.

Bridges, K., Harford, J., Ashwell, G., and Klausner, R. D. (1982). *Proc. Natl. Acad. Sci. U.S.A.* **79;** 350–354.

Ciechanover, A., Schwartz, A. L., Dantry, A., and Lodish, H. F. (1983). *J. Biol. Chem.* **258,** 9681–9689.
David, G. S. (1972). *Biochem. Biophys. Res. Commun.* **48;** 464–471.
David, G. S., and Reisfeld, R. A. (1974). *Biochemistry* **13,** 1014–1021.
Fallon, R. J., and Schwartz, A. L. (1986). *J. Biol. Chem.* **261;** 15081–15089.
Fraker, P. J., and Speck, J. C. (1978). *Biochem. Biophys. Res. Commun.* **80;** 849–857.
Goldstein, J. L., Basu, S. K., and Brown, M. S. (1983). *In* "Methods in Enzymology" (S. Fleischer and B. Fleischer, eds.), vol. 98; pp. 241–260. Academic Press, New York.
Greenwood, F. C., Hunter, W. M., and Glover, J. S. (1963). *Biochem. J.* **89;** 114–123.
Harford, J., Bridges, K., Ashwell, G., and Klausner, R. D. (1983). *J. Biol. Chem.* **258,** 3191–3197.
Levitzki, A. (1985). *In* "Endocytosis" (I. Pastan and M. C. Willingham, eds.), pp. 45–68. Plenum, New York.
Limbird, L. E. (1986). "Cell Surface Receptors: A Short Course on Theory and Methods." Martinus Nijhoff, Boston, Massachusetts.
Marchalonis, J. J. (1969). *Biochem. J.* **113;** 229–305.
Morell, A. G., and Ashwell, G. (1972). *In* "Methods in Enzymology" (V. Ginsburg, ed.), vol. 28; pp. 205–208. Academic Press. New York.
Owensby, D. A., Sobel, B. E., and Schwartz, A. L. (1988). *J. Biol. Chem.* **263,** 10,587–10,594.
Scatchard, G. (1949). *Ann. N.Y. Acad. Sci.* **51;** 660–672.
Schwartz, A. L., Rup, D., and Lodish, H. F. (1980). *J. Biol. Chem.* **255;** 9033–9036.
Schwartz, A. L., Fridovich, S. E., Knowles, B. B., and Lodish, H. F. (1981). *J. Biol. Chem.* **256;** 8878–8881.
Schwartz, A. L., Fridovich, S. E., and Lodish, H. F. (1982). *J. Biol. Chem.* **257,** 4230–4237.
Schwartz, A. L., Bolognesi, A., and Fridovich, S. E. (1984). *J. Cell Biol.* **98;** 732–738.
Shepherd, V. L., Freeze, H. D., Miller, A. L., and Stahl, P. D. (1984). *J. Biol. Chem.* **259,** 2257–2261.
Simmons, C. F., Jr., and Schwartz, A. L. (1984). *Mol. Pharmacol.* **26,** 509–519.
Van Lenten, L., and Ashwell, G. (1972). *In* "Methods in Enzymology" (V. Ginsburg, ed.), vol. 28; pp. 209–211 Academic Press, New York.
Wiley, H. S., and Cunningham, D. D. (1981). *Cell (Cambridge, Mass.)* **25;** 433–440.
Wiley, H. S., and Cunningham, D. D. (1982). *J. Biol. Chem.* **257;** 4222–4229.
Wolkoff, A. F., Klausner, R. D., Ashwell, G., and Harford, J. (1984). *J. Cell Biol.* **98,** 375–381.

Chapter 13

Expression and Analysis of the Polymeric Immunoglobulin Receptor in Madin–Darby Canine Kidney Cells Using Retroviral Vectors

PHILIP P. BREITFELD

Department of Pediatrics
University of Massachusetts
Medical School
Worcester, Massachusetts 01655

JAMES E. CASANOVA, JEANNE M. HARRIS, NEIL E. SIMISTER, AND KEITH E. MOSTOV

Whitehead Institute for Biomedical Research
Cambridge, Massachusetts 02139

I. Introduction

This chapter describes methods for studying the expression and transport of the polymeric immunoglobulin receptor (poly-IgR) in Madin–Darby canine kidney (MDCK) cells. This has been a very useful system

Copyright © 1989 by Academic Press, Inc.
All rights of reproduction in any form reserved.

for analyzing protein traffic and transport in polarized cells (Mostov and Blobel, 1982; Mostov and Simister, 1985; Mostov and Deitcher, 1986; Mostov *et al.*, 1986).

II. Expression of the Poly-IgR in MDCK Cells

A variety of methods have been used to express cloned cDNAs in MDCK cells, both transiently and in stable cell lines. We have primarily used a retroviral system developed by Richard Mulligan and colleagues at the Whitehead Institute (Mann *et al.*, 1983). Although the level of expression is modest, it has been adequate for our studies. Expression is very stable.

The system consists of an expression vector and a packaging cell line. Our earlier work utilized the vector, pDOL (Korman *et al.*, 1987). In this vector, the gene of interest (in this case the rabbit poly-IgR) can be cloned into either a *Bam*HI or *Sal*I site, both of which are driven by the viral long terminal repeat (LTR) promoter. The neomycin-resistance gene is driven by an internal Simian virus 40 (SV40) promoter. Recently we have obtained higher levels of expression (≤5-fold increased) in the unpublished pWE vector. In this vector, the gene of interest (cloned into the *Bam*HI site) is driven by an internal chicken β-actin promoter. The neomycin-resistance gene is driven by the viral LTR. The pWE vector probably does not give better expression than pDOL for most genes in most cell types.

The rabbit poly-IgR has internal *Bam*HI and *Sal*I sites (Mostov *et al.*, 1984). We have therefore added *Bgl*II linkers to the cDNA. This linked DNA can then be inserted into the *Bam*HI site, but the *Bgl*II and *Bam*HI sites are thereby lost. The *Bgl*II-linked DNA can also be converted to blunt end by Klenow fragment fill-in, and blunt ends ligated into the filled-in *Sal*I site of the M13 Mp8–19 series of vectors (Mostov *et al.*, 1986). This provides a convenient substrate for *in vitro*-mutagenesis experiments, which will not be described here. The cDNA can then be retrieved by cutting the M13 replicative form with *Bgl*II, as the *Bgl*II sites are retained by this construction.

Cloning with retroviral vectors should, if possible, be carried out in $RecA^-$ bacterial hosts. Even in such strains, these vectors (especially pWE) are subject to frequent deletions. Once a suitable construct has been made, plasmid DNA is purified by at least one round of CsCl centrifugation.

The ψAM packaging cells are obtained from Richard Mulligan (Cone and Mulligan, 1984). Cells are maintained in Dulbecco's minimal essential medium (DME) with 10% calf serum (not fetal bovine serum), 100 units/ml penicillin, and 100 μg/ml streptomycin in 5% CO_2. Cells are passaged with trypsin–EDTA every 4–7 days. For transfection a confluent 10-cm dish is divided 1 : 10 12–24 hours before use, so that cells are ~20% confluent when transfected.

Plasmid DNA, 10 μg in a volume of 5–20 μl is added to 0.5 ml of sterile HBS in a clear plastic tube. (HBS is prepared by combining 4 g NaCl, 0.185 g KCl, 0.05 g Na_2HPO_4, 0.5 g dextrose, 2.5 g HEPES in ~450 ml. The pH is adjusted to exactly 7.05 with NaOH. After bringing the volume to 500 ml, the solution is filter-sterilized.) Then 32 μl of sterile 2 *M* $CaCl_2$ are added and the tube gently flicked for 20 seconds. The tube is kept at room temperature for 45 minutes to allow a very faint, hazy precipitate to form. The medium is removed from a 10-cm plate of ψAM cells, and the DNA solution is added to the center of the plate. After 10 minutes at room temperature, the plate is gently rocked. After 10 additional minutes, 10 ml of medium are added and the plate placed in the 37°C CO_2 incubator for 4 hours. The medium is removed and 3 ml of a sterile mixture of 85% HBS–15% glycerol are added at room temperature. This is removed after 3.5 minutes and the dish gently washed three times with 10 ml of medium. Finally, 5 ml of medium are added and the dish placed at 37°C for 18 hours.

The virus produced during this incubation can potentially infect humans, although it is theoretically replication-incompetent. Replication-competent virus can be produced via recombination. All necessary biosafety precautions must therefore be observed, and gloves should be worn when handling the virus.

After 18 hours the medium, containing transiently produced virus, is removed. Polybrene (Sigma) is added to a final concentration of 8 μg/ml. (A polybrene stock of 0.8 mg/ml is prepared in H_2O, filter-sterilized, and kept at −20°C.) The virus stock can be frozen at −80°C, although each freeze–thaw cycle decreases the titer somewhat. The titer obtained varies from ~10 to 1000 colony-forming units per ml. Titer is determined by infecting the appropriate cells (in this case MDCK), and counting the number of neomycin-resistant colonies that result.

We have found considerable batch-to-batch variability in the concentration of the neomycin analog, G418, necessary to use with MDCK cells, G418, obtained from Gibco, is dissolved at 100 mg/ml in 0.2 *M* HEPES–NaOH (pH 7.9), sterile-filtered, and stored at −20°C. We have seen slow deterioration over several years of storage. It is necessary to determine the optimal concentration of G418 for each batch of drug.

MDCK cells are maintained in MEM with 5% fetal bovine serum (FBS), penicillin, and streptomycin, in 5% CO_2. (Calf serum can be used, but we prefer fetal serum, because it lacks IgA). A confluent 10-cm dish is split 1 : 10 into several 10-cm dishes. Various amounts of G418 are added to give 0.1–1 mg/ml final concentration. The media and drugs are changed after 7 days. The concentration of drug that kills all cells after 14 days should be used. Strain I (high-resistance) MDCK cells require less G418 than strain II.

We usually infect a 60-cm dish of strain II MDCK cells that is roughly 10% confluent. Then, 1–2 ml of transiently produced virus are added to the dish. After 3 hours, 5 ml of medium are added. In 3 days the cells should be confluent. Cells are trypsinized and taken up in 6.5 ml of medium. To a series of six 10-cm dishes we add 0.1, 0.2, 0.4, 0.8, 1.6, or 3.2 ml of cells. Medium is added to a final volume of 10 ml, and G418 is added. After 7 days, the media and drug are replaced. After 14 days medium can be replaced without adding G418. Colonies are visible after 10–14 days, and are picked around day 18–25.

Colonies should be picked from a plate containing a few well-separated colonies. Circle the desired colonies with a marker on the bottom outside of the plate. Remove all the medium from the plate. Suck the area around each colony dry with a Pasteur pipet and suction hose, using a separate pipet for each colony. Using sterile forceps, an 8-mm glass cloning ring (Bellco) is dipped in autoclaved Vaseline and then placed firmly over the colony. Then, 75 μl of concentrated tryspin–EDTA (0.5% trypsin, 5 m*M* EDTA) are added. The plate is incubated at 37°C for 5–10 minutes. Cells are monitored by phase-contrast microscopy. When the cells have rounded up, 75 μl of medium are added to each ring. Using a P200 Rainin pipetman and a sterile yellow tip, the cells are pipeted up and down a few times in the ring. Cells are then transferred. We generally put 80–90% of cells into a 35-mm dish for screening and the balance into a 25-cm^2 flask as a reserve. We usually screen by metabolic labeling and immunoprecipitation, which is described in a subsequent section. Usually we pick and screen six clones for a given construct.

III. Production of Antibody against Rabbit Secretory Component

Secretory component (SC) is a large proteolytic fragment of the poly-IgR and is easily purified from rabbit bile. Bile is purchased from Pel-Freez. Phenylmethylsulfonyl fluoride (PMSF) is added to 1 m*M*, and the bile is dialyzed against three 12-hour changes of 100 volumes 0.15 *M*NaCl. About 1 ml of bile is loaded onto each of six preparative sodium dodecyl sulfate (SDS)–7% polyacrylamide gels (20 × 20 × 0.15 cm) using the Laemlli system. No reducing agent is used. As a molecular-weight standard, 1 μl of whole serum can be run in a side lane. The gel is stained with Coomassie blue. The smeary complex of bands running slightly slower than serum albumin is SC and is excised. The gel slices are lyophilized and ground with a mortar and pestle.

We have found that guinea pigs produce excellent antibodies against this preparation of SC, and the antibodies bind well to protein A. Repeated bleeding of guinea pigs by cardiac puncture requires considerable skill, and so we have had our antibodies raised by Ribi Immunochemical Research under contract. The animals were injected with 50 mg of ground gel every 3 weeks for six injections, using the Ribi adjuvant. Animals were bled 7–10 days after each of the last three injections.

Whole serum can be used for immunoprecipitation. For many purposes, affinity-purified antibodies are desirable. An affinity resin is prepared by adjusting 20 ml of dialyzed bile to 2% SDS and boiling for 10 minutes. This is coupled to 10 ml of CNBr-activated Sepharose (Pharmacia) using the manufacturer's directions. Two milliliters of antiserum are diluted with 20 ml of mixed micelle buffer (1% Triton X-100, 0.2% SDS, 150 m*M* NaCl, 5 m*M* EDTA, 8% sucrose, 0.1% NaN_3, 10 units/ml Trasylol, and 20 m*M* triethanolamine-HCl, pH 8.6) and recirculated through the column overnight at 4°C. The column is then washed with three column volumes of mixed micelle buffer, followed by two volumes of final wash buffer (i.e., mixed micelle buffer lacking detergents), and finally two volumes of cold H_2O. Elution is with five volumes of 1% of triethylamine in H_2O (pH not adjusted). The eluate is neutralized with 1 *M* HEPES–NaOH, pH 7.4, and concentrated 10-fold with a Centricon concentrator. Typically 8 mg of antibody are recovered.

In our earlier studies we used human dimeric IgA as a ligand for uptake and transport studies. We have found that Fab fragments of the affinity-purified guinea pig anti-SC antibodies are a very useful ligand for the poly-IgR. They give a more reliable, higher signal and much lower background than dimeric IgA. Fab fragments of the affinity-purified antibody are produced using a kit from Pierce Chemical Co. We iodinate the Fab fragments using the iodine monochloride method, which is gentle but not very efficient. The iodine monochloride reagent is prepared as described (Goldstein *et al.*, 1983). Aliquots are stored at −20°C. Before use, the reagent is diluted 1 : 50 in 2 *M* NaCl. To a tube on ice we add, in order, 100 μl 1 *M* Tris-Cl (pH 8.0), 5 μl diluted reagent, 10 μl $Na^{125}I$ (Amersham, 100 mCi/ml), and 50 μg of Fab. After 10 minutes we add 100 μl of 5 g/liter NaI, and 100 μl of 10 mg/ml cytochrome *c* as a carrier. The reaction is dialyzed against three changes of 500 ml of phosphate-buffered saline (PBS) containing 5 g/liter NaI, over 24 hours at 4°C, and is stored at 4°C. It can be used for 1–2 months.

IV. Labeling of Cells Producing Poly-IgR and Immunoprecipitation

For screening clones, we usually label 35-mm dishes of cells. When cell are confluent or nearly confluent, the medium is removed and the monolayer rinsed with PBS. We add 0.5–0.6 ml of labeling medium containing 2–3 μl of [^{35}S]cysteine (Amersham, ~700–1100 Ci/mmol, 10–15 mCi/ml). Labeling medium is DME formulated without cysteine and supplemented with 5% FBS (dialyzed versus 0.15 *M* NaCl) and 20 m*M* HEPES–Na, pH 7.3. We label for 1–2 hours, and gently rock the plates every 15 minutes to keep the cells covered.

To harvest the cells, the labeling medium is removed. After this, 0.5 ml of SDS lysis buffer (0.5% SDS, 150 m*M* NaCl, 5 m*M* EDTA, 100 units/ml Trasylol, 20 m*M* triethanolamine-HCl, pH 8.1) is added, and the cells are scraped off the dish with a policeman. The cell lysate is transferred to a 1.5-ml tube and boiled for 2–5 minutes. After cooling at room temperature for at least 15 minutes, the cells are sonicated. We use a Branson sonicator with a cup horn attachment. The tube is sonicated at full power for two 30-second bursts, with intervening cooling period. Then, 0.5 ml of 2.5% Triton dilution buffer (2.5% Triton X-100, 100 m*M* NaCl, 5 m*M* EDTA, 100 units/ml Trasylol, 0.1% NaN_3, 50 m*M* triethanolamine-HCl, pH 8.6) is added, along with 30 μl of a 50% slurry of Sepharose CL-2B. The tube is mixed by inverting several times. The tube is centrifuged for 5 minutes in a microfuge, and the supernatant is transferred to a new tube. If any particulate material or "globs" of DNA are observed, a second preadsorption with Sepharose can be performed.

Antibody to SC is added to the supernatant. Generally we use 2 μl of a 1 : 10 dilution of whole serum. Immunoprecipitation is performed by placing the tubes on a gently rotating mixer for at least 90 minutes at room temperature, or overnight at 4°C. Twenty microliters of a 15% slurry of protein A–Sepharose (Pharmacia) are added and the tubes mixed for an additional 30 minutes. Beads are washed by brief (5 seconds) centrifugation, and resuspending in 1.4 ml of wash buffer. We usually perform four washes with mixed micelle buffer and one with final wash buffer. The protein A–Sepharose beads are sucked dry with a 50-μl Hamilton syringe. Then SDS–gel sample buffer is added directly to the beads, and the samples are boiled. The samples are analyzed on 7% polyacrylamide gels, which are then fluorographed.

V. Growth of Cells on Filters

Growth of MDCK cells on filters leads to increased cell polarity and allows separate access to the apical and basolateral surfaces. We have primarily used Millicell HA filters from Millipore. Some of the newer products, such as Millicell CM and Transwell (Costar Corp.) may offer lower nonspecific protein binding and/or improved optical properties.

We generally maintain cells on 10-cm tissue culture plates in medium containing 5% FBS. However, we use 10% serum when the cells are actually on the filters. To plate cells on filters we first place a Millicell in a 24- or 6-well tissue culture tray, depending on the size of Millicell. Sufficient medium (containing 10% serum) is added to both the inside and outside of the Millicell to wet the filter. Cells from a confluent 10-cm dish are trypsinized, gently pelleted by centrifugation, and resuspended in 10 ml of medium containing 10% serum. For a small (1.2 cm)-diameter Millicell, 0.4 ml of cells are pipeted into the Millicell, while for a large Millicell (2.4 cm diameter), 2 ml are used. Millicells are then placed in the 37°C incubator, and are generally used after 3 or 4 days. We usually change the medium after 1 or 2 days.

It is not easy to monitor directly the growth of cells on filters and formation of a confluent monolayer. A simple qualitative test relies on hydrostatic pressure. We fill the inside of the Millicell to the brim, and aspirate off the medium outside, so that only enough medium remains outside to cover the bottom of the filter. This results in a hydrostatic pressure pressing the cells down onto the filter. If there is a confluent monolayer, there should be no appreciable leakage of medium out of the Millicell during an overnight incubation. In contrast, incomplete monolayers rapidly leak, equilibrating the fluid level between inside and outside in less than an hour. We generally set up this leak test on every Millicell the night before our intended experiment. It is important during manipulations with Millicells to avoid an upward hydrostatic pressure, which can push the cells off the filter.

VI. Pulse–Chase Analysis of Cells on Filters

Filter-grown cells take up amino acids from the basolateral surface. To label cells we generally first wash the filters two or three times with PBS to remove medium. We usually label with cysteine because the poly-IgR is rich in cysteine and poor in methionine. Cysteine-free medium is assembled from individual components; 5–10% dialyzed serum is used. If labeling is for a short period, it is helpful to starve the cells for cysteine by incubating in cysteine-free medium for 10–15 minutes. Labeling is then accomplished by using cysteine-free medium supplemented with [^{35}S]cysteine (Amersham), generally at 0.5–1.5 mCi/ml. This is most economically accomplished by first stretching a piece of parafilm over a flat piece of plastic. A drop of radioactive medium (75 μl for small Millicells; 200 μl for large) is placed on the parafilm, and the Millicell placed over this. Then, 50–200 μl of nonradioactive, cysteine-free medium are added to the apical surface to keep the cells wet. Depending on the experiments, labeling can be from 5 to 60 minutes. To chase the cells, the Millicells are returned to 24- or 6-well trays, and rinsed once with complete medium. A convenient volume of complete medium is added to both the inside and outside of the Millicell, and the cells are incubated at 37°C for various times.

To harvest the cells on filters, the filter is cut out from the holder with a scalpel. The filter is placed in a 1.5-ml tube containing 0.5 ml of SDS lysis buffer and boiled for 5 minutes. The liquid is then transferred to a new tube and immunoprecipitated as described earlier.

To immunoprecipitate the medium, SDS is added to the medium to a final concentration of 0.8%, and the medium is boiled for 2 minutes. An equal volume of 5% Triton dilution buffer (same as 2.5% Triton dilution buffer, except with 5% Triton X-100) is added and the sample processed for immunoprecipitation in the same manner as the cell extract.

VII. Measurement of Transcytosis

We have previously used dimeric IgA as a ligand for transcytosis, but have recently switched to using iodinated Fab fragments of the antibodies against the receptor. A small Millicell is placed in a 24-well tray, and 0.4 ml of medium is added to both the inside and outside. Labeled Fab is added to either the inside or outside. Generally we use 0.1 μg (2×10^5 cpm) of Fab, but this can be varied. After several hours, the medium on the opposite surface is collected. One-tenth volume of 100% trichloroacetic acid (TCA) is added and the sample placed on ice for 30 minutes. After a 5-minute centrifugation in a microfuge, the supernatant is removed and the pellet counted. We usually find that transcytosis from basal to apical is 25- to 100-fold greater in cells expressing the receptor than in cells that do not make the receptor. Similarly, in cells that make the receptor, transcytosis from basal to apical is 25- to 100-fold greater than transcytosis from apical to basal.

REFERENCES

Cone, R. D., and Mulligan, R. C. (1984). *Proc. Natl. Acad. Sci. U.S.A.* **81,** 6349–6353.

Deitcher, D. L., Neutra, M. R., and Mostov, K. E. (1986). *J. Cell Biol.* **102,** 911–919.

Goldstein, J. L., Basu, S. K., and Brown, M. S. (1983). *In* "Methods in Enzymology" (S. Fleischer and B. Fleischer, eds.), Vol. 96, pp. 241–259. Academic Press, New York.

Korman, A. J., Frantz, J. D., Strominger, J. L., and Mulligan, R. C. (1987). *Proc. Natl. Acad. Sci. U.S.A.* **84,** 2150–2154.

Mann, R., Mulligan, R. C., and Baltimore, D. (1983). *Cell (Cambridge, Mass.)* **33,** 153–159.

Mostov, K. E., and Blobel, G. (1982). *J. Biol. Chem.* **257,** 11816–11812.

Mostov, K. E., and Deitcher, D. L. (1986). *Cell (Cambridge, Mass.)* **46,** 613–621.

Mostov, K. E., and Simister, N. E. (1985). *Cell (Cambridge, Mass.)* **43,** 389–390.

Mostov, K. E., Friedlander, M., and Blobel, G. (1984). *Nature (London)* **308,** 37–43.

Mostov, K. E., deBruyn Kops, A., and Deitcher. D. L. (1986). *Cell, (Cambridge, Mass.)* **47,** 359–364.

Chapter 14

Remodeling of Glycoprotein Oligosaccharides after Endocytosis: A Measure of Transport into Compartments of the Secretory Apparatus

MARTIN D. SNIDER

Department of Biochemistry
School of Medicine
Case Western Reserve University
Cleveland, Ohio 44106

I. Introduction

Nearly all proteins within the organelles involved in secretion and endocytosis are glycosylated. Polypeptides that are synthesized in the endoplasmic reticulum (ER) are extensively modified in an ordered set of reactions located in discrete compartments of the ER and Golgi complex. This glycosylation can be used to track the progress of a protein through the secretory apparatus. If the protein has been modified by an enzyme that resides in a given compartment, the protein must have passed through that compartment.

Copyright © 1989 by Academic Press, Inc.
All rights of reproduction in any form reserved.

This approach can be used in a similar way to study endocytosis of glycoproteins into compartments of the secretory apparatus. The major challenge results from the fact that cell surface glycoproteins have already passed through the secretory apparatus and have been modified. Therefore, it is necessary for the experimenter to create a situation where proteins in post-Golgi compartments have immature oligosaccharides that are substrates for enzymes in the secretory apparatus.

Experiments consist of three stages. (1) Glycoproteins with immature oligosaccharides are generated on the cell surface or other post-Golgi locations that are substrates for a particular modifying enzyme for which the location is known. This has been accomplished by incubating intact cells with glycosidases to digest surface proteins. Cells have also been treated with inhibitors of glycoprotein oligosaccharide synthesis, which results in the synthesis of glycoproteins with immature oligosaccharides. Pulse–chase labeling in the presence of inhibitor allows radioactive glycoproteins with immature oligosaccharides to reach the cell surface. (2) Cells are cultured under conditions in which membrane traffic occurs. (3) Cells are lysed, and individual glycoproteins are isolated and analyzed to determine whether they have been modified by the enzyme under study. If modification has occurred, then transport to the enzyme-containing compartment can be inferred.

This technique has a number of advantages in studying the entry of proteins into enzyme-containing compartments of the secretory apparatus. First, because most of the proteins in organelles of the secretory and endocytic apparatus undergo posttranslational modification, this technique can be used to study the behavior of a large number of proteins. Second, the posttranslational modifications of proteins occurs in a number of compartments in the ER and Golgi complex. The precise locations of many of these enzymes have been determined by cell fractionation and electron-microscopic studies (reviewed in Tartakoff, 1983; Farquhar, 1985; Roth, 1987). As a result, the entry of proteins into a number of different enzyme-containing compartments can be studied. Third, the signal in these experiments represents the transport through the enzyme-containing compartment since the beginning of the experiment. As a result, this method is more sensitive than cell fractionation or microscopic techniques, which only give the content of a given compartment at the moment the cells are examined. Finally, because covalent modifications are used to study transport, it is possible to follow proteins after they have passed through enzyme-containing compartments.

This approach has been used to study the behavior of both cell membrane and soluble glycoproteins in a number of Golgi compartments defined by different enzymes of glycoprotein oligosaccharide synthesis (reviewed in Snider, 1989). Work in our laboratory has concentrated on compartments defined by two of these enzymes: sialyltransferase and Golgi α-mannosidase I. Methods for studying the cycling of proteins through these compartments will be described. The behavior of cell surface transferrin receptor has been most extensively studied in our laboratory. Representative data will be shown from our experiments on this receptor.

In this approach, the behavior of glycoproteins with modified oligosaccharides is studied in cells that have been treated with glycosidases or inhibitors of glycosylation. Therefore, it is necessary to establish that the glycoproteins with modified oligosaccharides are adequate models for the unmodified glycoproteins. To show that this is the case we have compared the properties of the protein under study in treated and untreated cells. For transferrin receptor, we showed that transferrin binding and the rate of receptor internalization are unaffected by oligosaccharide modification (Snider and Rogers, 1985, 1986). In addition, the rate of transport of newly made molecules to the cell surface was not altered by inhibitors of glycosylation. Finally, we have shown that cell viability and rates of growth and macromolecular synthesis were unaffected in modified cells.

II. Transport of Glycoproteins to Sialyltransferase-Containing Compartments

The recycling of proteins from post-Golgi compartments through the secretory apparatus has been monitored most frequently using sialyltransferases. These enzymes have been localized to trans-Golgi cisternae and the trans-Golgi network. Evidence has been provided by studies using cell fractionation (Goldberg and Kornfeld, 1983), autoradiographic and cytochemical localization of the site of sialic acid incorporation (Bennett and O'Shaughnessy, 1981; Roth *et al.*, 1984), and immunocytochemistry using antisialyltransferase antibody (Roth *et al.*, 1985, 1986). Sialic acid residues are nearly always found at the nonreducing termini of N- and O-linked oligosaccharides. As a result, these residues can be readily removed from surface glycoproteins by neuraminidase treatment of intact cells. In addition, the sialic acid content of proteins can be easily monitored. Native gel electrophoresis and isoelectric focusing (IEF), which measure the influence of negatively charged sialic acid residues on the net charge of proteins (Regoeczi *et al.*, 1982, 1984; Snider and Rogers,

1985), as well as affinity chromatography on serotonin–agarose (Sturgeon and Sturgeon, 1982; Fishman and Fine, 1987) and exoglycosidase digestion (Duncan and Kornfeld, 1988) have been used.

Sialyltransferases have been used to monitor the transport both of cell surface receptors (Snider and Rogers, 1985, 1986; Duncan and Kornfeld, 1988) and of soluble proteins that enter cells via receptor-mediated endocytosis (Regoeczi *et al.*, 1982, 1984; Fishman and Fine, 1987). Work in our laboratory has concentrated on cell surface receptors; a method for studying the transport of these molecules to sialyltransferase-containing compartments is given here.

A. Labeling and Neuraminidase Treatment of Cells

Reagents:

PBS–BSA: Dulbecco's phosphate-buffered saline containing 1 mg/ml bovine serum albumin and 1 mg/ml glucose

PNE–BSA: (150 m*M* NaCl, 1 m*M* EDTA, 10 m*M* sodium phosphate (pH 7.4), 1 mg/ml BSA, 1 mg/ml glucose

IEF sample buffer I: 9.5 *M* urea, 2% dithiothreitol (DTT), 2% Nonidet P-40 (NP-40), 0.5% sodium dodecyl sulfate (SDS), and Polybuffer 96 and 74 (Pharmacia), each diluted 1 : 9

Recrystallized urea: A 10 *M* solution of urea is deionized with a mixed-bed ion exchange resin, crystallized at 4°C, and filtered. The crystals are washed with cold ethanol, dried, and stored desiccated at −20°C.

IEF sample buffer II: 1.1 g freshly recrystallized urea dissolved immediately before use in 1 ml of 1% 2-mercaptoethanol (2-ME), 3.7% NP-40, 3.7% Serva Isodalt 3-10 Ampholytes.

Cells are surface-labeled by lactoperoxidase-catalyzed iodination on ice using glucose oxidase and lactoperoxidase (Hynes, 1973). After washing at 4°C with Dulbecco's PBS, 10^7 cells/ml are iodinated in PBS containing 400 μCi/ml $Na^{125}I$ (carrier-free), 5 m*M* glucose, 2 U/ml lactoperoxidase (Calbiochem), and 50–100 mU/ml glucose oxidase for 20 minutes on ice. Then the mixture is made 1 m*M* in unlabeled NaI and washed three times in PBS–BSA. We have found that the ratio of lactoperoxidase to glucose oxidase activities must be at least 20. Lower ratios result in decreased cell viability and rapid degradation of labeled proteins. Under optimal iodination conditions the turnover rates of iodinated and metabolically labeled proteins are similar.

After iodination, 5×10^6 cells/ml are treated with 30 mU/ml *Vibrio cholerae* neuraminidase (Calbiochem) for 1 hour on ice in PBS–BSA.

Cells are then washed twice with PNE–BSA and once with PBS–BSA. Finally, the cells are cultured in growth medium under the desired conditions. After cells are lysed, the protein of interest is immunoprecipitated, and the sialic acid content of the isolated protein is assessed by IEF on polyacrylamide gels under denaturing conditions.

B. Isoelectric Focusing

1. Dissociating and Loading Samples

For some proteins, immunoprecipitates from 1–2.5 × 10^6 cells can be dissociated directly in 50 μl IEF sample buffer I by incubation of 30 minutes at 37°C. However, other proteins do not focus satisfactorily and do not survive freezing in this buffer. In this case, immunoprecipitates are dissociated in 1% SDS, 1% 2-ME, 50 m*M* Tris-HCl (pH 6.8) at 37°C for 1 hour. Samples can be stored at −20°C in this solution. Immediately before focusing, proteins are precipitated from supernatants by adding 100 μg mussel glycogen as carrier and then adding four volumes cold acetone and incubating for at least 2 hours at −20°C. The sample is then centrifuged; the pellet is dried in air to allow the acetone to evaporate, and then dissolved in 20 μl IEF sample buffer II.

2. Gels

Horizontal polyacrylamide slab gels are similar in composition to those used for the first dimension of two-dimensional gels (O'Farrell, 1975). To make a 12 × 15 × 0.08-cm gel, combine

12.375 g Urea
4.5 ml 10% NP-40 (w/v)
3.0 ml 28.4% Acrylamide, 1.6% methylene-bis-acrylamide (w/v)
4.5 ml Polybuffer 96 and 74 (the ratio of these two components is adjusted to give a gel of the desired pH range)
1.125 ml 0.1 mg/ml Riboflavin 5′-phosphate
9 μl 10% Ammonium persulfate
1.5 μl Tetramethylethylene diamine

The urea is dissolved in the first three solutions and degassed under vacumm. Then the final three components are added and the gel is poured into a commercially available casting apparatus modified by attaching a row of 6 × 6 × 0.6-mm pieces of polystyrene to one of the gel plates along one edge. When the gel plates are separated, these create sample wells in the upper surface of the gel. The gel is polymerized under fluorescent light for 15–20 minutes. Gel plates are then separated and two filter paper strips, one soaked in 1 *N* NaOH and the other in 1 *N* H_3PO_4, are placed at each edge of the gel, with the wells close to the basic end. Samples are loaded and platinum electrodes placed over the filter paper strips. A weight of ~200 g is placed on the electrodes to ensure even contact along the length of the gel. The gel is then run at 400 V for 16–20 hours at room temperature. Because the surface of the gel is exposed to air, urea crystals may form during focusing. Optimal results are obtained in gels where some crystallization has occurred. To keep gels from getting too dry, moistened strips of filter paper can be placed next to the gel in the electrophoresis apparatus. After focusing, the gels are fixed and dried, and labeled proteins are visualized by autoradiography or fluorography.

The results of an experiment with transferrin are shown in Fig. 1. Re-

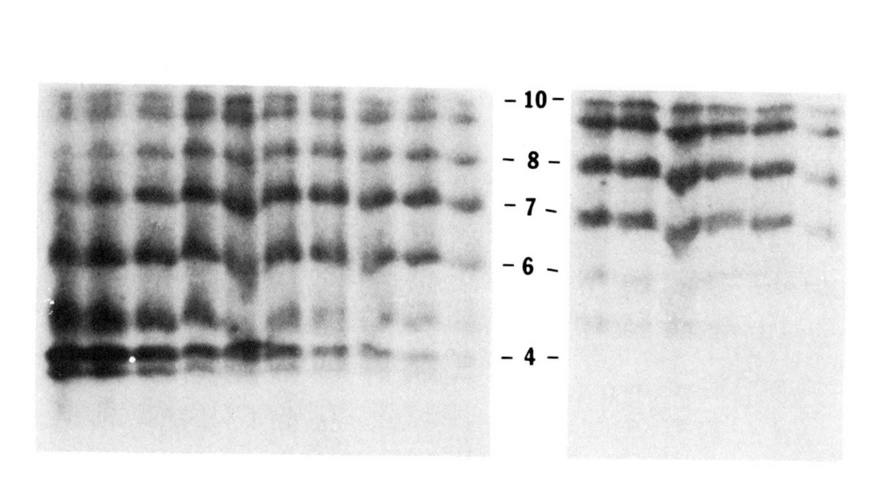

FIG. 1. Resialylation of cell surface asialotransferrin receptor. K562 human erythroleukemia cells were labeled with ^{125}I and then treated with neuraminidase. After culture in growth medium for the indicated times, transferrin receptor was immunoprecipitated and analyzed by IEF. Control cells were treated identically, except that the neuraminidase treatment was omitted. An autoradiograph of the dried gel is shown, with the acidic end at the top. Receptor species have been numbered from basic to acidic. Reproduced from Snider and Rogers (1985) by copyright permission of the Rockefeller University Press.

ceptor from control cells appears as a set of species at the acidic end of the focusing gel. When receptor is desialylated by neuraminidase treatment of the cell surface, these acidic species are lost and replaced by a new set of basic forms. During the reculture of neuraminidase-treated cells, the basic bands disappear and are replaced by acidic species, indicating that surface receptor has been transported to the sialyltransferase compartment. The half-time for this process was 2–3 hours, and nearly all of the receptor was resialylated after 20 hours.

III. Transport of Glycoproteins to Golgi α-Mannosidase I-Containing Compartments

Asparagine-linked oligosaccharides are synthesized from the precursor, $Glc_3Man_9GlcNAc_2$, which is added to glycoproteins in the ER. This high-mannose-type oligosaccharide is then extensively modified to a wide range of mature complex and hybrid-type structures (reviewed in Kornfeld and Kornfeld, 1985). After the Glc residues are removed in the ER, Golgi α-mannosidase I trims α-1,2-linked mannose residues from high-mannose oligosaccharides, yielding $Man_5GlcNAc_2$, which is the substrate for conversion to hybrid and complex species. While the exact location of mannosidase I is not known, it is the first Golgi enzyme to act on the asparagine-linked oligosaccharides of newly made glycoproteins. Because these glycoproteins traverse the Golgi complex in a cis–trans direction, mannosidase I is probably located in an early-Golgi compartment. The reversible inhibitor deoxymannojirimycin (dMM) has been used to study the recycling of glycoproteins through the compartment that contains this enzyme (Snider and Rogers, 1986; Duncan and Kornfeld, 1988; Neefjes *et al.*, 1988).

To study the transport of proteins to the mannosidase I compartment, cells are metabolically labeled and then chased in the presence of dMM to allow the transport of newly made glycoproteins through the Golgi complex. This results in glycoproteins in post-Golgi locations that have high-mannose-type oligosaccharides (primarily $Man_{8-9}GlcNAc_2$) instead of more highly processed ones. The drug is then removed and cells cultured in growth medium. Finally, the cells are lysed and glycoproteins analyzed to see if the immature oligosaccharides on the labeled glycoproteins have been converted to mature forms. A method that has been used in our laboratory to analyze the maturation of glycoprotein oligosaccharides in $[^3H]$mannose-labeled cells will be described here.

A. Labeling of Cells

Cells are pulse-labeled with 0.4–1.0 mCi/ml [2-^{3}H]mannose in minimal essential medium (MEM) with the glucose concentration reduced to 0.1 mg/ml glucose, dialyzed serum, and 1–2 m*M* dMM at a density of $\sim 10^7$ cells/ml for 1 hour. Following this labeling, the cells are washed and chased in growth medium containing dMM for 3–5 hours, a time that is long enough to allow the protein under study to pass through the Golgi complex. The rate of protein transport in the absence of drug is a reasonable guide, since dMM does not affect the intracellular transport of newly made proteins (Burke *et al.*, 1984; Elbein *et al.*, 1984; Snider and Rogers, 1986). Finally, the cells are cultured in dMM-free medium for periods of 1–20 hours. The cells are lysed and oligosaccharide processing analyzed.

B. Assessment of Oligosaccharide Processing

The most direct way to follow oligosaccharide processing is the chromatographic analysis of oligosaccharides and glycopeptides. Glycoproteins are digested with protease and the resulting glycopeptides analyzed by gel filtration (Snider and Rogers, 1986) or lectin-affinity chromatography (Duncan and Kornfeld, 1988). A method for gel filtration chromatography is given here.

For individual proteins, cells are lysed and the protein of interest is immunoprecipitated. The immunoprecipitate is then suspended in 100 μl of 0.1 *M* Tris-HCl (pH 8.0), 20 m*M* DTT, 1 m*M* NaN_3. Pronase (20 μl, 20 mg/ml) is then added and the samples incubated at 50°C for 36 hours, with additional aliquots of pronase added after 12 and 24 hours. Samples are then heated to 95°C for 5 minutes and equilibrated in 0.1 *M* sodium citrate (pH 5.5), 1 m*M* NaN_3 by centrifugation through 1.5-ml columns of Sephadex G-10 in this buffer. After mannosyl-glycoprotein endo-β-*N*-acetylglucosaminidase (Endo H, 30 ng) is added, the samples are incubated for 16 hours at 37°C and then heated for 5 minutes at 95°C. To analyze total glycoproteins, 0.2 mg ovalbumin is added as carrier to cell lysates in 1% NP-40 in PBS, and trichloroacetic acid (TCA, 10% final) is added to precipitate the proteins. The pellets are washed twice with ice-cold acetone, dried and digested with pronase and Endo H as described above.

The resulting glycopeptides and oligosaccharides are analyzed on 1 × 120-cm columns of Biogel P-4 (-400 mesh). This will not be discussed in detail, as it has been extensively described elsewhere (Yamashita *et al.*, 1982; Hubbard and Robbins, 1980). Species are identified by relative elution coefficients (K_d), which are calculated from the elution position and the excluded and included volumes, determined using serum albumin and mannose, respectively.

The results of such an analysis for transferrin receptor from K562 human erythroleukemia cells is shown in Fig. 2. Receptor from control cells has a mixture of high-mannose, hybrid, and complex oligosaccharides. In contrast, after pulse-labeling and chase in the presence of dMM, $Man_{8-9}GlcNAc_2$ are the principal oligosaccharides found on receptor (Fig. 2B). Following reculture without dMM, there has been extensive processing of the receptor oligosaccharides, indicating that receptor has returned to the mannosidase I compartment. Moreover, hybrid and complex oligosaccharides were synthesized during reculture, suggesting that receptors are exposed to other Golgi enzymes of oligosaccharide synthesis as well.

The chromatographic data can be quantitated by summing the radioactivity in each peak. These values are divided by the number of mannose residues in each oligosaccharide species, which yields a value that is proportional to the number of oligosaccharide chains. Converting the data in this way is necessary, because a large loss of radioactivity occurs during the reculture of dMM-treated cells, due to the trimming of Man residues from the labeled high-mannose oligosaccharides (compare Fig. 2B and C). This loss is not caused by a rapid degradation of labeled glycoproteins, because the loss of oligosaccharide chains during reculture is the same in control and dMM-treated cells.

The enzyme Endo H can also be used to assess the processing of glycoproteins during the reculture of dMM-treated cells. This enzyme cleaves high-mannose and hybrid-type oligosaccharides, but not complex species. The $Man_{8-9}GlcNAc_2$ species in dMM-treated cells are Endo H-sensitive, while highly processed complex-type oligosaccharides are not. It should therefore be possible to use the conversion of proteins from Endo H-sensitive to resistant forms during reculture as a test of oligosaccharide processing.

However, we have found that glycoproteins processed during reculture of dMM-treated cells acquire oligosaccharides that are different from those found on the glycoproteins in control cells. In particular, glycoproteins that ordinarily have Endo H-resistant complex oligosaccharides acquire Endo H-sensitive hybrid structures during reculture after dMM removal (compare transferrin receptor oligosaccharides in Fig. 2A and C). Consequently, the processing of these glycoproteins during reculture cannot be detected by Endo H (Neefjes *et al.*, 1988), even though it is readily observed by the direct analysis of oligosaccharides and glycopeptides (Snider and Rogers, 1986).

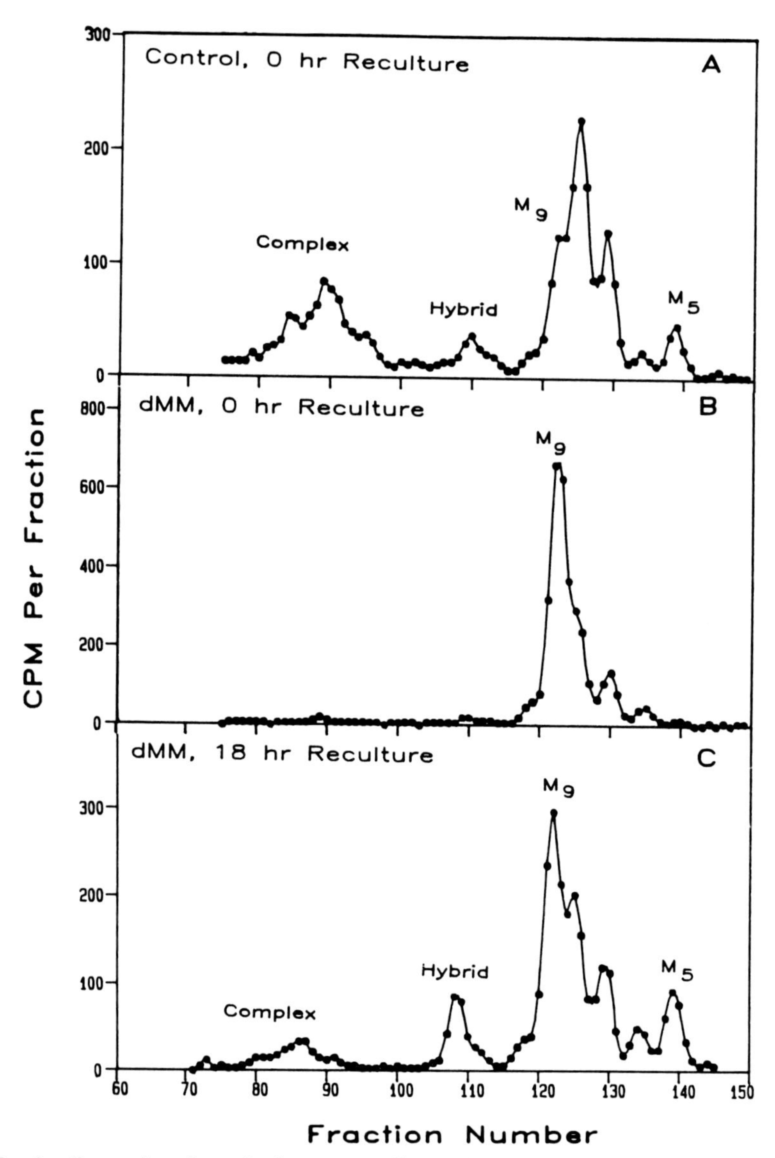

FIG. 2. Processing of transferrin receptor oligosaccharides after removal of dMM. K562 human erythroleukemia cells were labeled with [^{3}H]mannose and chased for 3 hours in the presence of dMM. Cells were lysed and transferrin receptor immunoprecipitated; glycopeptides were then prepared, treated with Endo H, and analyzed by gel filtration chromatography on Biogel P-4, as described in the text. The elution positions of complex glycopeptides, hybrid oligosaccharides, and the high-mannose species with five (M_5) and nine (M_9) mannose residues are shown. (A) Receptor from control cells at the end of the 3-hour chase. (B) Receptor from dMM-treated cells after the chase. (C) Receptor from dMM-treated cells after reculture for 18 hours in dMM-free medium. Reproduced from Snider and Rogers (1986) by copyright permission of the Rockefeller University Press.

IV. Conclusion

The methods described in this article have been used to test the transport of glycoproteins from post-Golgi regions into two compartments of the Golgi complex. Studies using these techniques have helped to demonstrate the entry of identified surface proteins into compartments of the secretory apparatus and to document the mixing of endocytic and exocytic traffic in these compartments. Moreover, it will be possible to extend these methods to test the entry of surface glycoproteins into additional compartments of the ER and Golgi complex defined by other enzymes of glycoprotein synthesis. Likely candidates include the ER glucosidases that process asparagine-linked oligosaccharides as well as galactosyltransferases and *N*-acetylgalactosaminyltransferases (Kozarsky *et al.*, 1988).

Acknowledgments

Work from the author's laboratory was supported by NIH grant GM38183 and a Pew Scholarship in the Biomedical Sciences to M.D.S.

References

Bennett, G., and O'Shaughnessy, D. (1981). *J. Cell Biol.* **88,** 1–15.

Burke, B., Matlin, K., Bause, E., Legler, G., Peyrieras, N., and Ploegh, H. (1984). *EMBO J.* **3,** 551–556.

Duncan, J. R., and Kornfeld, S. (1988). *J. Cell Biol.* **106,** 617–628.

Elbein, A. D., Legler, G., Tlusty, A. McDowell, W., and Schwarz, R. (1984). *Arch. Biochem. Biophys.* **235,** 579–588.

Farquhar, M. G. (1985). *Annu. Rev. Cell Biol.* **1,** 447–488.

Fishman, J. B., and Fine, R. E. (1987). *Cell (Cambridge, Mass.)* **48,** 157–164.

Goldberg, D. E., and Kornfeld, S. (1983). *J. Biol. Chem.* **258,** 3159–3165.

Hubbard, S. C., and Robbins, P. W. (1980). *J. Biol. Chem.* **255,** 11782–11793.

Hynes, R. O. (1973). *Proc. Natl. Acad. Sci. U.S.A.* **70,** 3170–3174.

Kornfeld, R., and Kornfeld, S. (1985). *Annu. Rev. Biochem.* **54,** 631–664.

Kozarsky, K., Kingsley, D., and Krieger, M. (1988). *Proc. Natl. Adad. Sci. U.S.A.* **85,** 4335–4339.

Neefjes, J. J., Verkerk, J. M. H., Broxterman, H. J. G., van der Marel, G. A., van Boom, J. H., and Ploegh, H. L. (1988). *J. Cell Biol.* **107,** 79–88.

O'Farrell, P. H. (1975). *J. Biol. Chem.* **250,** 4007–4021.

Regoeczi, E., Chindemi, P. A., Debanne, M. T., and Charlwood, P. A. (1982). *Proc. Natl. Acad. Sci. U.S.A.* **79,** 2226–2230.

Regoeczi, E., Chindemi, P. A., and Debanne, M. T. (1984). *Can. J. Biochem. Cell Biol.* **62,** 853–858.

Roth, J. (1987). *Biochim. Biophys. Acta* **906,** 405–436.

Roth, J., Lucocq, J. M., and Charest, P. M. (1984). *J. Histochem. Cytochem.* **32,** 1167–1176.

Roth, J., Taatjes, D. J., Lucocq, J. M., Weinstein, J., and Paulson, J. C. (1985). *Cell (Cambridge, Mass.)* **43,** 287–295.

Roth, J., Taatjes, D. J., Weinstein, J., Paulson, J. C., Greenwell, P., and Watkins, W. M. (1986). *J. Biol. Chem.* **261,** 14307–14312.

Snider, M. D. (1989). *In* "Intracellular Trafficking of Proteins" (J. Hanover and C. Steer, eds.). Cambridge Univ. Press, London and New York (in press).

Snider, M. D., and Rogers, O. C. (1985). *J. Cell Biol.* **100,** 826–834.

Snider, M. D., and Rogers, O. C. (1986). *J. Cell Biol.* **103,** 265–275.

Sturgeon, R. J., and Sturgeon, C. M. (1982). *Carbohydr. Res.* **103,** 213–219.

Tartakoff, A. M. (1983). *Int. Rev. Cytol.* **85,** 221–252.

Yamashita, K., Mizuochi, T., and Kobata, A. (1982). *In* Methods in Enzymology" (V. Ginsbury, ed.), Vol. 83, pp. 105–126. Academic Press, New York.

Chapter 15

A Flow-Cytometric Method for the Quantitative Analysis of Intracellular and Surface Membrane Antigens

JERROLD R. TURNER AND ALAN M. TARTAKOFF

Department of Pathology
Case Western Reserve University School of Medicine
Cleveland, Ohio 44106

MELVIN BERGER

Departments of Pediatrics and Pathology
Case Western Reserve University School of Medicine
Cleveland, Ohio 44106

Copyright © 1989 by Academic Press, Inc.
All rights of reproduction in any form reserved.

I. Introduction

In this chapter we describe a new method that allows discrimination between intracellular and plasma membrane antigens, and their quantitative measurement. A typical procedure for the study of human neutrophils is presented in Section IV, following consideration of the advantages, limitations, and potential technical obstacles of this approach.

The study of membrane traffic and sorting of membrane-associated proteins often focuses on the movement of proteins between the cell surface and intracellular compartments. In such cases, simultaneous measurement of intracellular and plasma membrane pools allows careful analysis of a protein's life cycle. Several quantitative immunological approaches have been used for such measurements. These include enzyme-linked and radioimmunoassays applied to cell populations, which compare antigen accessible to surface labeling on intact cells (i.e., surface antigen) to that accessible in detergent-permeabilized cells or extracts of detergent-solubilized cells (i.e., total cellular antigen) (Lipincott-Schwartz and Fambrough, 1986; Tse *et al.*, 1986; Unkeless and Healey, 1983). Intracellular antigen is then calculated as the difference between these two values. The information obtained from this type of assay is limited, since only average antigen content (surface or total) for a population of cells is measured. Thus, when measuring surface antigen, dead or broken cells may artifactually inflate the values obtained. Errors in cell counting or cell loss during washes may also render calculated average values unreliable. In addition, inefficiency of extraction or association of antigens with cytoskeletal elements or other insoluble structures may lead to underestimation of antigen pools. Finally, the existence of subpopulations of cells with distinct patterns of antigen distribution cannot be detected by methods that average the entire population.

To overcome these limitations, we have developed a new flow-cytometric approach for the measurement of total cellular antigen and plasma membrane (exposed) antigen (Turner *et al.*, 1988). Because flow cytometry measures each cell individually, subpopulations of cells with distinct antigen distributions are easily identified, and errors due to the calculation of average antigen content are eliminated. A further advantage of this method is that aliquots of the same samples used for flow cytometry may be inspected by fluorescence microscopy, thus permitting morphological examination of the intracellular antigen distribution as well as correlation with the quantitative results. Finally, the morphological observations may be extended to the ultrastructural level with minimal alterations of the immunostaining technique. The major requirements of this strategy are that the molecule studied must be immunoreactive, accessible to antibody, and not extracted or excessively altered by fixation and permeabilization.

We used this new approach to analyze trafficking of the complement C3b/C4b receptor (CR1) of human neutrophils (Turner *et al.*, 1988). While circulating neutrophils have only 5500 surface CR1 molecules (Fearon and Collins, 1983), translocation of intracellular CR1 pools results in a 6- to 10-fold increase in surface CR1 expression within minutes of neutrophil activation by a large number of compounds, including the synthetic chemoattractant *N*-formyl-Met-Leu-Phe (fMLP) or the calcium ionophore ionomycin (Fearon and Collins, 1983; Turner *et al.*, 1988). These increased levels of surface CR1 persist for at least 2 hours (Berger *et al.*, 1984), but it is not known whether such surface expression results from unidirectional recruitment to the cell surface or represents a new dynamic equilibrium between surface and intracellular pools. To characterize the redistribution of the total cellular pool of CR1 during neutrophil stimulation, we used monoclonal anti-CR1 antibody and flow cytometry to measure total detectable CR1 in fixed, saponin-permeabilized neutrophils, and surface CR1 on similarly fixed, but nonpermeabilized cells (Turner *et al.*, 1988).

II. Reagent Selection

The method described is a modification of routine indirect-immunofluorescence techniques commonly used for quantitative flow cytometry of surface antigens. The major technical problems encountered are likely to concern either loss of antigen reactivity during fixation or excessive nonspecific fluorescence, particularly after permeabilization. Thus, the selection of fixative, permeabilizing agent, and immune reagents is critical.

A. Antibodies

Because the flow cytometer measures all cell-associated fluorescence, it is essential that all of the immune reagents be highly purified. Thus, affinity-purified specific antibodies are preferable to whole antisera or immunoglobulin fractions. We have found that nonspecific fluorescence is variable and often indistinguishable from specific fluorescence when secondary-antibody conjugates are not affinity-purified. Cappel (West Chester, PA) and Boehringer Mannheim (Indianapolis, IN) have proved to be reliable sources of affinity-purified fluorescently tagged antibodies. Although we have generally used fluorescein-conjugated secondary antibodies, there is no reason to suspect that other fluorochromes will not perform satisfactorily, with one caveat. Due to the size of the phycobiliprotein component (100–250 KDa), immunoglobulin molecules conjugated to one or more phycobiliprotein molecules (e.g., phycoerythrin) may not penetrate saponin-permeabilized cells adequately to ensure uniform staining and washing.

Antibody dilutions must be determined empirically by preliminary titration experiments. Antibody should be saturating, but nonspecific fluorescence may interfere with analysis at excessive antibody concentrations (Jacobberger *et al.*, 1986). Nonspecific fluorescence can be minimized by using antibodies at concentrations just greater than those that give the maximum specific staining. If nonspecific staining persists, substitution of the 0.1% ovalbumin normally included in the staining solutions (Section IV,A, step 2) with 1–5% filtered serum of the same species as the second reagent (e.g., goat or sheep) may reduce nonspecific staining.

To remove aggregates that can preferentially stain surface antigen or cause artifactually bright cells, we generally prespin concentrated antibodies in a Beckman Airfuge (100,000 *g*) for 10 minutes immediately prior to use. The top one-third of the supernatant is diluted in phosphate-buffered saline (PBS) containing 0.1% ovalbumin and subsequently used for staining.

B. Fixatives

The choice of fixative is very important, since chemical derivatization and crosslinking reduce the immunoreactivity of many antigens. Of particular concern when using monoclonal antibodies (mAb) is the fact that some epitopes will not even tolerate 0.1% glutaraldehyde. In addition, crosslinking of cytoplasmic proteins by glutaraldehyde may restrict immunoglobulin penetration through the cytosol (Ohtsuki *et al.*, 1978). Buffered formaldehyde is gentler than glutaraldehyde, but it may be less effective in preventing antigen extraction and retaining ultrastructure. We have found the periodate–lysine–paraformaldehyde–fixative described by McLean and Nakane (1974) to be an excellent compromise. No matter what fixative is used, several control experiments must be done to verify that loss of antigen reactivity is minimal and does not interfere with the quantitative analysis. These controls include quantitative studies of surface antigen detection, and are described later (Section III,B).

C. Permeabilizing Agents

The procedure described uses saponin to permeabilize fixed cells. The saponins (sapogenin glycosides) are a heterogeneous group of plant glycosides. Commercially available saponin preparations are impure, and preparations of saponin from different suppliers vary significantly. We have found Sigma (St. Louis, MO) to be a reliable and consistent source with minimal lot-to-lot variation. Saponin binds cholesterol in biological membranes to form membrane pores (Elias *et al.*, 1979). Since the cholesterol content of different cell types and intracellular membranes may vary, the concentration to be used should be determined empirically as that which gives the maximum signal for the cell type and antigen studied. When using Sigma saponin, we have found concentrations of 0.04–0.06% to be suitable for immunofluorescence of most cells.

A variety of reagents other than saponin may be used for permeabilization, including Triton X-100 (TX-100), methanol, and lysolecithin (Goldenthal *et al.*, 1985; Jacobberger *et al.*, 1986; Schroff *et al.*, 1984). In fact, saponin permeabilization may be ineffective for staining of nuclear and mitochondrial antigens (Goldenthal *et al.*, 1985; Elias *et al.*, 1979). However, TX-100 may extract antigens, even from fixed cells (Goldenthal *et al.*, 1985), and cells permeabilized with other detergents may not be suitable for flow cytometry.

III. Control Experiments

A. Nonspecific Background Fluorescence

To evaluate nonspecific background fluorescence when using mAb, we generally omit primary antibody and stain only with the secondary fluorescent antibody. This practice was justified by preliminary experiments that demonstrated that the mean fluorescence of living, fixed, or fixed permeabilized neutrophils stained with secondary antibody alone was identical to that of similar samples stained with irrelevant primary antibody of the same isotype as the immune monoclonal and the same secondary fluorescent antibody. The equivalence of these values in our studies is probably due to the fact that neutrophils have only low-affinity Fc receptors and do not bind monomeric IgG. The validity of staining only with secondary antibody to assess nonspecific background fluorescence should not be assumed, particularly for cells that can bind monomeric IgG, but must be confirmed for each experimental system. The background fluorescence of unstained cells is not an adequate negative control, since the contribution of nonspecifically bound antibody to the total fluorescence must be considered. Nonspecific background fluorescence differed for living, fixed, and fixed permeabilized neutrophils. Thus, a separate nonspecific background fluorescence control should be included in each experiment for each condition used. Mean specific fluorescence of each sample is routinely calculated by subtracting the arithmetic mean fluorescence of the appropriate nonspecific control from the mean fluorescence of that sample. We use the mean specific fluorescence of the fixed permeabilized nonstimulated cells within each experiment as the reference point against which the specific fluorescence of other samples is normalized. This standard was chosen because it was the brightest and least variable sample in each experiment. Normalizing the values facilitates comparison of several experiments and controls for minor differences in cytometer settings, antibody dilutions, and cell handling.

B. Fixation

A number of control experiments are necessary to evaluate the effects of fixation and permeabilization. First, the decrease, if any, in antigen immunoreactivity during fixation must be determined. This is most easily accomplished by comparing aliquots of fixed cells to identical aliquots of living cells stained in parallel using the same reagents. (Living cells should be handled at 4°C in buffers containing 0.05% NaN_3 to avoid capping and endocytosis.) It is preferable to make this comparison over a wide range of surface antigen expression. In the context of the system studied, it may be possible to modulate the surface antigen expression physiologically. For example, we have upregulated the surface expression of CR1 to varying degrees by treatment with fMLP or ionomycin for several incubation durations (Turner *et al.*, 1988). If physiological regulation is not possible, graded protease treatment (before removing the aliquot to be washed and fixed) might provide samples with a broad range of surface antigen expression.

If each sample constitutes a single homogeneous population of cells, linear regression analysis of the mean specific fluorescence of fixed samples plotted as a function of the mean specific fluorescence of equivalent living samples should generate a line that passes through the origin. Ideally, all points should fall close to the line without a plateau at either extreme. If a plateau is present, fixation may be altering surface antigen reactivity nonlinearly. Points in the region of the plateau should be excluded from the analysis, with future measurements limited to the linear range. The difference between the slope of the regression line and 1.0 will reflect the loss of detectable antigen due to fixation. In our example (Fig. 1), the slope of the line determined by linear-regression analysis was 0.82, indicating a maximal loss of 18% of antigen reactivity due to fixation. The correlation coefficient *(r)*, 0.87 in our example, indicates the degree to which the relation between antigen detected on fixed versus living cells is linear.

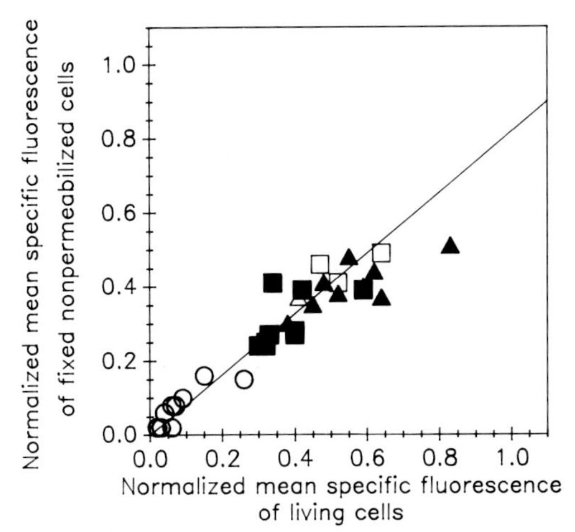

FIG. 1. Analysis of the effect of fixation on surface antigen detection. Isolated human neutrophils were incubated without stimuli at 0°C or 37°C for 0–60 minutes (○), with fMLP for 15 minutes (△) or 30–60 minutes (▲), or with ionomycin for 15 minutes (□) or 30–60 minutes (■). Linear-regression analysis of living and fixed nonpermeabilized aliquots stained for surface CR1 generated the solid line shown with slope of 0.82, r = 0.87. The mean specific fluorescence of each sample was normalized to the mean specific fluorescence of the fixed permeabilized nonstimulated cells from the same experiment as described in the text (Section III,A).

Since formaldehyde fixation may partially permeabilize some cells, care should be taken to evaluate whether fixation increases the levels of detectable antigen, that is, results in staining intracellular antigen. This might result in the mean specific fluorescence of fixed cells being greater than the mean specific fluorescence of living cells, and would be apparent in the graph prepared in Fig. 1 as a weak correlation and, possibly, a slope >1.0. Fluorescence microscopy is another effective way to evaluate staining of intracellular antigen in fixed nonpermeabilized cells.

C. Permeabilization

Once it has been established that fixation does not result in decreased detection of surface antigen, or that the decrease is linear throughout the range studied, the effects of saponin permeabilization must be evaluated. To determine whether surface antigen is extracted by saponin permeabilization of fixed cells, an aliquot of fixed nonpermeabilized cells should be stained by indirect immunofluorescence and analyzed by flow cytometry. The remainder of each sample should be treated with saponin and reanalyzed. If the antigen is not extracted by saponin, there should be no change in the mean specific fluorescence after saponin treatment.

Since intracellular antigen is not in the same environment as surface antigen, its loss of immunoreactivity following fixation may differ. Additionally, crosslinking of the surrounding protein matrix may make immunoreactive intracellular antigen inaccessible, even after permeabilization. The fixative's tonicity may influence the degree to which crosslinking limits intracellular diffusion of antibodies. Finally, the possibility that very close packing of antigen may result in excessive local fluorochrome concentrations and self-quenching of fluorescence should be considered. Thus, at least one control should examine the detectability of intracellular antigen in fixed permeabilized cells. This can be assessed in a variety of ways, some of which are only applicable to specific cell types and antigens (Turner *et al.,* 1988). The method chosen to evaluate intracellular antigen detectability in fixed saponin-permeabilized cells will depend on the system studied and the availability of alternate assays for intracellular antigen. Ideally, correlation of total cellular antigen measured by flow cytometry with independent measurements by an unrelated method should verify the results of each.

In our example, an independent measure of total cellular antigen was not available. Thus, to demonstrate that the measured changes in total detectable CR1 were the result of neutrophil activation and were not due to variables related to fixation or permeabilization, we compared total CR1 detected in fixed permeabilized neutrophils to surface CR1 detected on fixed nonpermeabilized neutrophils (Fig. 2). We were concerned that stimulated cells, with increased surface CR1, consistently had less total cellular CR1 than nonstimulated cells. To determine if this result was due to selective sensitivity of surface antigen to fixation, we estimated the maximum decrease in surface CR1 detection due to fixation as 18% of surface antigen (from Section III,B and Fig. 1). This is shown by the dashed line in Fig. 2, beginning at the theoretical point where 100% of the antigen is intracellular. The points representing stimulated neutrophils all fall well below the dashed line, indicating that selective destruction of surface CR1 during fixation cannot account for the observed loss of total cellular antigen. As in Fig. 1, we used neutrophils treated with different stimuli under various incubation conditions to achieve a broad range of surface expression. Since multiple values for total CR1 (fixed permeabilized cells) exist for any single level of surface expression (fixed nonpermeabilized cells), measured changes in total detectable CR1 are not likely to be related to surface CR1 expression. Furthermore, the data points tend to cluster according to the specific stimuli and incubation conditions, suggesting that the extent of total CR1 loss varied according to the type

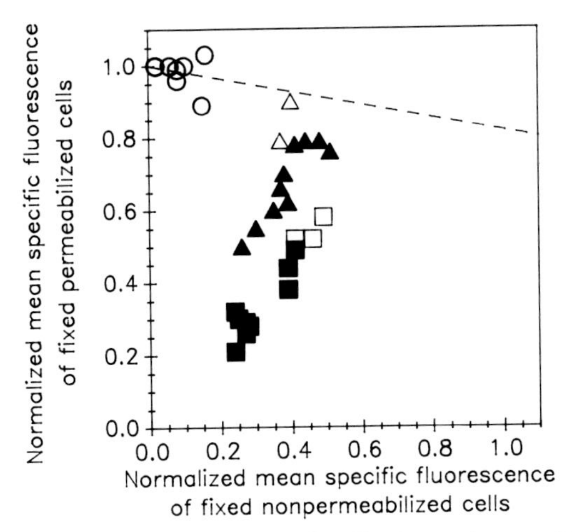

FIG. 2. Detection of intracellular antigen after fixation and permeabilization. Isolated human neutrophils were incubated without stimuli at 0°C or 37°C for 0–60 minutes (○), with fMLP for 15 minutes (△) or 30–60 minutes (▲), or with ionomycin for 15 minutes (□) or 30–60 minutes (■). After fixation, nonpermeabilized and permeabilized aliquots were stained for surface and total CR1, respectively. The dashed line estimates the maximum decrease in surface CR1 detection due to fixation as 18% of surface antigen (from Fig. 1). The mean specific fluorescence of each sample was normalized to the mean specific fluorescence of the fixed permeabilized nonstimulated cells from the same experiment as described in the text (Section III,A).

and duration of stimulation. These observations suggest that the decrease in total detectable CR1 was not due to fixation or permeabilization, but was the result of cellular processes and receptor trafficking pathways activated during stimulation of the neutrophils.

IV. Immunostaining and Flow Cytometry

A. Typical Immunostaining Procedure

1. To measure total and surface antigen, 2×10^6 cells should be prepared for each condition and washed once in 4°C PBS containing 0.05% NaN_3. The supernatant is discarded, and the pellet resuspended in 2–3ml of freshly prepared 4°C periodate–lysine–paraformaldehyde fixative (see Table I) for 20–30 minutes. The samples are kept on ice for the first 5–10 minutes of fixation, and removed from the ice to warm slowly to room

TABLE I

PREPARATION OF PERIODATE–LYSINE–PARAFORMALDEHYDE FIXATIVE[a]

Step 1: 50 m*M* Na–PO_4, 100 m*M* lysine, pH 7.4 buffer. 1.827 g of lysine-HCl is dissolved in 50 ml of distilled H_2O. The pH is adjusted to 7.4 with 0.1 *M* Na_2HPO_4. The total volume is brought to 100 ml with 0.1 *M* Na–PO_4 (pH 7.4), prepared by mixing 0.1 *M* Na_2HPO_4 and 0.1 *M* NaH_2PO_4.

Step 2: 8% Formaldehyde. 0.8g Paraformaldehyde is dissolved in 10 ml distilled H_2O by heating to 65°C and adding 25 μl 1 *N* NaOH. The solution should be filtered and allowed to cool before use. Formaldehyde may be stored at 4°C for 1 or 2 days.

Step 3: Ten milliliters of periodate–lysine–paraformaldehyde fixative can be prepared by dissolving 21.4 mg of $NaIO_4$ in a mixture of 7.5 ml of 50 m*M* Na–PO_4, 100 m*M* lysine (pH 7.4) buffer and 2.5 ml of 8% formaldehyde. The pH will be ~ 6.2. The fixative should be used within 30–60 minutes of preparation.

[a]From McLean and Nakane (1974).

temperature for the remainder of the fixation. All subsequent steps are at room temperature.

2. After fixation, the samples are diluted in 5 ml of room-temperature PBS containing 0.1% ovalbumin (PBS–ovalbumin) and pelleted by centrifugation at 200 *g* for 10 minutes. After fixation the cells may not form as compact a pellet as they had previously, so we generally fix in conical tubes to facilitate complete removal of the supernatant. The pellet is resuspended in 5 ml of PBS–ovalbumin and pelleted once again.

3. Each sample should now be resuspended in 3 ml of PBS–ovalbumin, and divided into two 1.5-ml aliquots in round-bottom tubes. These aliquots should be pelleted, and the first resuspended in 2 ml PBS–ovalbumin for staining of surface antigen. The second aliquot will be used to measure total cellular antigen and should be permeabilized by resuspension in 2 ml PBS–ovalbumin containing 0.04–0.06% saponin. (Saponin solutions should not be stored but must be prepared from powder each day.) After 10 minutes at room temperature, the samples should be pelleted once again. For permeabilized samples, saponin must be present throughout staining and washing.

4. The supernatant should be aspirated until only the pellet and ~80 μl of supernatant remain. The pellets can be resuspended by gentle tapping of the tubes (vigorous vortexing should be avoided). Primary antibody, appropriately diluted in PBS–ovalbumin (without saponin), should be added in a volume ≤20 μl. If >20 μl must be added, the saponin remaining in the supernatant will be diluted excessively, and separate dilutions of antibody should be prepared with and without saponin. The samples

should be incubated at room temperature for 30 minutes on a rotator or with frequent gentle agitation.

5. When the incubation with primary antibody is complete, the samples should be washed by dilution in 2 ml of PBS–ovalbumin with or without saponin, for permeabilized or nonpermeabilized samples, respectively. After centrifugation, the pellets should be washed two more times in PBS–ovalbumin with or without saponin.

6. Steps 4 and 5 should be repeated for staining with secondary antibody.

7. The samples should be washed once in PBS (without ovalbumin or saponin) before flow cytometry.

B. Flow Cytometry

We generally collect data on 10^4 volume-gated cells using a Becton-Dickinson FACS Analyzer (Mountain View, CA). The cytometer settings are similar to those used for living cells. However, volume gates should be examined carefully, since we have noticed a significant decrease in the electrically measured volume of saponin-permeabilized cells. Fixed nonpermeabilized cells had volume distributions that were indistinguishable from living cells. Right-angle scatter is affected by both fixation and permeabilization. Additionally, all fixed cells should be analyzed at slower rates (100–200 cells/second in the Becton-Dickinson FACS Analyzer) than living cells, since fixed cells tend to clog the analyzer when run at faster rates.

V. Microscopy

Flow cytometry is a very powerful approach that allows quantitative analysis of individual cells. However, the flow cytometer is blind to morphology, and measures all cell-associated fluorescence indiscriminately. Thus, a significant advantage of this method is the relative ease with which it can be extended to fluorescence and electron microscopy.

A. Fluorescence Microscopy

Cells stained exactly as described before are suitable for fluorescence microscopy. In fact, since 10^6 cells are stained, but only 10^4 are analyzed, the cells remaining in each sample after cytometry may be examined. The major limitation is that cells stained brightly enough for cytometry are not necessarily stained brightly enough for photomicroscopy. Mounting media that retard photobleaching may be effective solutions to this problem. We generally mount cells stained with fluorescein- or rhodamine-conjugated antibodies in 2% *n*-propyl gallate (Sigma) in glycerol–PBS (1 : 1), pH 9.0, to reduce photobleaching (Giloh and Sedat, 1982).

B. Electron Microscopy

Although the samples prepared for flow cytometry are not suitable for electron-microscopic examination, this procedure can be used for preembedding immunoperoxidase staining with minor modifications. These include reducing the saponin concentration and lengthening the times of fixation and antibody incubation. Additionally, the cells should be attached to a solid support such as poly-L-lysine-coated plastic in order to eliminate the centrifugation steps, which ultimately damage ultrastructural morphology. Finally, the fluorescent secondary antibody must be replaced with a peroxidase conjugate and incubation with peroxidase substrates followed by osmication. For a detailed discussion of this preembedding immunostaining approach, see Brown and Farquhar (Chapter 25, Vol. 31).

VI. Discussion

We have developed a new approach for the quantitative analysis of cellular antigens that allows concurrent measurement of total cellular and surface membrane antigen. As with CR1, this technique may facilitate the study of proteins with large intracellular pools and critically regulated surface expression, including the adipocyte glucose transporter (Karnieli *et al.*, 1981) and urinary epithelium H^+ transporters (Schwartz and Al-Awqati, 1986). Quantitative analysis of surface expression may also be simplified for proteins like the receptors for insulin (Krupp and Lane, 1982), epidermal growth factor (Beguinot *et al.*, 1984), and the Fc portion of IgG (Mellman *et al.*, 1983), which are degraded following ligand-mediated internalization. This method may also be used to distinguish epitopes of surface molecules that are exposed to the extracellular environment from those that are associated with cytoplasmic membrane domains (Kaplan *et al.*, 1988). Although similar information might be obtained by alternative approaches, flow cytometry has the additional advantage of measuring individual cells, thus permitting the analysis of heterogeneous samples. Finally, this immunofluorescent method simplifies correlation of quantitative flow-cytometric and qualitative microscopic approaches.

Acknowledgments

This work was supported by Grants AI-22687 (M. B.) and AI-21269 (A. M. T.) from the National Institutes of Health. J. R. T. is a trainee of the National Institutes of Health Medical Scientist Training Program GM-07250.

References

Beguinot, L., Lyall, R. M., Willingham, M. C., and Pastan, I. (1984). *Proc. Natl. Acad. Sci. U.S.A.* **81,** 2384–2388.

Berger, M., O'Shea, J., Cross, A. S., Folks, T. M., Chused, T. M., Brown, E. J., and Frank, N. M. (1984). *J. Clin. Invest.* **74,** 1566–1571.

Elias, P. M., Friend, D. S., and Goerke, J. (1979). *J. Histochem. Cytochem.* **27,** 1247–1260.

Fearon, D. T., and Collins, L. A. (1983). *J. Immunol.* **130,** 370–375.

Giloh, H., and Sedat, J. W. (1982). *Science* **217,** 1252–1255.

Goldenthal, K. L., Hedman, K., Chen, J. W., August, J. T., and Willingham, M. C. (1985). *J. Histochem. Cytochem.* **33,** 813–820.

Jacobberger, J. W., Fogleman, D., and Lehman, J. M. (1986). *Cytometry* **7,** 356–364.

Kaplan, D. R., Altman, A., Bergmann, C., and Gould, D. (1989). Submitted for publication.

Karnieli, E., Zarnowski, M. J., Hissin, P. J., Simpson, I. A., Salans, L. B., and Cushman, S. W. (1981). *J. Biol. Chem.* **256,** 4772–4777.

Krupp, M. N., and Lane, M. D. (1982). *J. Biol. Chem.* **257,** 1372–1377.

Lipincott-Schwartz, J., and Fambrough, D. M. (1986). *J. Cell Biol.* **102,** 1593–1605.

McLean, I. W., and Nakane, P. K. (1974). *J. Histochem. Cytochem.* **22,** 1077–1083.

Mellman, I. S., Plutner, H., Steinman, R. M., Unkeless, J. C., and Cohn, Z. A. (1983). *J. Cell Biol.* **96,** 887–895.

Ohtsuki, I., Manzi, R. M., Palade, G. E., and Jamieson, J. D. (1978). *Biol. Cell.* **31,** 119–126.

Schroff, R. W., Bucana, C. D., Klein, R. A., Farrell, M. M., and Morgan, A. C., Jr. (1984). *J. Immunol. Methods* **70,** 167–177.

Schwartz, G. J., and Al-Awqati, Q. (1986). *Annu. Rev. Physiol.* **48,** 153–161.

Tse, D. B., Al-Haideri, M., Pernis, B., Cantor, C. R., and Wang, C. Y. (1986). *Science* **234,** 748–751.

Turner, J. R., Tartakoff, A. M., and Berger, M. (1988). *J. Biol. Chem.* **263,** 4914–4920.

Unkeless, J. C., and Healey, G. A. (1983). *J. Immunol. Methods* **56,** 1–11.

Chapter 16

Control of Coated-Pit Function by Cytoplasmic pH

KIRSTEN SANDVIG AND SJUR OLSNES

Institute for Cancer Research
The Norwegian Radium Hospital
Montebello, 0310 Oslo 3, Norway

OLE W. PETERSEN AND BO VAN DEURS

Structural Cell Biology Unit
Department of Anatomy
The Panum Institute
University of Copenhagen
DK-2200 Copenhagen N, Denmark

I. Introduction

To investigate the role of endocytosis in cellular processes, methods to inhibit or alter the endocytosis are warranted. Thus, inhibition of the

Copyright © 1989 by Academic Press, Inc.
All rights of reproduction in any form reserved.

endocytic uptake of a certain ligand can provide information as to whether the uptake is required for its biological effect. Furthermore, selective inhibition of one kind of endocytosis can provide information about the type of endocytosis involved in the uptake of a certain ligand (see later). The best-studied kind of endocytosis occurs from coated pits that pinch off the plasma membrane to form coated vesicles. After removal of the clathrin coat, the vesicles fuse with endosomes. However, electron-microscopic studies of the uptake of certain ligands such as cholera toxin, tetanus toxin (Montesano *et al.*, 1982; Tran *et al.*, 1987), IgG-ferritin (Huet *et al.*, 1980), and insulin (Smith and Jarett, 1983) suggested that endocytic uptake occurs not only from coated pits but also from uncoated areas of the cell membrane. Also, studies on the uptake of ricin indicated that an alternative endocytic mechanism occurs (see later).

It has earlier been shown that depletion of cells for ATP inhibits both receptor-mediated and fluid-phase endocytosis (Steinman *et al.*, 1974). Methods have been developed to inhibit endocytic uptake of certain ligands but not that of others. Some cell types lose the coated-pit structures at the cell membrane when exposed to a hypotonic shock followed by potassium depletion (Larkin *et al.*, 1983). Upon such treatment, the endocytic uptake of ligands such as low-density lipoprotein (LDL), transferrin, and epidermal growth factor (EGF), which are all known to be endocytosed by the coated-pit, coated-vesicle pathway, is strongly inhibited. However, under the same conditions the toxic protein ricin is still endocytosed (Moya *et al.*, 1985) and capable of intoxicating the cells. This suggests that the toxin is endocytosed from uncoated areas of the membrane. Removal of coated pits by a method based on potassium depletion blocks infection of cells with the picornavirus poliovirus, whereas two other picornaviruses, rhinovirus 2 and encephalomycarditis virus, were still able to infect the cells (Madshus *et al.*, 1987b). At least in the case of human rhinovirus 2, there is evidence that endocytosis is required for infection.

Daukas and Zigmond (1985) have reported that high concentrations of salt blocked receptor-mediated uptake of a chemotactic peptide in leukocytes, whereas fluid-phase endocytosis was not blocked. Similar results were obtained by Oka and Weigel (1987), who reported that in rat hepatocytes hyperosmolarity blocked endocytosis of asialoorosomucoid, whereas uptake of lucifer yellow still occurred. We have tested the effect of high salt on endocytosis in Vero cells and found that uptake of transferrin and the protein toxins ricin and modeccin was inhibited at high external salt (data not shown). The results obtained so far with this method thus suggest that in some cells high salt has an inhibitory effect on endocytosis in general. It is not clear whether the method can be used to differentiate between different forms of receptor-mediated endocytosis.

The different methods used to interfere with the endocytic process have, as expected, a number of effects on the cells. When coated pits are removed from the cell surface by subjecting the cells to hypotonic shock and potassium depletion, not only does the membrane potential disappear, but there are also changes in the intracellular pH, the cells shrink, and there is a strong inhibition of chloride transport (Madshus *et al.*, 1987a). Other methods to regulate the endocytic uptake in cells are therefore warranted.

We have recently found that acidification of the cytosol inhibits endocytosis from coated pits, not by removal of the coated pits from the cell surface, but by inhibition of their pinching off to form coated vesicles (Sandvig *et al.* 1987, 1988). In addition, Davoust *et al.* (1987) and Heuser *et al.* (1987) have reported that low cytosolic pH blocks endocytosis from coated pits.

We also show that even though uptake of transferrin and EGF are strongly inhibited at the low pH, there is still ongoing endocytosis of ricin and of the fluid-phase marker lucifer yellow. This indicates that an alternate pathway of endocytosis exists.

II. Procedures for Measuring and Changing Cytoplasmic pH

To measure the cytosolic pH, we have determined the distribution of the radioactive weak acid [^{14}C]dimethyloxazolidine 2,4-dione ([^{14}C]DMO) as described by Deutsch *et al.* (1979). The cells were incubated with this compound (5 μCi/ml) for 5 minutes at 37°C, and then washed five times with ice-cold phosphate-buffered saline (PBS). The cell-associated radioactivity was then measured after extraction of the cells with 5% trichloroacetic acid (TCA). The extracellular space—that is, the extracellular liquid associated with the cells after they had been washed five times—was measured after incubation of the cells with [^{14}C]sucrose (5 μCi/ml). This space corresponded to 7% of the cell volume. The cell volume was calculated from the amount of ^3H_2O associated with cells after incubation with 1 μCi/ml ^3H_2O for 18 hours, and 7% of this value was subtracted to correct for extracellular ^3H_2O not removed by the washing.

We have used three different methods to alter the cytoplasmic pH.

1. In the method that is best tolerated by the cells, the cytoplasm is first loaded with NH_4^+ by preincubation of the cells with NH_4Cl. Upon

subsequent removal of NH_4Cl from the medium, ammonia diffuses out of the cells, and protons are left in the cytoplasm (Boron, 1977). It is important to remove all extracellular NH_4Cl, either by carefully removing the medium by suction or by washing the cells, to obtain maximal effect on the endocytic uptake.

The low pH activates the Na^+/H^+-exchanger present in the plasma membrane and, under normal conditions, the pH is rapidly brought back to neutrality (Moolenaar *et al.*, 1984; Sandvig *et al.*, 1986). However, the activity of this exchanger can be inhibited by removal of Na^+ from the external medium or by addition of the specific inhibitor, amiloride. Under these conditions, the pH remains low for a considerable period of time. The procedure most used in our work consists in preincubation of the cells for 30 minutes at 37°C with NH_4Cl and subsequent transfer of the cells to a Na^+-free buffer containing amiloride (see legend to Fig. 1). In spite of a standard protocol, the pH values obtained after such treatment varied from one experiment to another, possibly because of day-to-day variations in the permeability of the cell membrane.

2. Another method to acidify the cytoplasm involves incubation of cells with weak acids such as acetic acid. In its undissociated form the acid penetrates the membrane, dissociates in the cytoplasm, and thereby lowers the pH (Aubert and Motais, 1974; Rogers *et al.*, 1983a,b). In the experiment described here, the extracellular medium was adjusted to pH 5.0 to increase the amount of undissociated acid and to counteract the effect of the Na^+/H^+-exchanger.

3. A third method used has the advantage that it allows precise adjustment of the cytoplasmic pH. This method involves incubation of cells with isotonic KCl and the ionophores valinomycin and nigericin. Valinomycin ensures efficient potassium equilibration across the membrane. Since nigericin carries out electroneutral exchange of K^+ with H^+, the intracellular pH will be clamped at the same pH as the extracellular pH (Thomas *et al.*, 1979). In contrast to the two methods just described, such clamping of the cytoplasmic pH has a certain toxic effect on the cells (see Section IV). In spite of this, the method is a useful addition to the two other methods.

III. Effect of Cytoplasmic Acidification on Endocytosis

A. Endocytosis of Transferrin and Epidermal Growth Factor

At neutral pH transferrin is only bound to its receptor when saturated with iron. Iron-saturated transferrin (Sigma Chemical Co., St. Louis,

MO) was prepared essentially as described by Dautry-Varsat *et al.* (1983). Transferrin (1 mg/ml) was incubated with $FeNH_4$-citrate (0.1 mg/ml) in PBS for 3 hours at room temperature, and then dialyzed against the same buffer. To follow binding and endocytic uptake of transferrin and EGF, both ligands were iodinated by the Iodogen method as described by Fraker and Speck (1978).

The amount of endocytosed transferrin was determined after removal of the cell surface-bound ligand by pronase, as described by Ciechanover *et al.* (1983). After incubation of cells with ^{125}I-labeled transferrin (^{125}I-transferrin) at 37°C, the cells were washed rapidly three times with ice-cold PBS and then incubated for 1 hour at 0°C with serum-free medium containing 0.3% (w/v) pronase (protease from *Streptomyces griseus,* type XIV, Sigma). The cells and the medium were then transferred to Eppendorf tubes and centrifuged for 2 minutes. The radioactivity in the pellet and in the supernatant was measured, and the percentage of endocytosed ligand was calculated. Less than 5% of the total bound transferrin was associated with the cells when the exposure to ^{125}I-labeled ligand had been performed at 0°C.

In the case of EGF, cell surface-bound ligand was removed by treatment of the cells with low pH and high salt as described by Haigler *et al.* (1979). Cells that had been incubated with ^{125}I-labeled EGF (^{125}I-EGF) were washed five times with ice-cold PBS, incubated for 6 minutes at 0°C in 0.5 *M* NaCl, 0.2 *M* acetic acid (pH 2.5), and washed once in the same solution. The cells were dissolved in 0.1 *M* KOH and the cell-associated radioactivity was measured. The treatment with low pH and high salt very efficiently removed the cell surface-bound EGF. When the cells had been exposed to the ^{125}I-labeled ligand at 0°C, $<2\%$ was associated with the cells after the treatment with high salt and low pH.

Figure 1 shows the effect of acidification of the cytoplasm by NH_4Cl prepulsing on endocytosis of transferrin and EGF in different cell lines. All cell lines used were maintained as monolayer cultures in minimum essential medium (MEM) containing penicillin, streptomycin, and 10% (v/v) fetal calf serum (FCS) in an atmosphere containing 5% CO_2. The day before use the cells were seeded out in 24-well disposable trays, Petri dishes, or into flasks as indicated in the legends to the figures. In all the cell lines tested, the endocytosis of transferrin and EGF was blocked by low cytoplasmic pH. The actual pH after NH_4Cl prepulsing was measured with the radioactively labeled weak acid DMO as described earlier. The results indicated that the cytoplasmic pH must be reduced to values <6.5 to inhibit the endocytic uptake of transferrin and EGF (Sandvig *et al.*, 1987, 1988).

When Vero cells were acidified, more ^{125}I-transferrin became associated with the cells (Figs. 2 and 3). The Scatchard plot in Fig. 2 shows that

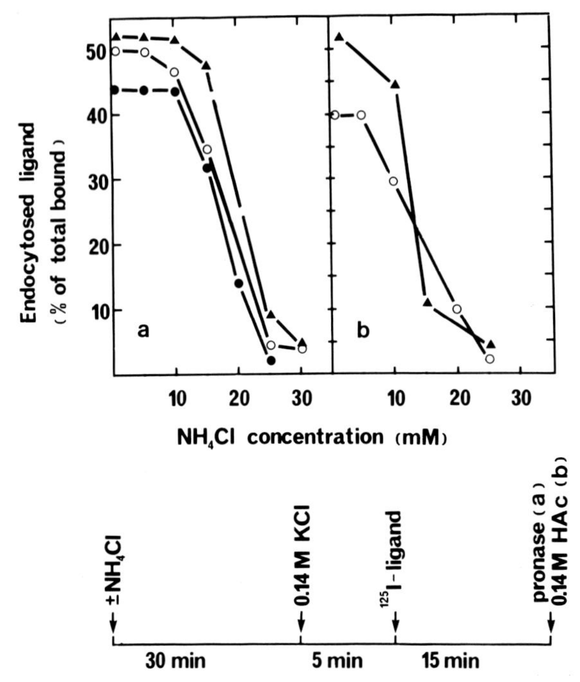

FIG. 1. Effect of NH_4Cl prepulsing on endocytic uptake of transferrin and EGF. Cells growing in 24-well disposable trays (a) or in Petri dishes (35-mm diameter) (b) were incubated for 30 minutes at 37°C with the indicated concentrations of NH_4Cl. The medium was then removed, and a buffer containing 0.14 *M* KCl, 2 m*M* $CaCl_2$, 1 m*M* $MgCl_2$, 1 m*M* amiloride, and 20 m*M* HEPES (pH 7.0), was added (0.2 ml in a, and 0.5 ml in b). After 5 minutes further incubation, ^{125}I-transferrin (200 ng/ml, 38,600 cpm/ng) (a) or ^{125}I-EGF (7 ng/ml, 15,000 cpm/ng) (b) was added. The endocytic uptake of transferrin and EGF was then measured after 10 and 15 minutes, respectively, as described in the text. Cells: A431 (○); Hep 2 (▲); Vero (●).

the increased binding is due to an increased number of transferrin receptors at the cell surface. There was also a slight increase in the amount of cell-associated EGF (Sandvig *et al.*, 1987, 1988). The increased binding of transferrin was dependent on Ca^{2+} in the medium, and could be prevented by the calmodulin antagonists trifluoperazine (Sigma) and W7 (Seikagaku Kogyo Co. Ltd., Tokyo, Japan), whereas W5, an inactive analog of W7, was without any effect. Also, serum-induced redistribution of transferrin receptors (Davis and Czech, 1986) and low pH-induced exocytosis of proton pumps in turtle bladder (van Adelsberg and Al-Awqati, 1986) were found to be dependent on Ca^{2+} in the medium. Furthermore,

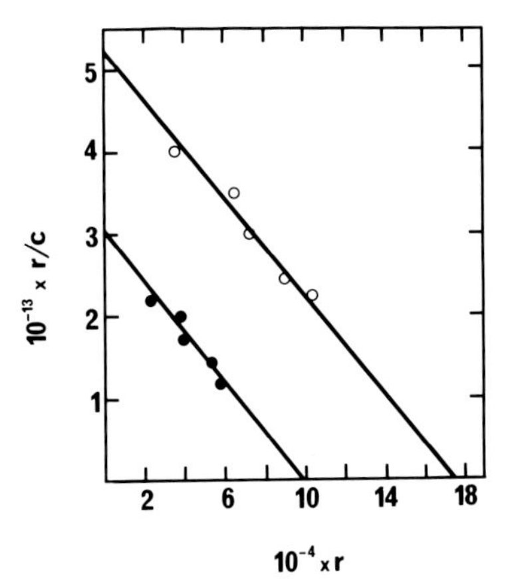

FIG. 2. Effect of NH_4Cl prepulsing on equilibrium binding of ^{125}I-transferrin to Vero cells. Vero cells growing in 24-well disposable trays were incubated for 30 minutes at 37°C with (○) and without (●) 25 m*M* NH_4Cl in HEPES medium, pH 7.0. The medium was then removed and a buffer containing 0.14 *M* KCl, 2 m*M* $CaCl_2$, 1 m*M* $MgCl_2$, 1 m*M* amiloride, and 20 mM HEPES (pH 7.0), was added. After 5 minutes at 37°C the cells were chilled to 0°C and increasing concentrations of ^{125}I-transferrin (6300 cpm/ng) were added. Unlabeled transferrin (100 μg/ml) was added to some of the wells. After 2 hours incubation at 0°C, the cells were washed three times in cold PBS, dissolved in 0.1 *M* KOH, and the cell-associated radioactivity was measured. The radioactivity associated with the cells in the presence of unlabeled transferrin was subtracted, and the data were plotted according to Scatchard (1949). *c*, Molar concentration of free transferrin; *r*, number of transferrin molecules bound per cell.

trifluoperazine and W7 have been found to inhibit exocytic processes in other systems as well (Garofalo *et al.*, 1983; Shechter, 1984), suggesting involvement of calmodulin or another Ca^{2+}-binding protein in the exocytic process.

The strong reduction of endocytosis in the acidified cells was rapidly reversible. When the cells were allowed to regulate the cytoplasmic pH back to neutrality by addition of Na^+, the endocytic uptake of transferrin was even larger than normal due to the increased number of transferrin receptors at the cell surface (data not shown).

The effect of acidification of cells by addition of acetic acid on binding and endocytosis of transferrin was similar to that of acidification by NH_4Cl prepulsing. As shown in Fig. 4, addition of increasing concentrations of acetic acid inhibited the endocytosis of transferrin to an increasing extent and, also in this case, higher amounts of transferrin were bound to the acidified cells than to cells with neutral cytosol.

The finding that low cytoplasmic pH inhibits endocytosis of transferrin

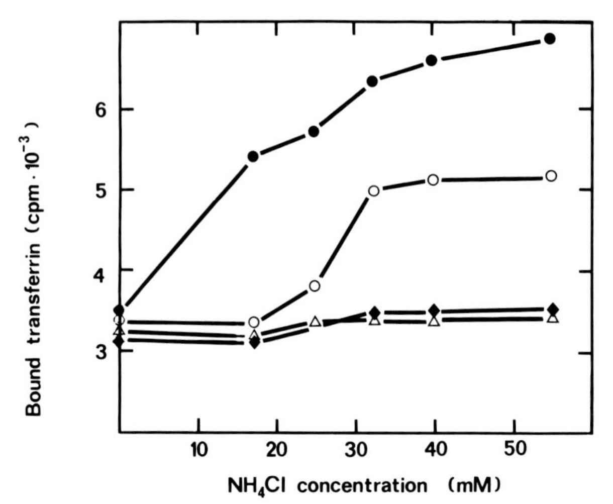

FIG. 3. Effect of calcium deprivation, trifluoperazine, and A23187 on the ability of low cytoplasmic pH to increase transferrin binding to Vero cells. Cells were preincubated with NH_4Cl and transferred to buffer without NH_4Cl as described in Fig. 1. Control (○). Where indicated (♦), trifluoperazine (10 μM) was present, or (△) the buffer contained no $CaCl_2$. The cells were then transferred to buffer without NH_4Cl. When indicated (●), A23187 (10 μM) was added upon removal of NH_4Cl. After 5 minutes incubation without NH_4Cl, the cells were chilled to 0°C, and ^{125}I-transferrin (200 ng/ml, 38,600 cpm/ng) was added in a $CaCl_2$-containing medium. After 1 hour at 0°C the cells were washed three times in ice-cold medium, dissolved in 0.1 *M* KOH, and the amount of bound radioactivity was measured.

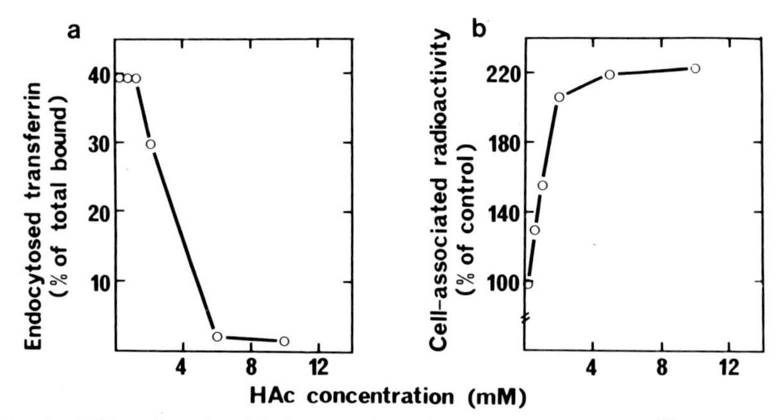

FIG. 4. Effect of acetic acid (HAc) on the endocytosis and binding of ^{125}I-transferrin to Vero cells. Vero cells growing in 24-well disposable trays were incubated for 5 minutes at 37°C in HEPES medium (0.2 ml, pH 5.0) with the indicated concentrations of acetic acid. ^{125}I-transferrin (200 ng/ml, 38,600 cpm/ng) was then added. After 10 minutes incubation at 37°C, endocytosis (a) and binding (b) of transferrin was measured as described in the text.

and EGF is supported by experiments where the internal pH was clamped at the same value as the extracellular pH by incubation of cells in isotonic KCl containing valinomycin and nigericin. As shown in Fig. 5, the endocytic uptake was strongly inhibited when the pH was reduced to values <6.5. The inhibition induced by this method was not as rapidly reversible as when the acidification was carried out by the two other methods described.

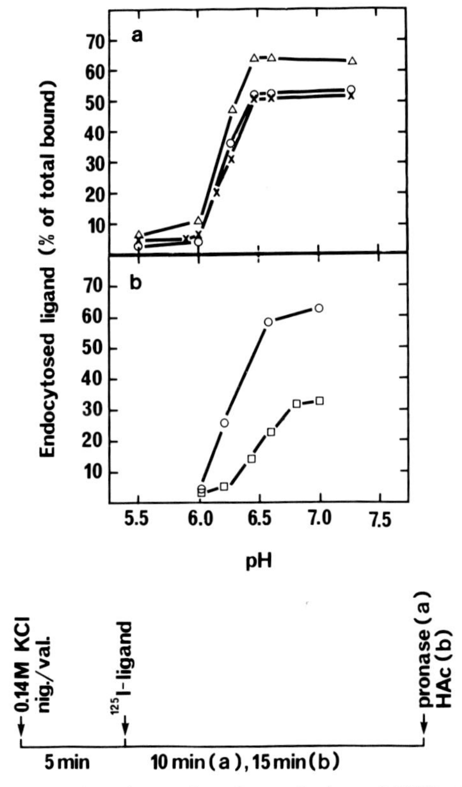

FIG. 5. Effect on the endocytic uptake of transferrin and EGF of pH clamping of the cytosol with nigericin and isotonic KCl. Cells growing in 24-well disposable trays (a) or in Petri dishes (35 mm diameter) (b) were incubated for 4 minutes in a buffer containing 0.14 *M* KCl, 2 m*M* $CaCl_2$, 1 m*M* $MgCl_2$, 5 μ*M* nigericin, 10 μ*M* valinomycin, and adjusted to the indicated pH values. Then increasing concentrations of (a) ^{125}I-transferrin (200 ng/ml, 38,600 cpm/ml) and (b) ^{125}I-EGF (7 ng/ml, 15,000 cpm/ng) were added, and the binding and endocytosis of the two ligands were measured after 10 and 15 minutes, respectively, as described in the text. Cells: A431 (□); Hep 2 (x); S3 (△); Vero (○).

B. Endocytosis of Ricin and Lucifer Yellow

The protein toxin ricin (purified from castor beans as previously described; see Olsnes and Pihl, 1973) binds to cell surface glycolipids and glycoproteins with terminal galactose residues. This binding is rapidly reversed upon addition of lactose (Olsnes *et al.*, 1974; Sandvig and Olsnes, 1979). Since washing the cells with lactose removes surface-bound toxin, endocytosed ricin can be measured as the amount of lactose-resistant ricin associated with the cells. Cells are incubated with ^{125}I-labeled ricin (^{125}I-ricin), the medium is removed, and the same medium containing 0.1 *M* lactose is added. The cells are incubated for 5 minutes at 37°C with this solution, washed three times, and dissolved in 0.1 *M* KOH; the radioactivity is then measured. The total amount of cell-associated ricin is determined by washing cells that were incubated with ricin rapidly three times in ice-cold medium, and then measuring the cell-associated radioactivity.

The effect of acidification of the cytosol on the endocytic uptake of ricin was measured in Vero cells, Hep 2 cells, MCF 7 cells, and in A431 cells. In all cell lines tested acidification induced only a slight reduction of the endocytic uptake of ricin. As shown in Table I, the reduction was strongest when the cells had been acidified by clamping the internal pH with nigericin and isotonic KCl. This is probably due to the slight toxic

TABLE I

EFFECT OF LOW CYTOPLASMIC pH ON ENDOCYTOSIS OF RICIN AND LUCIFER YELLOW

Treatment[a]	Endocytic uptake of ^{125}I-ricin[b] (% of control)	Endocytic uptake of lucifer yellow[c] (% of control)
Prepulsing with 25 m*M* NH_4Cl		
Vero	82 ± 15	82 ± 12
Hep 2	69 ± 17	
MCF 7	65 ± 10	
A431	79 ± 10	
5 m*M* Acetic acid, at pH 5.0 (Vero)	78 ± 8	
KCl–Nigericin–valinomycin, pH 6.0 (Vero)	30 ± 14	25 ± 9

[a]To lower the pH in the cytoplasm, cells were treated as described in the legends to Figs. 1–5.

[b]Endocytosed ricin was determined as described in the text.

[c]The uptake of lucifer yellow was determined fluorimetrically as described in the text.

effect of this method. That endocytic uptake of ricin occurred under conditions where uptake of transferrin and EGF was essentially blocked was confirmed by electron-microscopic studies (see later). Since the amount of NH_4Cl required to inhibit transferrin endocytosis varied from one day to another, it is important to note that endocytic uptake of ricin and transferrin were measured on the same cells the same day.

Endocytosis of lucifer yellow was measured essentially as described by Swanson *et al.* (1985). Cells were incubated with the dye (1 mg/ml) for 15 minutes at 37°C; they were washed eight times with ice-cold PBS containing bovine serum albumin (BSA; 0.1 mg/ml), and then dissolved in 0.05% Triton X-100 (v/v) containing BSA (0.1 mg/ml). Fluorescence was measured in a fluorescence spectrometer (model LS 5; Perkin-Elmer Corp., Norwalk, CT) with excitation at 430 nm (bandwidth, 10 nm) and emission at 540 nm (bandwidth, 10 nm). From the value obtained was subtracted the amount of lucifer yellow associated with cells incubated with the dye at 0°C to block endocytosis. Control experiments showed that incubation of the cells with metabolic inhibitors, which also block endocytosis, strongly reduced the uptake of lucifer yellow in the cells tested, indicating that the dye does not penetrate the membrane directly at 37°C.

As described earlier for ricin, the endocytic uptake of lucifer yellow also continued after acidification of the cytosol (Table I). This further indicates that there is an alternate pathway of endocytosis that could occur from uncoated areas of the plasma membrane.

IV. Metabolic Activity in Acidified Cells

None of the three methods strongly altered the level of ATP in the cells. Since the ATP content in cells must be reduced below 10% of the control level to affect endocytosis (Steinman *et al.*, 1974), it is clear that the inhibition of endocytosis at low cytoplasmic pH is not due to a reduced concentration of ATP.

To test if the acidification had any long-lasting effects on the cells we measured, after normalization of the cytoplasmic pH, the ability of the cells to endocytose transferrin as well as the rate of protein synthesis. The effect on endocytic uptake of transferrin of acidification by NH_4Cl prepulsing and by addition of acetic acid was rapidly reversible. However, cells that had been acidified by the pH-clamp method involving isotonic KCl in the presence of ionophores had a lower endocytic uptake than control cells even after normalization of the cytoplasmic pH. Also, when we measured the rate of protein synthesis 15 minutes after the cells had been acidified by the three different methods and then transferred to normal medium, there was a strong inhibition of protein synthesis in the cells that had been treated with KCl and ionophores, whereas there was a much smaller effect after acidification with the two other methods. In cells that had been treated with KCl and ionophores, the rate of protein synthesis was still somewhat reduced after 12 hours incubation in normal medium (Sandvig *et al.*, 1987, 1988).

V. Electron-Microscopic Observations

A. Preparation of Conjugates for Electron Microscopy

Conjugates with horseradish peroxidase (HRP) can be prepared by the SPDP (3-[2-pyridyldithio]-propionic acid *N*-hydroxysuccinimide ester) method as previously described (van Deurs *et al.*, 1986, 1987). The different fractions are filtered through a Sephacryl-200 column, and then analyzed by sodium dodecyl sulfate–polyacrylamide gel electrophoresis (SDS–PAGE). The monovalent conjugates (i.e., those containing one transferrin or ricin molecule and one HRP molecule) are selected for further experiments. This is important, since we found that polyvalent conjugates are transported efficiently only to the lysosomes, whereas monovalent conjugates of ricin as well as native ricin are transported also to the trans-Golgi network (van Deurs *et al.*, 1986, 1987, 1988).

The gold conjugates (particle size 5–10 nm) were prepared by the method of Slot and Geuze (1981). The amount of protein necessary to stabilize the colloidal-gold solution was determined as described by Horisberger and Rosset (1977). To test whether binding of the conjugates was specific we measured the ability of unconjugated ligand to block the binding of the conjugate to the cell surface. For ricin–gold conjugates the specificity of binding to galactose-terminating receptor sites could further be established by preincubation with 0.1 *M* lactose (van Deurs *et al.*, 1985).

B. Effect of Acidification on the Number of Coated Pits and on the Localization of Transferrin Receptors

The observation that acidification of the cytosol inhibits endocytosis of, for example, transferrin, which is a well-established "marker" of endocytosis from coated pits at the cell surface, could be explained in (at least) three ways: (1) acidification prevents formation of coated pits at the cell surface; (2) coated pits are present, but acidification prevents transferrin receptors from entering them; and (3) coated pits are present at the cell surface and contain a normal amount of receptors, but the coated pits do not pinch off (immobilized coated pits). These three possibilities were investigated by electron-microscopic studies.

1. Does Acidification Prevent Surface Coated-Pit Formation?

Quantification of coated pits at the cell surface was performed on low-magnification electron micrographs by determining with a map measurer the linear length of the plasma membrane, disregarding microvilli and other projections at the cell surface (Madshus *et al.*, 1987b; see also Larkin *et al.*, 1983, 1985). The number of coated pits at the same surface was counted at high magnification directly in the electron microscope. This technique may not be the most precise one, but it is simple and not very time-consuming.

For some years there was a discussion as to whether free, coated endocytic vesicles exist or whether they were all permanently surface-associated. However, serial section analysis studies documented that free coated vesicles indeed exist (Petersen and van Deurs, 1983). Moreover, quantitation of serial sections also demonstrated that by far most of the "free coated vesicular profiles" in a given section actually communicate with the cell surface at another plane of sectioning (Petersen and van Deurs, 1983). Therefore, when we counted the number of coated pits at the cell surface, we included not only those that were obviously pits, but also vesicular profiles less than three profile diameters away from the cell surface (Madshus *et al.*, 1987b) (Fig. 6). Those of these profiles that actually represent free, coated endocytic vesicles are most likely newly pinched off and will lose the coat soon.

Quantitations have been performed on both Hep 2 cells and Vero cells after acidification of the cytosol with the NH_4Cl prepulsing method (see earlier). However, in contrast to the result with hypotonic shock and K^+ depletion, acidified cells that showed a marked reduction in transferrin uptake had approximately the same amount of coated pits at the cell surface as control cells (Sandvig *et al.*, 1987, 1988) (Fig. 6). We therefore rule out the possibility that acidification—in the same way as K^+ depletion—prevents formation of coated pits at the cell surface.

Since, as described before, the concentration of NH_4Cl required to decrease the intracellular pH to values <6.5 varied from day to day even for the same cell line, the ultrastructural experiments were also carried

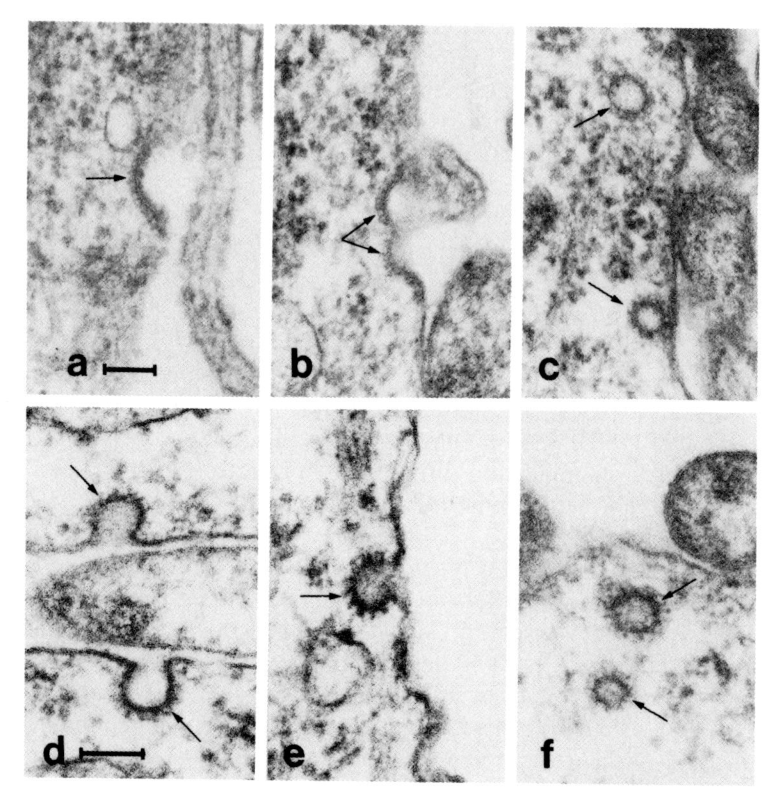

FIG. 6. Examples of coated pits in acidified cells. (a–c) Coated pits and vesicular profiles (arrows) in acidified HeLa S3 cells treated with 30 mM NH_4Cl. (d–f) Coated pits and vesicular profiles in acidified Vero cells treated with 40 mM NH_4Cl. In both cases the effect of the acidification procedure on transferrin uptake was checked on parallel cultures. (a–c) ×69,000; (d–f) ×86,000. Bars = 100 nm.

out after careful examination of the effect of various NH_4Cl concentrations (0, 10, 20, 30, and 40 mM) on the endocytosis of ^{125}I-transferrin in cells seeded out into 24-well disposable trays. Having established the concentration needed to abolish transferrin uptake almost completely, medium containing this concentration (as well as medium without NH_4Cl) was used on two parallel cultures in T-25 flasks. If similar results were obtained with respect to transferrin uptake, the cells in the T-25 flasks were considered acidified by the given NH_4Cl concentration, and the remaining T-25 cultures (i.e., six to eight flasks) were used for routine ultrastructural examinations, immunocytochemistry, and experiments with ligand conjugates.

2. Does Acidification Block Clustering of Transferrin Receptors in Coated Pits?

To test further whether transferrin receptors were present in coated pits of acidified cells, two ultrastructural approaches were used: detection of the receptor by labeling with ligand conjugates, and direct detection of the receptor by immunocytochemistry.

For ligand conjugates two types of probes can be used: particular probes such as colloidal gold, and enzymatic probes such as HRP (van Deurs *et al.*, 1985, 1986, 1987, 1988, 1989; Sandvig *et al.*, 1987). While ligand–gold conjugates are often very useful (for instance, they can be quantified), they are typically polyvalent; that is, they consist of several ligand molecules per gold particle, and may change the cell's way of handling the ligand (van Deurs *et al.*, 1986). Moreover, they may give problems with specificity. We experienced the latter problem when we tried to use a transferrin–gold conjugate to see whether the coated pits of acidified cells contained transferrin receptors. Thus, we could not prevent binding of some transferrin–gold aggregates by preincubating with unlabeled transferrin. With transferrin–HRP conjugates we did not have this problem (Sandvig *et al.*, 1987).

Upon incubation of Vero cells in monolayer culture at 4°C or 37°C with a monovalent transferrin–HRP conjugate, it turned out that 60–70% of the coated pits at the cell surface of both control cells and acidified cells were stained. Since the binding of transferrin–HRP could be prevented by addition of unlabeled transferrin, it is clear that the conjugate binds specifically to transferrin receptors, which thus are present in the coated pits of the acidified cells (Sandvig *et al.*, 1987).

As an alternative approach to study receptor expression in coated pits, we incubated Hep 2 cells (a human epidermal carcinoma cell line) at 4°C with a monoclonal anti-human transferrin receptor antibody and then with a secondary antibody conjugated to HRP (Sandvig *et al.*, 1987). By this approach we found that 75–80% of the coated pits in both control cells and acidified cells contained transferrin receptors (Fig. 7).

3. Do Coated Pits Fail to Pinch off?

Since we have shown that cytoplasmic acidification does not reduce the number of coated pits, and since the transferrin receptors still aggregated in coated pits under such conditions, the results obtained are thus in agreement with the posssibility that low cytoplasmic pH inhibits the pinching off of coated vesicles.

While a low cytoplasmic pH inhibits the coated-pit pathway, and thereby internalization of, for example, transferrin, the uptake of other

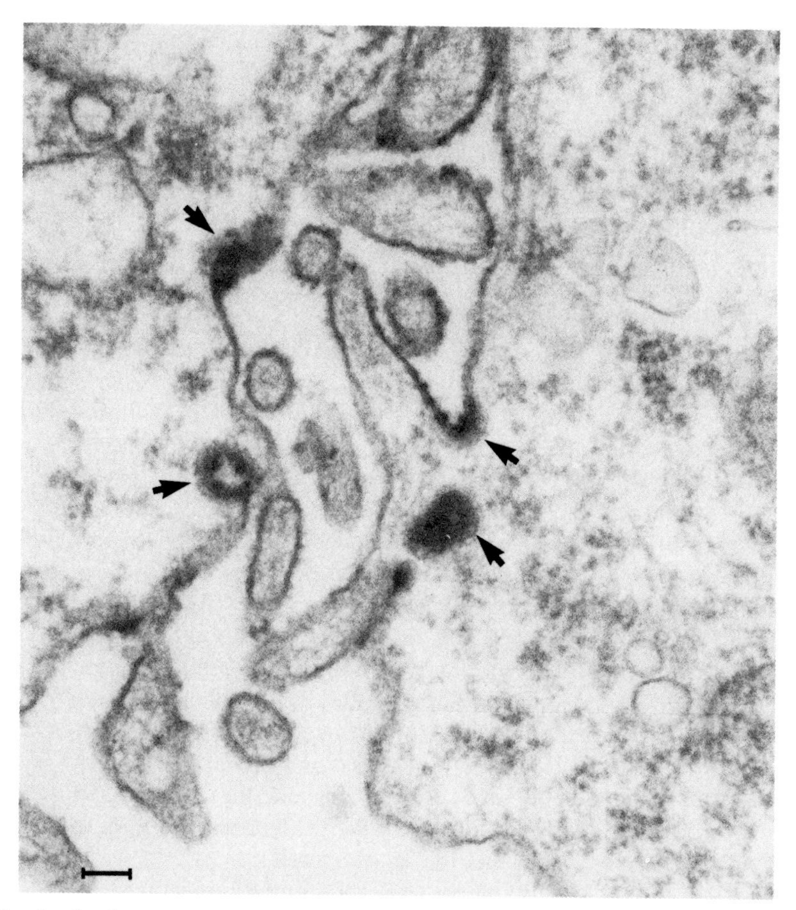

FIG. 7. Portions of two Hep 2 cells acidified by prepulsing with 40 m*M* NH_4Cl. In parallel cultures a reduction in endocytosis of transferrin (measured as internalized ^{125}I-transferrin after 10 minutes at 37°C in percentage of total cell-associated ligand) from 60% in controls to <5% in the acidified cells was obtained. The cells have been incubated at 4°C with a human anti-transferrin receptor antibody and thereafter with peroxidase-conjugated goat antimouse antibody before fixation, diaminobenzidine incubation, and processing for electron microscopy. Note that transferrin receptors are localized predominantly in the coated pits (arrows). ×69,000. Bar = 100 nm.

molecules such as ricin and lucifer yellow is only slightly reduced (see earlier). Thus, by using ricin–HRP and ricin–gold conjugates we have observed internalized ligand in endosomes of acidified cells (Sandvig *et al.*, 1987, 1988), to an extent comparable to that in nonacidified (control) cells (van Deurs *et al.*, 1985, 1986, 1987, 1989). The most likely explanation is that there exists an alternative, clathrin-independent endocytic pathway, and that this pathway is insensitive to low pH.

Acknowledgments

The work from our laboratories referred to in this review has been supported by the NOVO Foundation, the Danish Medical Research Council, the European Molecular Biology Organization (EMBO), and the Norwegian Cancer Society.

References

Aubert, L., and Motais, R. (1974). *J. Physiol (London)* **246,** 159–179.

Boron, W. F. (1977). *Am. J. Physiol.* **233,** C61–C73.

Ciechanover, A., Schwartz, A. L., Dautry-Varsat, A., and Lodish, H. F. (1983). *J. Biol. Chem.* **258,** 9681–9689.

Daukas, G., and Zigmond, S. H. (1985). *J. Cell Biol.* **101,** 1673–1679.

Dautry-Varsat, A., Ciechanover, A., and Lodish, H. F. (1983). **80,** 2258–2262.

Davis, R. J., and Czech, M. P. (1986). *J. Biol. Chem.* **261,** 8708–8711.

Davoust, J., Gruenberg, J., and Howell, K. (1987). *EMBO J.* **6,** 3601–3609.

Deutsch, C. J., Holian, A., Holian, S. K., Daniele, R. P., and Wilson, D. F. (1979). *J. Cell. Physiol.* **99,** 79–94.

Fraker, P. J., and Speck, J. C. (1978). *Biochem. Biophys. Res. Commun.* **80,** 849–857.

Garafoli, R. S., Gilligan, D. M., and Satir, B. H. (1983). *J. Cell Biol.* **96,** 1072–1081.

Haigler, H. T., Maxfield, F. R., Willingham, M. C., and Pastan, I. (1979). *J. Biol. Chem.* **255,** 1239–1241.

Heuser, J. E., Schlesinger, P. H., and Roos, A. (1987). *J. Cell Biol.* **105,** 91a.

Horisberger, M., and Rosset, J. (1977). *J. Histochem. Cytochem.* **25,** 295–305.

Huet, C., Ash, J. F., and Singer, S. J. (1980). *Cell (Cambridge, Mass.)* **21,** 429–438.

Larkin, J. M., Brown, M. S., Goldstein, J. L., and Anderson, R. G. W. (1983). *Cell (Cambridge, Mass.)* **33,** 273–285.

Larkin, J. M., Donzell, W. C., and Anderson, R. G. W. (1985). *J Cell. Physiol.* **124,** 372–378.

Madshus, I. H., Tønnessen, T. I., Olsnes, S., and Sandvig, K. (1987a). *J. Cell. Physiol.* **131,** 6–13.

Madshus, I. H., Sandvig, K., Olsnes, S., and van Deurs, B. (1987b). *J. Cell. Physiol.* **131,** 14–22.

Montesano, R., Roth, J., Robert, A., and Orci, L. (1982). *Nature (London)* **296,** 651–653.

Moolenaar, W. H., Tertoolen, G. J., and de Laat, S. W. (1984). *J. Biol. Chem.* **259,** 7563–7569.

Moya, M., Dautry-Varsat, A., Goud, B., Louvard, D., and Boquet, P. (1985). *J. Cell Biol.* **101,** 548–559.

Oka, J. A., and Weigel, P. H. (1987). *J. Cell Biol.* **105,** 311a.

Olsnes, S., and Pihl, A. (1973). *Eur. J. Biochem.* **35,** 179–185.

Olsnes, S., Refsnes, K., and Pihl, A. (1974). *Nature (London)* **249,** 627–631.

Petersen, O. W., and van Deurs, B. (1983). *J. Cell Biol.* **96,** 277–281.

Rogers, J., Hesketh, T. R., Smith, G. A., Beaven, M. A., Metcalfe, J. C., Johnson, P., and Garland, P. B. (1983a). *FEBS Lett.* **161,** 21–27.

Rogers, J., Hesketh, T. R., Smith, G. A., and Metcalfe, J. C. (1983b). *J. Biol. Chem.* **258,** 5994–5997.

Sandvig, K., and Olsnes, S. (1979). *Exp. Cell Res.* **121,** 15–25.

Sandvig, K., Tønnessen, T. I., Sand, O., and Olsnes, S. (1986). *J. Biol. Chem.* **261,** 11639–11644.

Sandvig, K., Olsnes, S., Petersen, O. W., and van Deurs, B. (1987). *J. Cell Biol.* **105,** 679–689.

Sandvig, K., Olsnes, S., Petersen, O. W., and van Deurs, B. (1988). *J. Cell. Biochem.* **36,** 73–81.
Scatchard, G. (1949). *Ann. N.Y. Acad. Sci.* **51,** 660–672.
Shecter, Y. (1984). *Proc. Natl. Acad. Sci. U.S.A.* **51,** 327–331.
Slot, J. W., and Geuze, H. J. (1981). *J. Cell Biol.* **90,** 533–536.
Smith, R. M., and Jarett, L. (1983). *J. Cell. Physiol.* **115,** 199–207.
Steinman, R. M., Silver, J. M., and Cohn, Z. A. (1974). *J. Cell Biol.* **63,** 949–969.
Swanson, J. A., Yirinee, B. D., and Silverstein, S. C. (1985). *J. Cell Biol.* **100,** 851–859.
Thomas, J. A., Buchsbaum, R. N., Zimniak, A., and Racker, E. (1979). *Biochemistry* **18,** 2210–2218.
Tran, D., Carpentier, J. L., Sawano, F., Gorden, P., and Orci, L. (1987). *Proc. Natl. Acad. Sci. U.S.A.* **84,** 7957–7961.
van Adelsberg, J., and Al-Awqati, Q. (1986). *J. Cell Biol.* **102,** 1638–1645.
van Deurs, B., Ryde Pedersen, L., Sundan, A., Olsnes, S., and Sandvig, K. (1985). *Exp. Cell Res.* **159,** 287–304.
van Deurs, B., Tønnessen, T. I., Petersen, O. W., Sandvig, K., and Olsnes, S. (1986). *J. Cell Biol.* **102,** 37–47.
van Deurs, B., Petersen, O. W., Olsnes, S., and Sandvig, K. (1987). *Exp. Cell Res.* **171,** 137–152.
van Deurs, B., Sandvig, K., Petersen, O. W., Olsnes, S., Simons, K., and Griffiths, G. (1988). *J. Cell Biol.* **106,** 253–267.
van Deurs, Sandvig, K., Petersen, O. W., and Olsnes, S. (1989). *In* "Trafficking of Bacterial Toxins" (C. B. Saelinger, ed.) (in press).

Part III. Selection and Screening of Vesicular-Transport Mutants of Animal Cells

A major motivation for increased emphasis on eukaryotic mutants defective in vesicular transport (Robbins and Myerowitz, 1981; Merion *et al.*, 1983; Robbins *et al.*, 1983, 1984; Klausner *et al.*, 1984; Marnell *et al.*, 1984; Nakano *et al.*, 1985; Roff *et al.*, 1986; Tufano *et al.*, 1987; Cain and Murphy, 1988; Nori and Stallcup, 1988; Krieger, 1986) comes from the desire to know the molecular basis of transport and the prospect for using nucleic acid-based strategies for analyzing and correcting such mutants. The greatest success, for the moment, has been obtained with yeast conditional transport mutants (Schekman, 1985). The precedent from studies of microorganisms argues persuasively for the importance of selection procedures for recovery of mutants. Outstanding applications of such procedures to higher eukaryotic cells are the selection of toxin- and lectin-resistant (Stanley, 1983) cells and the low-density lipoprotein (LDL) receptor mutants of Krieger (see Krieger *et al.*, Chapter 3, this volume). A recent review (Basch *et al.*, 1983) summarizes positive-selection procedures. Nevertheless, many mutant phenotypes are best recognized by screening protocols involving replica plating (Basilico and Meiss, 1974; Baker and Ling, 1978; Hohmann, 1978; Esko, 1986) and flow cytometry. Chapter 18 by Schindler *et al.* describes a powerful extension of conventional flow cytometry for adherent cells. Several general and technical references on somatic cell genetics are available (Thompson and Baker, 1973; Kao and Puck, 1974; Naha, 1974; Siminovitch, 1976; Puck and Kao, 1982; Hooper, 1985; Shay, 1982; Gottesman, 1987).

References

Baker, R., and Ling, V. (1978). Membrane mutants of mammalian cells in culture. *Methods Membr. Biol.* **9**, 337–384.

Basch, R., Berman, J., and Lakow, E. (1983). Cell separation using positive immunoselection techniques. *J. Immunol. Methods* **56**, 269–280.

Basilico, C., and Meiss, H. (1974). Methods for selecting and studying temperature-sensitive mutants of BHK-21 cells. *In* "Methods in Cell Biology" (D. M. Prescott, ed.), Vol. 8, pp. 1–22. Academic Press, New York.

Cain, C. C., and Murphy, R. (1988). A chloroquine-resistant Swiss 3T3 cell line with a defect in late endocytic acidification. *J. Cell Biol.* **106**, 269–277.

Esko, J. (1986). Detection of animal cell LDL mutants by replica plating. *In* "Methods in Enzymology" (J. Albers and J. Segrest, eds.), Vol. 129, pp. 237–253. Academic Press, Orlando, Florida.

M. Gottesman, ed., (1987). "Methods in Enzymology," Vol. 151. Academic Press, San Diego, California.

Hohmann, L. (1978). A simple replica plating and cloning procedure for mammalian cells using nylon. *In* "Methods in Cell Biology" (D. M. Prescott, ed.), Vol. 20, pp. 247–254. Academic Press, New York.

Hooper, M. (1985). "Mammalian Cell Genetics." Wiley, New York.

Kao, F.-T., and Puck, T. (1974). Induction and isolation of auxotrophic mutations in mammalian cells. *In* "Methods in Cell Biology" (D. M. Prescott, ed.), Vol. 8, pp. 23–40. Academic Press, New York.

Klausner, R., Renswoude, J., Kempf, C., Rao, K., Bateman, J., and Robbins, A. (1984). Failure to release iron from transferrin in a Chinese hamster ovary cell mutant pleiotropically defective in endocytosis. *J. Cell Biol.* **98,** 1098–1101.

Krieger, M. (1986). Isolation of somatic cell mutants with defects in the endocytosis of low-density lipoproteins. *In* "Methods in Enzymology" (J. Albers and J. Segrest, eds.), Vol. 129, pp. 227–236. Academic Press, Orlando, Florida.

Marnell, M., Mathis, L., Stookey, M., Shia, S.-P., Stone, D., and Draper, R. (1984). A Chinese hamster ovary cell mutant with a heat-sensitive, conditional-lethal defect in vacuolar function. *J. Cell Biol.* **99,** 1907–1916.

Merion, M., Schlesinger, P., Brooks, R., Moehring, J., Moehring, T., and Sly, W. (1983). Defective acidification of endosomes in Chinese hamster ovary cell mutants "cross-resistant" to toxins and viruses. *Proc. Natl. Acad. Sci. U.S.A.* **80,** 5315–5319.

Naha, P. (1974). Isolation of temperature-sensitive mutants of mammalian cells. *In* "Methods in Cell Biology" (D. M. Prescott, ed.), Vol. 8, pp. 41–46. Academic Press, New York.

Nakano, A., Nishijima, M., Maeda, M., and Akamatsu, Y. (1985). A temperature-sensitive CHO cell mutant pleiotropically defective in protein export. *Biochim. Biophys. Acta* **845,** 324–332.

Nori, M., and Stallcup, M. (1988). Temperature-sensitive transport of glycoproteins to the surface of a mouse lymphoma cell line. *Mol. Cell. Biol.* **8,** 833–842.

Puck, T., and Kao, F-T. (1982). Somatic cell genetics and its application to medicine. *Annu. Rev. Genet.* **16,** 225–271.

Robbins, A., and Myerowitz, R. (1981). The mannose 6-phosphate receptor of Chinese hamster ovary cells. Isolation of mutants with altered receptors. *J. Biol. Chem.* **256,** 10618–10622.

Robbins, A., Peng, S., and Marshall, J. (1983). Mutant Chinese hamster ovary cells pleiotropically defective in receptor-mediated endocytosis. *J. Cell Biol.* **96,** 1064–1071.

Robbins, A., Oliver, C., Bateman, J., Krag, S., Galloway, C., and Mellman, I. (1984). A single mutation in CHO cells impairs both golgi and endosomal functions. *J. Cell Biol.* **99,** 1296–1308.

Roff, C., Fuchs, R., Mellman, I., and Robbins, A. (1986). Chinese hamster ovary cell mutants with temperature-sensitive defects in endocytosis. I. Loss of function on shifting to the nonpermissive temperature. *J. Cell Biol.* **103,** 2283–2297.

Schekman, R. (1985). Protein localization and membrane traffic in yeast. *Annu. Rev. Cell Biol.* **1,** 115–143.

Shay, J., ed. (1982). "Techniques in Somatic Cell Genetics." Plenum, New York.

Siminovitch, L. (1976). On the nature of hereditable variation in cultured somatic cells. *Cell (Cambridge, Mass.)* **7,** 1–11.

Stanley, P. (1983). Selection of lectin-resistant mutants of animal cells. *In* "Methods in

Enzymology'' (S. Fleischer and B. Fleischer, eds.), Vol. 96, pp. 157–183. Academic Press, New York.

Thompson, L., and Baker, R. (1973). Isolation of mutants of cultured mammalian cells. *In* ''Methods in Cell Biology'' (D. M. Prescott, ed.), Vol. 6, pp. 210–282. Academic Press, New York.

Tufano, F., Snider, M., and McNight, S. (1987). Identification and characterization of a mouse cell line defective in the intracellular transport of glycoproteins. *J. Cell Biol.* **105,** 647–657.

Chapter 17

Replica Plating of Animal Cells

JEFFREY D. ESKO

Department of Biochemistry
Schools of Medicine and Dentistry
University of Alabama at Birmingham
Birmingham, Alabama 35294

I. Introduction

Replica plating as used in microbial genetics permits the identification of relatively rare mutants without resorting to direct selections based on resistance to drugs and inhibitors. In this procedure bacterial colonies established on the surface of an agar plate are transferred to a piece of velveteen cloth or filter paper, which is then used to duplicate the colony pattern onto a fresh plate (Lederberg and Lederberg, 1952). By preparing

Copyright © 1989 by Academic Press, Inc.
All rights of reproduction in any form reserved.

replica plates at different growth temperatures or in the absence of specific nutrients, temperature-sensitive mutants and nutrient auxotrophs can be isolated. Replica plating has proved most useful for isolating mutants of bacteria and yeast because they grow rapidly, form colonies of high cell density, and withstand the rigors of the duplication process.

Early attempts to adapt replica plating to animal cells met with limited success (Kuroki, 1975). In contrast to bacteria, animal cells are more fragile, adhere to each other and substrata, and produce colonies containing fewer cells. Enrichment and selection methods were devised, including 5-bromodeoxyuridine (BrdUrd) enrichment (Kao and Puck, 1974) and radiation suicide (Thompson, 1979) for the isolation of conditionally lethal mutants, and direct selections for mutants resistant to drugs, antibodies, or lectins (Baker and Ling, 1978; Stanley, 1985). In practice, specific selections do not exist for mutants with alterations in most enzymes of physiological and genetic interest, and available inhibitors are not entirely selective in their site of action.

Several methods for clonal analysis of animal cells are available, including single-cell dilution methods in microtiter plates (Goldsby and Zisper, 1969; Robb, 1970; Suzuki and Horikawa, 1973; Busch *et al.*, 1980) and manual replica plating of colonies with a glass rod (Kuroki, 1973; Saito *et al.*, 1977). In 1975, Stamato and Hohmann reported that animal cell colonies established on tissue culture plates adhere to nylon cloth. Because the transfer of colonies to nylon is based on cell adhesion, only 90% of the colonies on the plate transfer to the nylon cloth (Hohmann, 1978) and high-fidelity replica plates cannot be produced. Their findings stimulated a search for more reliable methods to replica-plate animal cells and led to the observation that animal cell colonies *grow* onto various supports, including filter paper (Esko and Raetz, 1978) and cloth of woven polyester (Raetz *et al.*, 1982). This strategy permits the identification of mutants by classic replica plating, *in situ* enzyme assays, and colony autoradiography. The purpose of this review is to describe in detail the replica plating of animal cells, emphasizing features of the technique that have emerged since its original description (also see Esko, 1986).

II. Replica Plating of Animal Cells

A. Chinese Hamster Ovary Cells

Replica plating of animal cells was originally devised for isolating mutants of Chinese hamster ovary (CHO) cells, an immortal cell line that has favorable genetic properties and growth characteristics. CHO cells grow attached to plastic surfaces and form colonies from single cells with high efficiency. They also grow in suspension culture, in soft agar, on top of agar, and as subcutaneous tumors in nude (athymic) mice (Gottesman, 1985). Their nutrient requirements are satisfied by a variety of synthetic media (Ham's F-12, α-MEM, DMEM, RPMI 1640) supplemented with 10% fetal calf serum or serum treated to remove small molecules (Kao and Puck, 1967), lipoproteins (Krieger *et al.*, 1981), total lipid (Esko and Matsuoka, 1983), or glycoconjugates (Kingsley *et al.*, 1986a; Esko *et al.*, 1987). CHO cells proliferate between 33° and 40°C, making possible the isolation of temperature-sensitive mutants.

CHO cells have a stable karyotype, and the phenotypes of mutants persist through many cell generations (Worton *et al.*, 1977). These features make it possible to isolate mutants clonally and to examine the biochemical and physiological effects of mutations *in vitro* and *in vivo*. Some investigators find CHO cell objectionable because the histological origin of the cells is not known and they exhibit few differentiated functions (Yergenian, 1985). However, their primitive nature makes them ideal for studying constitutive genes involved in "housekeeping" activities shared by all mammalian cells. Since CHO cells are amenable to gene transfer techniques (Abraham, 1985; Howard and McCormack, 1985), the expression of regulated genes can be studied as well. If, however, mutants of another cell line are desired, the replica-plating procedure developed for CHO cells is applicable to many other cell types, including contact-inhibited and nonadherent cells.

Treatment of a CHO cell monolayer with trypsin (0.25%) or EDTA (1 m*M*) produces a single-cell suspension that resumes growth when added to fresh dishes (Fig. 1A). The cells attach to the plate within a few hours and double every 12 hours at 37°C and every 24 hours at 33°C. After 8–16 days, depending on the temperature, macroscopic colonies appear on the bottom of the dish. Cells detach from the plate during each division cycle, diffuse, and reattach elsewhere on the plate, forming satellite colonies that eventually obscure the colony pattern (Fig. 2, left side). If the cultures are left undisturbed, the formation of satellite colonies is reduced but not entirely eliminated.

If a sterile disk of Whatman filter paper is placed over the cells (Fig. 1B) and a single even layer of glass beads is added to weight it against the bottom of the plate (Fig. 1C and D), the cells form colonies but satellite colonies do not arise (Fig. 2, right side). Changing the growth medium periodically maintains the cells in a rapidly proliferating state and does not disturb the colony pattern. Colonies that arise under the disk are much more uniform in size, and because of their high resolution they can be counted manually or with an electronic colony counter. This makes it

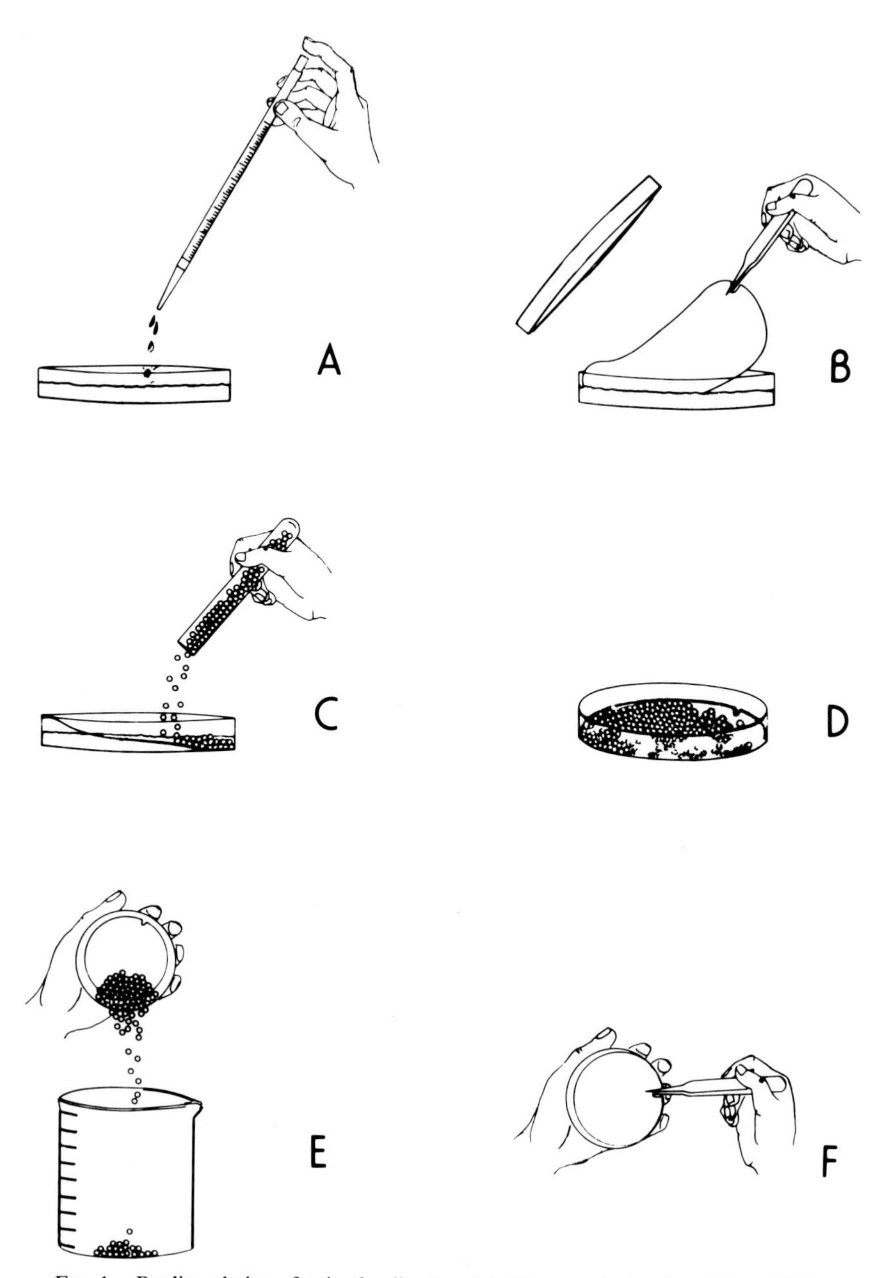

FIG. 1. Replica plating of animal cells. Reprinted by permission from Esko (1980).

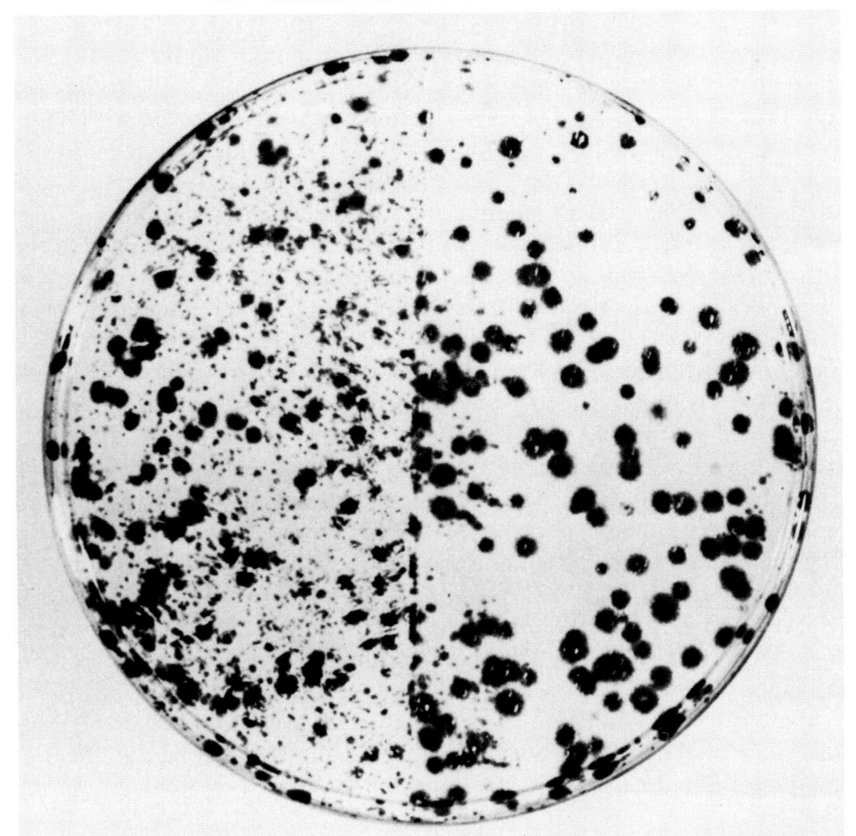

FIG. 2. Reduction of satellite colonies. The right-hand side of the plate was overlaid with a piece of Whatman no. 42 filter paper and glass beads. The left side of the dish was not overlaid. After 9 days, the paper and beads were removed and the plate was stained with Coomassie brilliant blue. Reprinted by permission from Esko (1980).

easier to determine the plating efficiency of the cells (number of colonies formed divided by the number of cells plated) and to monitor the response of mutants to drugs, nutrients, and temperature. Pure cell colonies can be isolated with glass or metal cloning cylinders and trypsin treatment (Jacobs and DeMars, 1977), or they can be scraped with a sterile toothpick or bacterial loop. Thus, cloning cells under filter paper circumvents the need for limiting dilution in microtiter plates.

B. Transfer of Colonies to Filter Paper

Animal cell replica plating is based on the observation that cells attached to plastic substrata grow into overlying disks of filter paper

(Esko and Raetz, 1978). When overlaid with grades of Whatman filter paper that have smooth surfaces (nos. 50, 52, 540, 541), cells from each colony grow up into the disk (Fig. 3). However, when overlaid with rough varieties of Whatman paper (nos. 1, 40, 42, or 3MM), the cells do not adhere as well to the overlay. The addition of 20 μg/ml of bovine pancreatic insulin to the growth medium improves cell transfer to filter paper, especially at 33°C. Enough cells from each colony remain attached to the dish so that the colony pattern is preserved on both supports. The efficiency of colony duplication on filter paper is easily determined by comparing the stained disks to the plates. The colonies are treated with 10% trichloroacetic acid (TCA) and stained with 0.05% Coomassie brilliant blue G (or R) prepared in 10% acetic acid (Esko and Raetz, 1978) or methanol–water–acetic acid (45 : 45 : 10, v/v) (Raetz *et al.,* 1982). The plates are destained by several rapid rinses in the latter solvent followed by a quick rinse in tepid water. The disks should be stained overnight for maximum sensitivity and destained until the background appears white by stirring them in a beaker on an orbital shaker. Other stains, including acid fuschins (Raetz *et al.,* 1982), eosin 2Y (Dantzig *et al.,* 1982), neutral red (Harvey and Bedford, 1988), Giemsa, methylene blue, and acridine orange, have been tried, but Coomassie brilliant blue is most sensitive. Filter paper disks can also be silver-stained.

Microscopic examination of a stained disk reveals that the cells not only have attached to the surface but also have crept along the fibers of the paper. Eventually, they appear on the other side of the disk, but the

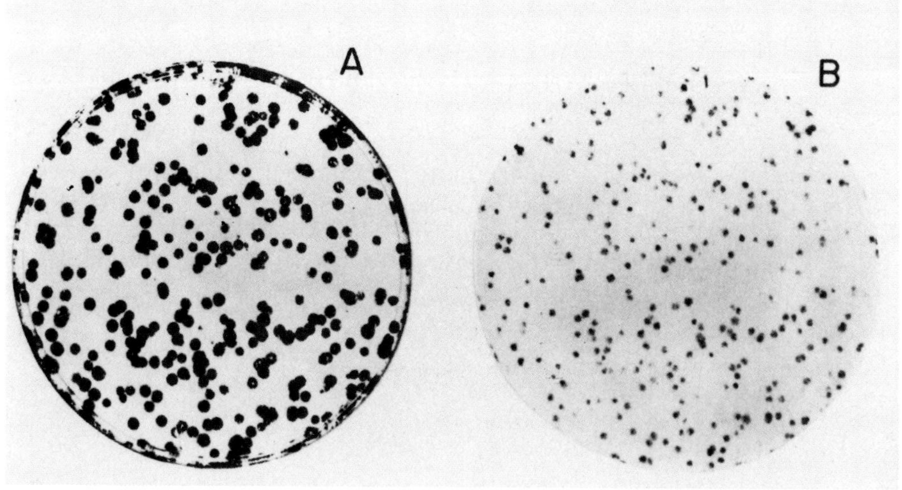

FIG. 3. Transfer of CHO cells to Whatman no. 50 filter paper. Reprinted by permission from Esko (1980).

colonies are more dense on the side that was against the plate. CHO cells overlaid at the single-cell stage need ~9 days at 37°C to achieve an adequate size for screening (Section IV). We refer to filter papers generated in this manner as 1 : 8 overlays; the number before the colon designates when the disk was applied, and the number after the colon defines the duration of the overlay in days. At 33°C, a 1 : 16 or 1 : 17 protocol is optimal. Under these conditions the colonies grow to ~2–4 mm in diameter on the dish and 1–3 mm in diameter on the disk (Fig. 3). If incubated longer, the centers of the colonies tend to die, leaving a halo of cells on the dish or the overlay. This phenomenon may be related to the tendency of the central regions of cell spheroids to become necrotic (Sutherland, 1988).

C. Transfer of Cells to Polyester

Although the transfer of CHO cells to filter paper is reliable and inexpensive, under centain conditions the transfer is not dependable. When grown in medium supplemented with delipidated serum, CHO cells are killed by a toxic substance in filter paper. This substance can be extracted by extensively washing the disks with phosphate-buffered saline (PBS) and ethanol or by presoaking the disks in growth medium supplemented with serum. Some CHO cell mutants (Raetz *et al.*, 1982) and differentiated cell lines, including $HSDM_1C_1$ fibrosarcoma cells (Neufeld *et al.*, 1984), mastocytoma cells (R. Montgomery and J. Esko, unpublished results), macrophage tumor cells, and antibody-producing hybridomas (Raetz *et al.*, 1982), are killed by filter paper overlays. Contact-inhibited mouse 3T3 fibroblasts proliferate poorly under filter paper, and the transfer lacks fidelity (R. LeBaron and J. Esko, unpublished results).

These observations led us to search for alternate replica-plating supports. CHO cells transfer to other overlays, including dialysis tubing, washed nitrocellulose and cellulose acetate membranes, Millipore filters, and nylon cloth (Esko, 1980), but polyester cloth has proved superior to all other supports (Raetz *et al.*, 1982). Transfer to polyester cloth is independent of the composition of the growth medium, and even small colonies which do not grow efficiently onto filter paper transfer well (Fig. 4). Other cell types, including FM3A mouse mammary carcinoma cells (Matsuzaki *et al.*, 1986), macrophage tumor cells, hybridoma and myeloma cells (Raetz *et al.*, 1982), fibrosarcoma cells (Neufeld *et al.*, 1984), C2 skeletal muscle cells (Black and Hall, 1985), mouse L cells (Gum and Raetez, 1983; Norcross *et al.*, 1984; Strazdis *et al.*, 1985), mastocytoma cells (R. Montgomery and J. Esko, unpublished results), 3T3 fibroblasts (R. LeBaron and J. Esko, unpublished results), MDCK epithelial cells

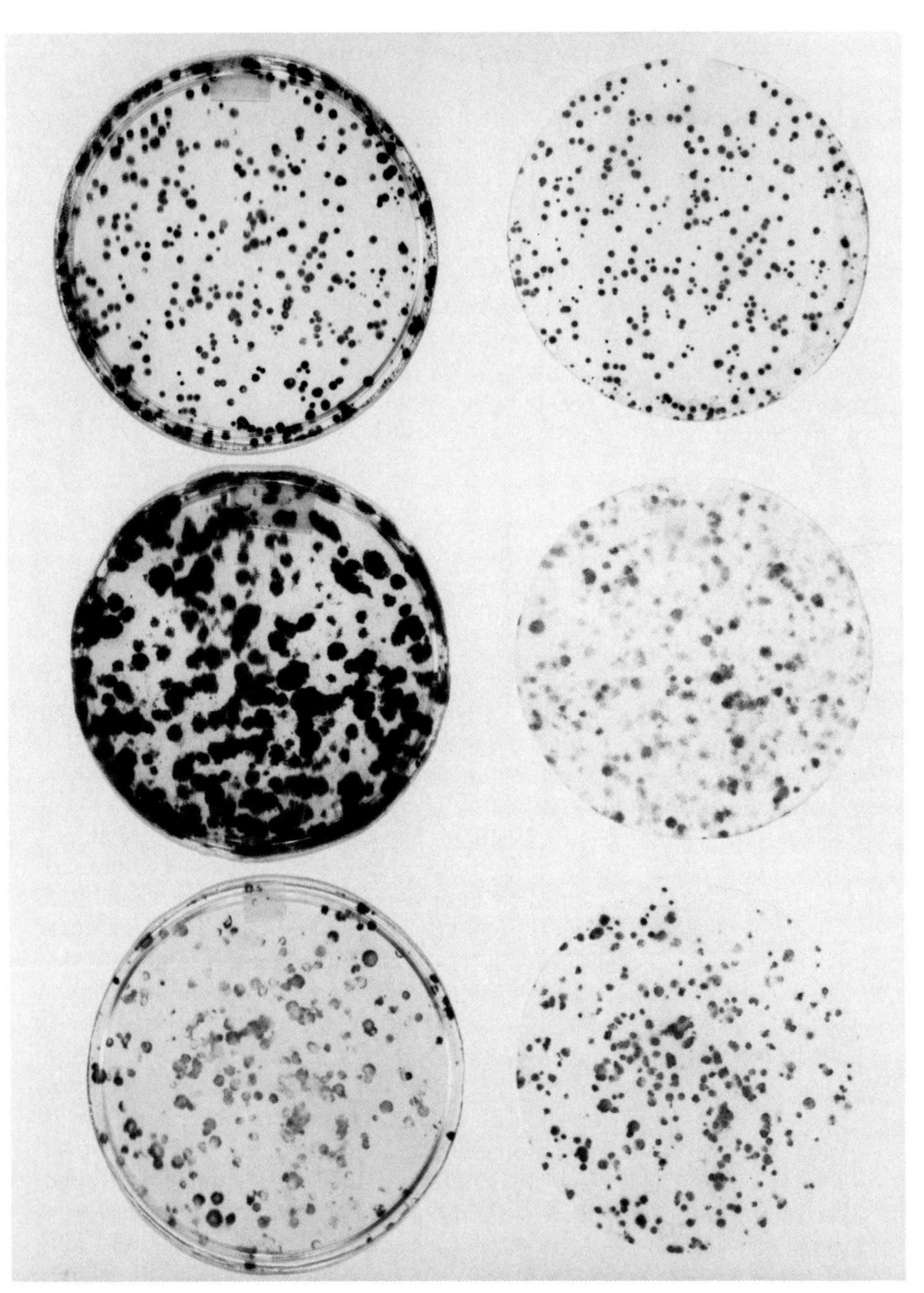

FIG. 4. Transfer of cells to polyester cloth. The polyester colony replicas on the right-hand side were generated from the dishes on the left-hand side. (Upper panels) CHO cells replica-plated to 17-μm pore diameter cloth (1 : 8); (middle panels) BALB/c 3T3 fibroblasts transferred to 17-μm pore diameter cloth (10 : 5); (lower panels) adherent mouse mastocytoma cells transferred to 10-μm pore diameter cloth (1 : 13).

(J. Mayne and R. Compans, personal communication), HeLa cells, and primary cultures of rat bone and mouse fibroblasts transfer to polyester. CHO cell hybrids formed by fusion with polyethylene glycol (Baker *et al.*, 1982), and cells derived from CHO tumors transfer efficiently (Esko *et al.*, 1988).

Some cell lines, like mouse 3T3 fibroblasts, are contact-inhibited. When plated at low density, single cells form colonies, but cells in the central region of each colony become highly organized and contact-inhibited, whereas those along the periphery continue to divide. Application of a polyester overlay to single 3T3 fibroblasts slows their growth, as if the presence of the overlay mimics contact inhibition. Replica plating 3T3 cells with a 6 : 4 protocol at 37°C or a 10 : 5 protocol at 33°C circumvents this problem (Fig. 4). Unlike CHO cells, 3T3 fibroblasts do not form satellite colonies and the colony pattern on the plates remains well resolved without an overlay.

Nonadherent cells also clone onto polyester cloth, but few cells remain attached to the dish. One way to circumvent this problem is to allow cells to settle to the bottom of a dish and overlay with two disks of polyester cloth and glass beads. As the cells form colonies, they grow through both disks, producing colony patterns on both disks. One disk serves as the master, while the other is screened for mutants. Another approach is to clone cells on top of agar. Colonies of nonadherent FM3A mouse mammary carcinoma cells established on the surface of soft agar plates (0.5% Difco purified agar prepared in growth medium) reliably replica-plate to polyester cloth (Matsuzaki *et al.*, 1986).

Mouse mastocytoma cells also do not grow attached to plastic, but an adherent subline (MRA1) has been selected which grows as a monolayer (Montgomery and Esko, 1988). The adhesive character of this subline appears to be adaptive rather than heritable, since release of the cells from selective conditions causes the population to revert to the nonadhesive phenotype. Precoating plates with cell adhesive compounds, such as poly-L-lysine (1 $\mu g/cm^2$), fibronectin (0.2 $\mu g/cm^2$), and fibrinogen (0.2 $\mu g/cm^2$) improves cell adhesion to the plastic. Mastocytoma cells also adhere better to Falcon Primaria plates compared to standard tissue culture plates. Colonies derived from the adherent subline replicate well from plastic to 1- or 10-μm pore diameter polyester cloth (Fig. 4).

Polyester cloth is available in a wide range of pore sizes, and some cell lines prefer specific varieties. Cells usually transfer well to 17-μm pore diameter discs, but mouse mastocytoma cells prefer 1- or 10-μm pore diameter cloth. Many cell lines proliferate upward through multiple layers (stacks) of polyester cloth allowing the formation of 3 to 4 high-resolution replicas of each colony (Fig. 5). The growth of CHO cells through polyester stacks occurs best with 17-μm pore diameter cloth (Raetz *et al.*, 1982). When CHO cells proliferate through polyester, they appear on the surface facing the growth medium and begin to shed and form satellite colonies. Application of a disc of 1-μm pore diameter cloth (Raetz *et al.*, 1982) or a piece of Whatman filter paper (Kuge *et al.*, 1986) over the polyester disk(s) circumvents this problem. Interestingly, transfer of mastocytoma colonies is best when the cells are overlayed with 27-μm cloth followed by two layers of 1-μm or one layer each of 10-μm and 1-μm pore diameter cloth (R. Montgomery and J. Esko, unpublished results). The cells adhere to the 27-μm disk and proliferate upwards into the second and third layers. The 27-μm disk can be left in place to obviate the formation of satel-

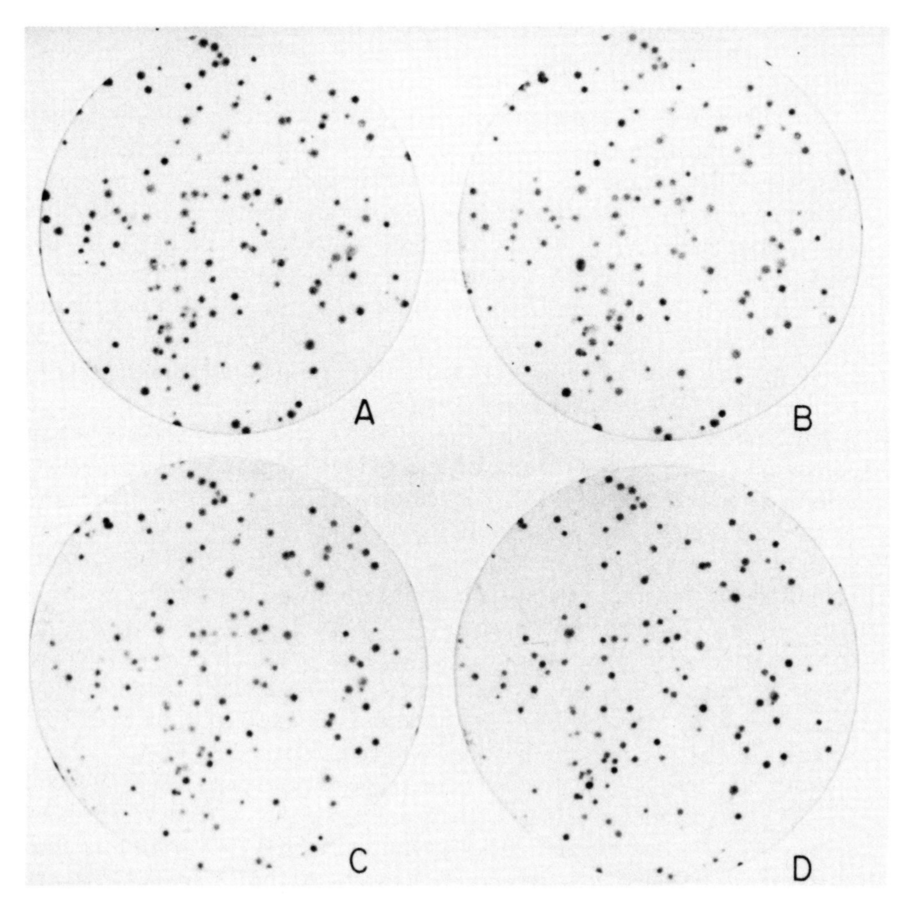

FIG. 5. Growth of CHO cells through polyester stacks. Reprinted by permission from Raetz *et al.* (1982).

lite colonies on the master plate while the analysis of the middle disk is underway. The top disk eliminates the formation of satellite colonies on the disks that arise as cells move through the stack.

III. Preparation of Disks, Beads, and Master Plates

A. Disks

Whatman filter paper fitted to the plates (82-mm diameter for a 100-mm diameter tissue culture dish) can be cut by hand or ordered directly from the manufacturer. Under normal growth conditions, filter paper disks do not require any prior treatment except that they should be notched for orientation, numbered with a soft lead pencil, and autoclaved. If necessary, the disks can be washed in PBS, deionized water, and ethanol to remove any toxic contaminants. Tetko, Inc. (Elmsford, NY) supplies polyester bolting cloth by the yard. Polyester disks of proper diameter (pore diameters: 17 μm, PeCap 7–17; 1 μm, PeCap 7–9) are traced with a metal template and cut with scissors. They should be soaked in PBS and deionized water, rinsed with ethanol, air-dried, interleaved with Whatman no. 1 filter paper (7-cm diameter), placed in a glass Petri dish, and autoclaved for 20 minutes. Residual moisture is removed from the disks by opening the dish in a laminar-flow hood for 1–2 hours. Stained polyester disks can be recycled by soaking them (10 ml/disk) at 50°C in two changes of 0.8% Terg-a-zyme (Alconox, NY) over a period of 24 hours, followed by extensive rinsing in hot tap water, deionized water, and ethanol. The original acid-washing method described by Raetz *et al.* (1982) tends to destroy the disks after repeated treatments, and for unknown reasons acid-washed disks when interleaved with Whatman filter paper are toxic to mastocytoma cells. Washed disks can be pressed with a household iron to remove wrinkles.

B. Beads

A single, even layer of 4-mm-diameter glass beads made of Pyrex or borosilicate glass weights the disk against the bottom of the plate. Beads varying from 2 to 6 mm in diameter have proved effective as well and are preferred over other types of ballasts because they conform to the contours of the plastic dish and allow the medium to percolate through the disk and around the cells. New beads are rinsed in deionized water, oven-dried in Pyrex dishes, and measured into test tubes so that each tube contains enough beads for one plate. The preferred method for recycling soiled beads is to soak them overnight in a 5% solution of Contrad 70 detergent (Curtin-Matheson), followed by extensive rinsing in tap water and deionized water in a pipet rinser. Glass rings also hold filter paper flush against the plate and completely eliminate the formation of satellite colonies, but cells transfer only at the points of contact between the ring and the paper, suggesting that cell transfer requires pressure (Esko, 1980). Since cell lines may be sensitive to "pressure contact," beads differing in diameter or density may prove advantageous.

C. Master Plates

The orientation of the notch on each disk should be marked on the side of the dish. After decanting the beads from the dish (Fig. 1E), the disks are removed (one at a time if stacked) with a pair of sterile tweezers (Fig. 1F), ensuring that the disk does not slide across the bottom of the plate. If the dishes are to be used as the master copies of the colonies, they are rinsed with growth medium or saline to remove cells loosened when the disks were removed. Several methods have been devised for storing the plates while screening the disks. The plates can be refilled with growth medium containing 100 U/ml of penicillin G, 100 μg/ml of streptomycin sulfate, 20 U/ml of Nystatin, and 2.5 μg/ml Fungizone (Gibco, NY) to inhibit bacterial and fungal growth (Esko and Raetz, 1978). Overlaying the cultures with Whatman no. 42 filter paper and glass beads prevents the formation of satellite colonies during storage. If the colonies contain few cells, the plates are first incubated for 2 or 3 days at 33°C to allow the colonies to fill in and then transferred to 28°C under an atmosphere of 5% CO_2 and 100% relative humidity. The cells divide about once each week and remain viable for at least 1 month. Master plates also can be stored at room temperature under CO_2 (Robbins, 1979) or frozen at −70°C, if filled with medium containing 10% dimethyl sulfoxide (DMSO; Baker *et al.*, 1982; Kuge *et al.*, 1985).

After the replicas have been screened and putative mutants identified, the medium in the master plates should be changed and the dishes incubated for a day under normal growth conditions. If stored frozen, thaw them quickly in a 37°C water bath. The desired colonies should be picked, treated with trypsin to produce a single-cell suspension, and diluted to 5 ml with growth medium. Because an average colony contains ~5000 cells, inoculation of three plates with 0.1, 0.3, and 1.0 ml of cell suspension should generate plates with 100–1000 colonies. These plates are overlaid and screened to ensure the purity and stability of putative mutants.

Adding the remainder of the cell suspension to a 25-cm^2 flask provides a backup cell stock in case the plates become contaminated. When the flask nears confluence, shift it to 28°C for long-term storage.

IV. Isolation of Mutants by Indirect Screenings

A. Mutagenesis

Unlike selection methods that permit the isolation of extremely rare mutants, replica plating has considerably lower capacity ($\leq 10^6$ colonies), and a mutagen is needed to raise the incidence of the desired mutant. However, if the incidence is $>10^{-6}$, then a mutagen should not be used because mutagenesis is likely to cause considerable DNA damage. To test if a mutagen is needed, a pilot screening should be conducted; if mutants are not observed, then a mutagen should be employed.

A variety of mutagens are available, including methyl methanesulfonate, ethyl methanesulfonate (EMS), *N*-methyl-*N'*-nitro-*N*-nitrosoguanidine, base analogs, γ irradiation and ultraviolet (UV) light (Shapiro and Varshaver, 1975). Retroviruses mutate the gene in which they insert and have the advantage that they mark the gene for subsequent cloning (Hartung *et al.*, 1986). Treatment of CHO cells with EMS has proved adequate for the identification of recessive and dominant mutants. Useful markers to score the effectiveness of mutagenesis include resistance to ouabain (OUA^R), a codominant mutation affecting the Na^+K^+-ATPase (Baker *et al.*, 1974); 6-thioguanine (TG^R), an X-linked recessive mutation in hypoxanthine phosphoribosyltransferase (Jacobs and Demars, 1977); and chromate (CHR^R), an autosomal recessive mutation affecting the plasma membrane anion exchanger (Esko *et al.*, 1987). Treatment of CHO cells with EMS increases the incidence of OUA^R mutants from $\leq 2 \times 10^{-7}$ to $\sim 10^{-4}$.

In this procedure, multiple 75-cm^2 flasks are seeded with 5×10^5 cells in regular growth medium at 37°C. Several cultures are usually mutagenized and kept separate, so that independent stocks of treated cells are available for subsequent screenings. Individual mutants found in different stocks therefore derive from independent mutational events and are not siblings. When the flasks have $1\text{–}2 \times 10^6$ cells, 15 ml of fresh medium containing 150–400 μg/ml of EMS (density = 1.17 g/ml) are added to each flask. Cap the flasks tightly, since EMS is carcinogenic and slightly volatile. After 16 hours at 37°C, remove the medium, rinse the flask, and add fresh medium. The cells are incubated at 33°C if temperature-sensitive strains are sought, and appropriate nutrients are added at this time to support the growth of auxotrophic strains. Supplementation of the growth medium with any deficient serum that may be used during a subsequent screening will ensure counterselection of cells that require other components absent from the serum. When the culture becomes confluent (usually 3–5 days), the cells are treated with trypsin and transferred to two 150-cm^2 flasks and grown until nearly confluent. The cells from each flask are then stored in ampules under liquid nitrogen (10^6 cells/ml of growth medium containing 8% glycerol or 10% DMSO, v/v), so that several caches of mutagen-treated cells are available for future screenings.

The plating efficiency and the incidence of drug-resistant cells is measured to ensure the effectiveness of mutagen treatment. After reviving an ampule, the cells are grown for 3–5 days until nearly confluent. Multiple 100-mm-diameter plates are then seeded with 300, 1000, 3000, and 10,000 cells in complete growth medium and incubated at both 33°C and 37°C. After 1 day, the cultures are overlaid with Whatman no. 42 paper (or polyester cloth) to determine the plating efficiency of the cells. Treatment of CHO cells with 400 μg/ml of EMS reduces the plating efficiency from ~90% to 20%. Another set of dishes are seeded with 10^5, 3×10^5, and 10^6 cells in medium containing 1 m*M* ouabain. After 3 days incubation at 37°C, the medium is changed to remove dead cells and the survivors are overlaid with Whatman no. 42 filter paper and glass beads. After 10 days, the colonies are counted and the incidence of ouabain-resistant mutants in the population is calculated after correction for plating efficiency. If the incidence of OUA^R clones is $<2 \times 10^{-4}$, discard the cells, since the likelihood of finding recessive mutants among 25,000 colonies (~100 plates) is low. Other cell lines, including hypotetraploid mouse 3T3 fibroblasts and mastocytoma cells, must be treated with a higher concentration of mutagen (600 μg/ml) or at lower cell density (3×10^5) in order to increase the incidence of OUA^R clones to the desired level.

B. Lederberg-Style Replica Plating

Indirect selection by classic replica plating is the technique of choice for isolation of temperature-sensitive strains and nutrient auxotrophs (Table I). To create authentic replica plates, a 1 : 8 colony replica prepared on filter paper at 37°C or a 1 : 16 replica prepared at 33°C is placed cell side-down in a fresh Petri dish filled with growth medium. All of the colonies on the disk transfer down to the plate within 3–5 days, depending on the temperature, and a single disk will generate four sequential replica plates with little loss of resolution (Esko and Raetz, 1978). Baker *et al.* (1982) reported that CHO cell colonies on filter paper will replicate to

TABLE I

ANIMAL CELL MUTANTS ISOLATED BY REPLICA PLATING

Type of mutant	Cell line	Support	References
Myo-inositol auxotrophs	CHO	Filter paper	Esko and Raetz (1978; 1980a)
	FM3A mouse mammary carcinoma	Polyester	Matsuzaki *et al.* (1986)
Lysophosphatidylcholine auxotrophs	CHO	Polyester	Raetz *et al.* (1982)
Phosphatidylserine auxotrophs	CHO	Polyester	Kuge *et al.* (1986)
Heat-sensitive	CHO	Polyester	Harvey and Bedford (1988)
UV-sensitive	CHO	Filter paper	Busch *et al.* (1980); Stefanini *et al.* (1982)

another piece of filter paper by placing the original filter cell side-down onto a fresh disk and covering with glass beads. Single CHO cells also form colonies when sandwiched between two pieces of filter paper, permitting the generation of as many as eight replica plates (Esko, 1980).

CHO cell mutants auxotrophic for *myo*-inositol were isolated by making two sequential replica plates (Esko and Raetz, 1978). After removing the medium from the dish, the beads were decanted by quickly inverting the plate over a beaker in a laminar-flow hood (Fig. 1E). The inverted dish was blotted on sterile absorbent paper to avoid drops of residual medium at the lip of the plate, which tend to carry contaminants into the dish when the plate is turned upright. The disk was carefully removed by its notch with a pair of sterile tweezers (Fig. 1F) and supported by surface tension against a metal or glass pan equipped with an aspirator line (Fig. 6). Thirty milliliters of medium were *forcibly* pipeted against the disk with an air-driven pipet aid (Drummond Scientific, Broomall, PA) to remove cells loosened during removal of the disk from the plate. The first replica plate was prepared in growth medium containing *myo*-inositol. The disks were washed again and placed cell side-up on top of an even layer of glass beads in a bacterial dish filled with just enough *myo*-inositol-deficient growth medium to cover the top of the beads. Incubation of the colonies for 2 days inhibited the growth of auxotrophic strains, and then a second replica plate was prepared in deficient medium. Colonies requiring *myo*-inositol for growth transferred to the first replica plate but not to the second (Fig. 7).

Another variation of replica plating, developed to isolate UV-sensitive mutants of CHO, uses a replica plate as the master (Busch *et al.*, 1980; Stefanini *et al.*, 1982). Mutagenized cells were cloned on filter paper, and

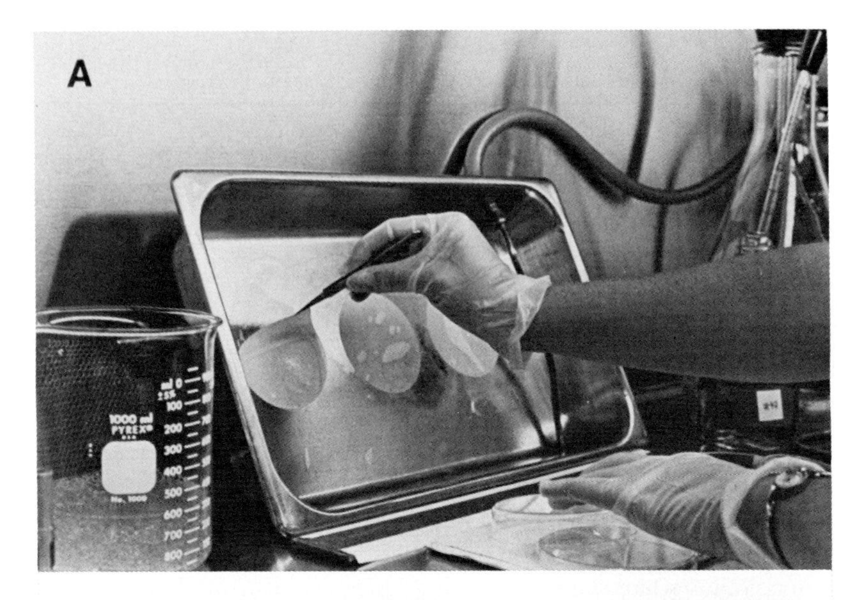

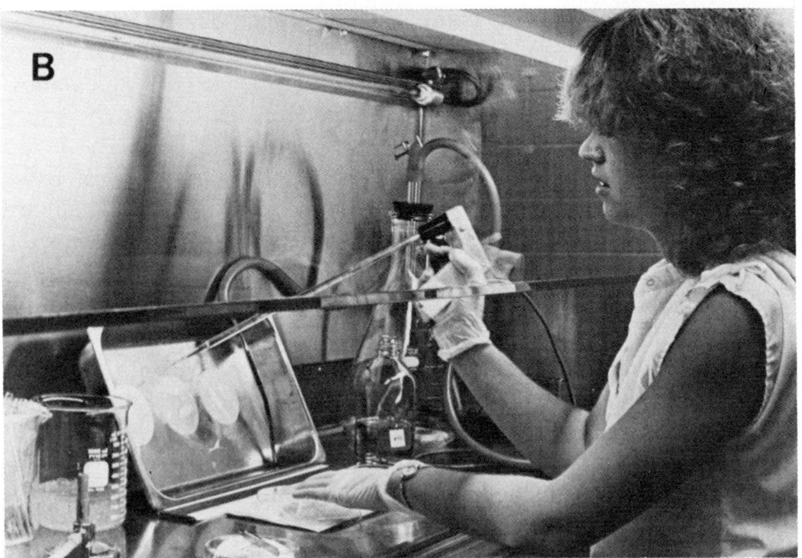

FIG. 6. Washing filter papers for replica plating.

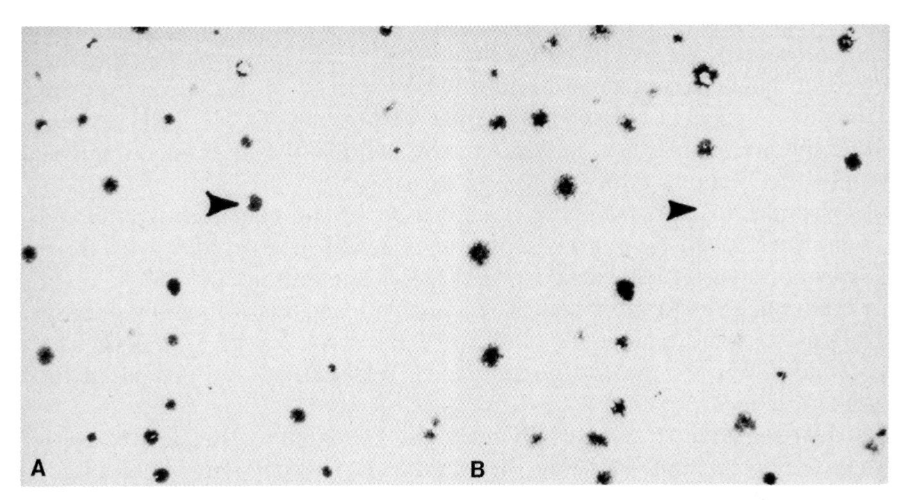

FIG. 7. Isolation of *myo*-inositol auxotrophs by replica plating. Reprinted by permission from Esko and Raetz (1978).

after removal of the disk the colonies were outlined on the bottom surface of the dish. The colonies on the plates were then UV-irradiated and grown for a few more days. Those that failed to increase in diameter on the master plate were judged sensitive to irradiation and cloned from replica plates generated from the filter papers.

Replica plates can also be created from polyester disks, but because CHO cells prefer to grow attached to polyester cloth (Fig. 4), the transfer back down to plastic is inefficient (Raetz *et al.*, 1982). Although the transfer to plastic improves by coating the dish with adhesive compounds (Section II), colonies transfer better from one polyester disk to another (Raetz *et al.*, 1982). Polyester replica disks are prepared by placing a fresh disk wetted with growth medium on top of a disk containing colonies. Glass beads are added to hold the disks together, and growth medium is added. Colonies transfer upward to the replica disk within 3 days at 37°C.

Lysophosphatidylcholine auxotrophs of CHO cells were isolated by cloning mutagen-treated cells onto two layers of polyester cloth in growth medium supplemented with phospholipids (Raetz *et al.*, 1982). The bottom disk (A) was incubated on a layer of beads in medium lacking lipid, while the second disk (B) was incubated in medium containing lipid. After 2 days, disks A and B were replicated to fresh disks (A′ and B′) in appropriate medium. The first replica disk (A′) was stained with 0.15% Ponceau G, R, 2R (acid red 26) prepared in methanol–water–acetic acid, 45 : 45 : 10 (v/v), and the second replica disk (B′) was stained with Coomassie brilliant blue. Superimposition and transillumination of the replicas (A′ over B′) presented mutants as blue colonies and wild type as purple colonies (red on blue). A similar procedure was described by Kuge *et al.* (1986) for the isolation of phosphatidylserine auxotrophs of CHO cells, except that the preincubation step was omitted and disk A was stained immediately after removal from the master dish.

Myo-inositol auxotrophs of nonadherent FM3A mammary carcinoma cells were found by overlaying colonies established on agar with three disks of polyester (Matsuzaki *et al.*, 1986). The bottom and middle disks were transferred to fresh agar plates, one with and one without *myo*-inositol, and two more layers of cloth were placed on top of each disk. The second disk from the top in each stack was stained and compared for mutant clones.

Heat-sensitive strains of CHO cells were isolated by transferring colonies to polyester and using the disk for the master copy (Harvey and Bedford, 1988). The original plates were photographed before and after heat treatment, and colonies that failed to increase in diameter were cloned from the the polyester copy. Colonies on the disks were stained with neutral red, and the desired clone was excised and treated with trypsin.

C. *In Situ* Enzyme Assays

Perhaps the most powerful feature of animal cell replica plating is that enzyme assays can be conducted in cell homogenates generated on filter paper disks (Esko and Raetz, 1978) as originally described for *Escherichia coli* colonies (Raetz, 1975). Although there have been two reports of enzyme assays on polyester disks (Gum and Raetz, 1983; Kuge *et al.*, 1985), cloth appears to be less effective than filter paper, possibly because of its open weave. To break open the cells, filter paper disks containing immobilized colonies are transferred to bacterial dishes containing 1 ml of 0.25 *M* sucrose, 20 m*M* Tris-HCl (pH 7.4), and protease inhibitors (1 μg/ml leupeptin, 0.5 μg/ml pepstatin A, and 20 μ*M* phenylmethylsulfonyl fluoride, or PMSF). Two or three cycles of freezing at −20°C (or −70°C) and rapid thawing at 37°C breaks open the cells. Interestingly, the enzymes remain localized, perhaps because they are trapped in the paper fibers. The disks are blotted on a paper towel to remove excess moisture and transferred to another bacterial dish containing 0.15–1.5 ml of a reaction mixture. The plates should be covered and equilibrated to the desired temperature in an incubator or by floating them in a water bath. After adequate time for products to accumulate, the reaction is stopped and products are detected colorimetrically, fluorimetrically, or by autoradiography, depending on the assay. Before attempting to perform an enzyme assay in immobilized colonies, it is wise to optimize test tube assays and devise detection methods for products on a piece of filter paper.

Assays based on acid precipitation of radioactive products have been developed for several lipid biosynthetic enzymes (Table II), including phosphatidylinositol synthase (Esko and Raetz, 1978), ethanolaminephosphotransferase (Polokoff *et al.*, 1981), choline and serine exchange reactions with phospholipids (Nishijima *et al.*, 1984; Kuge *et al.*, 1985), and dihydroxyacetonephosphate acyltransferase (Zoeller and Raetz, 1986). The assay of phosphatidylinositol synthase exemplifies the technique (Esko and Raetz, 1978). This enzyme transfers phosphatidic acid from CDP-diglyceride to *myo*-inositol. Colonies containing the enzyme convert *myo*-[^{14}C]inositol into phosphatidyl-[^{14}C]inositol, which precipitates when treated with a few milliliters of 10% TCA. Passing three 50-ml portions of 2% TCA through the disk on a Buchner funnel removes unincorporated, acid-soluble *myo*-[^{14}C]inositol. The dried disks are exposed to X-ray film (Kodak XAR-5 or Fuji RX medical X-ray film) and the time of autoradiography is adjusted so that wild-type colonies produce an image within the linear detection range of the film (Fig. 8). The disks are stained with Coomassie brilliant blue, and when the film is superimposed over the stained disks, mutants stand out as blue colonies lacking an autoradiographic halo.

We devised an *in situ* assay for heparan sulfate *N*-sulfotransferase based on the transfer of $^{35}SO_4$ from radioactive PAPS to the amino groups of glucosamine residues in N-desulfated heparin (Bame and Esko, 1988). In test tube assays addition of N-desulfated heparin causes a 20-fold stimulation of the reaction, if the products are collected by DEAE chromatog-

TABLE II

In Situ ENZYME ASSAYS

Enzyme reaction	Cell line	Support	References
Phosphatidylinositol synthase	CHO	Filter paper	Esko and Raetz (1978)
Ethanolaminephosphotransferase	CHO	Filter paper	Polokof *et al.* (1981)
Serine exchange activity with phospholipids	CHO	Filter paper	Kuge *et al.* (1985)
Choline exchange activity with phospholipids	CHO	Polyester	Nishijima *et al.* (1984)
Dihydroxyacetonephosphate acyltransferase	CHO	Filter paper	Zoeller and Raetz (1986)
Heparan sulfate N-sulfotransferase	CHO	Filter paper	Bame and Esko (1989)
Lysosomal glycosidases: α-mannosidase, β-galactosidase, β-glucuronidase, β-*N*-acetylhexosaminidase, α-L-fucosidase, α-L-iduronidase	CHO	Filter paper	Robbins (1979); Hall *et al.* (1986)
Alkaline phosphatase	L cells	Filter paper	Gum and Raetz (1983)

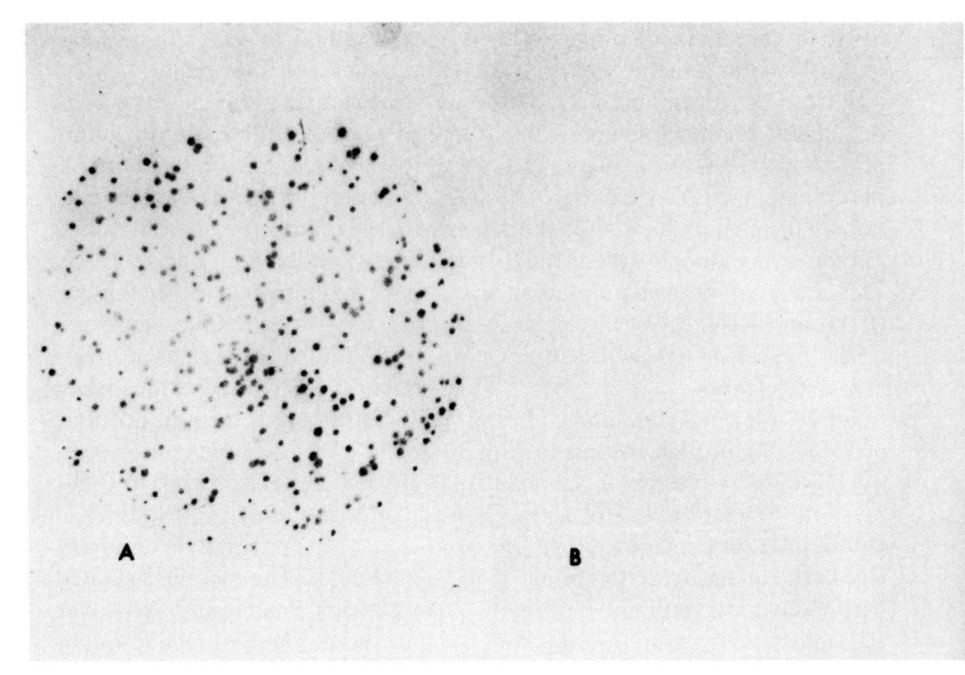

FIG. 8. *In situ* enzyme assay of phosphatidylinositol synthase. Disks were assayed in the presence (A) and absence (B) of the cosubstrate, CDP-diglyceride. Reproduced by permission from Esko and Raetz (1978).

raphy or through precipitation with cetylpyridinium chloride. On filter paper the products tend to diffuse, and addition of exogenous substrate causes a meager 2-fold stimulation of product formation. However, in the concentrated milieu of a colony homogenate immobilized on filter paper, enough endogenous substrate is present to form products detectable by autoradiography. Since almost 90% of the product consists of $^{35}SO_4$ attached to amino groups in heparan sulfate, the assay reliably measures *N*-sulfotransferase activity. However, when mutants were found some strains thought to lack enzyme activity actually had mutations in genes required for the formation of the substrate, not the product. These mutants lack heparan sulfate proteoglycans (PG) and accumulate chondroitin sulfate PG (Esko *et al.*, 1988).

Robbins and co-workers devised fluorometric assays for glycosidases based on the hydrolysis of nonfluorescent 4-methylumbelliferyl glycosides to fluorescent 4-methylumbelliferol (Robbins, 1979; Hall *et al.*, 1986). In this procedure, colony replicas on filter paper were rinsed with saline, frozen and thawed in 0.5–1.0 ml of 95% ethanol at −60°C, and blotted dry. The disks were incubated for 1.5–2 hours at 37°C in 0.15 ml of 1 m*M* 4-methylumbelliferyl α-mannoside at pH 4.4 and then photographed under long-wavelength UV light. Comparison of stained disks and the photographs revealed occasional nonfluorescent, α-mannosidase-deficient colonies. Robbins (1979) established similar assays for lysosomal β-galactosidase, β-*N*-acetylhexosaminidase, α-L-iduronidase, and α-L-fucosidase and suggested that Golgi α-mannosidase involved in glycoprotein oligosaccharide processing can be assayed by performing the reaction at pH 6. Hall *et al.* (1986) found that omission of the freeze–thaw step and ethanol treatment had no deleterious effect on the assay, suggesting that the cells were permeable to the glycosides under assay conditions. By screening both β-glucuronidase and α-L-iduronidase with a mixture of substrates, they found strains that bore a pleiotropic lesion affecting a total of seven lysosomal enzymes. Evidence suggested that the mutation altered the assembly of a dolichol-linked oligosaccharide precursor required for the synthesis of mature oligosaccharides on lysosomal enzymes.

Gum and Raetz (1983) reported a fluorometric assay for cAMP-inducible alkaline phosphatase. To circumvent the poor transfer of small colonies to filter paper, they cloned mutagen-treated cells onto polyester disks. They froze the disks at −80°C and soaked them in reaction mixture containing the fluorogenic substrate, 3-phospho-2-naphthoic acid 2,4-dimethylanilide. A disk of filter paper soaked in reaction mixture was placed on the bottom of a glass Petri dish; the polyester disk was placed on top of the filter paper, and a second soaked filter paper was laid over the polyester disk. After incubation at 37°C, the top filter was removed and photographed under UV light, and the polyester disk was stained. Comparison of the photographs to the stained disks revealed alkaline phosphatase-deficient clones as well as several clones that expressed markedly enhanced activity of the enzyme. The anilide substrate was superior to 4-methylumbelliferyl phosphate because the fluorescent product generated from the anilide substrate diffused to a lesser extent, increasing the sensitivity and resolution of the assay.

In situ assays for degradative enzymes are difficult to conduct without substrates that yield fluorescent products, but coupling degradative and biosynthetic reactions can solve this problem. Bulawa *et al.* (1981) devised a two-step assay for CDP-diglyceride hydrolase in which *E. coli* colonies established on filter paper converted α-[^{32}P]CTP and phosphatidic acid to [^{32}P]CDP-diglyceride. A second incubation carried out in the presence of EDTA stopped the synthetase reaction, and colonies containing hydrolase activity degraded [^{32}P]CDP-diglyceride generated in the first reaction. Mutants lacking the hydrolase appeared as radioactive "hot spots'' on autoradiograms. A similar approach yielded mutants in phosphatidic acid phosphatase and phosphatidylglycerophosphatase (Icho and Raetz, 1983). Adaptation of coupled assays to animal cell colonies has not been reported.

D. Colony Autoradiography

Animal cell colonies immobilized on filter paper and polyester cloth take up radioactive molecules from the growth medium and incorporate them into macromolecules. Uptake and metabolism of radioactive molecules is measured *in vivo* in the same manner as enzymatic assays are performed *in situ,* except that the colonies are not frozen and thawed. This strategy permits assays of entire metabolic pathways and makes possible the recovery of mutants defective in transport, intermediary metabolism, and macromolecular synthesis (Table III). Animal cell colonies also bind ligands and antibodies, permitting the detection of mutants with altered cell surface receptors and structural proteins of the cytoskeleton and extracellular matrix. It should also be possible to adapt nucleic hybridization methods to animal cell colonies, which would allow detection of RNA transcripts or foreign genes introduced by DNA transfection.

1. Uptake of Macromolecular Precursors

Colony autoradiography was originally devised to detect mutants in phosphatidylcholine synthesis (Esko and Raetz, 1980b). Colonies were cloned onto filter paper at 33°C and preincubated overnight at 40°C on top of glass beads (Section IV,B), in order to express any temperature-sensitive mutations. The disks were blotted, incubated at 40°C for 4 hours in 1 ml of growth medium containing [*methyl*-^{14}C]choline, and treated with TCA to precipitate newly made radioactive phosphatidylcholine. The assay is specific, since choline is the precursor of the head group of phosphatidylcholine and sphingomyelin, and degradation of [*methyl*-^{14}C]choline does not cause appreciable incorporation of radioactivity into other compounds. Superimposition of the autoradiograms over the stained disks revealed four mutants from ~20,000 colonies. One mutant possessed a thermolabile CDP-choline synthetase that rendered cells defective in phosphatidylcholine synthesis and temperature-sensitive for growth at 40°C (Esko *et al.*, 1981). Macromolecular synthesis continued for almost a day after phosphatidylcholine synthesis ceased, suggesting that enrichment schemes based on selective killing of proliferating cells would have counterselected the mutant. The other three mutants apparently bore defects in energy generation because DNA, RNA, and protein synthesis ceased rapidly under restrictive conditions.

We devised a similar autoradiographic assay for mutants defective in PG synthesis based on the incorporation of $^{35}SO_4$ into acid-precipitable material (Esko *et al.*, 1985). CHO cells use >95% of exogenous inorganic sulfate for PG synthesis, and because sulfate addition is the final step in the formation of the polysaccharide chains of PG (glycosaminoglycans, GAG), the screening permits the isolation of a broad range of mutants. Colony replicas were generated on polyester disks at 33°C to rescue mutants defective in essential genes. They were preincubated overnight in 10 ml of sulfate-free growth medium at 40°C to shorten the duration of autoradiography and to express any temperature-sensitive mutations. The next day, the disks were incubated at 40°C in 5 ml of sulfate-free growth medium containing 10–20 μCi/ml of $^{35}SO_4$. After 4 hours, the disks were treated with TCA and placed on top of a piece of filter paper in a Buchner funnel. Unincorporated $^{35}SO_4$ was removed by passing two 50-ml portions of 2% TCA through the disks. The disks were then exposed to X-ray film (Fig. 9).

Autoradiography of colonies with $^{35}SO_4$ has incredible sensitivity, since mutants with as little as 2-fold reduction in incorporation were detected. Many mutants were found (incidence ~0.1%), and 75 strains defective in PG synthesis have already been cloned. Biochemical studies showed that the collection contains 35 sulfate transport-deficient mutants (Esko *et al.*, 1986), 1 xylosyltransferase-deficient mutant (Esko *et al.*, 1985), 5 galactosyltransferase I-deficient mutants (Esko *et al.*, 1987), and 1 *N*-sulfotransferase-deficient mutant (Bame and Esko, 1989). Five heparan sulfate-deficient strains have also been found (Esko *et al.*, 1988), and 4 show a dramatic increase in chondroitin sulfate synthesis. Autoradiography of colonies with $^{35}SO_4$ has been adapted to a rat hepatoma cell line (E. Conrad, personal communication), mouse 3T3 fibroblasts (Keller *et al.*, 1988; R. LeBaron and J. Esko, unpublished results), mastocytoma cells (Montgomery and Esko, 1988), and Hela cells.

The large number of mutants derived from these screenings stimulated us to develop a rapid complementation test (Esko *et al.*, 1986, 1987). Pairs of mutants were coplated in 24-well plates, fused with polyethylene glycol, and 1 day later replated into 100-mm-diameter plates at 33°C. Colonies that arose on the plate were briefly labeled with $^{35}SO_4$, and newly made PG were precipitated with TCA. The sides of the plates were removed and the bottom of the plate was exposed to X-ray film. Mutants that bore defects in different genes complemented and gave rise to occasional colonies that incorporated as much $^{35}SO_4$ as wild-type colonies. Mutants that had defects in the same gene did not complement, and colonies incorporating wild-type levels of $^{35}SO_4$ did not arise. Colonies derived from fused cells also transfer to polyester cloth, permitting the isola-

TABLE III

MUTANTS ISOLATED BY COLONY AUTORADIOGRAPHY

Type of mutant	Cell line	Support	Precursor	References
CDP-choline synthetase	CHO	Filter paper	[^{14}C]Choline	Esko and Raetz (1980b); Esko *et al.* (1981, 1982)
CDP-ethanolamine synthetase	CHO	Filter paper	[^{14}C]Ethanolamine	Miller and Kent (1986)
Acyl-CoA synthetase	Fibrosarcoma	Polyester	[^{14}C]Arachidonate	Neufeld *et al.* (1984)
Proline transport	CHO	Filter paper	[^{14}C]Proline	Dantzig *et al.* (1982)
Glycine transport	CHO	Filter paper	[^{14}C]Glycine	Fairgrieve *et al.* (1987)
Sulfate transport	CHO	Polyester	$^{35}SO_4$	Esko *et al.* (1986)
Xylosyltransferase	CHO	Polyester	$^{35}SO_4$	Esko *et al.* (1985)
Galactosyltransferase I	CHO	Polyester	$^{35}SO_4$	Esko *et al.* (1987)
Heparan sulfate-deficient	CHO	Polyester	$^{35}SO_4$	Esko *et al.* (1988)
N-sulfotransferase	CHO	Polyester	$^{35}SO_4$	Bame and Esko (1989)
Folylpolyglutamate synthase	CHO	Polyester	[^{3}H]Deoxyuridine	Sussman *et al.* (1986)
Glycoprotein synthesis	CHO	Filter paper	[^{3}H]Fucose and [^{3}H]mannose	Hirschberg *et al.* (1981, 1982); Baker *et al.* (1982)
	CHO	Polyester	Secretion of [^{35}S]methionine-labeled protein	Nakano and Akamatsu (1985)

Mucin	Colon carcinoma	Polyester	^{125}I-Labeled antibodies	Kuan *et al.* (1987)
Acetylcholine receptors	C2 Muscle cells	Polyester	^{125}I-Labeled bungarotoxin	Black and Hall (1985); Black *et al.* (1987)
Man6P receptors	CHO	Filter paper	[^{35}S]Methionine in the presence of a ricin conjugate of Man6P	Robbins *et al.* (1981)
LDL receptors	CHO	Polyester	^{125}I-LDL	Esko (1986)
LDL endocytosis	CHO	Polyester	^{125}I-LDL	Esko (1986)
Cholesterol auxotrophs	CHO	Polyester	^{125}I-LDL	Esko (1986)
Endocytosis	CHO	Filter paper or polyester	Uptake of [^{35}S]methionine-labeled CHO cell secretions that contain Man6P; [^{35}S]methionine in the presence of diphtheria toxin	Robbins *et al.* (1983, 1984); Klausner *et al.* (1984); Roff *et al.* (1986); Robbins and Roff (1987)
Secretion	CHO	Polyester and nitrocellulose	Secretion of [^{35}S]methionine-labeled proteins	Nakano *et al.* (1985)

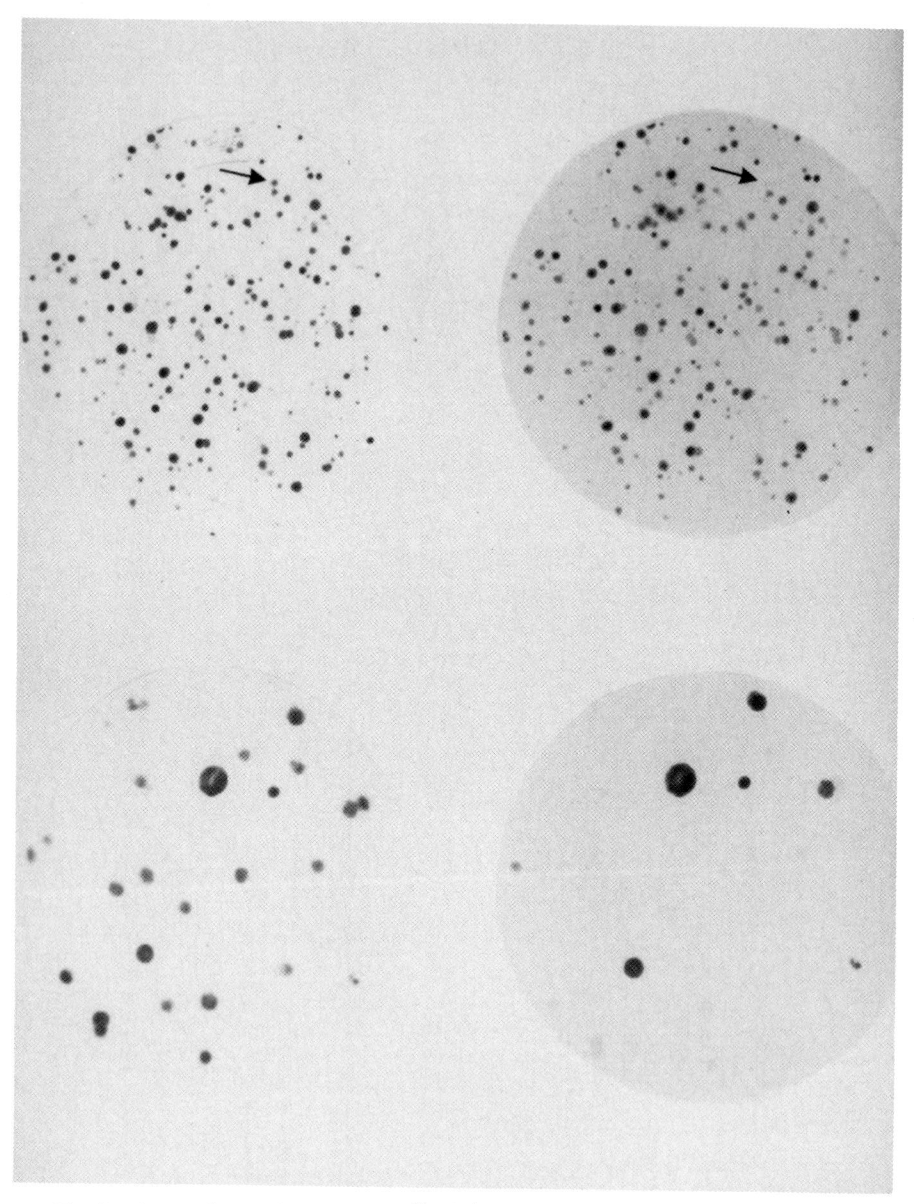

FIG. 9. Autoradiographic detection of $^{35}SO_4$ incorporation into proteoglycans. The upper two panels are the stained disk and the autoradiogram from a primary screening of mutagen-treated colonies. The lower panels are from the repurification of the deficient mutant designated by the arrow in the upper panels. Stained disks are on the left side and the corresponding autoradiograms are on the right side. Reprinted by permission from Esko *et al.* (1985).

tion of hybrid strains without construction of sublines containing drug selection markers.

Most of the strains isolated by $^{35}SO_4$ colony autoradiography grew normally, suggesting that PG is not required for cell division *in vitro*. About 10% of the mutants were also temperature-sensitive for growth, but the PG deficiency did not co-revert in temperature-resistant subclones selected at 40°C, suggesting that the altered growth properties resulted from secondary mutations in the strains. A few temperature-sensitive mutants were found in which reversion to temperature resistance restored PG synthesis. That the PG deficiency rendered the cells temperature-sensitive must be considered, but since $^{35}SO_4$ incorporation is energy-dependent, mutants altered in ATP synthesis would also behave in this manner.

An *in situ* method for measuring folylpolyglutamate synthase activity based on the folate cofactor requirement of thymidylate synthase was developed by Sussman *et al.* (1986) A CHO mutant lacking folylpolyglutamate synthase does not accumulate folate and therefore fails to incorporate [6-^{3}H]deoxyuridine into DNA. This screening could potentially detect mutants in dihydrofolate reductase as well, but CHO cells apparently carry two copies of this gene (Urlaub *et al.*, 1983). Thus, dihydrofolate reductase-deficient mutants would have to be isolated in two steps, the first to obtain heterozygotes that incorporate 2- to 3-fold less [6-^{3}H]deoxyuridine and the second to obtain sublines dramatically deficient in [6-^{3}H]deoxyuridine incorporation.

2. Ligand and Antibody Blotting

Ligand-blotting and antibody-binding reactions conducted on disks provide a direct route for obtaining mutants in receptors and structural proteins lacking enzymatic activity. In general, blotting assays are more variable than assays based on the incorporation of radioactive precursors, and a 2- to 3-fold variation in the intensity of the autoradiographic signal can occur. Thus, blotting assays should be biased toward colonies with dramatic alterations. Polyester replicas work better than filter paper replicas in blotting reactions because the cloth has lower nonspecific binding.

Black and Hall (1985) cloned C2 muscle cell colonies onto 10-μm polyester disks under conditions optimized for CHO cells. Incubation of transferred colonies in special growth medium induced myotube fusion and expression of acetylcholine receptors. ^{125}I-Labeled α-bungarotoxin bound to the colonies, and rinsing the cloth four times in PBS containing 0.2% albumin removed unbound toxin. The colonies were fixed in 1% formaldehyde and 0.8% glutaraldehyde, stained with copper ferrocyanide, and autoradiographed. Screening ~20,000 colonies yield several mutants that failed to produce acetylcholine receptors because of faulty fusion. Two other mutants fused normally but did not synthesize acetylcholine receptors because of deficient production of the α subunit (Black *et al.*, 1987), while another strain made normal amounts of receptor but failed to export it to the cell surface (Black and Hall, 1985).

Incubation of fixed muscle cell colonies with rabbit antiserum against acetylcholine receptors discriminated receptor-deficient strains from the wild type (Black and Hall, 1985). Coating the disks with a 1 : 40 dilution of normal goat serum prevented nonspecific binding of second antibody. Biotinylated goat antirabbit immunoglobulin reacted with primary antibody bound to the colonies, and after inactivation of endogenous peroxidase activity, bound second antibody was detected with an avidin–horseradish peroxidase conjugate assayed with diaminobenzidine.

A human colon cancer cell line (LS174T) was cloned under two layers of 1-μm polyester cloth, and acetone-fixed colonies were treated with an antiserum against cell surface mucins (Kuan *et al.*, 1987). The treated disks were rinsed with saline and reacted with ^{125}I-labeled protein A. After autoradiography and staining, several strains were found with altered levels of mucin.

Blotting intact CHO colonies with ^{125}I-labeled low-density lipoproteins (LDL) successfully yielded mutants altered in LDL metabolism (Esko, 1986). Preincubation of colonies for several days in medium containing lipoprotein-deficient serum increased the amount of LDL binding and uptake, significantly enhancing the autoradiographic signal on the film. The original mutagen-treated population had not been preincubated in lipoprotein-deficient medium in order to obtain strains that might require LDL or other lipids for growth. Four mutants identified by LDL blotting lacked LDL receptors, whereas nine strains required LDL for growth. Subsequent studies showed that the nine mutants contained defects in de novo cholesterol synthesis, and that cholesterol satisfied the auxotrophic requirement for LDL. Two other mutants took up 3-fold more radioactive LDL than other colonies (Fig. 10) and appeared as autoradiographic "hot spots." These mutants contain normal receptors, and LDL accumulates in the interior of the cell in a smooth membrane fraction (J. Esko, unpublished results).

3. Enrichment Methods

When desirable mutants are rare and have not been found in screenings of 10^4–10^5 colonies (100–300 plates), it may prove useful to combine an enrichment procedure with replica plating. Various enrichment schemes that selectively kill dividing cells have been devised based on BrdUrd

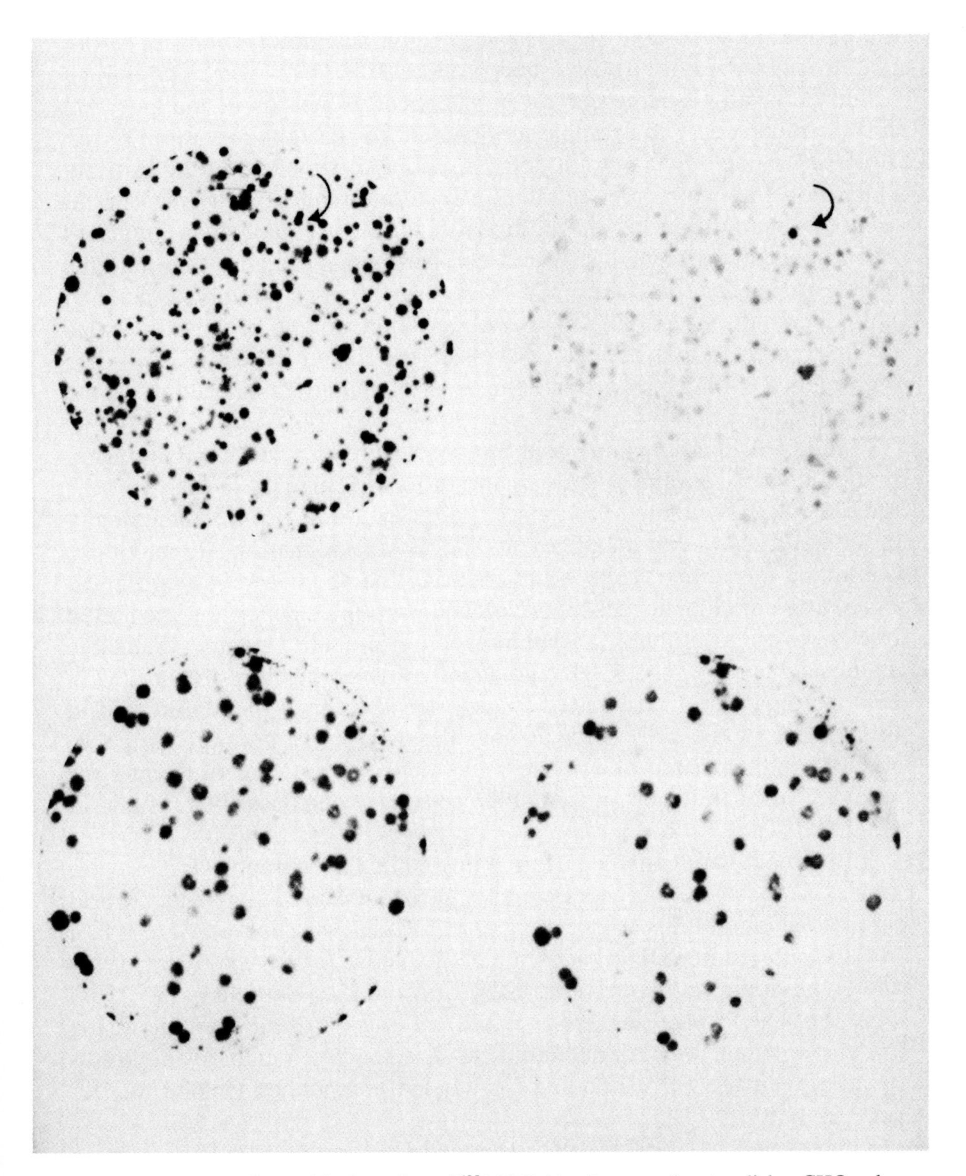

FIG. 10. Autoradiographic detection of ^{125}I-LDL blotting on polyester disks. CHO colonies cloned on polyester disks were incubated with ^{125}I-LDL, washed, and exposed to X-ray film. The upper two panels are the stained disk and the autoradiogram showing a mutant that takes up more LDL than other colonies. The lower panels represent the disk and the autoradiogram obtained from the repurification of the mutant indicated by the arrow in the upper panels. Stained disks are on the left side and the corresponding autoradiograms are on the right side.

incorporation (Kao and Puck, 1974), tritium suicide (Thompson, 1979), and toxin sensitivity (Baker and Ling, 1978). However, enrichment schemes usually assume certain biochemical properties of mutants that may be incorrect. For example, enrichment for nutrient auxotrophs with BrdUrd assumes that mutant cells stop DNA synthesis during nutrient starvation and can be recovered after restoration of the nutrient. Enrichments of mutants based on antibody-dependent, complement-mediated cell lysis or fluorescence-activated cell sorting also require that mutants in essential genes can be rescued. In contrast, replica-plating methods do not depend on reversibility of the mutant's phenotype because a replica of each colony is analyzed. Nevertheless, enrichment procedures coupled with replica plating may provide the order of magnitude needed to isolate rare mutants.

Mutants blocked in amino acid transport were enriched from mutagen-treated cells by tritium suicide, using proline or glycine of high specific radioactivity (Dantzing *et al.*, 1982; Fairgrieve *et al.*, 1987). Dantzing *et al.* (1982) discussed in detail how to use ^{3}H suicide methods to maximize the killing of parental cells while ensuring the viability of mutant cells. Following enrichment, cells were cloned onto filter paper and incubated briefly with [^{14}C]proline, [^{14}C]aminoisobutyric acid (a nonmetabolizable amino acid analog), or [^{14}C]glycine. Gently rinsing the disks six times with cold isotonic buffer terminated transport and trapped the amino acid in the interior of the cell. After drying, the disks were autoradiographed, stained, and superimposed in the usual manner. A similar procedure for assaying sugar transport in LLC-PK1 cells has been described (Mullin *et al.*, 1984).

Enrichment with tritiated sugars followed by fluorographic assays were employed to isolate mutants defective in fucose and mannose incorporation into glycoproteins (Hirschberg *et al.*, 1981; 1982; Baker *et al.*, 1982). Mutants altered in CDP-ethanolamine synthetase (Miller and Kent, 1986) and arachidonic acid-specific acyl-CoA synthetase (Neufeld *et al.*, 1984) were enriched from mutagen-treated cells by tritium suicide using [^{3}H]ethanolamine or [^{3}H]arachidonic acid, respectively. Colony autoradiography with the corresponding ^{14}C-labeled compounds yielded the desired mutants.

Robbins and co-workers (1981) isolated CHO mutants defective in receptor-mediated uptake of mannose 6-phosphate (Man6P)-containing ligands through a two-step procedure involving enrichment and replica plating. Treatment of cells with ω-(6-phospho)-pentamannose oligosaccharides coupled to ricin killed 99.9% of the cells because uptake through the Man6P receptor delivered ricin to the interior of the cell. The survivors were replica-plated to filter paper and incubated with the toxic conjugate and [^{35}S]methionine to measure protein synthesis. Since ricin inhibits protein synthesis, colonies expressing the Man6P receptor took up the toxin and did not incorporate [^{35}S]methionine. Toxic conjugates of growth factors may aid in the isolation of mutants altered in growth factor receptors as well.

Diphtheria toxin enrichments and subsorting by replica plating produced mutants altered in receptor-mediated endocytosis (Robbins *et al.*, 1983, 1984; Klausner *et al.*, 1984; Roff *et al.*, 1986; Robbins and Roff, 1987). Diphtheria toxin enters cells from an acidic subcellular compartment and inhibits protein synthesis. Therefore, mutants altered in toxin binding, internalization, or acid-dependent penetration of endosomes might survive enrichment. Survivors were replica-plated and incubated with [^{35}S]methionine-labeled CHO cell secretions enriched in lysosomal hydrolases that bear Man6P terminated oligosaccharides. Washing the disks with growth medium and saline solutions stopped uptake, and after fixation with TCA and staining, the disks were fluorographed. Many mutants that failed to incorporate labeled secretions had a pleiotropic mutation affecting the uptake of toxin, lysosomal hydrolases, and transferrin, and the maturation of viral glycoproteins through the Golgi was altered. These mutants have normal endocytic receptors, but endosome acidification is dramatically altered.

A modification of this procedure with stacked polyester cloth replicas identified temperature-sensitive mutants altered in endocytosis and endosome acidification (Roff *et al.*, 1986; Robbins and Roff, 1987). In this procedure, cells were shifted to 39°C, treated with diphtheria toxin, and recovered at 34°C. Survivors were subsequently cloned under five layers of polyester disks, and one replica was incubated with ^{35}S-labeled secretions at the permissive temperature, whereas another was labeled at the restrictive temperature. Desirable mutants appeared positive on the first disk and negative on the second. Two more disks were incubated with [^{35}S]methionine and diphtheria toxin at 34°C and 39°C to assess receptor-mediated uptake of toxin.

Enrichments with diphtheria toxin coupled with replica plating also yielded CHO cell mutants defective in secretion (Nakano *et al.*, 1985). Cells were extensively treated with trypsin to remove cell surface proteins and incubated at 39.5°C in tissue culture plates. After 7 hours they were treated with diphtheria toxin, and nonadherent cells were collected 1 hour later, under the assumption that the procedure would enrich for cells that did not restore surface expression of diphtheria toxin receptors and proteins that mediate cell adhesion. The survivors were replica-plated to polyester cloth, and replicas were soaked in [^{35}S]methionine to label cellular proteins. Next, the disk was overlaid with a layer of polyester cloth and a disk of nitrocellulose wetted with labeled growth medium, and the stack was incubated at 39.5°C for 15 hours. The intervening layer of polyester cloth prevented transfer of cells to the nitrocellulose, but permitted secreted ^{35}S-labeled proteins to reach the membrane. One mutant labeled well on the polyester disk and produced a weak signal on the nitrocellulose membrane, suggesting that it bore a defect in protein export. Another strain isolated in this manner was judged defective in glycoprotein oligosaccharide synthesis (Nakano and Akamatsu, 1985). The use of secondary overlays composed of nitrocellulose or other specialized filters provides a useful method to trap secretory compounds.

4. Gene Transfer and Replica Plating

Reversion of mutants to the normal phenotype by DNA transfer provides the opportunity to isolate the corresponding wild-type gene and to examine the expression of foreign genes in the transformed mutant. Animal cells incorporate DNA fragments after electroporation (Potter *et al.*, 1984) or from mixtures of DNA with calcium phosphate (Graham and van der Eb, 1973), dextran sulfate (Warden and Thorne, 1968), or polybrene (Chaney *et al.*, 1986). If we consider that the human genome contains $\sim 3 \times 10^9$ bp of DNA and each cell that takes up DNA incorporates a 50-kbp segment, then we calculate that $\sim 6 \times 10^4$ transformants must be analyzed to ensure that at least one cell receives the desired gene. Experiments in which selection or enrichment schemes were used indicate that the recovery of revertants occurs at a frequency of $\sim 10^{-4}$–10^6 (Pellicer *et al.*, 1980; Westerveld *et al.*, 1984; Kingsley *et al.*, 1986b). However, the observed reversion frequency depends critically on the efficient expression of the transferred gene. Some cells that received the gene but did not express it efficiently may have been lost because the dose of drug or concentration of nutrient employed for enrichment was incorrect. Replica plating can solve this problem, since the screening assays generally have great sensitivity.

A ligand-blotting assay for detecting epidermal growth factor (EGF) receptors was used to sort mouse L cells for EGF receptors after transfection of genomic A431-cell DNA (Strazdis *et al.*, 1985). Colony replicas of transfected cells were generated on polyester cloth and screened with ^{125}I-labeled EGF. Surprisingly, ~1–4% of transfected cells scored positive for EGF receptors, a number unusually high for gene transfer experiments. Transfer of HeLa-cell DNA resulted in 10-fold less transfectants that scored positive for EGF receptor activity, suggesting that the gene is amplified in A431 cells. This study demonstrates that replica plating has adequate capacity for finding revertants corrected by DNA transfection.

Replica plating also permits sorting of cells that express different levels of a desired gene product. A class II major histocompatibility gene (I-A^d) on a plasmid was introduced into mouse L cells by calcium phosphate coprecipitation (Norcross *et al.*, 1984). Transfectants were replica-plated to polyester disks and incubated with a monoclonal antibody against I-A^d. The colonies were then reacted with ^{125}I-labeled goat antimouse IgG and the clone giving the highest autoradiographic signal was picked for further studies.

It should also be possible to hybridize nucleic acid probes to cellular DNA or RNA in colonies, using suitable hybridization membranes for replica plating. Attempts to replica-plate CHO cells to Gene Screen Plus (New England Nuclear) or Zeta Probe (BioRad) membranes have not proved successful, but replica plating to nitrocellulose merits further development. Conceivably, CHO colonies establised on polyester disks will transfer to these membranes electrophoretically or through passive-transfer techniques equivalent to those used for Southern blot analysis (Maniatis *et al.*, 1982). Nucleic acid hybridization in colonies would simplify the detection of foreign genes introduced by DNA transformation or viral infection and permit identification of mutants with altered transcriptional responses to environmental factors.

V. Conclusions

In this review I have tried to describe the many uses of animal cell replica plating. The original publication on replica plating of animal cells on filter paper was in 1978 (Esko and Raetz, 1978), and polyester replica plating was devised in 1982 (Raetz, *et al.*). During this period, dozens of variations of the original procedure have appeared, and more than 35 classes of mutants have been isolated, mostly from CHO cells (Tables I–III). These mutants account for over one-third of the strains obtained through selection, enrichment, and indirect-screening schemes (as collated by Gottesman, 1985). Replica plating is inexpensive, relatively easy to employ, and adapts well to the individual characteristics of different cell lines and biochemical systems.

Mutants obtained by replica plating have been used to analyze lipid and PG biosynthesis, transport, secretion, and endocytosis, and in some cases they have provided models for in-born errors of metabolism in humans (Esko, 1986; Zoeller and Raetz, 1986; Esko *et al.*, 1987; Kresse *et al.*, 1987). Their characterization has provided insights into the organization and regulation of metabolic pathways, often correcting erroneous conclusions deduced from the analysis of enzymic activities in cell-free extracts (Nishijima *et al.*, 1984; Kuge *et al.*, 1985, 1986). The most exciting facets of mutant isolation are the unanticipated discoveries that arise through serendipity (Esko *et al.*, 1988).

Acknowledgments

I thank R. Montgomery, R. LeBaron, K. Bame, K. Rostand, J. Weinke, and G. vanden Heuval for their contributions to the replica-plating technique and their editorial comments. This work was supported by National Institutes of Health grant GM33063 and grant BC-605 from the American Cancer Society.

References

Abraham, I. (1985). *In* "Molecular Cell Genetics: The Chinese Hamster Cell" (M. M. Gottesman, ed.), pp. 181–210. Wiley, New York.

Baker, R. M., and Ling, V. (1978). *Methods Membr. Biol.* **9,** 337–384.

Baker, R. M., Brunette, D. M., Mankovitz, R., Thompson, L. H., Whitmore, G. F., Siminovitch, L., and Till, J. E. (1974). *Cell (Cambridge, Mass.)* **1,** 9–21.

Baker, R. M., Hirschberg, C. B., O'Brien, W. A., Awerbuch, T. E., and Watson, D. (1982). *In* Methods in Enzymology" (V. Ginsburg, ed.), Vol. 83, pp. 444–458. Academic Press, New York.

Bame, K. J., and Esko, J. D. (1989). *J. Biol. Chem.* **264,** 8059–8065.

Black, R. A., and Hall, Z. W. (1985). *Proc. Natl. Acad. Sci. U.S.A.* **82,** 124–128.

Black, R. A., Goldman, D., Hochschwender, S., Lindstrom, J., and Hall, Z. W. (1987). *J. Cell Biol.* **105,** 1329–1336.

Bulawa, C. E., Ganong, B. R., Sparrow, C. P., and Raetz, C. R. H. (1981). *J. Bacteriol.* **148,** 391–393.

Busch, D. B., Cleaver, J. E., and Glaser, D. A. (1980). *Somat. Cell Genet.* **6,** 407–418.

Chaney, W. G., Howard, D. R., Pollard, J. W., Sallustio, S., and Stanley, P. (1986). *Somat. Cell Mol. Genet.* **12,** 237–244.

Dantzig, A. H., Slayman, C. W., and Adelberg, E. A. (1982). *Somat. Cell Genet.* **8,** 509–520.

Esko, J. D. (1980). "Membrane Phospholipid Mutants of Animal Cells." University of Wisconsin, Madison. Available from author.

Esko, J. D. (1986). *In* "Methods in Enzymology" (J. J. Albers and J. P. Segrest, eds.), Vol. 129, pp. 237–253. Academic Press, Orlando, Florida.

Esko, J. D., and Matsuoka, K. Y. (1983). *J. Biol. Chem.* **258,** 3051–3057.

Esko, J. D., and Raetz, C. R. H. (1978). *Proc. Natl. Acad. Sci. U.S.A.* **75,** 1190–1193.

Esko, J. D., and Raetz, C. R. H. (1980a). *J. Biol. Chem.* **255,** 4474–4480.

Esko, J. D., and Raetz, C. R. H. (1980b). *Proc. Natl. Acad. Sci. U.S.A.* **77,** 5192–5196.

Esko, J. D., Wermuth, M. M., and Raetz, C. R. H. (1981). *J. Biol. Chem.* **256,** 7388–7393.

Esko, J. D., Nishijima, M., and Raetz, C. R. H. (1982). *Proc. Natl. Acad. Sci. U.S.A.* **79,** 1698–1702.

Esko, J. D., Stewart, T. E., and Taylor, W. H. (1985). *Proc. Natl. Acad. Sci. U.S.A.* **82,** 3197–3201.

Esko, J. D., Elgavish, A., Prasthofer, T., Taylor, W. H., and Weinke, J. L. (1986). *J. Biol. Chem.* **261,** 15725–15733.

Esko, J. D., Weinke, J. L., Taylor, W. H., Ekborg, G., Rodén, L., Anantharamaiah, G., and Gawish, A. (1987). *J. Biol. Chem.* **262,** 12189–12195.

Esko, J. D., Rostand, K. S., and Weinke, J. L. (1988). *Science* **241,** 1092–1096.
Fairgrieve, M., Mullin, J. M., Dantzig, A. H., Slayman, C. W., and Adelberg, E. A. (1987). *Somat. Cell Mol. Genet.* **13,** 505–512.
Goldsby, R. A., and Zisper, E. (1969). *Exp. Cell Res.* **54,** 271–275.
Gottesman, M. M. (1985). *In* "Molecular Cell Genetics: The Chinese Hamster Cell" (M. M. Gottesman, ed.), pp. 139–154. Wiley, New York.
Graham, F. L., and van der Eb, A. J. (1973). *Virology* **52,** 456–467.
Gum, J. R., Jr., and Raetz, C. R. H. (1983). *Proc. Natl. Acad. Sci. U.S.A.* **80,** 3918–3922.
Hall, C. W., Robbins, A. R., and Krag, S. S. (1986). *Mol. Cell. Biochem.* **72,** 35–45.
Hartung, S., Jaenisch, R., and Breindl, M. (1986). *Nature (London)* **320,** 365–367.
Harvey, W. F., and Bedford, J. S. (1988). *Radiat. Res.* **113,** 526–542.
Hirschberg, C. B., Baker, R. M., Perez, M., Spencer, L. A., and Watson, D. (1981). *Mol. Cell. Biol.* **1,** 902–909.
Hirschberg, C. B., Perez, M., Snider, M., Hanneman, W. L., Esko, J. D., and Raetz, C. R. H. (1982). *J. Cell. Physiol.* **111,** 255–263.
Hohmann, L. K. (1978). *In* "Methods in Cell Biology (D. M. Prescott, ed.), Vol 20, pp. 247–253. Academic Press, New York.
Howard, B. H., and McCormack, M. (1985). *In* "Molecular Cell Genetics: The Chinese Hamster Cell" (M. M. Gottesman, ed.), pp. 211–234. Wiley, New York.
Icho, T., and Raetz, C. R. H. (1983). *J. Bacteriol.* **153,** 722–730.
Jacobs, L., and DeMars, R. (1977). *In* "Handbook of Mutagenicity Test Procedures" (B. J. Kilbey, M. Legator, W. Nichols, and C. Ramel, eds.), pp. 193–230. Elsevier, Amsterdam.
Kao, F.-T., and Puck, T. T. (1967). *Genetics* **55,** 513–524.
Kao, F.-T., and Puck, T. T. (1974). *In* "Methods in Cell Biology" (D. M. Prescott, ed.), Vol. 8, pp. 23–39. Academic Press, New York.
Keller, K. M., Brauer, P. R., and Keller, J. M. (1988). *Exp. Cell Res.* **179,** 137–158.
Kingsley, D. M., Kozarsky, K. F., Hobbie, L., and Krieger, M. (1986a). *Cell (Cambridge, Mass.)* **44,** 749–759.
Kingsley, D. M., Sege, R. D., Kozarsky, K. F., and Krieger, M. (1986b). *Mol. Cell. Biol.* **6,** 2734–2737.
Klausner, R. D., van Renswoulde, J., Kempf, C., Rao, K., Bateman, J. L., and Robbins, A. R. (1984). *J. Cell Biol.* **98,** 1098–1101.
Kresse, H., Rosthøj, S., Quentin, E., Hollmann, J., Glössl, J., Okada, S., and Tønnesen, T. (1987). *Am. J. Hum. Genet.* **41,** 436–453.
Krieger, M., Brown, M. S., and Goldstein, J. L. (1981). *J. Mol. Biol.* **150,** 167–184.
Kuan, I.-F., Byrd, J. C., Basbaum, C. B., and Kim, Y. S. (1987). *Cancer Res.* **47,** 5715–5724.
Kuge, O., Nishijima, M., and Akamatsu, Y. (1985). *Proc. Natl. Acad. Sci. U.S.A.* **82,** 1926–1930.
Kuge, O., Nishijima, M., and Akamatsu, Y. (1986). *J. Biol. Chem.* **261,** 5790–5794.
Kuroki, T. (1973). *Exp. Cell Res.* **80,** 55–62.
Kuroki, T. (1975). *In* "Methods in Cell Biology" (D. M. Prescott, ed.), Vol. 9, pp. 157–177. Academic Press, New York.
Lederberg, J., and Lederberg, E. M. (1952). *J. Bacteriol.* **63,** 399–406.
Maniatis, T., Fritsch, E. F., and Sambrook, J. (1982). "Molecular Cloning: A Laboratory Manual." Cold Spring Harbor Lab., Cold Spring Harbor, New York.
Matsuzaki, H., Yamauchi, M., and Shibuya, I. (1986). *Cell Struct. Funct.* **11,** 75–80.
Miller, M. A., and Kent, C. (1986). *J. Biol. Chem. 261,* 9753–9761.
Montgomery, R. I., and Esko, J. D. (1988). *J. Cell Biology* **107,** 157a.
Mullin, J. M., Adelberg, E. A., and Slayman, C. W. (1984). *J. Gen. Physiol.* **84,** 33a–34a.
Nakano, A., and Akamatsu, Y. (1985). *Biochim. Biophys. Acta* **845,** 507–510.

Nakano, A., Nishijima, M., Maeda, M., and Akamatsu, Y. (1985). *Biochim. Biophys. Acta* **845,** 324–332.

Neufeld, E. J., Bross, T. E., and Majerus, P. W. (1984). *J. Biol. Chem.* **259,** 1986–1992.

Nishijima, M., Kuge, O., Maeda, M., Nakano, A., and Akamatsu, Y. (1984). *J. Biol. Chem.* **259,** 7101–7108.

Norcross, M. A., Bentley, D. M., Margulies, D. H., and Germain, R. N. (1984). *J. Exp. Med.* **160,** 1316–1337.

Pellicer, A., Robins, D., Wold, B., Sweet, R., Jackson, J., Lowy, I., Roberts, J. M., Sim, G. K., Silverstein, S., and Axel, R. (1980). *Science* **209,** 1414–1422.

Polokoff, M. A., Wing, D. A., and Raetz, C. R. H. (1981). *J. Biol. Chem.* **256,** 7687–7690.

Potter, H., Weir, L., and Leder, P. (1984). *Proc. Natl. Acad. Sci. U.S.A.* **81,** 7161–7165.

Raetz, C. R. H. (1975). *Proc. Natl. Acad. Sci. U.S.A.* **72,** 2274–2278.

Raetz, C. R. H., Wermuth, M. M., McIntyre, T. M., Esko, J. D., and Wing, D. C. (1982). *Proc. Natl. Acad. Sci. U.S.A.* **79,** 3223–3227.

Robb, J. A. (1970). *Science* **170,** 857–858.

Robbins, A. R. (1979). *Proc. Natl. Acad. Sci. U.S.A.* **76,** 1911–1915.

Robbins, A. R., and Roff, C. F. (1987). *In* "Methods in Enzymology" (V. Ginsbury, ed.), Vol. 138, pp. 458–470. Academic Press, Orlando, Florida.

Robbins, A. R., Myerowitz, R., Youle, R. J., Murray, G. J., and Neville, D. M., Jr. (1981). *J. Biol. Chem.* **256,** 10618–10622.

Robbins, A. R., Peng, S. S., and Marshal, J. L. (1983). *J. Cell Biol.* **96,** 1064–1071.

Robbins, A. R., Oliver, C., Bateman, J. L., Krag, S. S., Galloway, C. J., and Mellman, I. (1984). *J. Cell Biol.* **99,** 1296–1308.

Roff, C. F., Fuchs, R., Mellman, I., and Robbins, A. R. (1986). *J. Cell Biol.* **103,** 2283–2297.

Saito, Y., Chou, S. M., and Silbert, D. F. (1977). *Proc. Natl. Acad. Sci. U.S.A.* **74,** 3730–3734.

Shapiro, N. I., and Varshaver, N. B. (1975). *In* "Methods in Cell Biology" (D. M. Prescott, ed.), Vol. 10, pp. 209–234. Academic Press, New York.

Stamato, T. D., and Hohmann, L. K. (1975). *Cytogenet. Cell Genet.* **15,** 372–379.

Stanley, P. (1985). *In* "Molecular Cell Genetics: The Chinese Hamster Cell" (M. M. Gottesman, ed.), pp. 745–772. Wiley, New York.

Stefanini, M., Reuser, A., and Bootsma, D. (1982). *Somat. Cell Genet.* **8,** 635–642.

Strazdis, J., Lanahan, A., Johnson, D. E., Bothwell, M., and Kucherlapati, R. (1985). *Int. J. Neurosci.* **26,** 129–140.

Sussman, D. J., Milman, G., Osborne, C., and Shane, B. (1986). *Anal. Biochem.* **158,** 371–376.

Sutherland, R. M. (1988). *Science* **240,** 177–184.

Suzuki, F., and Horikawa, M. (1973). *In* "Methods in Cell Biology" (D. M. Prescott, ed.), Vol. 6, pp. 127–142. Academic Press, New York.

Thompson, L. H. (1979). *In* "Methods in Enzymology" (W. B. Jakoby and I. H. Pastan, eds.), Vol. 58, pp. 308–322. Academic Press, New York.

Urlaub, G., Käs, E., Carothers, A. M., and Chasin, L. A. (1983). *Cell (Cambridge, Mass.)* **33,** 405–412.

Warden, D., and Thorne, H. V. (1968). *J. Gen. Virol.* **3,** 371–377.

Westerveld, A., Hoeijmakers, J. H. J., Van Duin, M., de Wit, J., Odijk, H., Pastink, A., Wood, R. D., and Bootsma, D. (1984). *Nature (London)* **310,** 425–429.

Worton, R. G., Ho, C. C., and Duff, C. (1977). *Somat. Cell Genet.* **3,** 27–45.

Yergenian, G. (1985). *In* "Molecular Cell Genetics: The Chinese Hamster Cell" (M. M. Gottesman, ed.), pp. 3–36. Wiley, New York.

Zoeller, R. A., and Raetz, C. R. H. (1986). *Proc. Natl. Acad. U.S.A.* **83,** 5170–5174.

Chapter 18

Analysis, Selection, and Sorting of Anchorage-Dependent Cells under Growth Conditions

MELVIN SCHINDLER, LIAN-WEI JIANG, AND
MARK SWAISGOOD

Department of Biochemistry
Michigan State University
East Lansing, Michigan 48824

MARGARET H. WADE

Meridian Instruments, Inc.
Okemos, Michigan 48864

I. Introduction

The introduction of flow-cytometric analysis and cell sorting to cell biology opened a research door for the investigation of cellular individuality

METHODS IN CELL BIOLOGY, VOL. 32

Copyright © 1989 by Academic Press, Inc.
All rights of reproduction in any form reserved.

(Kruth, 1982; Muirhead *et al.*, 1985; Melamed *et al.*, 1986). Previously, analysis of cellular function and biochemical composition was technically constrained to bulk measurements of mixed cell populations. Although cell subpopulations, in particular those comprising lymphoid cells, could be monitored and observed using conventional microscopy and staining, the acquisition of sufficient quantities of pure subpopulations for biochemical and biophysical analysis and subsequent cloning remained a difficult task. Flow cytometry, as represented by a variety of fluorescence-based analytical instrumentation (Kruth, 1982; Muirhead *et al.*, 1985; Melamed *et al.*, 1986), in conjunction with monoclonal antibodies (Gasterson *et al.*, 1985; Dumont *et al.*, 1985) and lectins (Steinkamp and Kraemer, 1979; Kraemer *et al.*, 1973)—both reagents that bind to cell type-specific antigens—offered a means for rapid analysis and sorting of lymphoid subpopulations. To a more limited extent, time domain measurements were also exploited for analysis (Martin and Swartzendruber, 1980; Sklar *et al.*, 1984). Clearly, because of the flow properties of the analytical system, a number of inherent constraints must be considered in the design of experimental and selection protocols for flow equipment. In general, flow measurements require that cells be maintained in single-cell suspension; in addition, the cells must be capable of withstanding moderate shear forces generated by rapid flow through narrow apertures (Kruth, 1982; Muirhead *et al.*, 1985; Melamed *et al.*, 1986). Although multiple-laser or split-beam systems have been utilized to permit multiparametric detection and analysis in flow cytometers (Melamed *et al.*, 1986; Bauer *et al.*, 1986), the rapid movement of cells past detectors not only severely limits the types of time domain analyses that can be performed on a single cell, but equally significant, precludes the multi-time point analysis of cellular variations on an individual cell over a range of physiological times, for examples, milliseconds to days.

During the past 15 years, lymphoid cells, particularly lymphocytes, have been employed for flow-cytometric analysis. Significant insights have been obtained into the cell cycle (Gray and Coffino, 1979; Fox *et al.*, 1985), mechanisms of mitogen stimulation (Nuesse and Kramer, 1984; Marcus *et al.*, 1987), and cellular immune response (James *et al.*, 1986; Loken and Stall, 1982). Flow cytometers have been used clinically to monitor tumors (Velet *et al.*, 1984) and for bone marrow transplantation (Loken *et al.*, 1987). Successful attempts to utilize anchorage-dependent tissue-forming cells for analysis and sorting in a flow cytometer, for the most part, have been limited because of a requirement of cellular anchorage to extracellular matrix (ECM) for maintenance of cell viability and preservation of differentiated phenotype and function (Bissell and Barcellas-Hoff, 1987; Gospodarowicz *et al.*, 1980; Ingber and Jamieson, 1985).

Similar difficulties have been encountered in utilizing whole-plant cells because of cell wall-mediated aggregation and clumping of these cells in suspension culture (Ho *et al.*, 1986). Bacterial analysis has also been hampered by clumping problems. Clearly, any instrumental approaches for analysis and sorting of nonlymphoid cell types must be capable of both monitoring cellular activity under optimum growth conditions and performing appropriate cellular separation or sorting without perturbing cellular attachment or cell wall organization. Although a number of physicochemical procedures have been pursued for cell sorting of anchorage-dependent cells, they are usually limited to a binary selection strategy and will not be further discussed. Since anchorage-dependent cells must maintain their surface adherence, the construction of an anchored-cell analyzer must be based on the principle of bringing the means of perturbation and detection to the adhering cells. Two significant advantages of this nonflow approach are that (1) the same cell may be probed and monitored in a number of ways over a long period of time, resulting in the elucidation of one or more selection criteria by which cells may be segregated or sorted; and (2) new modes of selection and analytical criteria may now be pursued based on cell architecture (e.g., cytoskeletal organization: Mitchinson and Kirschner, 1984; Lazarides, 1987) and long-term dynamic properties (e.g., endocytosis: Pastan and Willingham, 1983; van Renswoude *et al.*, 1982), vesicle transport (Lasek and Brady, 1985), and membrane lateral mobility (Jacobson *et al.*, 1987).

The goals of this chapter will be to discuss analytical instrumentation for analysis, selection, and sorting; to examine the types of noninvasive cellular experimentation that may be performed and subsequently used for selection criteria; and to present anchorage-dependent cell-sorting strategies.

II. Instrumental Design—Components for Anchored-Cell Analysis, Selection, and Sorting

The construction of an instrument to analyze, select, and sort anchored cells has in general terms proceeded using the following components:

1. A microscope for visual inspection that also serves as a focusing element for fluorescence excitation and subsequent fluorescence emission from cells containing probe fluorophores
2. A tunable laser to provide a coherent, high-intensity source of light capable of exciting a wide variety of fluorescent dyes and stains

3. Optical detector(s) to convert emitted fluorescence to electronic pulses
4. An acousto-optical modulator and/or controllable series of neutral-density filters to modulate light intensity
5. High-speed two-dimensional stages to locate and address individual cells in tissue or tissue culture
6. A computer to provide system control and management, experimental organization, data collection, and data manipulation and analysis.

A schematic view of the ACAS 470 Interactive Laser Cytometer, an instrument designed to image fluorescence distributions in cells for analysis, selection, and sorting, is presented in Fig. 1. A more thorough examination of the system components has been previously described (Schindler *et al.*, 1987; Schindler and Jiang, 1987).

Recent work with high-resolution computer imaging and video frame grabbers (devices that digitize video images for display on computer terminals) now permit simultaneous overlays of fluorescent and phase images of cells. Such methods permit a direct colocalization of morphological structure and emitted fluorescence as observed for lysosomes and coincident acridine orange staining (Fig. 2). Multiple fluorescent probes may now be monitored and then examined for coincidence with intracellular structures.

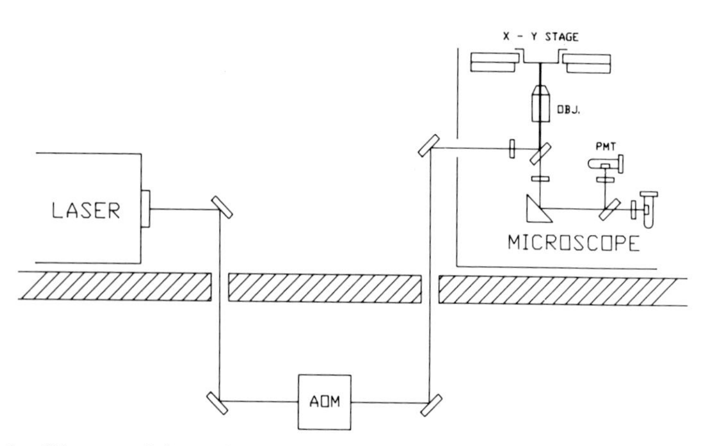

FIG. 1. Diagram of the optical path of a representative fluorescence-imaging system (ACAS 470 Interactive Laser Cytometer). The argon laser beam intensity is controlled by passage through an acousto-optic modulator. The amplitude-modulated beam is then directed into an inverted microscope containing a dichroic element. The excited fluorescence in the sample, which rests on an X–Y scanning stage, is captured by a photomultiplier tube (PMT) and digitized by an analog-to-digital converter. An image of fluorescence distribution is prepared by computer and displayed on a terminal screen.

III. Single-Cell Analytical Parameters

The advantage of single-cell analysis of anchorage-dependent cells is that, for the most part, it removes the demands of rapid data capture as a consideration in experimental design and multiparametric analysis. Cells may be maintained under growth conditions and optically probed in a continuous manner. Under computer control, experiments may be performed sequentially on hundreds or thousands of cells with results presented as histograms in a manner analogous to flow-cytometric analyses (Kruth, 1982; Muirhead *et al., 1985; Melamed et al.,* 1986). These histograms may be inspected for cellular subpopulations. Computer- or user-defined subpopulations of anchorage-dependent cells may be listed and prepared for segregation and sorting, as will be described in Sections IV, A and B. Considering that the cells can be maintained in a fully differentiated state under growth conditions and, if desired, under conditions of maximum cell–cell contact, a significant number of biochemical properties may be exploited for understanding cellular structure, function, and individuality. The following section will serve to illustrate the types of single-cell-based experiments that can be performed on large populations of cells with the multiple goals of analysis, detection of anchored cells with altered functions or responses, and cell isolation and cloning.

A. Intracellular pH

Alteration of cytoplasmic pH has been demonstrated to be a cellular response to mitogens and oncogenic transformation. (Moolenaar *et al.,* 1986; Cassel *et al.,* 1986). Variations in pH of intracellular compartments (e.g., lysosome, endosome, plant vacuole) have been demonstrated to have dramatic consequences for transmembrane signal processing and membrane biosynthesis (Gonzalez-Noriega *et al.,* 1980; Steinman *et al.,* 1983). The control of pH occurs through the intervention of proton pumps in the plasma membrane and subcellular organelles (Mellman *et al.,* 1986). Posttranslational modification or specific mutation are a means by which the biological activity of the proteins may be altered with significant consequences for cellular function. Marlin and Lindquist (1975) demonstrated that the pH sensitivity of fluorescein emission could be utilized as a probe for intracellular pH. Using this principle, fluoresceinated dextrans may be utilized to target fluorescein to intracellular compartments to measure specifically endosomal and lysosomal pH (Ohkuma and Poole, 1978; Tycko and Maxfield, 1982). Experiments to explore vesicular pH may be performed in the following manner.

A working calibration curve can be created by fluorescence-imaging equipment to give the pH dependence of 40- or 64-kDa fluorescently derivatized dextrans [fluorescein isothiocyanate (FITC)–dextrans]. Fluorescence intensity of FITC–dextrans dissolved in a series of buffers (pH 4.3–7.0) is measured at an emission wavelength of 530 nm. Excitation is performed with an argon laser at 457 and 488 nm, respectively. Figure 3A shows a sample standard curve of the measured-intensity ratio $R = (I_{488} - I_{B488})/(I_{457} - I_{B457})$ for 40-kDa FITC–dextran in buffers at the indicated pH values. The fluorescence-intensity ratio of scanned cellular images may be determined and compared to values obtained on the standard curve. This correlation permits a direct reading of intralysosomal pH. In this manner, the pH of individual intracellular compartments may be determined. Uptake of FITC–dextran into lysosomes can be performed as follows. Cells are exposed to FITC–dextran in the medium or Hank's balanced salt solution (HBSS)–HEPES (FITC–dextran concentration 1–2 mg/ml) and incubated at 37°C for 20–30 hours followed by intensive washing. The dextrans have been demonstrated to reside in the lysosomal fraction (Ohkuma and Poole, 1978). Figure 3B shows typical FITC–dextran lysosomal distribution patterns in 3T3-1 fibroblasts.

An example of the analysis for compartmental pH measurements is presented in Fig. 3C. Panel 1 gives an intensity ration of 0.71, yielding a calculated lysosomal compartmental pH of 4.7 for 3T3-1 fibroblasts. Panel 2 represents an intensity ratio of 1.4 or a pH of 6.2 for 3T3-1 cells in the presence of 30 μM chloroquine (4–6 hours incubation).

Using mutagenized or virally transformed cells, variations in compartmental pH may be explored by these methods in large populations of cells leading to mutant selection and clonal isolation and analysis.

B. Intracellular Calcium (Ca^{2+})

The significant and diverse roles of calcium in cellular activity have been amply documented (Tsien *et al.*, 1982; Poenie *et al.*, 1985). Cellular mechanisms of transformation, signal propagation, cell–cell recognition and communication, differentiation, and proliferation all share a requirement for a transient cytoplasmic change in Ca^{2+} concentration. Such a change may trigger specific enzymatic cascades as a result of amplitude- or frequency-modulated Ca^{2+} release from sequestration in endoplasmic

FIG. 2. (A) Phase-contrast image of acridine orange-stained 3T3 cells gathered using a video camera attached to the imaging system described in Fig. 1. (B) The fluorescent image of these stained cells. In conjunction with a video frame grabber, the images in (A) and (B) are now overlaid, resulting in a demonstration that the nuclei and lysosomes are truly labeled by acridine orange (C).

A

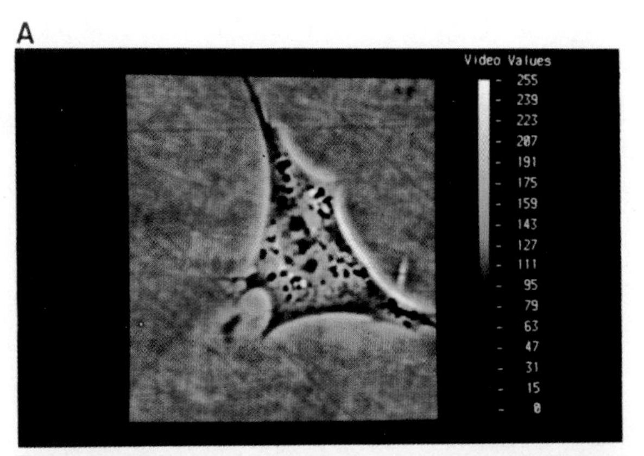
Video Values
255
239
223
207
191
175
159
143
127
111
95
79
63
47
31
15
0

B

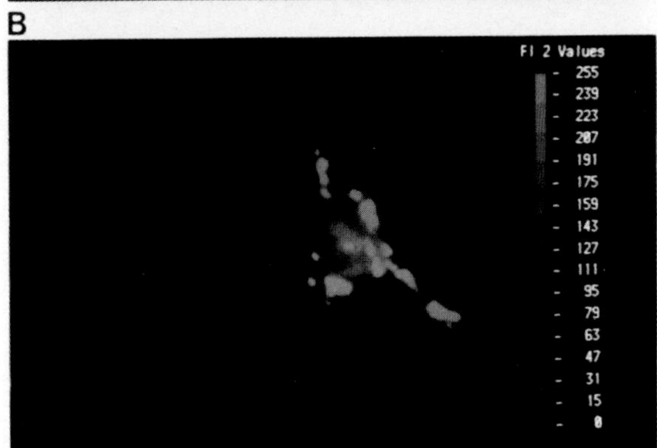
Fl 2 Values
255
239
223
207
191
175
159
143
127
111
95
79
63
47
31
15
0

C

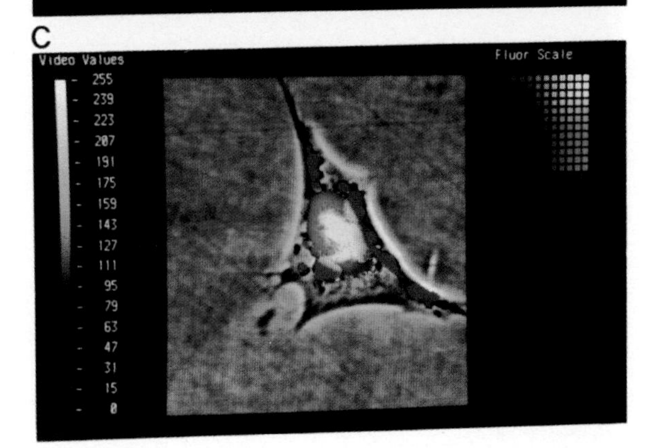
Video Values
255
239
223
207
191
175
159
143
127
111
95
79
63
47
31
15
0
Fluor Scale

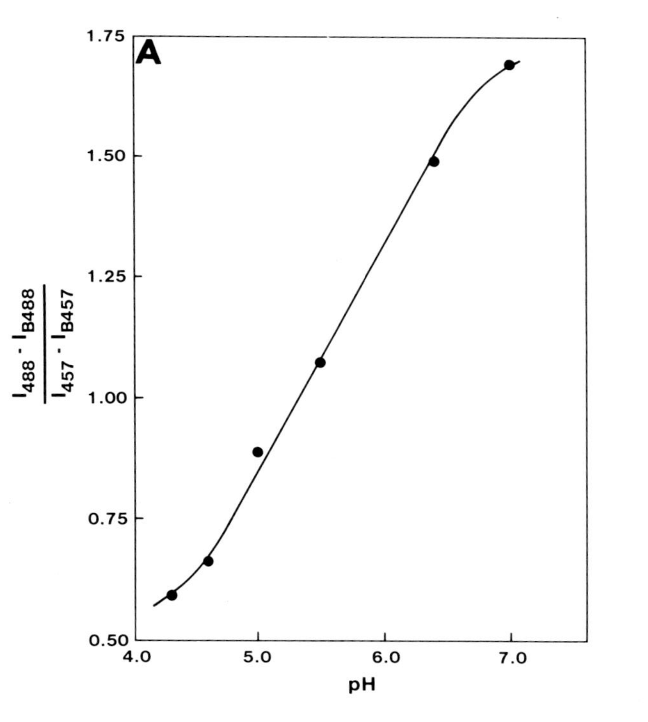

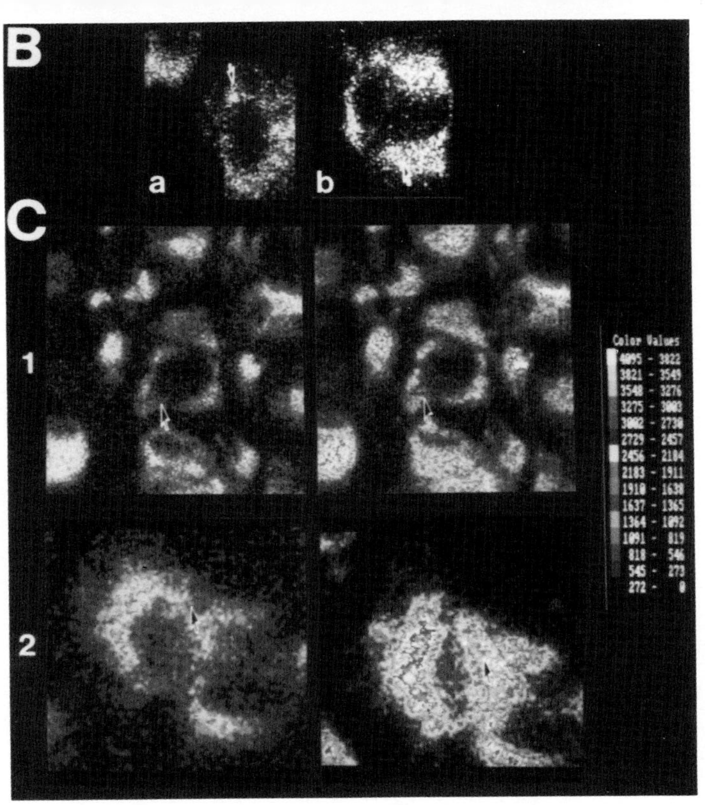

FIG. 3. (A) A working standard curve of fluorescence intensity ratio $R = (I_{488} - I_{B488})/(I_{457} - I_{B457})$ versus pH for 40-kDa FITC–dextran in buffers of varying pH. I_{488} and I_{457} represent the 530-nm emission intensity at the excitation wavelengths of 488 and 457 nm, respectively. I_{B488} and I_{B457} are background-intensity corrections at each excitation wavelength. (B) Typical images of lysosome fluorescence patterns are observed for (a) 40-kDa FITC–dextran in 3T3-1 cells and (b) 40-kDa FITC–dextran in 3T3-1 cells incubated with 30 μM chloroquine. (C) Fluorescent images of (panel 1) FITC–dextran-loaded 3T3-1 cells with $R = 0.71$, pH = 4.7 and (panel 2) 3T3-1 cells incubated with 30 μM chloroquine for 4–6 hours with $R = 1.4$, pH = 6.2.

reticulum, or enhanced Ca_{2+} transport across the plasma membrane. A number of fluorescent Ca^{2+}-chelation probes have been developed, each having inherent advantages and disadvantages (Grynkiewicz *et al.*, 1985; Haugland, 1983). These probes may all be incorporated into the cell cytoplasm and have the property that free Ca^{2+} causes concentration-dependent changes in their fluorescence emission. Employing appropriate standard curves for Ca^{2+} concentration $[Ca^{2+}]$, fluorescence of these chelators may be directly related to intracellular free $[Ca^{2+}]$. In this manner, cellular contour maps of free $[Ca^{2+}]$ changes may be prepared. A typical $[Ca^{2+}]$ analysis may be performed as follows.

Human teratocarcinoma (HT) cells are labeled with 1 μM Indo-1, acetoxymethyl ester (Indo-1AM) for 1 hour at 37°C in medium. The cell-permeant form of the Ca^{2+} chelator enters the cell and is cleaved by esterases to the fluorescent Indo-1 free acid (Grynkiewicz *et al.*, 1985). Unbound Indo-1 emits at 485 nm while Ca^{2+}-bound Indo-1 emits at 405 nm (Grynkiewicz *et al.*, 1985). By exciting at 360 nm with a 5-W argon laser and monitoring both emissions simultaneously, a ratio of emission wavelengths, 405/485, can be prepared. The ratio method is very powerful in that it is concentration and to some extent, photobleaching-independent.

The result of such an analysis on serum-starved cells is presented in Fig. 4. The upper panel of Fig. 4 represents the fluorescent images recorded using dual detectors for emissions at 485 nm (detector 1) and 405 nm (detector 2). The center panel (Fig. 4B) demonstrates the computer-generated image of the pixel-by-pixel dual-detector ratioed fluorescence-emission image; Fig. 4C represents the ratio in the same group of cells following stimulation by fetal calf serum (10%). A plot of total cellular $[Ca^{2+}]$ as a function of time is shown in Fig. 4D. Although two-dimensional images of Ca^{2+} distribution are demonstrated, single-point or single-dimension scans of cells may be performed that can significantly decrease the time scale of Ca^{2+} transient analysis to the range of milliseconds. In this manner, many cells may be examined to pursue the isolation of cell variants that are altered in their Ca^{2+} response.

C. Endosomal and Lysosomal Compartments

The endosomal and lysosomal compartments serve as processing centers for plasma membrane ligand–receptor complexes as elements of the

FIG. 4. Serum-starved HT cells were labeled with Indo-1, and (A) the fluorescence at 405 and 485 nm was measured simultaneously. (B) A ratio of 405/485 at time 0 is depicted in the left center panel and (C) the ratio after the addition of serum is shown in the right center panel. (D) A time plot of the calcium change over time is shown in the lower panel. The ratio can be correlated to free intracellular $[Ca^{2+}]$ using a standard curve.

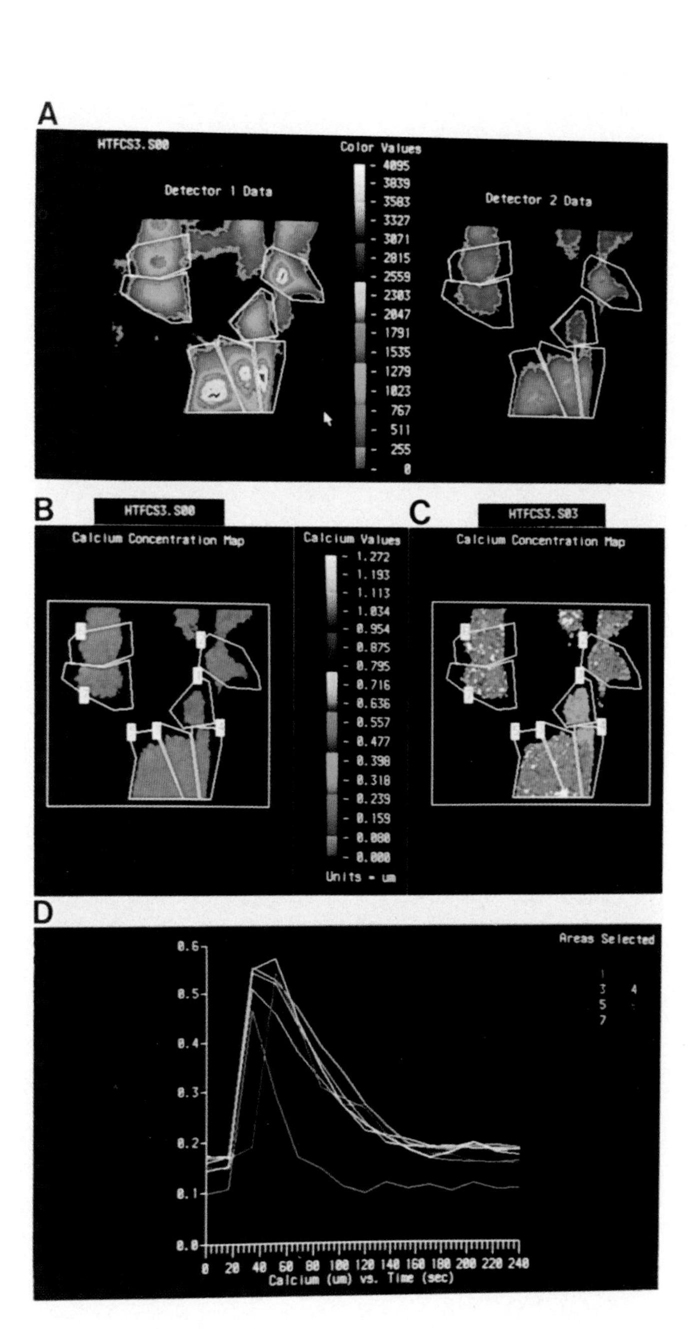
A
HTFCS3.S00
Color Values
Detector 1 Data
Detector 2 Data
4095
3839
3583
3327
3071
2815
2559
2303
2047
1791
1535
1279
1023
767
511
255
0
B
HTFCS3.S00
Calcium Concentration Map
Calcium Values
1.272
1.193
1.113
1.034
0.954
0.875
0.795
0.716
0.636
0.557
0.477
0.398
0.318
0.239
0.159
0.080
0.000
Units = um
C
HTFCS3.S03
Calcium Concentration Map
D
Areas Selected
1
3
4
5
7
0.6
0.5
0.4
0.3
0.2
0.1
0.0
0 20 40 60 80 100 120 140 160 180 200 220 240
Calcium (um) vs. Time (sec)

recycling and degradative pathways (Steinman *et al.*, 1983; Mellman *et al.*, 1986). Polypeptide growth factors bind to plasma membrane receptors, triggering receptor aggregation and internalization into endosomes, which then target receptors for recycling or lysosomal degradation. Evidence has accumulated that normal cytoskeletal organization is required for the functioning of the endosomal–lysosomal pathway (Brown *et al.*, 1980; Mellman *et al.*, 1986), suggesting that all dynamic experiments of this pathway should attempt to maintain normal anchorage-dependent architecture (i.e., adhering and spread). Such conditions would also be of benefit in a search of cell variants in mutagenized or transformed cells at various stages of the endocytotic pathway.

An example of such diversity in lysosomal distribution is observed in Fig. 5. The distribution of lysosomes (as determined by acridine orange staining) in Swiss 3T3 fibroblasts (Fig. 5A) is compared to Kirsten murine sarcoma virus-transformed (KMSV) fibroblasts (Fig. 5B). It is readily apparent that the intracellular distribution and number of lysosomes is significantly different. The dramatic relocation of lysosomes to the growth tips of fibroblast processes in transformed cells may suggest a distinct redistribution of cellular-response mechanisms to these specific sites of enhanced biosynthetic activity. Such spatial differences would be lost if these cells were to be analyzed by cell suspension methods (Murphy and Roederer, 1986). In a similar manner, using fluorescently derivatized transferrin, α_2-macroglobulin, or growth factors (Tycko and Maxfield, 1982; Mellman *et al.*, 1986), the endosomal pathway may be analyzed with regard to quantitative uptake, intracellular targeting, and degradative or recycling time course. Newly identified selection criteria may then be employed to isolate endosomal pathway variants.

D. Transport Dynamics—Fluorescence Photobleaching As a Noninvasive Perturber

Intracellular and intercellular communication is mediated in general by a mixture of molecular diffusion and energy-driven flow processes. Molecules synthesized at one cellular site generally transit to other cellular compartments for subsequent processing, metabolism, or activation. Measurements of particular patterns or rates of movement may serve as important clues to specific cellular synthetic and activational pathways as well as providing information about specific cellular structures that can influence movement, that is, cytoskeletal organization, transmembrane channels, and plasma membrane organization and continuity. In recent years, the dominant technique for quantitating two- and three-dimensional transport in plant and animal cells has been the noninvasive optical technique of fluorescence redistribution after photobleaching (FRAP)

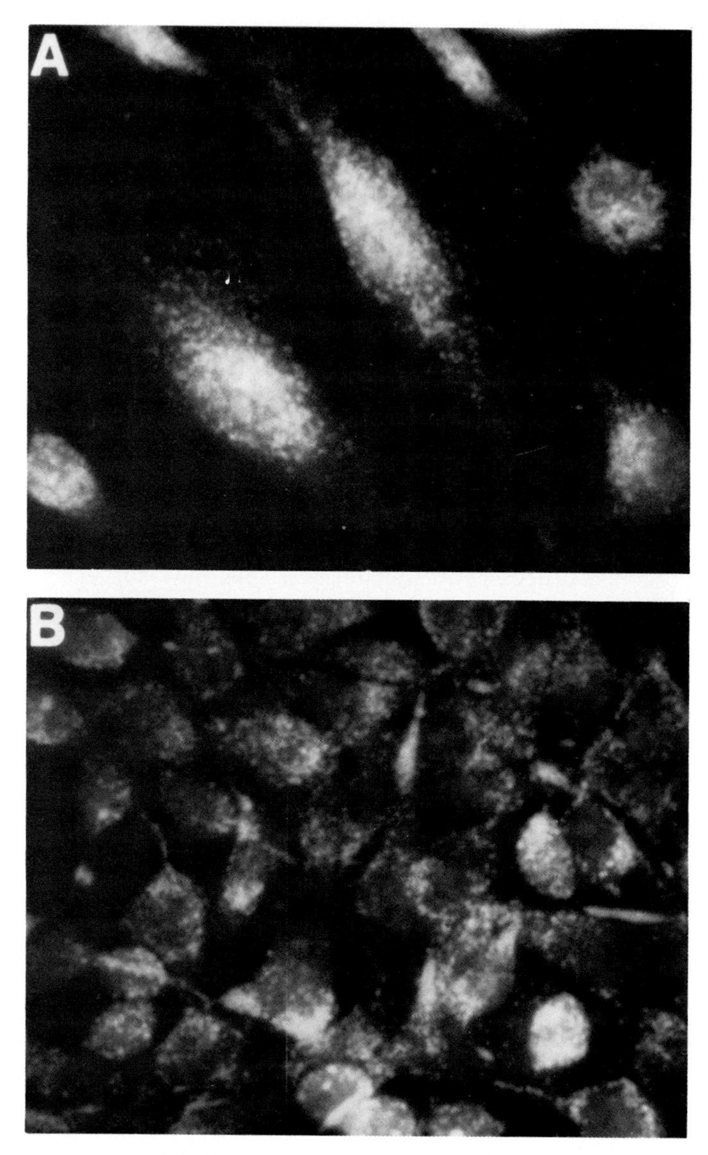

FIG. 5. Lysosome distribution in 3T3 fibroblasts stained with acridine orange showing lysosomes. (A) BALB/c 3T3 fibroblasts; (B) Kirsten murine sarcoma virus (KMSV)-transformed BALB/c 3T3 fibroblasts.

(Kapitza and Jacobson, 1986). Fluorescent probes (e.g., lipids), antibodies to membrane proteins (glycolipids), are introduced into cells to serve as reporter groups sampling the biochemical and biophysical environment of the nonderivatized species. Following the establishment of a probe equilibrium, a short, intense burst of laser light may be employed to photochemically destroy fluorescence on a given surface or in a volume within the cell membrane, cytoplasm, or an intracellular organelle. The dissipation of the induced spatial gradient may then be monitored over time to examine the return of nonphotobleached fluorescent molecules. Mathematical analysis of recovering fluorescence results in a measure of molecular diffusion or flux, while the degree of recovery may be interpreted in terms of structural hindrances to molecular movement, such as cytoskeletal pore diameter (Luby-Phelps *et al.*, 1986) and membrane continuity (Baron-Epel *et al.*, 1988a). In this manner, a dynamic measurement may be utilized to map not only cellular architecture but also the structural elements involved in intercellular communication.

1. Intercellular Communication

Coordination and synchrony of cellular metabolism and growth are essential for the normal integration of cells into functioning tissue structures. Gap junctions in animal cells and plasmodesmata in plants serve as the specialized channel structures through which low molecular weight ($\leq$1700) hydrophilic substances may partition between contacting cells (Baron-Epel *et al.*, 1988a). Changes in the patency of these structures have major consequences for ultimate tissue differentiation in embryogenesis, tissue viability, and oncogenic transformation (Fraser *et al.*, 1987; Azarnia *et al.*, 1988). Clearly, measurements of intercellular communication must of necessity be performed on aggregates of contacting cells. Such experiments would preclude the use of flow-cytometric procedures but are quite amenable to fluorescence imaging and photobleaching approaches. An example of such a measurement on soybean cells grown in suspension culture is presented in Fig. 6. Briefly, soybean [*Glycine max* (L.) Merr cv. Mandarin] root cells (SB-1 cell line) grown in 1B5C medium (Metcalf *et al.*, 1983) in suspension culture were intracellularly stained with fluorescein as described (Baron-Epel *et al.*, 1988a). A uniform pattern of cytoplasmic staining was observed as represented in Fig. 6A. Labeled samples are placed on the stage of an ACAS 470 Interactive Laser Cytometer. The automated stage moves the sample in 1.5-μm steps in a two-dimensional pattern past a microscope objective that focuses the excitation beam (488 nm) from an argon ion laser to a 1-μm-diameter beam on the sample. Emitted intensities at each point are captured by a PMT, col-

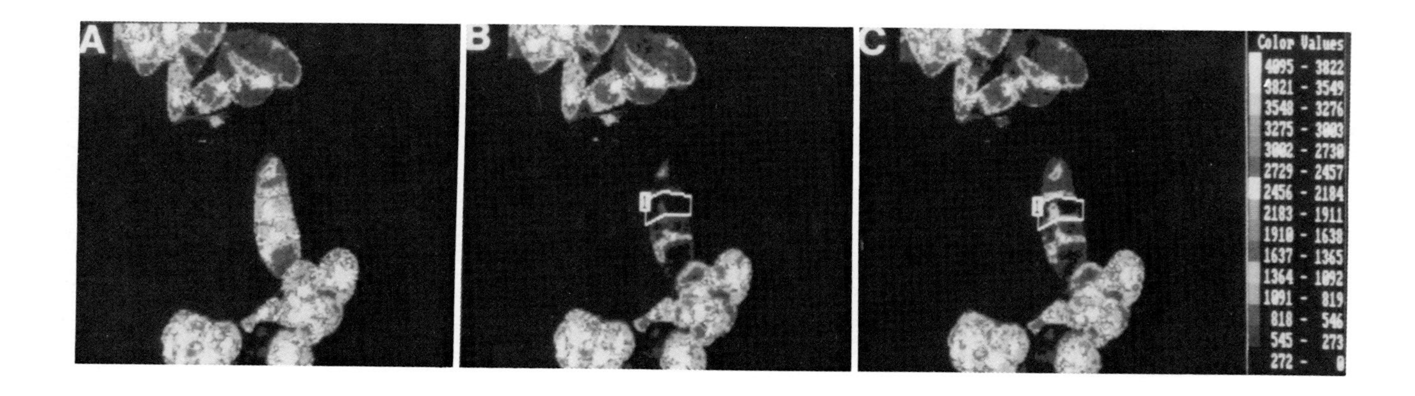

FIG. 6. Recovery of fluorescence in photobleached soybean tissue culture cells labeled with fluorescein. (A) Before photobleaching there is a strong fluorescence intensity prior to bleach distribution. (B) This is reduced after photobleaching the cell delineated by the polygon. (C) Fluorescence intensity returns to the boxed cell 10 minutes after photobleaching.

or-coded, and displayed on a cathode ray tube as false color images or shades of gray (Fig. 6) of cellular fluorescence (Schindler *et al.,* 1987). Photobleaching of cells results in cellular fluorescent images as observed immediately (Fig. 6B) and 10 minutes after photobleaching (Fig. 6C).

2. Intracellular Transport

The movement of molecules and vesicles through the cytomatrix may be rate-limited by (1) the polymeric organization of the cytoplasm (Webster *et al.,* 1978; Luby-Phelps *et al.,* 1986), (2) recognition and attachment to cytoskeletal structures (Masters, 1984; Kachar *et al.,* 1987), and (3) availability of translocator molecules such as kinesin (Vale *et al.,* 1985). In addition, movement of molecules between nucleoplasmic and cytoplasmic compartments are dependent on the state and availability of nuclear pore complexes (Jiang and Schindler, 1986). Changes in cell architecture, or proliferative or differentiated state have all been demonstrated to modify significantly all aspects of intracellular macromolecular and vesicular transport. Consequently, measurements of these dynamic properties for anchorage-dependent cells must be performed under conditions of preserved differentiated structure. Technical approaches to these measurements for macromolecular movement through the cytomatrix (Luby-Phelps *et al.,* 1986; Wojcieszyn *et al.,* 1981) and through the nuclear pore complex (Jiang and Schindler, 1986; Schindler and Jiang, 1987) have relied on mildly perturbing conditions to introduce fluorescently derivatized macromolecules into the cytoplasm, while fluorescence photobleaching is employed to induce spatial gradients whose dissipation kinetics may be monitored to obtain diffusion coefficients or flux rates. Procedures for cellular incorporation of probe macromolecules and the specific experimental detail of performing these photobleaching experiments are fully presented elsewhere (Schindler *et al.,* 1987; Schindler and Jiang, 1987).

3. Membrane Lateral Mobility—A Dynamic Phenotype for Mutant Selection

Lateral mobility of membrane proteins is strongly influenced by membrane and submembranous cortical cytoskeletal organization (Edelman, 1976; Koppel *et al.,* 1981; Tank *et al.,* 1982). Changes in membrane transport phenomena may be indicative of significant reorganizations in cytoskeleton or ECM. Accordingly, measurements of plasma membrane protein lateral mobility may be utilized as selection criteria for isolating subpopulations of mutagenized or transformed cells with altered dynamic phenotypes. We have pursued such a dynamic selection in the following manner.

It was previously observed that a significant population (~30%) of transformed KMSV-3T3 fibroblasts when maintained in the spherical state demonstrated "fast" mobilities ($\geq 1 \times 10^{-9}$ cm^2/second) for wheat germ agglutinin (WGA) membrane receptors (M. Swaisgood and M. Schindler, unpublished results). To explore these differences, clonal populations of fast-mobility KMSV-3T3 fibroblasts were identified using FRAP. Rhodamine- or fluorescein-labeled WGA was used to monitor protein lateral mobility as described (Schindler *et al.*, 1985a). Cells were labeled for 30 minutes at 4°C with 100 μg/ml rhodamine– or fluorescein–WGA in HBSS–HEPES. A fast-mobility population was derived through a series of subclonings on 96-well tissue culture plates from single cells. Clones having fast WGA receptor lateral diffusion were further subcloned. After a total of three subclonings on 96-well plates and two selections from soft agar, the fast-mobility population was obtained and further maintained through common tissue culture practice.

As observed in Fig. 7, fast membrane diffusion cells may indeed be observed following the FRAP selection technique.

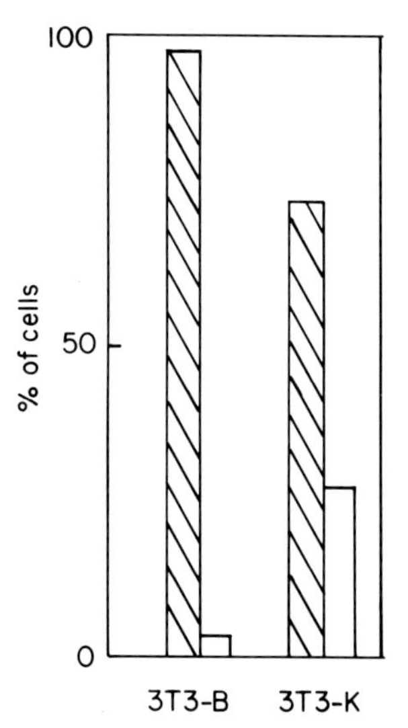

FIG. 7. Histogram of lateral mobility in BALB/c 3T3 and KMSV-transformed BALB/c 3T3 fibroblasts. Cells with WGA receptor diffusion coefficients in the range of 10^{-10} cm^2/second (cross-hatched bars) and 10^{-9} cm^2/second (open bars).

4. Cell Wall Permeation

Prokaryotes and plants are cell wall-containing organisms. Cell walls serve as osmotic containers, macromolecular permeation barriers, and organizers for cellular proliferation (Albersheim *et al.*, 1973). Information concerning their structure is primarily compositional; little is known about three-dimensional organization. In plants, the thickness and composition of the wall varies among cell types and plant species. Since the plant wall serves as the first point of cellular interaction with the environment, variations in its permeability properties may greatly influence the type of biochemical signaling molecules that may ultimately initiate transmembrane signals at the plasma membrane. Experiments designed to tackle such questions must work with whole-plant cells. In culture, such cells generally form filamentous structures. As with anchorage-dependent cells, experimentation must be designed to bring the means of analysis to the cell aggregates under growth conditions. Fluorescence-imaging and photobleaching methods can be utilized to perform measurements of transwall macromolecular permeation (Baron-Epel *et al.*, 1988b). In this manner, a catalog of rates of transit for macromolecules differing in Stokes radii and charge may be prepared correlating transwall mobility to physiological state or differentiated function. Selection criteria may then be established based on these transit properties of cell walls to use sorting techniques to be discussed for isolating cell wall variants. A typical experiment may be performed in the following manner.

Soybean (cv. Mandarin) root cells (SB-1 cell line) grown in 1B5C medium in suspension culture (Ho *et al.*, 1986) were employed for porosity measurements. After 72–96 hours of growth following transfer to fresh medium, cells at the logarithmic stage of growth were centrifuged, and resuspended in 10 m*M* 2-amino-2(hydroxymethyl)-1,3-propanediol (Tris) (pH 5.5), 10 m*M* $CaCl_2$, and 0.5 or 0.7 *M* mannitol to induce plasmolysis. Fluorescein-derivatized dextrans of 4.1 kDa, 9.0 kDa, 17.9 kDa, 41.0 kDa, 65.2 kDa, and 156.9 kDa in mean molecular mass were employed. Labeled dextrans were added to cell suspensions and the cultures incubated for various times at room temperature. A 7- to 10-μl aliquot of the cell suspension containing the fluorescent dextrans or lectins was placed on a slide, and a coverslip was placed on top of the sample. Melted paraffin was used to seal the coverslip to the slide. Following probe equilibration, the FRAP technique was used to measure the translocation rate of fluorescent-labeled dextrans across the plant wall. Photobleaching experiments were essentially done as described by Peters (1986), with a number of technical modifications. A scanning rather than stationary Gaussianprofile laser beam (~2–4 μm in diameter) was moved across the soybean cell. The scanning technique was as described in Schindler *et al.* (1985a). In these experiments, the focused laser beam is repeatedly scanned across the volume between the cell wall and the plasma membrane that results following plasmolysis. The fluorescein-derivatized dextran located in the volume between the wall and the retracted protoplast membrane is photobleached by a high-intensity burst of laser light as previously described (Jiang and Schindler, 1986; Wade *et al.*, 1986). Following photobleaching, the redistribution of unbleached and bleached FITC–dextrans is monitored as recovery of fluorescence in the cell space between the protoplast plasma membrane and the wall. When the data are analyzed as previously described for nucleocytoplasmic transport (Jiang and Schindler, 1986) and cell–cell communication (Wade *et al.*, 1986), rate constants for transwall dye transport may be derived and cells with variant cell wall permeation rates may be identified.

IV. Anchorage-Dependent Cell Sorting

In the previous sections we have attempted to demonstrate that anchorage-dependent cells have physical properties that require analytical measurements to be pursued under conditions of attachment and cell–cell integration. An instrumental configuration was presented that brings the sensitivity and noninvasive advantages of fluorescence analysis to anchorage-dependent animal cells and plant cells under growth conditions. As discussed, removal of time constraints for analysis permits a wide range of dynamic experimentation and enables multiple experiments to be performed on the same cell with the capability for long-term monitoring. A consequence of these analytical procedures is that new biochemical and biophysical criteria may be explored for the selection of atypical cells. In pursuing such a strategy, it must be stressed that not only must unique cells or cell types in anchorage-dependent cell populations be analyzed under appropriate conditions of spreading, differentiation, and growth, but cell segregation or sorting must also be performed under conditions that maintain these properties.

A. Cell Ablation Sorting

An advantage in using a laser source for fluorescence excitation is that a beam focused to 1 μm still maintains significant incident irradiation power. Cell ablation sorting takes advantage of two modes of laser illumination. The first mode is a low-intensity nondestructive illumination utilized for fluorescence excitation and photobleaching as previously described (Section III,D). For cell ablation sorting, the effective laser intensity is increased to serve as a cell-killing beam. Cells to be analyzed and sorted are grown on coverslips or tissue culture surfaces that have previously been coated with a heat-absorptive material (Schindler *et al.*, 1985b). This dark surface may now be utilized to convert high-intensity spot laser illumination into heat. Cells anchored to the growth surface that are to be excluded from selection are literally boiled, while nearby cells that have been spared from killing-intensity illumination are preserved and may proliferate (Schindler *et al.*, 1985b). A typical cell ablation and sorting experiment is presented in Fig. 8.

Two populations of CHO cells, one of which binds an FITC–lectin, were mixed and plated onto a heat-absorptive coverslip (Fig. 8A). After 24 hours, the cells were labeled and a selection and sort was performed over a small area. The cells were imaged (Fig. 8B), and the fluorescent cells were selected for saving by setting a threshold value above which cells were saved, and below which cells were killed. The distance around the saved cells (border) and the frequency of "kill lines" was usually set at 10 μm for the border and 8 μm for the kill distance (Fig. 8C). Figure 8D shows the same area after 24 hours. After 2 days growth (Fig. 8E), the area was stained again and imaged (Fig. 8F), showing proliferation of the fluorescent population.

Such selections may in fact be positive or negative as demonstrated in Fig. 9. Cells, in this case fluorescent derivatized beads, that are labeled are imaged (Fig. 9A) and spared from the subsequent laser-killing pattern (Fig. 9B). In a similar manner, labeled cells may be killed, sparing unlabeled cells for proliferation and analysis (Fig. 9C).

B. Cookie Cut Sorting

Another anchorage-dependent cell-sorting mode that utilizes the high intensity of the laser beam to isolate single cells has been termed the "cookie cutter" approach (Schindler *et al.*, 1985b). This method has significant advantages for isolating single-cell variants within very large populations of undesired cells. In outline, cells are cultured on a thin heat-absorptive film liner made to adhere to the bottom of a tissue culture plate. Should selection be based on antigen detection, appropriate fluorescent antibodies are employed and fluorescent detection analysis proceeds. Each fluorescently positive cell is noted by computer with regard to an *X–Y* coordinate. Following detection, the monitoring beam is switched to higher intensity and proceeds to circumscribe the desired

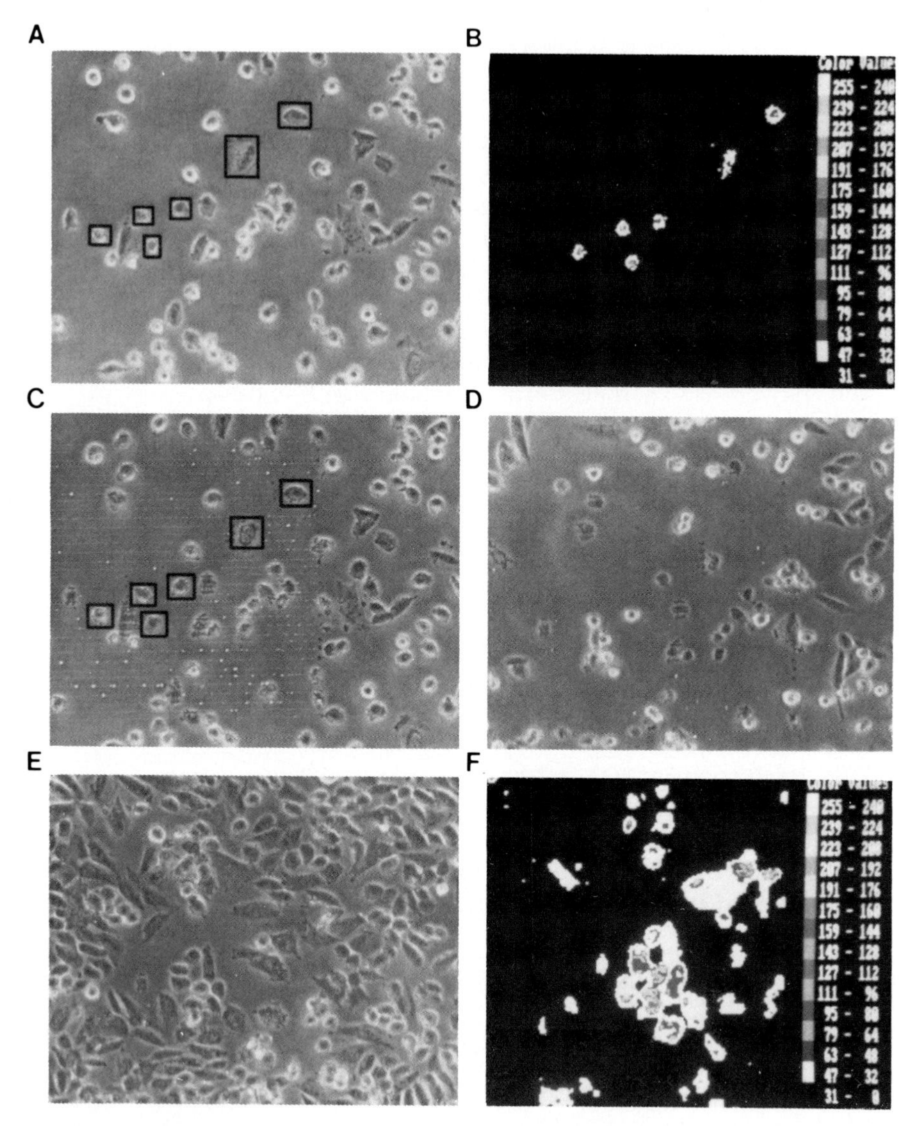

FIG. 8. Mixed population of Chinese hamster ovary (CHO) cells in which one type of cell population is differentially labeled with an FITC–lectin. In this example, the fluorescent cells (B; outlined in A and C) are saved and the rest of the area ablated by increased laser power. (C) The area immediately after killing; (D) the same area after 24 hours. After 2 days growth, the cells were restained and imaged, showing that the number of fluorescent cells had increased [E (phase) and F].

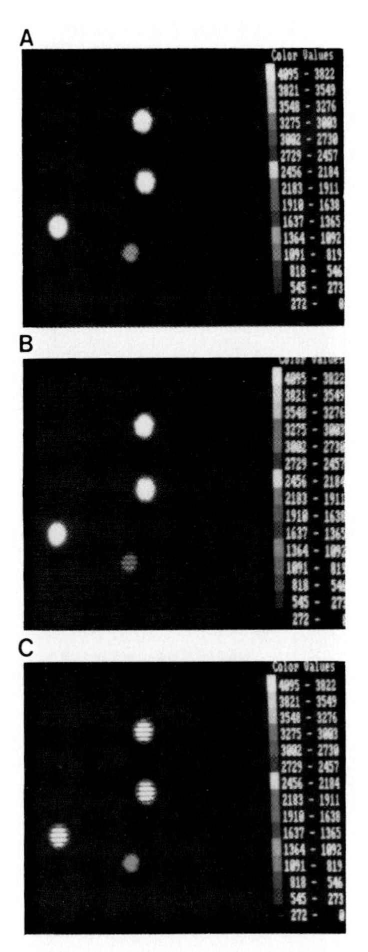

FIG. 9. (A) Fluorescent image of 10-μm beads showing positive and negative sorting protocols. (B) The highly fluorescent beads are saved; (C) the highly fluorescent beads are killed.

cells. The outermost octagon welds the film to the culture dish; optional killing octagons can be used to eliminate unwanted cells. Laser intensity for the interior octagons is not at sufficient laser power to weld. The result is a small "cookie" to which the cells maintain their attachment. Unwanted cells may now be removed by physically pulling the film off the plastic growth plate, leaving unperturbed cookies with attached anchored cells. A sequence of these events is presented in Fig. 10. A demonstration of viability following cell isolation by this method is presented in Fig. 11. These cells were monitored over a period of 7 days, and they demonstrate that if the laser intensity required for welding is kept at a sufficient distance from the desired cells, then the cells remaining on the cookie will proliferate, overgrowing the welded points onto the clear tissue culture surface.

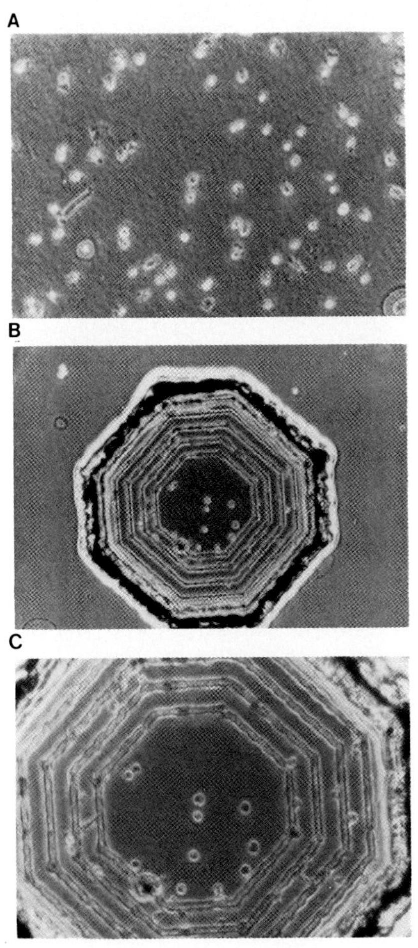

FIG. 10. CHO cells were plated on film-lined dishes (A), and an area containing the cell(s) was circumscribed with the laser. (B) The etched film containing unwanted cells is peeled away, and (C) the remaining cookie containing the desired cell is left behind. (B) × 100; (C) × 400.

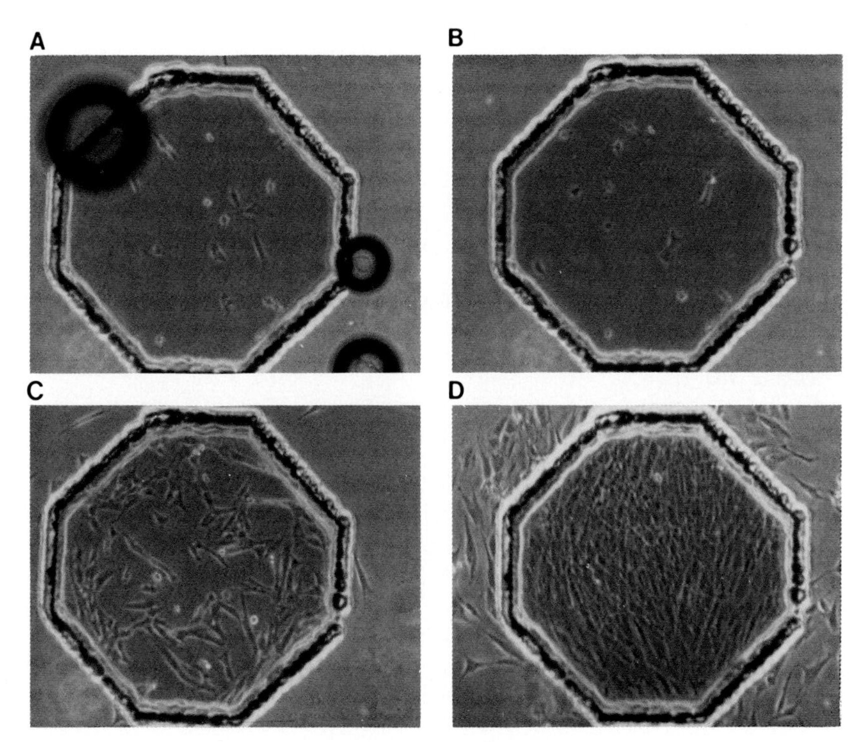

FIG. 11. (A) Myoblasts were plated on the film-lined dishes and a cookie was isolated. (B) After 18 hours; (C) 3 days and (D) 7 days, growth can be seen.

V. Conclusion

An attempt has been made to design a number of instrumental approaches to extend the benefits of automated single-cell analysis and sorting from lymphoid to anchorage-dependent animal and plant cells. We believe that these types of instrumental and procedural methods offer major advantages for real-time temporal and spatial biochemical analysis and subsequent cellular isolation of anchorage-dependent cells. Applications for such methods will be found across the spectrum of cell biological investigations.

ACKNOWLEDGMENTS

Melvin Schindler wishes to express his thanks to Dr. Y. Reisner of the Department of Biophysics, Weizmann Institute of Science for extending his kindness and hospitality during the writing of this chapter at the Institute. Some work described in this chapter was supported by grant G30158 from the National Institutes of Health. The authors also wish to thank T. Sneider for excellent secretarial assistance.

References

Albersheim, P., Bauer, W. D., Keegstra, K., and Talmadge, K. W. (1973). *In* "Biogenesis of Plant Cell Wall Polysaccharides" (F. Loewus, ed.), pp. 117–147. Academic Press, New York.

Azarnia, R., Reddy, S., Kmiecik, T. E., Shalloway, D., and Loewenstein, W. R. (1988). *Science* **239,** 396–400.

Baron-Epel, O., Hernandez, D., Jiang, L.-W., Meiners, S., and Schindler, M. (1988a). *J. Cell Biol.* **106,** 715–721.

Baron-Epel, O., Gharyal, P. K., and Schindler, M. (1988b). *Planta* **175,** 389–395.

Bauer, K. D., Clevenger, C. V., Williams, T. J., and Epstein, A. L. (1986). *J. Histochem. Cytochem.* **34**(2), 245–250.

Bissell, M. J., and Barcellas-Hoff, M.-H. (1987). *J. Cell Sci., Suppl.* **8,** 327–345.

Brown, K. D., Friedkin, M., and Rozengurt, E. (1980). *Proc. Natl. Acad. Sci. U.S.A.* **77,** 480–484.

Cassel, D., Rothenberg, P., Whiteley, B., Mancuso, D., Schlessinger, P., Reuss, L., Cragoe, E. J., and Glaser, L. (1986). *Curr. Top. Membr. Transp.* **26,** 157–173.

Dumont, F. J., Coker, L. Z., Habbersett, R. C., and Trefinger, J. A. (1985). *J. Immunol.* **134**(4), 2357–2365.

Edelman, G. M. (1976). *Science* **192,** 218–226.

Fox, M. H., Read, R. A., and Bedford, J. S. (1985). *Radiat. Res.* **104**(3), 429–442.

Fraser, S. E., Green, C. R., Bode, H. R., and Gilula, N. B. (1987). *Science* **237,** 49–55.

Gasterson, B. A., Mcilhinney, R. A., Shashhikant, P., Knight, J., Monaghan, P., and Ormerod, M. D. (1985). *Differentiation* **30**(2), 102–110.

Gonzales-Noriega, A., Grubb, J. H., Talkad, V., and Sly, W. S. (1980). *J. Cell Biol.* **85,** 839–852.

Gospodarowicz, D., Delgado, D., and Vlodovsky, I. (1980). *Proc. Natl. Acad. Sci. U.S.A.* **77,** 4094–4098.

Gray, J. W., and Coffino, P. (1979). *In* "Methods in Enzymology" (W. B. Jakoby and I. H. Pastan, eds.), Vol. 58, pp. 233–248. Academic Press, New York.

Grynkiewicz, G., Poenie, M., and Tsien, R. Y. (1985). *J. Biol. Chem.* **260,** 3440–3450.

Haugland, R. P. (1983). *In* "Excited State of Biopolymers" (R. Steiner, ed.), pp. 29–58. Plenum, New York.

Ho, S.-C., Malek-Hedayat, S., Wang, J. L., and Schindler, M. (1986). *J. Cell Biol.* **103,** 1043–1054.

Ingber, D. E., and Jamieson, J. D. (1985). *In* "Gene Expression during Normal and Malignant Differentiation" (L. C. Andersson, C. G. Gahmberg, and P. Ekblom, eds.). Academic Press, Orlando, Florida.

Jacobson, K., Ishihara, A., and Inmon, R. (1987). *Annu. Rev. Physiol.* **49,** 163–175.

James, S. P., Fiocchi, C., Graeff, A. S., and Strober, W. (1986). *Gastroenterology* **91**(6), 1483–1486.

Jiang, L.-W., and Schindler, M. (1986). *J. Cell Biol.* **102,** 853–858.

Kachar, B., Bridgman, P. C., and Reese, T. S. (1987). *J. Cell Biol.* **105,** 1267–1271.

Kapitza, H.-G., and Jacobson, K. A. (1986). *In* "Techniques for the Analysis of Membrane Proteins" (C. I. Ragan and R. J. Cherry, eds.), pp. 345–375. Chapman & Hall, London.

Koppel, D. E., Sheetz, M. P., and Schindler, M. (1981). *Proc. Natl. Acad. Sci. U.S.A.* **78,** 3576–3580.

Kraemer, P. M., Tobey, R. A., and Van Dilla, M. A. (1973). *J. Cell. Physiol.* **81,** 305–314.

Kruth, H. S. (1982). *Anal. Biochem.* **125,** 225–242.

Lasek, R. J., and Brady, S. T. (1985). *Nature (London)* **316,** 645–647.

Lazarides, E. (1987). *Cell (Cambridge, Mass.)* **51,** 345–356.

Loken, M. R., and Stall, A. M. (1982). *J. Immunol. Methods* **50,** R85–R112.

Loken, M. R., Shah, V. O., Dattilio, K. L., and Civin, C. I. (1987). *Blood* **69**(1), 255–263.

Luby-Phelps, K., Taylor, D. L., and Lanni, F. (1986). *J. Cell Biol.* **102,** 2015–2022.

Marcus, D. M., Dustira, A., Diego, I., Osovitz, S., and Lewis, D. E. (1987). *Cell. Immunol.* **104**(1), 71–78.

Marlin, M. M., and Lindquist, L. (1975). *J. Lumin.* **10,** 381–390.

Martin, J. C., and Swartzendruber, D. E. (1980). *Science* **202,** 199–201.

Masters, C. (1984). *J. Cell Biol.* **99,** 2225–2255.

Melamed, M. R., Mullaney, P. F., and Mendelsohn, M. L., eds. (1986). "Flow Cytometry and Sorting," 2nd ed. Wiley, New York.

Mellman, I., Fuchs, R., and Helenius, A. (1986). *Annu. Rev. Biochem.* **55,** 663–700.

Metcalf, T. N., III, Wang, J. L., Schubert, K. R., and Schindler, M. (1983). *Biochemistry* **22,** 3969–3975.

Mitchinson, T., and Kirschner, M. (1984). *Nature (London) 312,* 237–242.

Moolenaar, W. H., Defize, L. H. K., van der Saag, P. T., and de Laat, S. W. (1986). *Curr. Top. Membr. Transp.* **26,** 137–156.

Muirhead, K., Horan, P. K., and Poste, G. (1985). *Bio/Technology* **3,** 337–356.

Murphy, R. F., and Roederer, M. (1986). *In* "Applications of Fluorescence in the Biomedical Sciences" (D. L. Taylor, A. S. Waggoner, R. F. Murphy, F. Lani, and R. R. Birge, eds.), pp. 545–566. Liss, New York.

Nuesse, M., and Kramer, J. (1984). *Cytometry* **5** (1), 20–25.

Ohkuma, S., and Poole, B. (1978). *Proc. Natl. Acad. Sci. U.S.A.* **75,** 3327–3331.

Pastan, I., and Willingham, M. C. (1983). *Trends Biochem. Sci.* **8,** 250–254.

Peters, R. (1986). *Biochim. Biophys. Acta* **864,** 305–359.

Poenie, M., Alderton, J., Tsien, R. Y., and Steinhardt, R. A. (1985). *Nature (London)* **315,** 147–149.

Schindler, M., and Jiang, L.-W. (1987). *In* "Methods in Enzymology" (P. M. Conn and A. R. Means, eds.), Vol. 141, pp. 447–458. Academic Press, Orlando, Florida.

Schindler, M., Holland, J. F., and Hogan, M. (1985a). *J. Cell Biol.* **100,** 1408–1415.

Schindler, M., Allen, M. L., Olinger, M. R., and Holland, J. F. (1985b). *Cytometry* **6,** 368–374.

Schindler, M., Trosko, J. E., and Wade, M. H. (1987). *In* "Methods in Enzymology" (P. M. Conn and A. R. Means, eds.), 141, pp. 439–447. Academic Press, Orlando, Florida.

Sklar, L. A., Finney D. A., Oades, Z. G., Jesiatis, A. J., Painter, R. G., and Cochrane, C. G. (1984). *J. Biol. Chem.* **259,** 5661–5669.

Steinkamp, J. A., and Kraemer, P. M. (1979). *In* "Flow Cytometry and Sorting" (M. R. Melamed, P. F. Mullaney, and M. L. Mendelsohn, eds.), pp. 457–501. Wiley, New York.

Steinman, R. M., Mellman, I., Muller, W. A., and Cohn, Z. A. (1983). *J. Cell Biol.* **96,** 1–27.

Tank, D. W., Wu, E.-S., and Webb, W. W. (1982). *J. Cell Biol.* **92,** 201–212.

Tsien, R. Y., Pozzan, T., and Rink, T. J. (1982). *Nature (London* **295,** 68–70.

Tycko, B., and Maxfield, F. R. (1982). *Cell (Cambridge, Mass.)* **28,** 643–651.

Vale, R. D., Reese, T. S., and Sheetz, M. P. (1985). *Cell (Cambridge, Mass.)* **42,** 39–50.

van Renswoude, J., Bridges, K., Harford, J., and Klausner, R. D. (1982). *Proc. Natl. Acad. Sci. U.S.A.* **79,** 6186–6190.

Velet, G., Warnecke, H. H., and Kahle, H. (1984). *Blut* **49,** 37–43.
Wade, M. H., Trosko, J. E., and Schindler, M. (1986). *Science* **232,** 525–528.
Webster, R. E., Henderson, I., Osborn, M., and Weber, K. (1978). *Proc. Natl. Acad. Sci. U.S.A.* **75,** 5511–5515.
Wojcieszyn, J. W., Schlegel, R. A., Wu, E.-S., and Jacobson, K. A. (1981). *Proc. Natl. Acad. Sci. U.S.A.* **78,** 4407–4410.

Chapter 19

Positive and Negative Liposome-Based Immunoselection Techniques

LEE LESERMAN, CLAIRE LANGLET, ANNE-MARIE SCHMITT-VERHULST, AND PATRICK MACHY

Centre d'Immunologie
INSERM-CNRS de Marseille-Luminy
13288 Marseille CEDEX 9, France

I. Introduction

Liposomes are vesicles composed of one or several phospholipid bilayers surrounding a closed aqueous compartment. They may be synthesized in sizes ranging from that of small viruses to large bacteria, and can be made to contain a number of different kinds of molecules stably encapsulated in their aqueous spaces or associated with their lipid bilayers, including chromophores, drugs, proteins and nucleic acids. Inter-

Copyright © 1989 by Academic Press, Inc.
All rights of reproduction in any form reserved.

est in liposomes for purposes of cell selection derives from their large capacity and polyvalent nature, and from the possibility of coupling various ligands to the surface of the liposomes, notably monoclonal antibodies, which permit selective targeting of the liposomes to cells bearing surface molecules for which the antibodies are specific. Depending on the target molecule and the cell type, the liposomes may be taken up and release the transported molecule intracellularly.

The application of liposome technology for cell selection may be divided into four areas, presented in order of development and of technical complexity:

1. Liposomes containing fluorescent reagents as cell markers, in conjunction with cell sorting
2. Liposomes as transporters of cytotoxic drugs, for specific elimination of targeted cells
3. Liposomes as transporters of "antidotes" for cytotoxic drugs present in the medium, for specific rescue of targeted cells
4. Transport of genes conferring drug resistance

Numerous techniques exist for liposome formation, depending on the application and the material to be liposome-associated. The literature on liposomes is extensive, and a complete discussion of different technical possibilities is beyond the scope of this paper (for more information, see Gregoriadis, 1984, 1988; Ostro, 1987; Machy and Leserman, 1987). We here limit the presentation to those methods in routine use in our laboratory, with technical differences relevant to different applications indicated.

II. Preparation of Protein-Coupled Liposomes

Liposomes are formed with the compound of interest encapsulated within the enclosed space (for reagents soluble in aqueous medium), or as part of the component phospholipids (for lipid-soluble drugs or phospholipid derivatives of drugs). A fraction of the phospholipid used for the liposome preparation is a derivative of phosphatidylethanolamine, modified with a heterobifunctional crosslinking reagent, SPDP (see Fig. 1) (Carlson *et al.*, 1978), so that it expresses a protected thiol group. In a second step, protected thiol groups are introduced into the proteins to be coupled to the liposomes. The thiol groups on the protein are activated by mild reduction and incubated with the liposomes, resulting in a thiol disulfide exchange reaction, coupling the protein covalently to the liposome (Leserman *et al.*, 1980a; Barbet *et al.*, 1981). Conceptually, the protected disulfide of the modified phosphatidylethanolamine of the liposomes could be reduced, followed by incubation with the nonreduced, SPDP-modified protein. In fact, when the coupling reaction is performed in this direction it works much less efficiently, for reasons we have not explored in detail.

The steps in the coupling reaction are presented schematically in Fig. 1.

A. Reagent Preparation and Purification

A typical liposome preparation is made using dipalmitoyl phosphatidylcholine (DPPC) or dimyristoyl phosphatidylcholine (DMPC), and cholesterol (which serves to stabilize the liposomes against leakage of contents) in mole ratios of 1 : 1 or 2 : 1, together with 1 or 2% (molar) dipalmitoyl or dimyristoyl phosphatidylethanolamine (DPPE or DMPE) modified

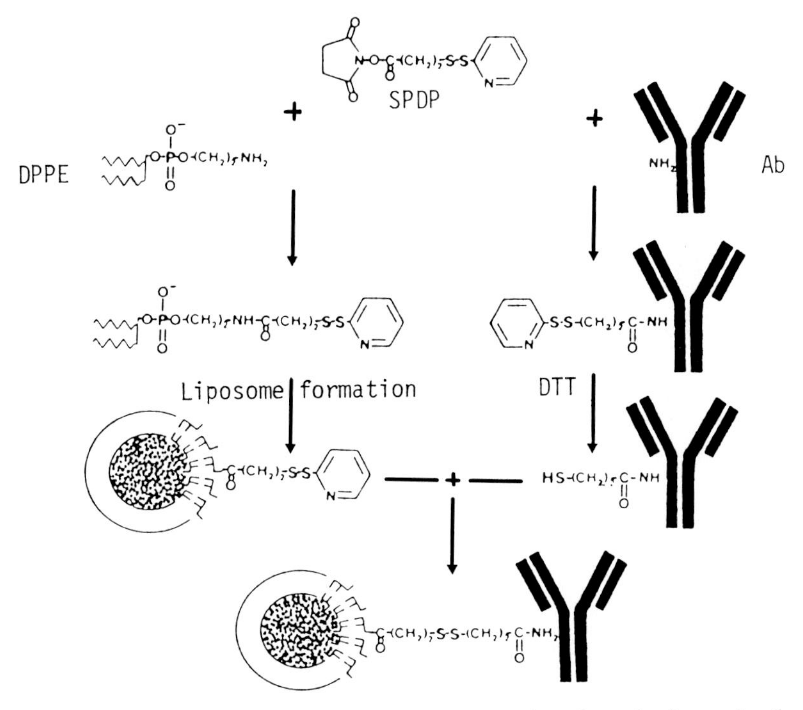

FIG. 1. Schematic representation of the covalent coupling of proteins (monoclonal antibodies or protein A) to preformed liposomes. SPDP, *N*-succinimidyl-3-(2-pyridyldithio) propionate; DPPE, dipalmitoyl phosphatidylethanolamine; Ab, antibody; DTT, dithiothreitol.

with *N*-succinimidyl-3-(2-pyridyldithio) propionate (SPDP). We have used lipids from Sigma Chemical Co. (St. Louis, MO), Avanti Polar lipids, Inc. (Birmingham, AL), and other major suppliers with equivalent results. Our experience with SPDP is limited to that obtained from Pharmacia (Pharmacia LKB Biotechnology, Piscataway, NJ); it is also supplied by Sigma and Pierce Europe BV (Aud-Beijerland, The Netherlands).

1. Synthesis of Derivatized Phospholipids

Dipalmitoyl (or dimyristoyl) phosphatidylethanolamine 3-(2-pyridyldithio) propionate (DPPE or DMPE-DTP) are synthesized as follows: to 10 μmol of DPPE (DMPE) in 700 μl of chloroform–methanol (9 : 1) are added 12 μmol of SPDP in 300 μl of methanol and 20 μmol triethylamine. Because SPDP hydrolyzes on storage in methanol, it should always be freshly prepared from desiccated powder. The mixture is stirred at room temperature in a small glass vial for 2 hours. A volume of water equal to that of the organic solvent is then added. After mixing, the two phases are separated by centrifugation (in a benchtop centrifuge), and the organic (lower) phase is aspirated, remixed with water, and the operation repeated two times for the purpose of removing all unreacted SPDP and triethylamine. During this process the organic phase will evaporate to some extent and should be replaced by the addition of chloroform–methanol; an emulsion will form at the chloroform–water interface, which should be recovered to the extent possible. The organic phase should then be evaporated and resuspended in fresh chloroform–methanol. We have made larger batches of this reagent (100 μmol) by multiplying the quantity of the reagents indicated by 10, with equivalent results. The reagent is stable upon storage at −20°C. The concentration of the final product, PE-DTP, may be determined by the spectrophotometric measurement of the release of pyridine-2-thione by dithiothreitol (DTT) reduction. To do so, the A_{343} of an aliquot of the PE-DTP evaporated from chloroform–methanol is measured in aqueous solution before and after the addition of DTT (to a final concentration of 10 m*M* from a stock solution of 500 m*M* in methanol). The difference in the A_{343} divided by the E_M of the pyridine-2-thione (8080) is the molar concentration of PE-DTP.

2. Modification of Protein by SPDP

The bulk of our experience is with modification of protein A from *Staphylococcus aureus* (Pharmacia), or of monoclonal or polyclonal antibodies. A detailed examination of the modification of other proteins by

SPDP is presented in the original publication on this reagent (Carlson *et al.*, 1978).

Because SPDP is an *N*-succinimide, which reacts with α-amino groups (of terminal amino acids or the α-amine of phosphatidylethanolamine) and the ε-amine of lysines, it is necessary to perform all reactions in solutions free of competing amines (i.e., Tris, and $(NH_4)_2SO_4$, or NaN_3). We use phosphate or HEPES buffers. For a typical reaction, 0.01 μmol of protein (~500 μg protein A or 1.5 mg of antibody) in 250–1000 μl 0.15 *M* NaCl buffered to pH 7.4 with 10 m*M* HEPES is reacted in a glass tube with a 5- to 10-fold molar excess of SPDP, introduced as a 10–20 m*M* solution (10–20 μl) in methanol. The preparation is rapidly mixed by vortexing and the reaction allowed to continue for 30 minutes at room temperature. The choice of the quantity of SPDP depends on the protein in question; protein A is unaffected by a 10-fold molar excess of SPDP, while some antibodies are damaged at that level. We routinely modify antibodies at a 5-fold molar excess of SPDP. For routine preparations of protein A, for which we have ample experience, we then add DTT to a final concentration of 50 m*M* from a stock solution of 500 m*M* in water and, after 15 minutes additional incubation at room temperature, separate the protein from the excess DTT over a small column of Sephadex G25, preequilibrated with a suitable buffer at pH 8–8.5. The addition of DTT serves to break the protected disulfide of the dithiopyridine, leaving a reactive sulfhydryl (SH) group. The protein is then ready for incubation with the liposome preparation, which results in a thiol disulfide exchange with the protected SH previously coupled covalently to phosphatidylethanolamine (Fig. 1). It is important that the protein be completely free from residual DTT when it is incubated with the liposomes; otherwise the protected disulfide groups on the lipid will be cleaved by the DTT before they can participate in the reaction with the protein.

The protein may also be dialyzed or separated by gel filtration without addition of DTT and reduced at a later time. The SPDP-modified protein is as stable on storage as the unmodified protein. Separation of the protein from unreacted SPDP before the addition of DTT also has the advantage of permitting the amount of protein-associated 2-thiopyridone liberated by the reduction to be evaluated spectrophotometrically, and thus the calculation of the number of molecules of the crosslinking reagent introduced per molecule of protein, verifying that the protein in question has been successfully modified by SPDP. This is done as discussed previously for PE-DTP, by measuring in aqueous solution the concentration of thiopyridone liberated at OD_{343} following reduction of an aliquot of the modified protein, divided by the protein concentration as determined by its absorbance at OD_{280}. The concentration of the protein measured at 280 nm is artifically enhanced by the absorbance of the heterobifunctional crosslinking reagent at that wavelength, and may be corrected according to the formula (concentration of 2-thiopyridone released on reduction) × 5100 = A_{280} due to the 2-pyridyl disulfide group (Carlson *et al.*, 1978).

Alternatively, and especially for proteins such as protein A for which the E_M at OD_{280} is low, a small quantity of that protein, radiolabeled to a known specific activity, may be introduced with the protein to be modified, and the concentration of 2-thiopyridone measured spectrophotometrically is compared to the concentration of the protein measured by its radioactivity. The introduction of a radiolabel additionally facilitates collection of the modified protein following column chromatography and the verification of protein–liposome association following coupling.

Protein A contains no cysteine groups, and as such is not susceptible to reduction by DTT. It can therefore be treated by DTT to activate the thiol introduced by SPDP without change of pH. Antibodies are disulfide-linked, and these bonds can be broken by DTT at neural pH, but if the protein is brought to pH 4.5 by gel filtration or dialysis it can be incubated with DTT, at which pH the pyridyldithio groups are much more susceptible to reduction than the disulfide groups of the antibody (Carlson *et al.*, 1978).

B. Liposome Formation

Liposomes are made by techniques separable into those requiring or not requiring detergent. As the majority of our liposome preparations are made for the purpose of encapsulating water-soluble molecules, we have emphasized those techniques in which detergent is not used, in order to avoid the problems associated with contamination of the preparation by residual detergent and an increased leakage of contents. These techniques start either by the formation of large multilamellar vesicles by vortexing or by freezing and thawing the lipid preparation mixed with the solute to be encapsulated, or by the formation of reverse micelles in organic solvent, followed by the removal of the solvent to obtain large unilamellar liposomes. These liposomes are heterogeneous in size, and the majority are too large to be taken up by nonphagocytic cells (Machy and Leserman, 1983). Reduction of the mean size of the liposomes by sonication is a simple technique, but the efficiency of sonication depends on the size of the probe, the volume to be sonicated, and the geometry of the vessel in which the sonication takes place. As such, instructions typically suggest that a preparation be sonicated until it clarifies, usually followed by ultracentrifugation to remove tungsten from the sonicator tip and aggregated lipid. An alternative method for liposome preparation, that of extrusion, has been described by Hope *et al.* (1985). This technique is more reproducible than sonication, and is useful for the encapsulation of macromolecules (and especially DNA), since no shearing stress is applied during preparation of small liposomes. Reverse-phase evaporation liposomes (Szoka and Papahadjopoulos, 1978) can encapsulate large quantities of solutes.

Liposomes may be prepared containing low molecular weight water-soluble molecules such as carboxyfluorescein (CF) or sulforhodamine (both from Molecular Probes, Inc, Eugene, OR, or Kodak, Rochester, NY); the CF from Kodak, while much less expensive, requires two purification steps to be sufficiently pure to be usable (Ralston *et al.*, 1981). Methotrexate (MTX), and N^5- formyl-tetrahydrofolate (F-THF) (both from the Division of Cancer Treatment, NCI, NIH, Bethesda, MD) are commercially available from Sigma. The constituent phospholipids and cholesterol are dissolved in chloroform–methanol (9 : 1) at a concentration of 50–100 m*M*. After solubilization they should be stored in tightly capped glass vials fitted with Teflon cap liners at −20°C. In a glass vial suitable for sonication are mixed 20 μmol phospholipid (DPCC or DMPC), 20 μmol cholesterol, and 0.8 μmol PE-DTP. The organic solvent is removed by evaporation with nitrogen or on a rotary evaporator in such a way that it is evenly distributed as a fine lipid film on the sides of the vial. Residual solvent is removed under vacuum. To the dried lipid film is added the appropriate concentration of the solute to be encapsulated in 1–3 ml total volume. Liposomes are normally made containing a fluorescent compound, or if the purpose of the preparation is that of drug transport they are mixed at a known molar ratio with a fluorescent marker such as CF. This permits subsequent quantification of the encapsulated reagent by fluorescence, and also permits determination of the number of liposomes that become cell-associated, by fluorometry or with the fluorescence-activated cell sorter (see later). Because of the low encapsulation efficiency of liposomes, it is desirable to include drugs at concentrations near their maximum solubility. We routinely encapsulate MTX at 25 m*M*, and THF at 50 or 100 m*M* in an alkaline solution, such as 100 m*M* $NaHCO_3$. Complete solubilization of these drugs at high concentration may require mild heating. The solution to be encapsulated is added to the vial containing the lipid film, and the mixture is alternatively frozen (a container with a few pieces of dry ice in alcohol is convenient for this) and thawed in a water bath preheated to 50°C, with occasional vortexing of the warm solution.

Liposomes made of pure phospholipids will not form at temperatures below that of the sol–gel transition of the phospholipid, which in the case of DPPC is 41°C and for DMPC is 27°C. This temperature requirement is reduced to some extent, but not eliminated, by the addition of cholesterol. The choice of the phospholipid is determined by the heat resistance of the material to be encapsulated, and by the possible presence of other lipid components, such as constituents designed to destabilize liposomes taken up in endocytic vesicles (Yatvin *et al.*, 1978; Connor *et al.*, 1984). In the case of the solutes presented here, heat sensitivity is not an issue and DPPC can be used. The mixture should gradually form a turbid solution with no residual lipid adherent to the walls of the container. At this point, the liposomes are large and primarily multilamellar. Their size may be reduced by sonication or extrusion.

1. Sonication

Liposomes are sonicated for 10–15 minutes in a probe-type sonicator equipped with a small tip, with the glass container partially immersed in a water bath preheated to 50°C. We routinely introduce a fine stream of nitrogen gas into the sonication vessel via a thin catheter during sonication to minimize oxidation of the lipids or the solute. The clarified solution is centrifuged at 100,000 *g* for 30 minutes to eliminate aggregated material and larger liposomes. The material remaining in solution should be clear, exhibiting only the faint opalescence of Rayleigh scattering, and their mean diameter is <1000 Å.

2. Extrusion

Small liposomes may also be formed using a device such as the Extruder (Lipex Biomembranes Inc., Vancouver, BC, Canada), according to published references (Hope *et al.*, 1985) and the manufacturer's instructions, by repeatedly forcing multilamellar or large unilamellar liposomes at high pressure through polycarbonate membranes with pores of defined sizes. The body of the device is immersed in a water bath to maintain a temperature of 50°C. A bottle of nitrogen gas is used as the pressure source. The mean size of the liposomes is then determined by the pore size of the membrane; membranes of pore sizes 500–6000 Å are available.

3. Reverse-Phase Evaporation

For the encapsulation of large macromolecules such as DNA plasmids, the classic technique is reverse-phase evaporation (Szoka and Papahadjopoulos, 1978), although extrusion methods with large-pore filters may also be used. In addition to the lipids described for the aforementioned procedures, we add 10% (molar) of phosphatidylserine; we use the synthetic product supplied by Avanti. Lipids are disolved in 3 ml of a mixture of diisopropyl ether and chloroform (2 : 1), and mixed with 1 ml of 1 mg/ml of linearized DNA plasmid in HEPES-buffered saline. It is necessary to form an emulsion of this material without sonication, which would break the DNA; a convenient technique is repeatedly to pass the solution between two Luer-lock syringes joined by a double-hubbed needle (Machy and Leserman, 1983). The organic solvent is then removed under vacuum using a rotary evaporator, resulting in the formation of large liposomes, the heterogeneous size of which may be rendered more uniform by passage over polycarbonate filters. A detailed description of reverse-phase evaporation may be found in the original reference (Szoka and Papahadjopoulos, 1978).

C. Protein Coupling

After liposome formation, the liposomes are coupled to the freshly activated protein at room temperature, usually by mixing the protein and liposomes and dialyzing against buffered saline at pH 8–8.5 for several hours. The initial quantity of PE-DTP in the preparation is ~0.8 μmol, and the initial quantity of protein ~0.01 μmol; hence, assuming roughly equivalent loss of both reagents during the preparation, the final ratio of protein to PE-DTP is ~1 : 80, although only about half of these PE-DTP-groups are present in outer leaflet of the liposome, and thus in contact with the protein solution. We routinely couple 10–30 molecules of protein per liposome, as calculated from the mean number of phospholipid molecules per liposome and the coupling efficiency, as determined by the use of radiolabeled proteins.

Encapsulation of material in small liposomes is inefficient: only a small percentage of the solute is entrapped in the liposomes. It thus may be desirable to recover the nonencapsulated material prior to the coupling reaction by passage over a column of Sephadex (Pharmacia). As the apparent molecular weight of the liposomes is several million, their separation from the nonencapsulated low molecular weight solute is quite rapid, and is facilitated if they contain a colored solute. This column step is desirable if the encapsulated material is expensive or difficult to prepare and is to be recovered, if the excess of noncoupled protein is to be recovered free of the encapsulated solute, or if the encapsulated material would interfere chemically with the protein–liposome coupling. There is no such interference with materials routinely encapsulated in our laboratory. In the case of DNA plasmids, we separate unencapsulated material by gel filtration over a Sepharose column after treatment with DNase. Encapsulated DNA is protected from the enzyme, while unencapsulated DNA is degraded. Alternatively, it is possible to recover unencapsulated DNA on a sucrose gradient.

Coupled liposomes are usually separated from nonbound protein on columns of Sepharose 2 or 4B. If the protein is trace-labeled, radioactivity measurements of fractions should show two peaks, the first being the coupled and the second being the free protein. Other techniques for confirming coupling include the following:

1. Rechromatography of the liposome peak should confirm the stable association of the coupled protein, which can be cleaved from the liposomes by reduction of the disulfide bond, as with DTT.
2. Flotation of the liposomes on a sucrose or other density gradient should also result in flotation of the attached protein.
3. Centrifugation of the liposomes with protein A-bearing Sepharose beads should precipitate protein A-binding antibodies coupled to liposomes, together with the liposome contents, which can easily be detected if they are colored or fluorescent. Alternatively, protein A-bearing liposomes will be precipitated by relevant antibodies bound to beads (Barbet *et al.*, 1981).

Following protein coupling, liposomes can be sterilized by filtration through 0.45 μm filters, and may be stored in a refrigerator for several months without loss of activity.

III. Measurement of Liposome–Cell Association and Drug Effects

A. Fluorescence Measurements

Liposomes containing fluorescent markers can be used, in conjunction with a fluorescence-activated cell sorter, for the identification and collection of a subpopulation of cells whose binding characteristics differ from the bulk population. The advantage of liposomes over conventional fluorochrome-marked antibodies or protein A include a higher fluorescence signal and lower background. Since the encapsulation procedure is essentially the same for each soluble fluorophore, the protocol does not have to be changed according to the technique used for conjugation of the marker in question. The use of fluorophores with potentially interesting spectral properties that cannot easily be modified to accommodate a prosthetic group necessary for their linkage to protein is not precluded (Truneh and Machy, 1987). Finally, liposomes also offer the possibility of combining phenotypic selection (selection based on the expression or lack of expression of a marker) with selection pressure (liposomes that will kill or protect the selected cells), when the fluorophore is coencapsulated with a drug (Machy and Leserman, 1984) (see later). Liposomes containing CF, sulforhodamine, or other fluorophore and coupled to protein A are a convenient probe for the evaluation of the level of expression of a number of cell surface determinants for which protein A-binding antibodies are available, since a single preparation of liposomes can be used. Alternatively, cells can be directly coupled to an antibody or other ligand with affinity for a cell surface determinant.

Cells (0.5×10^6–10^6) are incubated in 100 μl of culture medium containing serum in the presence of 10–20 μg/ml of antibody at 4°C for 1 hour. The liposomes can be added together with the antibody, as the liposomes express multiple protein A molecules, which increases their avidity for cell-bound antibody despite the competition of soluble antibody. Alternatively, the cells can be washed and liposomes incubated in a second step. After washing to remove nonbound liposomes, the cells can then be evaluated for fluorescence by fluorometry, by fluorescence microscopy, or by the fluorescence-activated cell sorter. In order to optimize the fluorescence signal for fluorescence microscopy and for flow cytometry, it is useful to incorporate the fluorochromes in liposomes at a concentration where concentration-dependent quenching is not important—that is, in the range of 1–10 m*M*. When the cells are to be evaluated by fluorometry, which is usually done after detergent lysis, the concentration of the fluorophore can be increased to amplify the fluorescence signal, as the fluorophore released by lysis is diluted in the large volume of the cuvet. This increase in fluorescence constitutes a sensitive test for the integrity of liposome preparations (Weinstein *et al.*, 1984). The uptake by cells of liposomes containing high concentrations of fluorophore, and the subsequent release of the fluorophore into the cytoplasm also dilutes the fluorophore and increases the cells' fluorescence. This phenomenon is energy-dependent, and has been used as a measure of the rate of internalization of different cell surface determinants to which the liposomes were bound (Truneh *et al.*, 1983).

An example of the use of liposomes for marking cell surface determinants expressed at a low level is shown in Fig. 2. In this experiment, equivalent amounts of protein A, in the form of protein A-bearing liposomes of 800 Å diameter and containing ~1500 molecules of CF per liposome, or FITC-labeled protein A (2.5 molecules fluorescein per molecule protein) were incubated with RDM4 murine thymoma cells, in the presence of isotype-matched control (antihuman) or antibodies specific for the cell surface molecules H-2D or H-2K, which are expressed at levels, respectively, of 2500 or 25,000 molecules per cell, based on Scatchard plots obtained using the same antibodies. The cells were analyzed by flow cytometry, and the mean fluorescence of cells plotted. The H-2K molecule is easily seen with fluorescent protein A, but the H-2D molecule

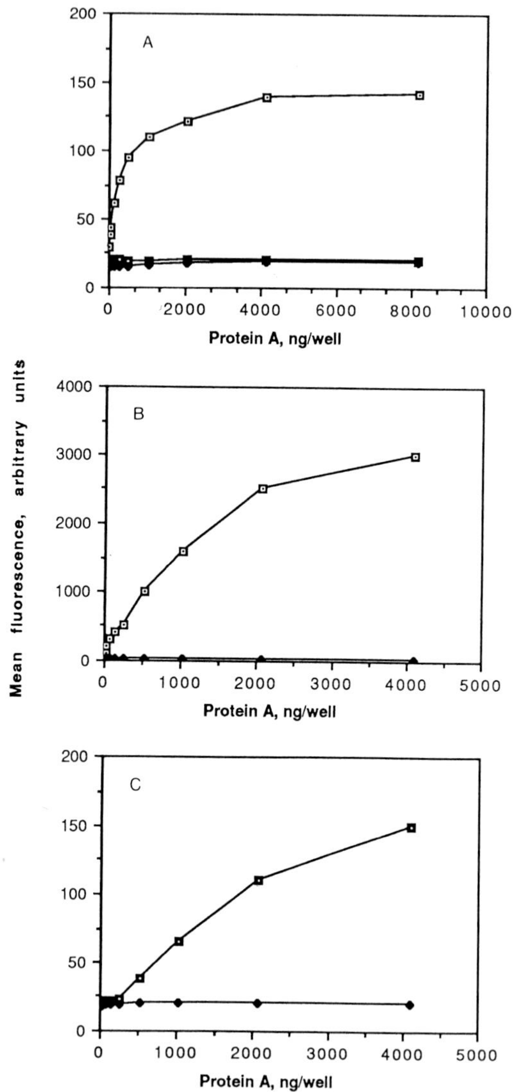

FIG. 2. Comparison by flow cytometry of the fluorescence signal of RDM4 cells incubated with different antibodies (controls, ◆; anti-H-2K, ⊡; anti-H-2D, ■) and with (A) protein A coupled to FITC or (B and C) liposomes of 800 Å diameter containing CF, as described in the text. The level of protein A refers to the concentration of protein A bearing FITC or liposomes added per well. The figure shows the mean fluorescence of 10^4 cells. The intrinsic fluorescence of cells incubated without fluorescent reagents is ~20. Note the difference in scale for (B). Reprinted, in modified form, from Truneh and Machy (1987), with permission.

could not be resolved from background (Fig. 2A). In contrast, the signal from protein A-bearing liposomes bound to the same determinant was significantly higher (Fig. 2B; note the difference in the scale). These liposomes permitted clear resolution of the H-2D molecule, for an equivalent level of background binding as fluorescent protein A (Fig. 2C) (Truneh and Machy, 1987).

B. Short-Term Assays for the Evaluation of Liposome Effects on Cells

In order for drugs in liposomes to have an effect on target cells, it is necessary for the cells to express a sufficient level of the cell surface molecule for which the liposomes are specific and to internalize that molecule, so that the liposome-encapsulated drug enters into the cell. Although it was originally thought that cell-bound liposomes were capable of fusion, it is clear that this is a rare phenomenon. One of the ways in which this was shown was to incubate cells with various antibodies directed at different cell surface molecules and a single preparation of protein A-bearing liposomes containing MTX. Although the cells were sensitive to free MTX, and to liposomes bound to some cell surface determinants, they were insensitive to liposomes bound to other molecules, expressed at identical or even higher levels (as measured by fluorometry of cell-bound liposomes) (Leserman *et al.*, 1981; Machy *et al.*, 1982 a,b). Since delivery of drugs by fusion should depend only on the concentration of the bound liposomes, this experiment suggested that fusion was not the major mechanism of entry of cell-bound liposomes. This test was performed by the uptake of radiolabeled deoxyuridine, which is one of two relatively rapid procedures for the determination of the effect of drug-containing liposomes on cells.

1. Uptake of Radiolabeled Deoxyuridine

This technique depends on the specific blockade by MTX of the action of the cytoplasmic enzyme dihydrofolate reductase, whose action is necessary for the reduction of dihydrofolate into tetrahydrofolate, a necessary cofactor in thymidine synthesis (Jolivet *et al.*, 1983). Cells are incubated with various concentrations of free MTX, or liposomes containing encapsulated MTX, either directly coupled to an antibody with affinity for a cell surface molecule, or with protein A-bearing liposomes in the presence of relevant and control antibodies. These tests are conveniently performed in 96-well flat-bottom microtiter plates in 0.1–0.2 ml of medium, in which are suspended 10^4–10^5 mitogen-stimulated lymphocytes, tumor, or other proliferating cells. After 3–4 hours, exposure to free or encapsulated MTX at 37°C, 0.5–1 μCi d[^{3}H-6]Urd is added, and after an additional 12–16 hours of incubation the cells are harvested with an automatic harvester and radioactivity incorporated in DNA determined by liquid scintillation. Results are expressed as the percentage of incorporation of radioactivity as compared to a relevant control, such as cells incubated in medium alone, in the presence of the targeting antibody but not liposomes, or in the presence of antibody and liposomes made without MTX (Leserman *et al.*, 1980b, 1981; Machy *et al.*, 1982a,b). A typical experiment of this type is presented in Fig. 3. CBA spleen cells were separated into B and T cells and stimulated with the mitogens lipopolysaccharide or concanavalin A, respectively. Then 10^5 cells were incubated with free MTX or with dilutions of a stock solution of protein A-bearing liposomes and an antibody specific for the H-2K molecule, richly expressed on both cell types, or an isotype-matched control antibody that does not bind to these cells. Despite equivalent sensitivity to free MTX, and comparable levels of specific liposome binding, as confirmed by flow cytometry (not shown), the liposomes were effective at significantly inhibiting d[^{3}H-6]-Urd incorporation only in the case of the T-cell blasts. B cells are, however, sensitive to the action of liposomes of the same composition bound to other cell surface determinants (Machy *et al.*, 1982a,b). The capacity of a determinant to act as a target for liposome-mediated drug delivery thus is a property both of the target cell and of the cell that expresses it.

2. Direct Measure of Cell Proliferation with MTT

There are instances in which d[^{3}H-6]Urd incorporation cannot be used, such as for mutant cell lines that lack the enzyme thymidylate kinase, and so cannot incorporate exogenous d[^{3}H-6]Urd into thymidine. In addition, the reduction in d[^{3}H-6]Urd incorporation reflects a metabolic insult to a single enzyme pathway, and not the absence of the possibility of proliferation. This can be determined by counting the cells, either directly (Machy and Leserman, 1984) or by a procedure that measures the enzymatic activity of living cells, such as the MTT test of Mosmann (1983). This test is performed according to the same basic format as the d[^{3}H-6]Urd incorporation test, except that cells are incubated with the liposomes or free drug for 2 days, following which 3-(4,5-dimethylthiazoyl-2)-2,5 diphenyl tetrazolium bromide (MTT), dissolved in saline at 5 mg/ml is added to a final concentration of 500 μg/ml. After 2–3 hours, incubation at 37°C plates are centrifuged, the supernatant removed, and the cells resuspended in propanol-2. which solubilizes the blue formazan precipitate formed as a consequence of the intracellular reduction of MTT, which is an energy-dependent process. The resulting blue color is directly proportional to the number of living cells in the well (Schmitt-Verhulst *et al.*, 1987), and may be measured spectrophotometrically with an automatic

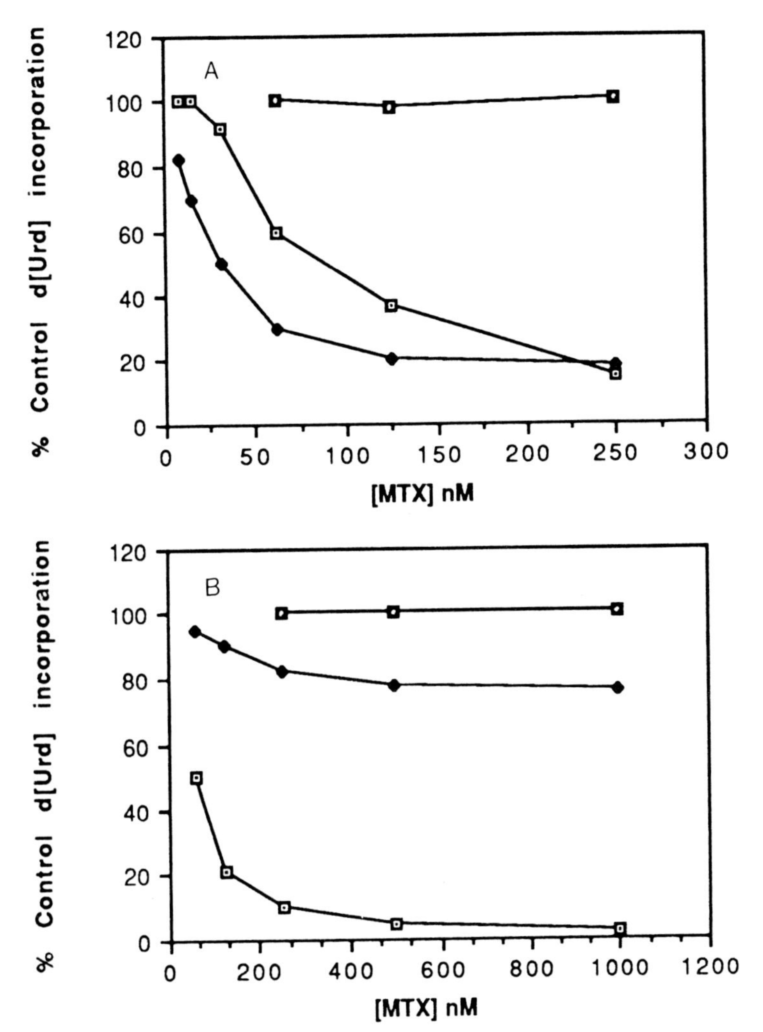

FIG. 3. Differential effect of targeted liposomes on the same cell surface determinant expressed by different cell types. T and B lymphocytes were obtained from spleen cells after enrichment as described (Machy *et al.*, 1982a,b). They were then stimulated with (A) concanavalin A for T cells and with (B) lipopolysaccharide for B cells in order to stimulate DNA synthesis. After 3 hours incubation with various concentrations of free or protein A-bearing liposome-encapsulated MTX, in the presence or absence of antibodies, 0.5 μCi d[^{3}H-6]Urd was added, and after 16 hours additional incubation the radioactivity incorporated in DNA was determined by liquid scintillation. Free MTX (⊡); anti-H-2K (◆); control antibody (■). Reprinted, in modified form, from Machy *et al.* (1982b), with permission.

tissue culture plate reader using a test wavelength of 570 nm and a reference wavelength of 630 nm.

C. Long-Term Assay for the Evaluation of Liposome Effects on Cells

We have already discussed two short-term procedures that permit the investigation of the entry of liposomes into target cells and have indicated the importance of the cell type and the target determinant for specific drug delivery. We have also verified, in long-term assays, that cells that were sensitive to the targeted drug were effectively killed. This has been evaluated with different cell types and antibodies. Murine (RDM4) or human (A431) tumor cells were incubated for 2 days with free MTX, or with 5 μg/ml of mouse-specific (anti-H-2K) or human specific (anti-HLA-B and C) antibodies and MTX-containing liposomes coupled to protein A, in 0.5 ml culture medium. The concentration of the encapsulated drug used for each cell type was the same as the free drug—the level of these being the level of free MTX that gave 60–80% inhibition of incorporation of d[^{3}H-6]Urd in a short-term assay (for RDM4, 60 n*M;* for A431, 250 n*M*). After 2 days the medium was aspirated and fresh medium was added. After 1 week cells expressing the target determinant could not be observed when incubated with free drug or with liposomes and the relevant antibody, but their growth was unaffected by liposomes in the presence of the nonbinding antibody (Fig. 4) (Machy and Leserman, 1984).

It was also possible to eliminate a subpopulation of cells expressing the target determinant without affecting the proliferation of nontargeted cells (Machy and Leserman, 1984). The ability to eliminate cells expressing a given determinant suggested the use of this technique for the selective isolation of spontaneous or induced variant cells lacking the target determinant (Fig. 5). This was investigated in two model systems, using both tumor and normal cells.

IV. Model Systems

A. Negative Selection

1. MHC Deletion Variants of the Murine Thymoma RDM4

Murine thymoma RDM4 cells (50,000) were resuspended in RPMI 1640 medium supplemented with 5% fetal calf serum, and incubated in a volume of 0.5 ml in 24-well tissue culture plates (Costar). To some wells was

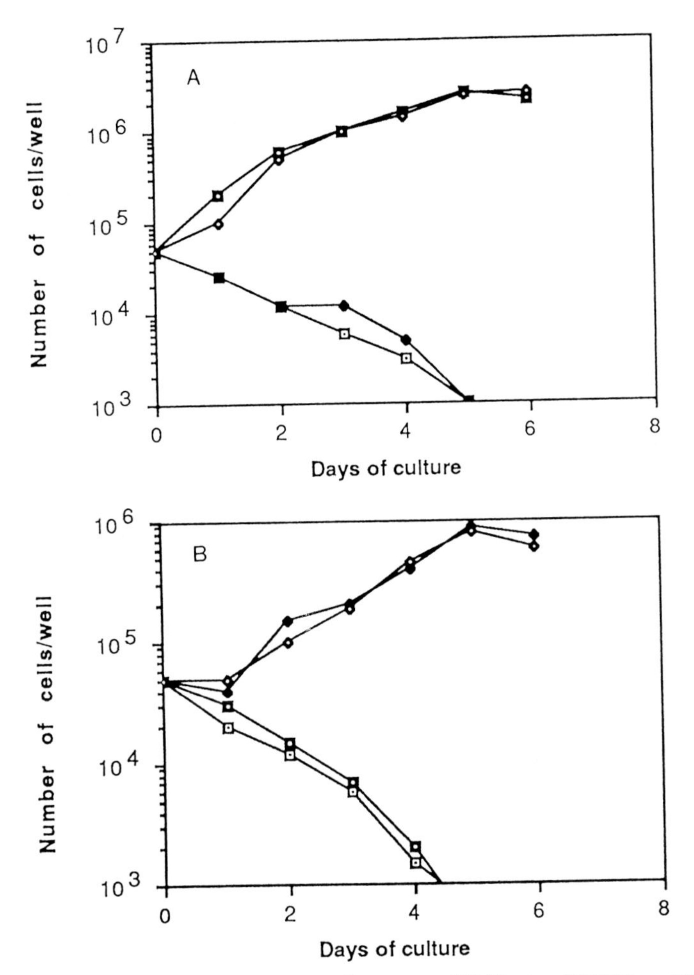

FIG. 4. Specificity of drug delivery from liposomes. (A) Murine (RDM4) or (B) human (A431) tumor cells were incubated for 2 days with free MTX, or with mouse- or human-specific antibodies and MTX-containing liposomes coupled to protein A. Cell counts were performed daily. Anti-HLA (■); no drug (◆); other symbols as in Fig. 3. The same antibodies bound to protein A-bearing liposomes made without MTX had no effect on cell growth (not shown). Reprinted, in modified form, from Machy and Leserman (1984), with permission.

added a monoclonal antibody (H-100-5/28, final concentration of 5 μg/ml), specific for the major histocompatibility complex (MHC)-encoded H-2K molecule, which is expressed by these cells (see Section III,A). Protein A-bearing liposomes containing 20 m*M* MTX were added to a final MTX concentration of 60 n*M;* this concentration was chosen on the basis of a screening test, which showed a 60–80% inhibition of the incorporation of

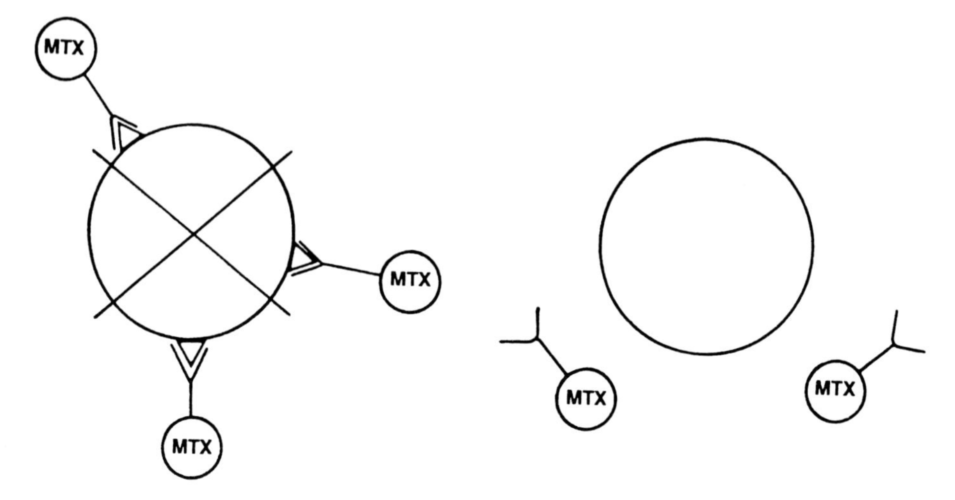

FIG. 5. Negative selection. Cells expressing the target determinant are killed by liposomes that bind to the determinant and are endocytosed. Cells with low levels of, or lacking, the target determinant survive.

d[^{3}H-6]Urd for this cell type with this antibody and concentration of protein A-bearing MTX liposomes. After 2 days of culture the medium was aspirated and fresh medium was added without antibody or liposomes. Cells that escaped targeted liposomes could be detected after 2 weeks. They were then tested for their sensitivity to free MTX and MTX-containing targeted liposomes and were analyzed by cytofluorography and in a radioimmunoassay (RIA) for their capacity to bind the selecting monoclonal anti-H-2K antibody (H-100-5/28). The selected cells remained sensitive to free MTX at a concentration of 60 n*M* but were insensitive to MTX–liposomes targeted by the antibody H-100-5/28, even at a concentration of the drug as high as 250 n*M*.

Examination of the selected cells using the antibody H-100-5/28, or other FITC-labeled antibodies specific for the H-2K molecule with the fluorescence-activated cell sorter indicated that the cells failed to express measurable levels of the H-2K molecule (not shown). A RIA using the selected cells and a radiolabeled anti-H-2K antibody demonstrated that the selected cells expressed 1500–2000 H-2K molecules, as compared with the starting cell population, which expressed 25,000–30,000 H-2K molecules per cell as calculated from Scatchard plots. Limiting dilution experiments of similar design but with graded numbers of RDM4 cells per well indicated that the H-2K expression variants appeared at a frequency of ~1 cell in 10^4 (Machy and Leserman, 1984).

2. Selection of T-Cell Receptor-Negative Variants of a Nontransformed Cell Line

In addition to the selection of the tumor variant discussed above, we selected a variant, negative for the expression of the T-cell receptor for antigen, of the nontranformed and nonmutagenized, functional cytolytic T-cell line called KB5-C20. This molecule is expressed by wild-type cells at levels of 100,000 molecules per cell. The protocol was modified slightly with respect to the RDM4 protocol, in that the culture medium was supplemented with the growth factor interleukin-2 (IL-2), on which the cells remain dependent, and the selection was performed on an initial population of 10^6 cells, using two 24-hour incubations with 60 n*M* MTX in protein A-bearing liposomes in the presence of 1 μg/ml of a protein A-binding antibody (Désiré-1) specific for the T-cell receptor of that cell line. The selections were separated by 3 days of incubation in medium without liposomes. Following expansion of the surviving cells in medium without liposomes, we observed a loss of the targeted T-cell receptor, but no effect on the level of other normally expressed cell surface molecules (Schmitt-Verhulst *et al.*, 1987) (Fig. 6). The difference in the protocol for selection of KB5-C20, as compared to RDM4 cells, is based on our subjective impression that the KB5-C20 cells were dying more rapidly than RDM4, and thus should be incubated without liposomes during a brief rest period. We have not rigorously evaluated different periods of incubation with liposomes with respect to selection of determinant-negative variants, which will undoubtedly depend on the cell type and determinant in question. As an example, using the protocol just given we have been unable to obtain H-2K-negative variants from 10^6 KB5-C20 cells (unpublished results) or from L cells (Machy and Leserman, 1984), although these are obtained at high frequency from RDM4 cells (see earlier). The possiblilty of obtaining these rare variants from larger numbers of KB5-C20 cells or L cells has not yet been studied.

B. Positive Selection

This technique depends on the observation that N^5-formyl-tetrahydrofolate (F-THF), which is an analog of the product (methyl-tetrahydrofolate), the synthesis of which is blocked by MTX, may reverse the toxicity of MTX. This combination of reagents is used in cancer chemotherapy, since the folate requirements of tumor cells often exceed those of normal cells, permitting preferential rescue of the latter (Jolivet *et al.*, 1983). In this *in vitro* application, MTX is present in the culture medium at a level just sufficient to kill the cells (determined for each cell type in a d[^{3}H-6]-

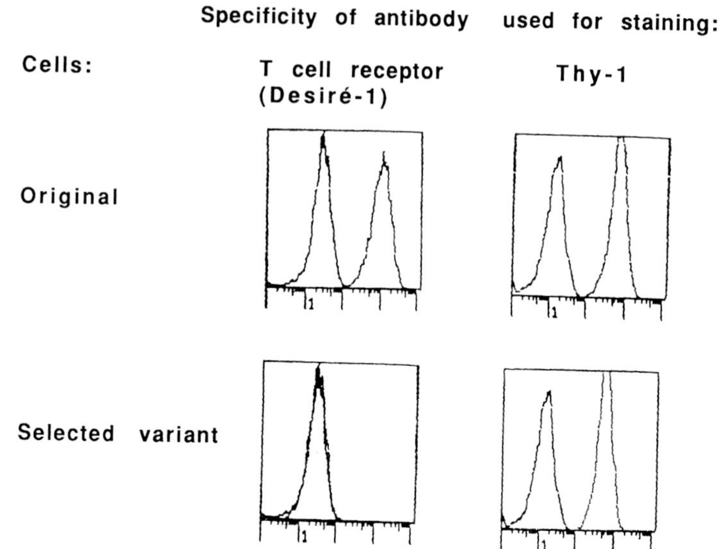

FIG. 6. Negative selection of cells expressing the T-cell receptor for antigen in a cytotoxic T-cell line, KB5-C20. Surviving cells were evaluated for the expression of the targeted and a nonselected cell surface determinant, the Thy-1 molecule. Histograms of the fluorescence of 10^4 cells by flow cytometry show the level of binding of fluorescent protein A-bearing liposomes and the selecting antibody (Desiré-1, left panel), or an antibody specific for the nonselected Thy-1 molecule (right panel) for both the variant and original cell populations. The histogram at the left of each panel is the binding of fluorescent liposomes in the presence of a control antibody that does not bind to the cells. The control and experimental histograms superimpose for the T-cell receptor of the variant population. This result confirmed loss of the targeted T-cell receptor, without effect on the level of the Thy-1 molecule. Northern dot–bolt analysis indicated that the variant lacked mRNA for the α chain of the T-cell receptor. From Schmitt-Verhulst *et al.* (1987). Reprinted with permission.

Urd assay), and antibody- or protein A-bearing liposomes containing F-THF are added to the culture. Cells expressing a determinant for which the liposomes are specific will take up sufficient F-THF to overcome the MTX effect, and thus will grow, while cells lacking the target determinant will die (Fig. 7) (Machy and Leserman, 1984).

1. ENCAPSULATION OF F-THF IN LIPOSOMES

N^5-Formyl-tetrahydrofolate is prepared for injection as the Ca^{2+} salt, which cannot easily be encapsulated because it causes aggregation and precipitation of liposomes. It was therefore passed through a 5-ml Chelex 100 (Biorad) column pretreated with 1 *N* HCl and then 0.5 *N* NaOH, to

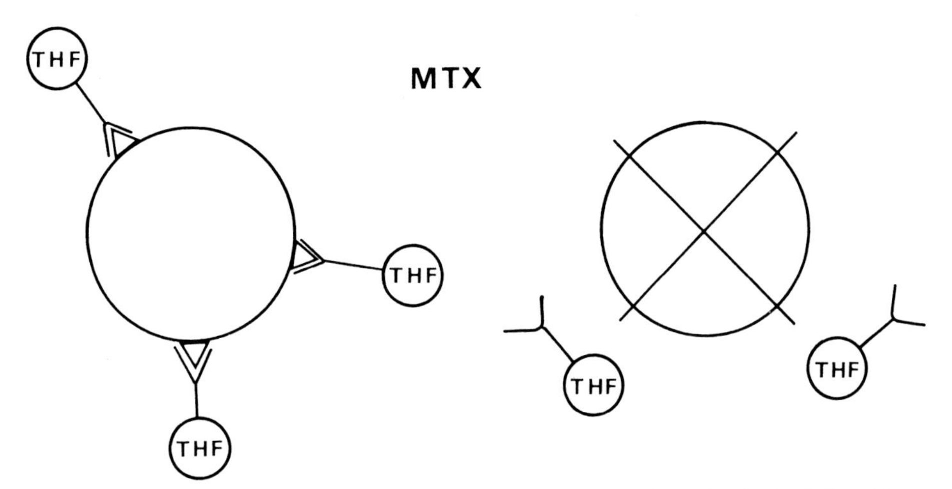

FIG. 7. Positive selection. Methotrexate, which is taken up by cells via the folic acid transport system, will prevent the synthesis of tetrahydrofolate (THF). Tetrahydrofolate in ligand-bearing liposomes will "rescue" cells expressing the target determinant; cells not expressing the target determinant will die.

obtain the sodium salt, which was encapsulated in liposomes at a concentration of 100 m*M*.

2. HIGH-MHC EXPRESSION VARIANTS OF THE MURINE THYMOMA RDM4

In our study of the expression of H-2 molecules by RDM4 cells, we had noted the low level of expression of H-2D molecules in the nonselected, wild-type population (Fig. 2). We attempted to select for cells expressing increased levels of H-2D molecules from this population. Cells were initially incubated in 0.5 ml culture medium in 24-well cluster plates at 10^5 cells/ml in the presence of an antibody specific for the H-2D molecule (5 μg/ml), 125 n*M* MTX, and protein A-bearing liposomes diluted such that the final concentration of F-THF in culture was 2 μ*M*. After 2 days, when cell proliferation was noted, the medium was changed and MTX added to a final concentration of 250 n*M*, without addition of liposomes. After 2 days cessation of cell growth was noted, the medium was replaced, and liposomes, antibody, and MTX were added at their initial concentrations. The cycle of selection with liposomes and MTX, followed by a double concentration of MTX without liposomes was repeated three times; then cells were incubated in normal medium for expansion. The cells were then tested in a cellular RIA for their capacity to bind the selecting anti-H-2D antibody. As compared to the starting population, the mean level of H-2D expression was increased by a factor of 4–5 (Machy and Leserman, 1984).

3. Recovery of Transfected Cells from a Mixed Population

To study the properties of a molecule, it is often useful to transfect the gene encoding that molecule into a cell population in which it is not normally expressed. This leads to the problem of isolating the transfected from the majority of nontransfected cells.

We have evaluated the possibility of saving a mouse cell population ($LMtk^-$) transfected with a gene encoding a human cell surface molecule (the HLA molecule Cw3) using targeted liposomes containing F-THF, while killing the nontransfected cells by free MTX in the medium. In a model experiment demonstrating this possibility we mixed the HLA-transfected $LMtk^-$ cell line TRH42 with various numbers of nontransfected $LMtk^-$ cells. The mixed cell population was incubated with 500 n*M* free MTX and an anti-HLA antibody plus F-THF-containing protein A-bearing liposomes at a concentration of 4 μ*M* F-THF. On days 3, 7, and 11 of culture, free MTX (500 n*M*) was added. On days 4, 8, and 12, the medium was changed and the selection repeated as for the first day. On or about day 15, when active cell proliferation was noted, the medium was aspirated and MTX was added at a concentration of 1000 n*M*. Liposomes were not added. After one additional day, medium was aspirated and an additional cycle of selection (liposomes plus antibodies, plus 500 n*M* MTX) was applied. Cells were subsequently grown in flasks—initially in the presence of 4 μ*M* free F-THF for 3 or 4 days, afterward in normal medium. Growing cells were analyzed for HLA expression and were found to express this determinant, even when initially cultured at a level of 1 TRH42 cell incubated with 100,000 $LMtk^-$ cells (Fig. 8) (Machy and Leserman, 1984).

C. Transport of Genes Conferring Drug Resistance

As discussed in the previous section, targeted liposomes can be used to select transfected cells from a mixed population. When liposomes contain DNA. they can also be used for the purpose of transfecting a specific cell type. We used large unilamellar liposomes (4000 Å), coupled to protein A, and containing a plasmid that confers resistance to mycophenolic acid to demonstrate gene transfer into specific lymphoid cells. Human B lymphoblastoid cells (BJAB 10^7) were washed with phosphate buffer and incubated in 0.5 ml of this buffer on ice with 5 μg of linearized plasmid

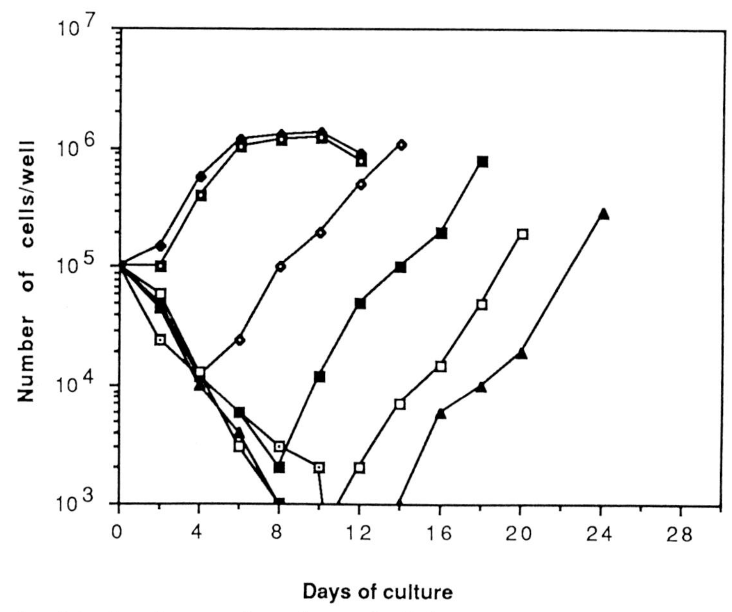

FIG. 8. Selection of a rare subpopulation of cells from mixed culture. All cells were killed by free MTX (⊡), but survived when an excess of free F-THF was added (◆). Cell growth was normal in the presence of free MTX and liposomes targeted to a murine determinant (H-2K) expressed by all cells (◘). When various numbers of cells transfected with a human (HLA) molecule were added (◇, 1 : 100; ■, 1 : 1000; □, 1 : 10,000; ▲, 1 : 100,000) and the cells were incubated with antibodies specific for the HLA determinant, cell proliferation was delayed. The surviving cells were found to express the HLA determinant (data not shown). Reprinted, in modified form, from Machy and Leserman (1984), with permission.

contained in liposomes targeted to the cells via an antibody specific for an expressed cell surface determinant, or a control antibody. The cells were left on ice for 1 hour to permit liposome–cell binding, and were then electroporated (Potter *et al.*, 1984) using a LKB 2197 power supply (2500 V, 0.9 mA). For liposome-encapsulated DNA, both the relevant antibody and electroporation was essential, since only cells incubated in this manner survived in a selective medium containing mycophenolic acid (Fig. 9) (Machy *et al.*, 1988). We have reported (Machy and Leserman, 1983) that large liposomes cannot be taken up by nonphagocytic cells such as lymphocytes. Since the plasmid we used for the transfection cannot be encapsulated in small liposomes, we do not yet know whether this specific transfection procedure would be efficient in the absence of electroporation if small liposomes could be used.

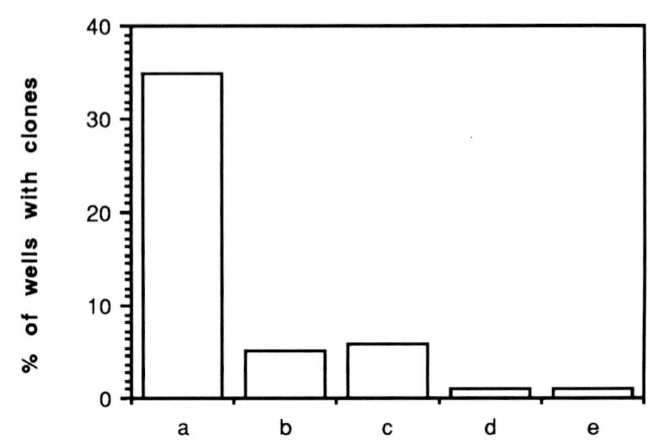

FIG. 9. Transfection of a human B lymphoma by DNA-containing, protein A-bearing liposomes and electroporation. Cells were incubated with or without antibodies and liposomes. The liposomes contained 5 μg of DNA. Transfectants obtained after (a) electroporation of cells preincubated with an anti-HLA antibody; (b) with a control anti-H-2 antibody; (c) without antibody plus protein A-bearing liposomes; (d) with anti-HLA antibody and liposomes without electroporation of cells; (e) cells electroporated after incubation with anti-HLA antibody and liposomes made without DNA. From Machy *et al.* (1988).

V. Conclusions and Perspectives

Liposomes are useful for marking, killing, or rescuing cell populations. The use of liposomes as fluorescent reagents, notably for the cell sorter, offers high signal with low background. High specificity of action is also the case for the use of liposomes for negative selection, which requires, however, that the target determinant is endocytosed by the cell (a criterion shared with immunotoxins), in contrast to the action of antibody and complement, for which expression of the molecule in question is sufficient for the cell to be killed. Positive selection with liposome-encapsulated protective reagents, though studied in a small number of model systems, is seriously challenged only by the fluorescence-activated cell sorter, which has been used for this application in only a few laboratories highly experienced in its use. Despite the advantages noted, liposomes have been little used for these purposes, probably because of lack of a commercial supplier of "ready-to-use" liposomes, and the normal hesitation of most individuals to initiate new technologies for which more or less satisfactory alternatives exist. We hope that, as these applications are perfected, and new uses are found, that these interesting reagents will find additional enthusiasts.

References

Barbet, J., Machy, P., and Leserman, L. D. (1981). *J. Supramol. Struct. Cell. Biochem.* **16,** 243–258.

Carlson, J., Drevin, H., and Axen, R (1978). *Biochem. J.* **173,** 723–737.

Connor, J., Yatvin, M. B., and Huang, L. (1984). *Proc. Natl. Acad. Sci. U.S.A.* **81,** 1715–1718.

Gregoriadis, G. (1984). "Liposome Technology." CRC Press, Boca Raton, Florida.

Gregoriadis, G. (1988). "Liposomes as Drug Carriers: Recent Trends and Progress." Wiley, Chichester.

Hope, M. J., Bally, M. B., Webb, G., and Cullis, P. R. (1985). *Biochim. Biophys. Acta* **812,** 55–65.

Jolivet, J., Cowan, K. H., Curt, G. A., Clendeninn, N. J., and Chabner, B. A. (1983). *N. Engl. J. Med.* **309,** 1094–1104.

Leserman, L. D., Barbet, J., Kourilsky, F. M., and Weinstein, J. N. (1980a). *Nature (London)* **288,** 602–604.

Leserman, L. D., Weinstein, J. N., Blumenthal, R., and Terry, W. D. (1980b). *Proc. Natl. Acad. Sci. U.S.A.* **77,** 4089–4093.

Leserman, L. D., Machy, P., and Barbet, J. (1981). *Nature (London)* **293,** 226–228.

Machy, P., and Leserman, L. D. (1983). *Biochim. Biophys. Acta* **730,** 313–320.

Machy, P., and Leserman, L. D. (1984). *EMBO J.* **3,** 1971–1975.

Machy, P., and Leserman, L. (1987). "Liposomes in Cell Biology and Pharmacology." John Libbey Eurotext Ltd./INSERM, Paris.

Machy, P., Barbet, J., and Leserman, L. D. (1982a). *Proc. Natl. Acad. Sci. U.S.A.* **79,** 4148–4152.

Machy, P., Pierres, M., Barbet, J., and Leserman, L. D. (1982b). *J. Immunol.* **129,** 2098–2102.

Machy, P., Lewis, F., McMillan, L., and Jonak, Z. (1988). *Proc. Natl. Acad. Sci. U.S.A.* **85,** 8027–8031.

Mosmann, T. (1983). *J. Immunol. Methods* **65,** 55–63.

Ostro, M. (1987). "Liposomes: From Biophysics to Therapeutics." Dekker, New York.

Potter, H., Weir, L., and Leder, P. (1984). *Proc. Natl. Acad. Sci. U.S.A.* **81,** 7161–7165.

Ralston, E., Hjelmeland, L. M., Klausner, R. D., Weinstein, J. N., and Blumenthal, R. (1981). *Biochim. Biophys. Acta* **649,** 133–137.

Schmitt-Verhulst, A.-M., Guimezanes, A., Boyer, C., Poenie, M., Tsien, R., Buferne, M., Hua, C., and Leserman, L. (1987). *Nature (London)* **325,** 628–631.

Szoka, F., and Papahadjopoulos, D. (1978). *Proc. Natl. Acad. Sci. U.S.A.* **75,** 4194–4198.

Truneh, A., and Machy, P. (1987). *Cytometry* **8,** 562–567.

Truneh, A., Mishal, Z., Barbet, J., Machy, P., and Leserman, L. D. (1983). *Biochem. J.* **214,** 189–194.

Weinstein, J. N., Ralston, E., Leserman, L. D., Klausner, R. D., Dragsten, P., Henkart, P., and Blumenthal, R. (1984). *In* "Liposome Technology" (G. Gregoriadis, ed.), Vol. 3, pp. 183–203. CRC Press, Boca Raton, Florida.

Yatvin, M. B., Weinstein, J. N., Dennis, W. H., and Blumenthal, R. (1978). *Science* **202,** 1290–1293.

INDEX

A

B

D

E

M

N

O

P

R

S

T

V

CONTENTS OF RECENT VOLUMES

Volume 27

Echinoderm Gametes and Embryos

Volume 28

Dictyostelium Discoideum: Molecular Approaches To Cell Biology

Volume 29

Fluorescence Microscopy of Living Cells in Culture

Volume 30

Fluorescence Microscopy of Living Cells in Culture

Part B. *Quantitative Fluorescence Microscopy—Imaging and Spectroscopy*

Volume 31

Vesicular Transport

Part A.

Part I. *Gaining Access to the Cytoplasm*